sponsors, not only from France and Europe but from the United States and from Japan too. The three chairmen want to express here their gratitude to all the sponsors of the workshop: *The French Centre National de la Recherche Scientifique, the American National Science Foundation; the French agency ANDRA, the oil companies Elf Aquitaine, Schlumberger, Total, and a group of 10 Japanese building companies, namely Fujita, Hazama-gumi, Kajima, Kumagaï-gumi, Maeda, Obayashi, Satoh, Shimizu, Taisei, Takaneka.*

We are very grateful too, to the members of our international organising committee, who have been in charge of finding sponsorship in their country: Professor Oka for Japan, Professor Drescher for the United States, doctor Mühlhaus for Australia, and have been extremely efficient in that essential role. The scientific organisation has been done with the help of these scientists, together with the members of the advisory committee, Professors Berest, Darve, Gudehus, Mróz, Tatsuoka, Vermeer, and Muir Wood. We want to thank all of them for their very profitable advises.

René Chambon
Jacques Desrues
Ioannis Vardoulakis

Preface

Localisation and other related instability phenomena are known to play a crucial role in the ultimate behaviour of soils and rocks at and during rupture and constitute therefore the basis of a continuum theory of rupture. Proper modelling of strain localisation and other bifurcation phenomena can be very significant in several fields, among which one can cite: Environmental Engineering & Disaster Prevention: Slope stability analysis, monotonous and cyclic behaviour, liquefaction phenomena; Petroleum Engineering: Borehole stability, sand production, drilling; Mining Industry: Rock bursting, tunnel instability; Civil Engineering: Stability of foundations, embankments, excavations, retaining walls; and Powder & Grains Engineering: Storage and controlled flow of granular masses.

Since the middle of the 70's, strain localisation in Geomaterial has become a more and more topical subject among the scientists involved in the geomechanics field. At the beginning the localisation community was rather small over the world, but it increased progressively. The first workshop on localisation in granular bodies was held in Karlsruhe, Germany, on February 22-25, 1988. It was followed by a second in Gdansk, Poland, on September 25-30, 1989. At the end of the latter, René Chambon and Jacques Desrues, from Grenoble, were asked to consider the possibility of organising the next event in France, which they accepted gratefully. Ioannis Vardoulakis kindly accepted to join them to constitute a team of three chairmen for the third Localisation Workshop. The first two events emphasised more the fundamentals of localisation and bifurcation theory as applied mostly to soil mechanics and were sponsored by the German and Polish Research Councils. With the third event, the organisers wanted to extend the scope of the Workshop so as to accommodate the interest of mining, petroleum and building industry.

The *Third International Workshop on Localisation and Bifurcation Theory for Soils and Rocks* was held in Grenoble, France, on September 6-9, 1993. The venue of the Workshop was a Colloquium center of the French Scientific Research Centre CNRS in Aussois, a nice mountain village in the French Alps.

Following the appreciation of the scientific committee and the participants, the Workshop has been a great success. We have had about 82 scientific participants, with 45 communications and three discussion sessions in 4 days. Participants were coming from most of the developed countries in the world, Europe, America, Japan, Australia.

The present book contains a selection of papers presented during the Workshop. All the papers have been reviewed by members of the Scientific Committee, namely P.Berest, F.Darve, A.Drescher, G.Gudehus, Z.Mróz, H.-B.Mühlhaus, D.Muir Wood, F.Oka, F.Tatsuoka, P.Vermeer and R.Chambon, J.Desrues, I.Vardoulakis (editors).

This worldwide meeting of reputed specialists involved in localisation studies would not have been possible without the strong support of both governmental agencies and private industrial

The Third International Workshop on Localisation and Bifurcation Theory for Soils and Rocks, held in Grenoble, France, on September 6-9, 1993, was sponsored by: The French Centre National de la Recherche Scientifique, the American National Science Foundation, the French company Elf Aquitaine, the French agency ANDRA, Schlumberger, Total, and a group of 10 Japanese building companies, namely Fujita, Hazama-gumi, Kajima, Kumagaï-gumi, Maeda, Obayashi, Satoh, Shimizu, Taisei, Takaneka.

Le 3ième Atelier international sur la localisation et la théorie de la bifurcation dans les sols et les roches, qui s'est tenu à Grenoble, France du 6 au 9 septembre 1993, a été co-financé par: Le CNRS (Centre National de la Recherche Scientifique), la NSF (National Science Foundation américaine), Elf Aquitaine, l'ANDRA, Schlumberger, Total, et un groupe de 10 compagnies de construction japonaises: Fujita, Hazama-gumi, Kajima, Kumagaï-gumi, Maeda, Obayashi, Satoh, Shimizu, Taisei, Takaneka.

Scientific Committee/Comité scientifique:

P.Berest, F.Darve, A.Drescher, G.Gudehus, Z.Mroz, H.B.Mühlhaus, D.Muir Wood, F.Oka, F.Tatsuoka, P.Vermeer and R.Chambon, J.Desrues, I.Vardoulakis (editors/éditeurs).

Thanks are due to the French Rock Mechanics Committee CFMR, who sponsored the edition of this book.
Les éditeurs remercient le Comité Français de Mécanique des Roches pour son aide à l'édition de cet ouvrage.

Cover: Strain localization in plane strain biaxial test on dense Hostun RF sand, incremental displacement and strain fields, experimental characterization using stereophotogrammetry, after M.Mokni, Thèse de doctorat de l'Université Joseph Fourier, Grenoble 1992.

The texts of the various papers in this volume were set individually by typists under the supervision of each of the authors concerned.

Authorization to photocopy items for internal or personal use, or the internal or personal use of specific clients, is granted by A.A.Balkema, Rotterdam, provided that the base fee of US$1.50 per copy, plus US$0.10 per page is paid directly to Copyright Clearance Center, 222 Rosewood Drive, Danvers, MA 01923, USA. For those organizations that have been granted a photocopy license by CCC, a separate system of payment has been arranged. The fee code for users of the Transactional Reporting Service is: 90 5410 511 9/94 US$1.50 + US$0.10.

Published by
A.A.Balkema, P.O.Box 1675, 3000 BR Rotterdam, Netherlands (Fax: +31.10.413.5947)
A.A.Balkema Publishers, Old Post Road, Brookfield, VT 05036, USA (Fax: +1.802.276.3837)

ISBN 90 5410 511 9
© 1994 A.A.Balkema, Rotterdam
Printed in the Netherlands

PROCEEDINGS OF THE THIRD INTERNATIONAL WORKSHOP ON LOCALISATION AND BIFURCATION THEORY FOR SOILS AND ROCKS
GRENOBLE (AUSSOIS) / FRANCE / 6-9 SEPTEMBER 1993

Localisation and Bifurcation Theory for Soils and Rocks

Edited by
R.CHAMBON & J.DESRUES
Laboratoire Sols, Solides, Structures (3S), UJF – INPG – CNRS, Grenoble, France
I.VARDOULAKIS
National Technical University, Athens, Greece

A.A.BALKEMA / ROTTERDAM / BROOKFIELD / 1994

COMPTES RENDUS DU TROISIEME ATELIER INTERNATIONAL SUR LA LOCALISATION
ET LA THEORIE DE LA BIFURCATION DANS LES SOLS ET LES ROCHES
GRENOBLE (AUSSOIS) / FRANCE / 6-9 SEPTEMBER 1993

Localisation et Théorie de la Bifurcation dans les Sols et les Roches

Edité par
R.CHAMBON & J.DESRUES
Laboratoire Sols, Solides, Structures (3S), UJF – INPG – CNRS, Grenoble, France
I.VARDOULAKIS
National Technical University, Athens, Grèce

A.A.BALKEMA / ROTTERDAM / BROOKFIELD / 1994

LOCALISATION AND BIFURCATION THEORY FOR SOILS AND ROCKS
LOCALISATION ET THEORIE DE LA BIFURCATION DANS LES SOLS ET LES ROCHES

Préface

La localisation et les phénomènes d'instabilité qui y sont liés jouent un rôle crucial dans le comportement à la rupture des sols et des roches, et sont ainsi nécessairement à la base d'une théorie de la rupture dans le cadre de la mécanique des milieux continus. Une modélisation correcte de la localisation de la déformation et des autres phénomènes de bifurcation est un enjeu majeur dans plusieurs domaines d'application, parmi lesquels on peut citer l'Environnement et les Risques naturels: Stabilité de pentes, comportement monotone et cyclique en rapport avec le phénomène de liquéfaction; le Génie Pétrolier: Stabilité de parois de trou de forage, venue de sable, forage; les Mines: Écaillage, stabilité de parois de tunnels; le Génie Civil: Stabilité des fondations, remblais, excavations, murs de soutènement ainsi que l'Ingénierie des Poudres et des Grains: Stockage et transport des matières pulvérulentes.

A partir du milieu des années 70, la localisation de la déformation est devenue un sujet de plus en plus important pour les chercheurs en géomécanique. La communauté intéressée, assez restreinte au début, s'est étendue rapidement. Le premier atelier sur la localisation dans les milieux granulaires s'est tenu à Karlsruhe, en Allemagne, du 22 au 25 février 1988. Un second atelier fut organisé à Gdansk, en Pologne, du 25 au 30 septembre 1989. A l'issue de ce dernier, René Chambon et Jacques Desrues, de Grenoble, furent sollicités pour organiser l'édition suivante en France. A leur invitation, Ioannis Vardoulakis se joignit à eux pour constituer une équipe de trois organisateurs pour le troisième atelier sur la localisation.

Le *troisième Atelier international sur la localisation et la théorie de la bifurcation dans les sols et les roches* s'est tenu à Grenoble, en France du 6 au 9 septembre 1993. L'atelier a eu lieu au Centre du CNRS 'Paul Langevin', situé à Aussois, village des Alpes françaises.

Suivant les avis recueillis auprès du comité scientifique et des participants à l'issue de l'atelier, celui-ci a été un grand succès. Suivi par 82 participants, il a été l'occasion de 45 communications et de 3 sessions de discussion réparties sur 4 journées. Les participants provenaient de tous les pays industrialisés, notamment d'Europe, des Etats-Unis, du Japon et d'Australie.

Le présent ouvrage contient une sélection d'articles présentés pendant l'atelier; tous ont été soumis pour avis à un comité de lecture constitué de MM. P.Berest, F.Darve, A.Drescher, G.Gudehus, Z.Mróz, H.-B.Mühlhaus, D.Muir Wood, F.Oka, F.Tatsuoka, P.Vermeer et de R.Chambon, J.Desrues, I.Vardoulakis (éditeurs).

Cette réunion mondiale de spécialistes réputés dans le domaine de la localisation n'aurait pas été possible sans le soutien vigoureux de plusieurs institutions gouvernementales et de plusieurs groupes industriels, non seulement français et européens mais aussi américains et japonais. Les trois organisateurs expriment ici leur gratitude à tous les co-financeurs de l'atelier: *Le CNRS (Centre National de la Recherche Scientifique français), la NSF (National Science Foundation américaine), Elf Aquitaine, l'ANDRA, Schlumberger, Total, et un groupe de 10 compagnies de construction japonaises: Fujita, Hazama-gumi, Kajima, Kumagaï-gumi, Maeda, Obayashi, Satoh, Shimizu, Taisei, Takaneka.*

Notre gratitude va aussi aux membres du Comité international d'organisation, qui ont eu notamment la charge de solliciter les financements dans leur propre pays: le Professeur Oka au Japon, le Professeur Drescher aux Etats-Unis, le docteur Mühlhaus en Australie. L'organisation scientifique s'est appuyée sur ces scientifiques, ainsi que sur les membres du comité d'experts, constitué des Professeurs Berest, Darve, Gudehus, Mróz, Tatsuoka, Vermeer, et Muir Wood. A tous nous exprimons nos remerciements pour leurs avis éclairés.

René Chambon
Jacques Desrues
Ioannis Vardoulakis

Table of contents
Table de matières

1 General theories and considerations
Théories et considérations générales

Localisation in granular bodies – Position and objectives 3
La localisation dans les milieux granulaires – Position du problème et objectifs à atteindre
G.Gudehus

Localised consolidation 13
Consolidation localisée
D.Kolymbas

Post-critical response of soils and shear band evolution 19
Comportement post-critique des sols et évolution des bandes de cisaillement
Z.Mróz & J.Maciejewski

2 Localisation and bifurcation conditions
Localisation et critères de bifurcation

Existence and uniqueness for B.V. problems involving CLoE model 35
Existence et unicité dans des problèmes aux limites avec le modèle CLoE – Études des solutions avec bandes de cisaillement
D.Caillerie & R.Chambon

Bifurcation conditions within hypoplastic constitutive theory 41
Conditions de bifurcation dans le contexte de la théorie hypoplastique
Z.Sikora & S.Rybicki

A dynamical interpretation of flutter instability 51
Une interprétation dynamique de l'instabilité de flottement
D.Bigoni & J.R.Willis

Stability problems related to static liquefaction of loose sand 59
Problèmes de stabilité en relation avec la liquéfaction statique dans le sable lâche
C.di Prisco & R.Nova

3 Localisation and constitutive modelling
Localisation et lois de comportement

Stability and uniqueness in geomaterials constitutive modelling — 73
Stabilité et unicité dans la modélisation rhéologique des géomatériaux
F. Darve

A new effective non-local strain-measure for softening plasticity — 89
Une nouvelle mesure non-locale de déformation pour la plasticité avec adoucissement
P.A. Vermeer & R.B.J. Brinkgreve

Shear moduli identification versus experimental localisation data — 101
Identification des modules des cissaillement en fonction de résultats expérimentaux
R. Chambon, J. Desrues & D. Tillard

Beyond invertibility surface in granular materials — 113
Au-delà de la rupture dans les milieux granulaires
W. Wu & A. Niemunis

Localised failure analysis using damage models — 127
Étude de la rupture localisée avec des modèles d'endommagement
A. Dragon, F. Cormery, T. Désoyer & D. Halm

Essential features of a Cosserat continuum in interfacial localisation — 141
Localisation d'interface dans les milieux de Cosserat
P. Unterreiner, I. Vardoulakis, M. Boulon & J. Sulem

4 Experiments
Expérimentation

Some observations of zones of localisation in model tests on dry sand — 155
Observations de zones de localisation dans un sable dense
D. Muir Wood & K.J.L. Stone

Shear banding in sands observed in plane strain compression — 165
La déformation des bandes de cisaillement dans les sables soumis à une compression en déformation plane
T. Yoshida, F. Tatsuoka, M.S.A. Siddiquee, Y. Kamegai & C.-S. Park

Deformation of shear zone in sedimentary soft rock observed in triaxial compression — 181
Déformation des zones de cisaillement dans les roches sédimentaires tendres observées en compression triaxiale
F. Tatsuoka & Y.-S. Kim

Experimental observations of strain localisation in plane strain compression of a stiff clay — 189
Observations expérimentales de la localisation de la déformation dans des essais de compression
en déformation plane sur une argile raide
G. Viggiani, R.J. Finno & W.W. Harris

5 Micromechanics of granular media
Micromécanique des milieux granulaires

A gradient elasticity model for granular materials — 201
Un modèle élastique à gradient pour les milieux granulaires
H.-B. Mühlhaus & F. Oka

Analytical solutions of deformation in gradient dependent model 211
Solutions analytiques de déformation avec des modèles dépendant du gradient
F. Oka & H.-B. Mühlhaus

6 Numerical modelisation
Modélisation numérique

Numerical modelling for the behaviour of an elastic medium in the presence 219
of a discontinuity for geotechnical applications
Modélisation numérique du comportement d'un milieu élastique en présence de discontinuités,
pour application dans le cadre de la géotechnique
E. Sakellariadi & G. Scarpelli

Instability of a viscoplastic model for clay and numerical study of strain localisation 237
Instabilité d'un modèle élasto-viscoplastique pour l'argile et étude numérique de la localisation
de la déformation
F. Oka, A. Yashima, I. Kohara & T. Adachi

A strain localisation analysis of frozen sand by elasto-viscoplastic softening model 249
Modélisation de la localisation de la déformation dans un sable gelé avec un modèle élasto-viscoplastique
adoucissant
T. Adachi, F. Oka, A. Yashima & L. L. Chu

Numerical study on localised deformation in a Cosserat continuum 257
Étude numérique de la déformation localisée dans un milieu continu de Cosserat
J. Tejchman

Author index 275
Index des auteurs

1 General theories and considerations
  Théories et considérations générales

Localisation in granular bodies – Position and objectives
La localisation dans les milieux granulaires – Position du problème et objectifs à atteindre

G. Gudehus
Institute for Soil Mechanics and Rock Mechanics, Karlsruhe University, Germany

ABSTRACT: The importance of localization in granular bodies is demonstrated by a series of examples with loading or excavation, loss of equilibrium, flow and penetration. A rather simple hypoplastic constitutive relation is shown to describe the granular behaviour with an extension for polar and capillary effects. Some boundary value problems are recommended for research: Shearing of a thin layer, shifting of plugs and plates, channel flow, and penetration. As localization is likewise important for stability of equilibrium, deformation and flow, the present paper tends towards a unified concept.

Cet article illustre l'importance de la localisation dans les milieux granulaires par une série d'exemples concernant des chargements, des excavations, des pertes d'équilibre, des écoulements et des pénétrations. Une loi de comportement hypoplastique assez simple est présentée, capable de décrire le comportement du milieu granulaire avec une extension pour la prise en compte des effets polaires et capillaires. On suggère des problèmes recommandés pour des travaux de recherche: cisaillement d'une couche mince, déplacement de plaques et d'ancrages, écoulement en canal, pénétration. Comme la localisation est également importante pour la stabilité de l'équilibre, la déformation et l'écoulement, le présent article vise à déboucher sur un concept unifié.

1 INTRODUCTION

Localization in granular bodies is known - without using modern terms - since Coulomb's times. It is of importance for numerous cases of civil, mining, and chemical engineering and also for structural geology. This is shown by a series of examples dealing with problems of deformations caused by loading or excavation, loss of equilibrium, flow or penetration. Localization with shearing and extension, which can lead to slip or separation, explains the inadequacy of many presently used calculation models.

The following section on constitutive behaviour is restricted to so-called reversible states of granular materials. Elastoplastic and hypoplastic constitutive equations are briefly discussed using the concept of state variables. A crucial argument is the definition and determination of material parameters. Polar and non-local approaches are discussed with the intention to achieve the simplest physically justified regularization.

The subsequent series of boundary value problems is recommended for research prior to tackling the more practical problems outlined before. Shearing along rigid boundaries is considered in order to determine material parameters first. Shifting of plugs and plates past granular bodies follows. Some cases of avalanches and penetration belong to the same group of problems. Brief remarks help to better understand the influence of grain properties and shear reversals.

Shearing inside granular bodies is first discussed for so-called element tests. Penetration and flow with sharp rigid edges follow. The sequence of deformation, stability, and flow, is considered. Separation is briefly discussed for cases of so-called reversible cohesion. Some remarks on the numerical treatment of localization, especially on interfaces, are added.

Closing remarks deal with some limitations and possible extensions. Various coupled problems appear with the penetration of tools and with grouting into granular bodies. Extensions of present concepts will also be needed for very loose or cemented granular bodies.

2 PRACTICAL CASES WITH RELEVANT LOCALIZATION

2.1 Deformation

Shallow foundations on granular ground have to be designed with sufficiently high bearing capacity and low displacements. Consider the every-day example with inclined and eccentric load. Even with small allowable displacements, localization of deformations in the ground is rather inevitable: There is bottom shear, penetration of the edge with shear jumps, and separation behind the leeward edge. It is not surprising, therefore, that predictions with various calculation models are often far from reality even for rather uniform granular soil. It is frequently overlooked also that the state prior to loading is the result of previous deformations with localizations: Natural ground is rich of shear and crack patterns, improved and articifical ground contains further localizations (as outlined below).

Deep foundations with piles or anchors are even more influenced by localizations. Their placement involves excavation, penetration and grouting with various localizations. The shaft resistance is strongly influenced by the grain size due to localized shear, and lateral loading leads to leeward separation from the ground. Almost all the conventional calculation models neglect these factors and are therefore far off reality. Loading tests are required, but much more use could be made of them with the aid of calculation models allowing for localization. Likewise, the analysis of pile and anchor groups cannot proceed without considering localizations.

The importance of localizations is more evident for retaining structures. Gravity or L-shaped walls are influenced by localizations as they have shallow foundations. In addition, the fill undergoes shear localization during placement and subsequent service states, and separation behind the wall top appears due to cohesion. Sheetpile or concrete walls reaching deeper than the bottom of excavation involve strong localizations during their placement. Subsequent shearing of the ground past the surface is inevitable. A strong discontinuity occurs at the foot if this is shifted against the ground. Further localization effects arise in case of supporting anchors or piles behind or below retaining structures.

Displacements caused by neighbouring excavations are also often influenced by localizations. Opening a cut can reactivate localized shearing along formerly active slip surfaces or tectonic faults, and also localizations due to ground improvement. Digging or driving a tunnel involves localized penetration and flow which determines the ground displacements. Localizations are also of importance for dams and offshore structures where finite element methods are widely used: the contact surface of structural and granular bodies will inevitably undergo localized shear, and often also separation.

The common feature of these cases is evident: the serviceability of engineering structures on and inside granular bodies is strongly influenced by localizations arising prior to and during the service time. Consequently, conventional elastic and elastoplastic calculation models cannot match the reality. Recurrence to load tests and field observations with back analysis cannot overcome these shortcomings if the latter is again made without allowance for localizations.

2.2 Loss of equilibrium

is rather easily understood for a shallow foundation with one degree of freedom and dead load in granular ground. A collapse occurs if the dead load reaches the peak value of the load-displacement curve. In other words, even a minute disturbance then leads to accelerations, i.e. an excess of kinetic energy. In conventional limit state analysis only the equilibrium of load and ground resistance is considered, which is a necessary condition. In order to decide whether an equilibrium is stable, indifferent, or unstable, one has to consider the excess of kinetic energy, which is normally not done in the analysis. Localization can play an important role then which is frequently overlooked. The transfer of these statements to piles and anchors is rather straightforward.

Similar statements hold for geotechnical structures with more degrees of freedom. For slender structures under dead load geometrical non-linearity is of importance. Foundation bodies connected with stiff superstructures obtain forces which change substantially with displacements and rotations. These changes of foundation loads are interrelated with effects of localization inside the ground. As they are not allowed for in most of the presently used limit equilibrium and finite element calculations, one is still very far from a reliable prediction of collapse.

Similar statements can be made for collapse due to excavation. A slope failure is actually progressive, so that a grain size effect due to shear localization arises. This is not principally different with retaining structures or soil improvement, but more complicated due to induced further localizations. The same is true for the roof or face collapse of tunnels. The recurrence to field measurements and stepwise prediction in the sense of an observational method requires a better mechanical understanding, however. Otherwise one is tempted to make smooth extrapolations of observed trends without being aware of bifurcations.

The considered cases can be strongly influenced by the pore fluid. Spontaneous or successive liquefaction of loose granular masses is a very dramatic example. The delayed collapse of fine - grained soils (creep rupture) is likewise important. One has to admit that the conventional effective or total stress analysis is frequently far off reality so that the confidence into calculation models is dwindling. The role of localized fluid pressure dissipation is not even qualitatively well understood.

2.3 Flow and penetration

Flow and penetration with rigid walls lead to related problems of granular mechanics. Consider the penetration of rods, plates, or tubes. Shear localization occurs along the contact among a rod or plate and a granular body during relative displacement. The same is true for hollow cylinders (shafts or tunnel tubes), where inside flow has to compensate the outside penetration. The latter cases can involve also localization inside the granular body depending on the kind of removal of material. Localization inside granular bodies is produced by flow and penetration with sharp edges. A moving shovel produces repeated localization patterns and thus obtains a pulsating resistance (Mróz and Maciejewski, 1994). Similarly, discharge from a hopper causes repeated localization patterns and pulsating wall pressures. Grain size effects are qualitatively known for such cases but not yet allowed for in calculation models.

Various kinds of localization can occur in a layered ground with pore fluid. Originally horizontal quaternary layers can be disturbed with modes of shearing or bulging with temporary local pore pressure increase (Eissmann, 1987). Grouting of various fluids into granular ground also leads to localized bulging, shearing and cracking. These mechanisms are yet poorly understood so that predictions are scarcely possible.

Other features of granular flow and penetration may briefly be mentioned here. Rather regular patterns of shear bands and density fluctuations can occur in natural ground in various scales and should be essentially understood for a better interpretation of ground investigations. The flow can be rather chaotic under other boundary conditions. Regular humming vibrations or rather irregular shocks - such as silo- or earthquakes - seem to be inevitable. Flexible penetrating bodies tend to deviate from the desired trajectory even in initially uniform ground. Very rapid penetration of solid bodies, explosions and impacts cause further kinds of localization with strong inertial effects (Kolymbas, 1994). All the implied stability problems are far outside the range of presently used calculation models.

3 CONSTITUTIVE RELATIONS

3.1 Granular interactions

It is helpful to consider idealized groups of grains in order to obtain qualitative restrictions for constitutive relations (Fig. 1).

The well-known considerations of grain quadruplets (a) support the following assumptions for gra-

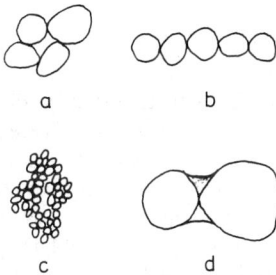

FIGURE 1. Idealized granular assemblies: quadruple (a), chain (b), cluster (c), pair with capillary bridge (d)

nular materials: deformations are mostly plastic due to slip, static limitations are imposed by dry friction, the range of void ratios is limited, dilatancy and contractancy occur, viscosity can be neglected. A chain of grains (b) suggests that each of them interacts only with the nearest neighbours, and that the interaction can be switched out due to loss of contact. Considering a group of grain clusters (c) one can imagine that there are considerable fluctuations of void ratio, grain rotations, and dislocations. Capillary fluid among a pair of grains (d) acts like an internal prestressing which increases with surface energy and the inverse of the grain diameter. This prestress is evidently zero for a dry and a fully saturated granular assembly so that there is a maximum in between. With relative velocities of grain skeleton and fluid, viscous resistance arises.

3.2 Simple materials

Granular materials are mostly assumed as simple in the sense of Truesdell and Noll, i.e. the symmetric stress tensor is required to be a functional of the history of the deformation gradient. More specifically, rate type relations of the kind

$$\mathring{\mathbf{T}} = \mathbf{F}(\mathbf{T}, \mathbf{D}, \mathbf{S}) \qquad (1)$$

are used, wherein $\mathring{\mathbf{T}}$ denotes an objective stress rate, $\mathbf{D}$ the rate of stretching, i.e. the symmetric part of the velocity gradient, and $\mathbf{S}$ a tensorial internal state variable. The function $\mathbf{F}$ is homogeneous of degree 1 with respect to $\mathbf{D}$ in case of rate independence. $\mathbf{F}$ must be non-linear in $\mathbf{D}$: ($\mathbf{F}(-\mathbf{D}) \neq -\mathbf{F}(\mathbf{D})$) in order to achieve irreversible deformations. This non-linearity is achieved in elastoplastic relations by sector-wise linear representations for $\mathbf{F}$ with switch functions for the transitions. $\mathbf{F}$ can also be defined by hypoplastic relations (Kolymbas, 1991).

A rather simple hypoplastic relation (Gudehus, 1993) can be written as

$$\mathring{\mathbf{T}} = h_g f_o [\mathbf{L}(\hat{\mathbf{T}}, \mathbf{D}) + f_d \mathbf{N}(\hat{\mathbf{T}}) ||\mathbf{D}||] \qquad (2)$$

wherein $\hat{\mathbf{T}} = \mathbf{T}/\mathrm{tr}\mathbf{T}$ denotes the stress direction tensor, and $\mathbf{L}$ a linear function of $\mathbf{D}$. The required non-linearity is achieved by the second term which is proportional to the norm of $\mathbf{D}$. This representation assures that the response envelope (Gudehus, 1979) is always elliptic with the reference point not lying outside. A kind of statical limit condition is obtained for zero stress rates and volume changes associated with certain stress states and directions of rates of stretching.

The factor f_d depends on the relative void ratio, i.e. the position of e within the bounding values e_f and e_c. Stationary flow is connected with the maximum e_f and an angle of granular friction (φ_g), determined by the functions $\mathbf{L}$ and $\mathbf{N}$. By cyclic shearing e can be reduced to e_c. By keeping f_d within the limits 0 and 1 for $e = e_c$ and e_f, respectively, dilatancy and contractancy cannot lead outside an allowed range of void ratios.

The limiting void ratios decrease with mean stress from maximum values to zero. This dependency is most easily described by

$$e = e_o \exp(\mathrm{tr}\,\mathbf{T}/h_g) \qquad (3)$$

wherein h_g denotes a granular hardness. The reference values for zero pressure are principally the same as the familiar limits of void ratio of a coarse-grained material. The scalar factor f_o, which depends on the void ratio, controls the position of peak stress ratios and is determined from assuming the pore pressure decrease for proportional loading following the equation above. It turns out that rather simple representations for the functions $\mathbf{L}, \mathbf{N}, f_d$ and f_o suffice for a wide group of granular materials. Consequently, only four material parameters are required:

h_g = granular hardness
e_{fo} = void ratio for granular flow at zero stress
e_{co} = minimum void ratio for cyclic shearing at zero stress level
φ_g = angle of a granular friction.

This consitutive relation describes the main features of granular materials. It covers a wide range of void ratios, pressures, and deformations. The four basic material parameters can be estimated from granulometric properties and calibrated from element tests.

Hypoplastic relations have been criticized because of the lack of bound theorems and the inability to describe shakedown. Bound theorems are very useful for engineering design, but they depend on ductility and normality, properties which granular materials unfortunately do not generally have. Bound theorems for the asymptotic behaviour of granular bodies are certainly desirable but they should be based on more realistic constitutive relations with due consideration of localizations. A shakedown, which implies a linear dependence of $\mathbf{F}$ on $\mathbf{D}$, is achieved with the hypoplastic relation outlined above only for $f_d = 0$, i.e. $e = e_c$. Various elastoplastic relations predict elastic ranges for certain finite stress oscillations, whereas a real shakedown is only observed asymptotically for vanishing cyclic stress amplitudes.

The most important state variable is the void ratio e. Capillary cohesion can additionally be allowed for by adding an isotropic capillary stress

$$p_c = \frac{U_f}{d_g} f(e, S_r) \qquad (4)$$

to $\mathbf{T}_g$ in our constitutive equation. A characteristic grain size d_g and a fluid surface tension U_f are needed. The dependence of p_c on the degree of saturation, S_r, and the void ratio can be described by suitable functions $f(e, S_r)$ having a maximum with respect to S_r between 0 and 1. A similar isotropic internal stress can be used for other kinds of cohesion which are uniquely determined by the void ratio. This kind of cohesion may be called reversible in the sense that it is determined by the arrangement and properties of grains and pore fluid only.

Stress-induced anisotropy is already modelled by the dependence of f on $\mathbf{T}$ in hypoplastic relations, whereas rather complicated hardening rules are needed in elastoplastic relations. Inherent anisotropy requires additional tensorial hidden variables represented by a structure tensor $\mathbf{S}$.

A main idea of our hypoplastic relation is the possibility to reconstitute states of granular materials. Therefore, the basic parameters can be determined from the asymptotic behaviour of disturbed samples which remain unchanged after arbitrary deformation histories. States are assumed to be sufficiently characterized by stress tensor and void ratio, and additionally the degree of saturation with pore fluid. This approach admittedly excludes various kinds of macropores (such as cracks or crumps in agricultural soil), strong fluctuations of void ratio, various packings of flat particles, strongly adsorbed pore fluid, and cementation among grains. It is proposed for studies of localization so that this is not veiled by numerous other factors.

3.3 Higher order materials

The need of regularization arose from analytical and numerical studies yielding localization. Viscosity can be left aside for dry granular materials. It has to be introduced later for fine-grained materials with weak drainage or capillary pore fluid. An elastoplastic polar granular material has been proposed by Mühlhaus (1987) motivated by consideration of grain assemblies yielding a realistic estimation of the shear zone thickness. This relation has been used numerically and supported experimentally by Tejchman (1989). The stress tensor and the rate of stretching tensor are no more symmetric. An additional constitutive relation appears for the momentum stress depending on the curvature tensor. Additional boundary conditions depending on wall roughness are required which replace the conventional constitutive relations for an interface between wall and granular body. Noticeable Cosserat effects, i.e. deviations from a symmetric stress tensor and additional rotations, appear only within narrow shear zones the thickness of which depends also on the boundary conditions.

As outlined by Tejchman in this volume, the Cosserat approach can be incorporated into hypoplastic relations. One can retain the constitutive relation outlined above by inserting the non-symmetric stress and rate of stretching tensors instead of $\mathbf{T}$ and $\mathbf{D}$ and introducing the grain size d_g so that mixed invariants are commensurable (this is analogous to Mühlhaus' proposal). An additional constitutive equation is required for the momentum stress tensor in formal analogy with the equation above. For testing various representations the 'sandwich' example outlined in the next section can be used. As the test for this case is principally simple, a rather easy and almost direct determination of the additional Cosserat parameters is possible. It appears that only one additional dimensionless parameter suffices.

Localizations with extension leading to separation cannot adequately be modelled by a polar approach but by gradient dependent formulations. Cracking of granular bodies requires cohesion. Consider an extension with a cohesion as described by the above equation for p_c: the increase of void ratio leads to a reduction of cohesion approaching zero for a certain value of e. The thickness of a zone of extension localization is proportional to the grain size as this enters the equation for p_c. Thus, a gradient approach is not necessary. The combination with shear localization is straightforward.

4 Boundary value problems

4.1 Shearing along rigid boundaries

A rather simple 'sandwich' system (Fig.2a) is proposed for the beginning.

It consists of two rough plates moving parallel to

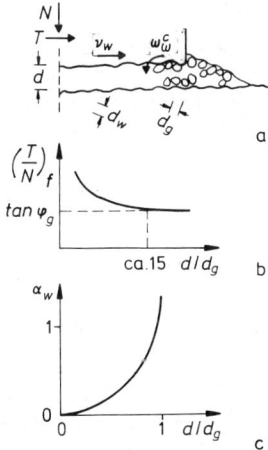

FIGURE 2. 'Sandwich' test (a), result for stationary flow (b), factor for Cosserat boundary condition

each other and a granular layer of variable thickness and void ratio in between. The rim is stress-free so that constraints as in other shearing devices cannot occur. Only ratios of forces and velocities are considered as they are decisive for particle slip. These quantities are rather constant along the shear zone. One can consider either constant void ratio and shear zone thickness or constant mean normal stress. The ratios of grain size d_g, surface roughness d_w, and zone thickness d are variable.

The state of stationary shear flow (subscript f) is

most instructive and can also be achieved in the experiment. With polar elastoplastic and hypoplastic approaches (Tejchman 1994) a decrease of the flow shear force ratio with increasing relative shear zone thickness is predicted (Fig. 2b). For very rough interfaces, i.e. $d_w \geq d_g$, back analysis of such shear tests yields the additional constitutive parameter of the polar material. With less rough interfaces, i.e. $d_w \leq d_g$, the wall parameter α_w in the boundary condition

$$\omega_w^c/v_w = \alpha_w/d_g, \qquad (5)$$

wherein ω_w and v_w denote rates of Cosserat rotation and shear displacement along the boundary, respectively, can be determined. It was found that α_w typically depends on d_w/d_g as plotted in Fig. 2c.

Angles of wall friction for stationary flow, thickness of shear zone and void ratio distributions can be calculated and compared with experimental observations. The reduction of flow stress ratio with the increase of zone thickness (Fig.2b) seems to contradict the additional freedom of rotations (ball bearing effect); it is explained by the additional resistance against them (momentum stress). It is an open question whether unconfined shearing in narrow zones leads to higher void ratios than under shearing without localization. This question cannot be clarified by element tests as a shear localization is inevitable. Only theoretical and experimental microscopic considerations can possibly give an answer, although further problems with fluctuations and dislocations will occur. It is only clear that for rapid shearing inertial effects lead to a substantial increase of void ratio and decrease of flow friction (as outlined by Jenkins in this volume).

One should then consider the shifting of plugs and plates (Fig. 3).

The initial field of stress components and void ratios has to be varied within realistic limits (including Janssen's state or earth pressure at rest) as it is not completely swept out by the transition to stationary flow.

Strong discontinuities, which properly lead to shear localization into the interior of the granular body, can be neglected at the beginning in case of slender plugs and plates. Predicted zone thickness and force ratios can be compared with experimental observations in order to check constitutive assumptions and parameters. Thus the consistency of the model can be partly verified. The results can also help to establish simplified engineering approaches for forces on the walls of silos and buried plates. These examples analysed already by Tejchman (1989) show that interface elements in numerical calculations are not necessary; they should at best be deduced from comparative calculations with a polar constitutive approach.

A next case of interest is shearing with variable direction. The 'sandwich' of Fig.2a can be sheared along an arbitrary curve within the plane. Shearing to the side can first be studied in order to obtain objective formulations, including asymptotic stationary flow. For constant normal force, the typical contractancy after an abrupt change of shearing direction can be calculated and compared with observations. Various shearing cycles can also be studied in order to clarify maximum densification in narrow shear zones.

A related case is the combination of axial and torsional shearing of a cylinder. Different realistic initial state fields have to be assumed. Again the asymptotic behaviour is of primary interest: pure torsion after axial shearing and vice versa, and cyclic shearing with small amplitudes. Such studies can also establish consistent and objective interface relationships for simplified numerical calculations in order to avoid mesh refining along shear zones.

Penetration of rods and plates should be studied first for two idealized cases (Fig.4).

Horizontal driving parallel to the ground surface (a) can lead to stationary resistance if shear localization inside the granular body is avoided by a sufficient ratio of overburden pressure and compressibility. Internal

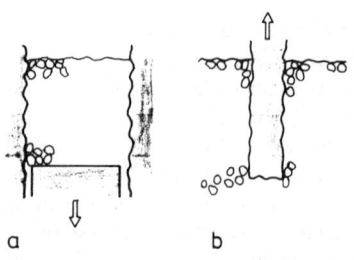

FIGURE 3. Shifted plug (a) and plate (b)

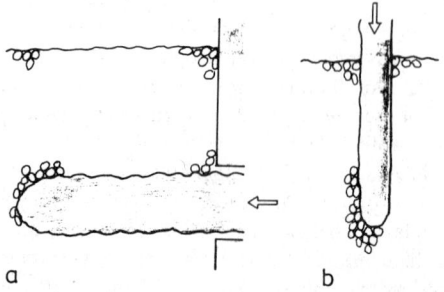

FIGURE 4. Horizontal (a) and vertical (b) penetration of plate or rod

shear localization can also be suppressed in front of a vertically penetrating body (b), but the resistance increases monotonically if initial stress ratios and void ratios are constant. Shearing along the interface is to be treated as in the cases treated above. Rather high pressure gradients arise in front of the penetrating body; they are covered by the proposed hypoplastic relation but may cause numerical problems because of high gradients of incremental stiffness. Large relative displacements require a mixed Euler-Lagrange formulation.

Different extensions of these test examples may be briefly indicated. One can introduce a pore fluid and consider various drainage situations. Drainage along the interfaces has to be specially considered (cf. the contribution by Kolymbas in this volume). By introducing some kind of elastic resilience in the support of rigid bodies in contact with the granular mass, one can model various kinds of self-induced vibrations. Elasticity of the granular skeleton or the pore fluid can also be introduced for this goal. The bottom conditions for avalanches can also be derived from shear localization.

4.2 Shearing inside a granular body

It is advisable to start with idealized biaxial tests (Fig. 5).

Imperfections of initial void ratio - indicated by I in the drawings - are randomly distributed. In case of lateral external fluid pressure (a) a single plane shear zone can result. More shear zones are obtained if they are successively confined by end plates. This example serves again to better understand bifurcation in granular bodies. Mesh refinements and internal interface elements for numerical modelling of the growth of shear zones can also be clarified.

A pattern of shear bands can be obtained for the conditions of a displacement - controlled biaxial test (b). A collapse of the system is excluded, so that notions of statical instability should not be used. If the granular body tends to dilate, the increase of the shear zone thickness with grain size causes an increase of externally measured mean stress, i.e. stiffness and strength are overestimated.

It is not yet advisable to analyse idealized triaxial tests in a similar manner. It appears that shear localization then causes a loss of axial symmetry so that the analysis becomes far more difficult. It appears more promising to analyse the penetration of rigid edges from the surface (Fig.6).

Two symmetric shear zones develop from a vertically penetrating wedge (a), whereas only one zone develops for inclined penetration (b). Because of large displacements a mixed Euler-Lagrange formulation is needed. For an embedded foundation (c) under inclined loads, inertial stress and void ratio fields have to be introduced, and shear zones can grow from one edge only. The passive trap door problem (d) is similar.

As outlined by Mróz in this book, new shear zones can develop after large displacements in such cases, accompanied by fluctuations of the driving force. This is a kind of repeated hardening and softening of the whole system as shear zones are first dilated and then blocked by the translated rigid body. Amplitude and wave length of force fluctuations will increase with the grain size. Experimental verifications are not very difficult. The importance for various practical cases and the formulation of simplified calculation models for them is evident.

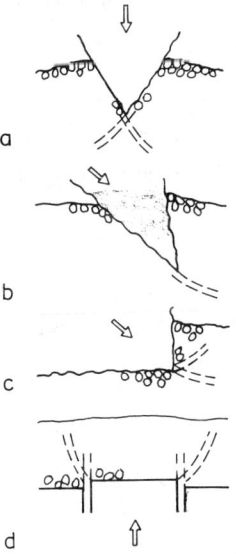

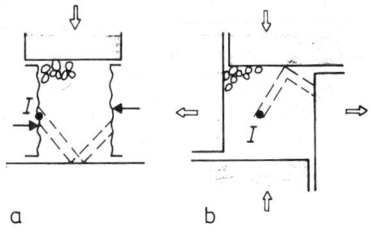

FIGURE 5. Biaxial tests with lateral pressure (a) or displacement (b) control

FIGURE 6. Penetration of sharp edges: wedges (a & b), foundation (c), trap door (d)

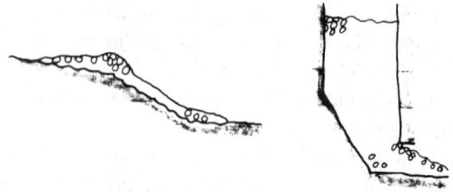

FIGURE 7. Granular flow on a slope (a) and in a hopper (b)

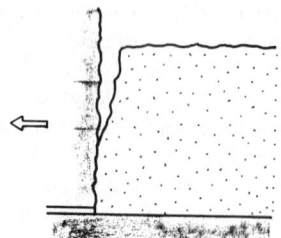

FIGURE 9. Tension crack behind a shifted wall

Granular flow along confining walls with edges should also be studied (Fig. 7).

One can start with quasistatic flow on a slope (a) or from a hopper (b) caused by slow removal of granular material at the foot. Shear zones grow repeatedly into the granular body if this tends to dilate so that the flow then is pulsating. Mixed Euler-Lagrange formulations with adaptive mesh refinement are again necessary. Grain size effects will arise as before. It is especially interesting to look for asymptotic stationary solutions, produced by feeding the granular material removed at the foot to the top again. It is not even sure whether such solutions exist, however.

Some extensions of such examples may be mentioned only without further drawings. By introducing pore fluid and full saturation, one can study the influence of drainage conditions and velocity. An additional internal length enters due to viscosity and certainly influences the localizations. Various extensions with inertial forces are thinkable, but only few of them will be numerically and experimentally feasible in near future.

4.3 Separation

Uniaxial extension (Fig. 8) should be studied in order to understand the proposed constitutive relation for cohesion.

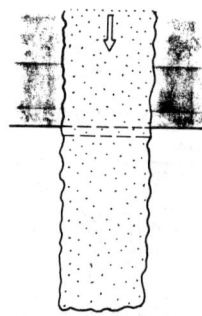

FIGURE 8. Extrusion of a wet granular string or plate

Consider a vertically extruded string of wet or moist fine silt. With the increase of length, and weight correspondingly, the void ratio increases at the upper end so that the cohesion decreases there. This clearly leads to tension fracture. In terms of our hypoplastic formulation, the intrinsic capillary pressure p_c is reduced by an increase of e. Separation means zero granular pressure, i.e. balance of intrinsic pressure and external traction after a certain maximum extension. As the grain size enters the proposed relationship for intrinsic pressure, it also influences the thickness of the zone of extension (without using higher order gradients). Thus the zone of localization is also grain - size dependent. Experiments can help to check and improve this approach.

A similar case is the development of tension cracks behind moving walls (Fig. 9).

There is an increase of void ratio along the wall up to zero effective stress prior to cracking. The relevance for retaining walls and associated numerical models is evident. The same approach should be applicable to the cracking of granular bodies due to shrinking or rapid injection of gas or fluid. In all these cases the rate of pore fluid flow influences the localization via the specific seepage force.

More complicated is the arching of cohesive granular bodies. A static case is achieved by slow lowering of a trap door. Localizations with shearing *and* extension start from sharp edges, meet and lead to separation. The grain size enters both via Cosserat terms and cohesion (when using the above equation for p_c). New localizations appear after large displacements, and periodic patterns can be achieved. Calculations of this kind will help to extend simplified models as proposed by Mróz in this volume. By including inertial forces one could even explain and quantify silo and earth quakes.

REFERENCES

Eissmann, L. 1987. Lagerungsstörungen im Lockergebirge. *Geophys. Veröff. KMU Leipzig.* 111, 4, 7-77.s

Gudehus, G. 1979. A comparison of some constitutive laws under radially symmetric loading and unloading. *Proc. 3th Conf. on Num. Meth. Geomech..* 1309-1324. Rotterdam: Balkema.

Gudehus, G. 1993. A comprehensive constitutive equation for granular materials. Submitted to *Soils and Foundations, Jap.Soc. Soil Mech.*

Kolymbas, D. 1991. An outline of hypoplasticity. *Arch. Appl.Mech.* 61: 143-154.

Kolymbas, D. 1994. Localized Consolidation. In this book.

Mróz, Z. & Maciejewski, J. 1994. Post-critical response of soils and shear band evolution. In this book.

Mühlhaus, H.-B. 1987. Berücksichtigung von Inhomogenitäten im Gebirge im Rahmen einer Kontinuumstheorie. *Veröff. Inst. Boden- u. Felsmech., Univ. Karlsruhe.* 106.

Tejchman, J. 1989. Scherzonenbildung und Verspannungseffekte in Granulaten unter Berücksichtigung von Korndrehungen. *Veröff. Inst. Boden- u. Felsmech., Univ. Karlsruhe.* 117.

Localised consolidation
Consolidation localisée

D. Kolymbas
Institute of Soil Mechanics and Rock Mechanics, University of Karlsruhe, Germany

ABSTRACT: Dynamic excitation of loose watersaturated sand gives rise to peculiar phenomena of segregation of pore fluid from the grain matrix and corresponding compaction of the soil. In the present paper this compaction is considered as a localized consolidation. In contrast to Terzaghi's consolidation theory, localized consolidation accounts for rapid compaction since the drainage path of water leads across a moving discontinuity surface and is, thus, extremely short. Consideration of the kinematics allows to determine the velocity of propagation of the density discontinuity.

Dans les sables lâches saturés sousmis à certaines sollicitations dynamiques, on peut observer des phénomènes particuliers d'expulsion d'eau de la matrice granulaire avec densification correspondante. Dans le présent article on considère cette densification comme une consolidation localisée. Contrairement avec la théorie de Terzaghi, la consolidation localisée permet une densification rapide car les chemins de drainages à considérer ne traversent qu'une surface de discontinuité mobile, et sont donc extrêmement courts. Des considérations cinématiques permettent de déterminer la vitesse de propagation de la discontinuité de densité.

1 INTRODUCTION

In this paper some localization phenomena are discussed which appear only at multiphase media such as a mixture of sand and water. There, the multiphase character is relevant only if the solid particles have a different motion than the water. This is the case at any process of compaction (or consolidation) which means a densification of the grain skeleton and a segregation of water from the grains. Up to now, the consolidation of a water saturated granular medium is treated by means of the theory of TERZAGHI or by some more refined theories mainly based on TERZAGHI. The hart of this theory is the heat equation which governs the gradual dissipation of the overpressure of the pore fluid.

There are, however, some indications that the consolidation and water-segregation can also take place in a much more quick and violent way than according to TERZAGHI's theory. Such compaction modes are observed in loose water-saturated sands and are released by dynamic actions such as impacts or earthquakes.

In this paper is presented a theory which tries to explain these quick compaction modes by moving discontinuities of the deformation field which can be considered as *compaction waves* or *moving vortex sheets* or as *phase transition waves*. The reader may wonder how a discontinuity can propagate in a problem goverved by the parabolic *heat equation*, which smooths any discontinuity. The answer is that TERZAGHI's parabolic equation is not valid for the processes considered here, because the latter are based on a completely different mechanism. In the case of the considered discontinuities, the drainage path (the length of which controls the relaxation time) is extremely short, as it only leads across the discontinuity.

Perhaps the most intricate property of granulates is that their density can vary within a relatively large range no matter what the corresponding pressure is. Whereas the density of gases, fluids and elastic bodies is uniquely determined by the pressure, this is not the case with granulates. The transition to a denser state can occur either as a deformation process (by increasing the pressure or by shearing) or as a phase transition (see Figure 1). The dotted curve reminds the behaviour of a VAN DER WAALS fluid. A phase transition into a denser state can be associated with a moving discontinuity surface.

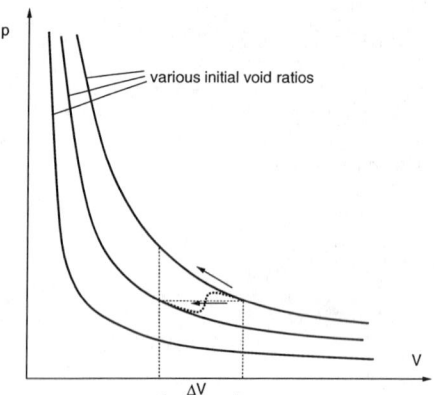

Figure 1: Typical compression diagram for a soil with various initial densities

2 DYNAMIC COMPACTION OF SOIL

Very peculiar phenomena accompany dynamic soil compaction. Following Japanese authors [22] "sand boils and sand volcanoes ... during the earthquake were reported". Several minutes after underground explosions in loose sand deposits, water eruptions and formation of craters are usually observed.

Considering dynamic compaction, *contractancy* plays an important role: As known, undrained shear leads to build up of pore pressure. As a consequence, the stiffness and the strength of soil reduce. This process is often called *liquefaction*.

The point now is, that liquefaction is necessary but not sufficient in explaining dynamic compaction. Shear and build up of pore pressure take place at undrained conditions, whereas the compaction sets on subsequently. Of course, one can try to explain the compaction by means of Terzaghi's theory. If complete liquefaction is assumed, the initial pore overpressure is $h(z) = \gamma' \cdot z$, where z is a depth-coordinate and γ', the unit weight of the submerged grain skeleton or boyant unit weight, equals $\gamma - \gamma_w$. Herein, γ is the unit weight of the mixture. According to Terzaghi's theory this pore overpressure dissipates and eventually (strictly speaking for $t \to \infty$) a final settlement s_∞ will be obtained. This settlement can be calculated if the compressional stiffness of the soil, E_s, is known. Taking into account that in the considered case a *reloading* takes place, the stiffness E_s is rather high and, consequently, the obtained settlement is low. For realistic cases a far too low settlement will be obtained (as compared to measurements). In addition, the so obtained compaction will increase linearly with depth, i.e. no compaction will be obtained at the unloaded surface. This, also, appears not realistic. Therefore the observations cannot be explained by the constitutive behaviour of soil (i.e. the stress-strain-relation) and another mechanism of compaction should be considered.

A striking similarity exists between the phenomena of magma segregation and *volcanism,* on the one hand, and porewater segregation from loose soil deposits due to dynamic actions on the other hand. Melting of rock proceeds in such a way (see [16]) that the melt is confined to those places where three grain boundaries meet. Thus, the melt plays the role of the porewater in the soil. The isomorphism between these phenomena is supported from the fact that related soil mechanics reports use words such as 'volcanoes' and 'cratering' in describing the phenomena observed in watersaturated loose soil subjected to dynamical actions. In fact, the observed expelling of water, which sets on several minutes after an underground explosion, strongly reminds the eruption of lava, and the remaining craters in the soil surface look like veritable *minivolcanoes*. The corresponding theories developed in the field of geophysics [14], [15], [16] also express this similarity. Notice that tha magma migration is a slow process. The ascent through the earth crust requires thousands of years. Several publications treat the question whether the ascending magma is a soliton or not. Following [16] it is not.

3 WATERSATURATED SOIL AS A MIXTURE

Watersaturated soil can be considered as a two-component mixture. Introducing the volume fractions $\alpha_i := V_i/V$ for each constituent i we can consider quantities referring to these uniformly distributed constituents. Such quantities will be denoted by superposed indices. For watersaturated soil with $i = s$ for solids and $i = w$ for water we have:

$$\alpha_s + \alpha_w = 1 \ .$$

Of course, α_w equals the porosity n defined as V_{voids}/V_{total}. The porosity n should be contrasted with the void ratio e defined as V_{voids}/V_{solids}. e and n are related by the equation $e = n/(1-n)$. The partial densities ϱ^s and ϱ^w read as follows:

$$\varrho^s = \alpha_s \varrho_s \quad , \quad \varrho^w = \alpha_w \varrho_w$$

The total density of the mixture is ϱ :

$$\varrho = \alpha_s \varrho_s + \alpha_w \varrho_w$$

The well-known jump relation expressing the conservation of mass across discontinuities reads in the one-dimensional case

$$[\varrho^i(U - v^i)] = 0, \qquad (1)$$

with U being the propagation speed of the discontinuity.

Note that for each constituent i a seperate jump relation holds, whereas the conservation of momentum can only be expressed globally (i.e. for the total stress)

$$\sum_i [\sigma^i] = \sum_i [\varrho^i v^i (v^i - U)].$$

As usually, $[x]$ means the jump $x_1 - x_2$ of any quantity x across the discontinuity. In this paper the momentum jump relation will not be considered further since the involved velocities are too small for inertia to be of any importance. In the equations state above the one-dimensional case is considered. For the three-dimensional case the velocities have to be considered as vectors and the stress as tensor.

4 TERZAGHI'S CONSOLIDATION THEORY

When a saturated soil is suddenly loaded, the additional load acts in the beginning only upon the water, the compressional stiffness of which is much higher than the one of the grain skeleton. A pore overpressure is thus immediately created and subsequently dissipates slowly as the water is squeezed out from the pores. The rate of water volume which is being squeezed out from a volume element amounts div v. Herein, v is the flux of the porewater relatively to the grains. Introducing DARCY's law, $\mathbf{v} = k \,\mathrm{grad}\, h$, we obtain the volumetric strain rate $\dot{\varepsilon}_v$ of the grain skeleton

$$\dot{\varepsilon}_v = k \,\mathrm{div}\,\mathrm{grad}\, h$$

Herein, h expresses the overpressure p as the height of a water column: $p = \gamma_w \cdot h$. The permeability k depends on the viscosity of the pore fluid and on the void ratio. The linearity follows from the assumption that $E_s \cdot k$ remains constant during consolidation. Actually, with decreasing void ratio E_s increases and k decreases (according to $k \propto e^3/(1+e)$ or $k \propto e^2$). In a first approximation it can be argued that both tendencies counterbalance eachother.

On the other hand, the volumetric strain rate, i.e. the volume change rate of the grain matrix, can be related to the change of the stress acting upon the soil skeleton. Considering the one-dimensional case ($\varepsilon_v \equiv \varepsilon_1$), we have

$$\dot{\varepsilon}_v = -\dot{\sigma}_1/E_s = \dot{p}/E_s = \gamma_w \cdot \dot{h}/E_s \ .$$

Herein, E_s is the stiffness modulus of the soil for one-dimensional compression. σ_1 is the corresponding normal effective (i.e. intergranular) stress component. In the above equation TERZAGHI's principle has been taken into account. In the quasistatic case the vertical total stress component must be constant. Thus we obtain

$$\frac{\partial^2 h}{\partial z^2} = \frac{\gamma_w}{E_s \cdot k} \cdot \frac{\partial h}{\partial t} \qquad (2)$$

5 DISCONTINUOUS CONSOLIDATION MODE

In the model presented here, sand grains and water are assumed to be incompressible. Thus the compaction of the soil skeleton proceeds only by means of rearrangement of grains and squeeze out of porewater. This is also assumed in TERZAGHI's theory.

5.1 Sedimentation wave

Consider again the differential equation 2. Its left part cannot be wrong, since it merely expresses the volume change rate by means of DARCY's law and the assumption $k =$const. So, we have to modify the right part, which relates the volume change rate with the behaviour of the grain skeleton, such that the volume decrease should now not depend on the increase of compressional stress. For this purpose we assume $E_s = \infty$, i.e. we neglect any consolidation according to TERZAGHI's theory. In reality, of course, both modes co-exist, as this has been considered by Scott (1968). Here I consider only a compactional wave which travels through the medium, say from the bottom to the top (see also Ivanov (1967), Cowin and Comfort (1982), Scott (1986), Nigmatulin (1987, Chapter 8)). This wave can be conceived as a zone of vanishing thickness in the interior of which the compaction takes place. Such a wave can be mathematically described by means of DIRAC's δ-function, $\delta(x - ct)$, which denotes a discontinuity moving with the velocity c. Thus, equ. 2 is now replaced by

$$k \frac{\partial^2 h}{\partial z^2} = b \cdot \delta(x - ct) \ . \qquad (3)$$

b, the volume change rate within the discontinuity, will

be specified below. $x = H - z$ is a co-ordinate which counts in the direction of the wave propagation. c is the wave velocity. It can be determined if we consider the jump relation expressing the conservation of mass. Denoting by e_1 the void ratio and by n_1 the corresponding porosity ($n = \frac{e}{1+e}$) of the original (loose) soil deposit, and by e_2 and n_2 the corresponding quantities of the compacted soil, we can express the mass conservation of water by means of equ. 1 as follows:

$$[n(c - v^w)] = 0 \qquad (4)$$

Herein, v^w is the velocity of water. Similarly, equ. 1 yields the mass conservation of solid particles as

$$[(1-n)(c - v^s)] = 0 \qquad (5)$$

with v^s being the velocity of the solids. Setting the positive directions of v^s and v^w upwards, respectively, we obtain from DARCY's law (with $i := \partial h / \partial z$) as follows:

$$v^w - v^s = k \cdot i \quad ,$$

Now, when we eliminate v^w from equ. 4 and v^s from equ. 5 and substitute into DARCY's law, we obtain the following equation for the propagation velocity c:

$$c = \alpha \cdot k \cdot i \qquad (6)$$

with the abbreviation

$$\alpha = \frac{e_1}{e_1 - e_2} \cdot \frac{1 + e_2}{1 + e_1} .$$

For the case of *total* liquefaction (i.e. vanishing effective stresses), the 'hydraulic gradient' i reads as follows:

$$i = i_{tot} = \left. \frac{\partial h}{\partial z} \right|_{t=0} = \frac{\gamma'}{\gamma_w} \approx 1.$$

According to the measured values of e_1 and e_2 for fine sand the factor α can be found to have a value of ca. 50. We see thus, that

$$c \approx 50 \cdot k$$

holds true. This relation has been experimentally corroborated by a simple laboratory test, which can be carried out in the following way. Take a glas tube and fill it with water-saturated loose sand having the void ratio e (see Figure 2). The compaction wave can be triggered by a hit on the lower part of the tube. A fine sand deposit of the initial height $H = 60$ cm can thus settle by ca. 2 cm within a time lapse of less than 2

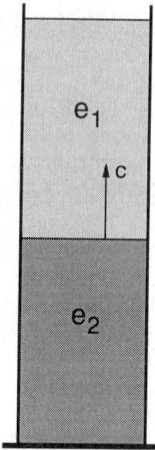

Figure 2: Glas tube with compacted (lower) and loose (upper) parts.

seconds. An instantaneous picture of the compaction wave can be obtained by a photo taken in such a way that the particles in its upper part appear hazy due to their motion.

The time lapse needed for complete consolidation is

$$t_1 = \frac{1}{\alpha} \cdot \frac{\gamma_w}{\gamma' k} \cdot H \quad .$$

H is the thickness of the consolidated stratum.

Of course, k refers here to the uncompacted region of loose sand. t_1 is considerably shorter than according to TERZAGHI's theory, which predicts an infinite time lapse for 100 % consolidation. For 90 % consolidation it predicts

$$t_2 \approx \frac{\gamma_w}{E_s \cdot k} \cdot H^2 \quad .$$

Note that t_2 is proportional to H^2, which is typical for diffusion processes, whereas t_1 is proportional to H.

The constant b appearing in equ. 3 can be determined in the following way: The integral of equation 3 reads:

$$h(x,t) = \frac{b}{k}\langle x - ct \rangle + C_1(t) \cdot x + C_2(t) \qquad (7)$$

with $\langle x \rangle = x$ for $x \geq 0$ and $\langle x \rangle = 0$ for $x < 0$.

The following boundary conditions are introduced:
(i) $\partial h(x = 0, t)/\partial x = 0$, i.e. an impermeable underlying stratum. It follows: $C_1(t) = 0$
(ii) $h(x = H, t) = 0 \longrightarrow C_2(t) = \frac{b}{k}(H - ct)$.

Thus, we obtain:

$$h(x,t) = \frac{b}{k} \cdot (\langle x - ct \rangle - H + ct) \quad (8)$$

Equ. 3 requires that the distribution of h be linear on both sides of the discontinuity. Thus, the distribution of $h(x,t)$ for $x > ct$ equals to the initial distribution $h(x, t = 0)$. The latter reads

$$h(x, t = 0) = \psi \cdot (H - x). \quad (9)$$

Herein, ψ is a coefficient denoting the degree of liquefaction, such that $0 < \psi \le 1$, and $\psi = 1$ for complete liquefaction. From equations 8 and 9 we obtain

$$\frac{b}{k} = -\psi. \quad (10)$$

In other words, k is the maximum value of the compaction rate b. Thus, it follows:

$$|b| \le k \quad .$$

The initiation or nucleation of the considered waves is not analyzed here. It it merely assumed that pre-existing discontinuities of the density start propagating. This effect is triggered by the generation of an overpressure in the porewater, say by an underground explosion or some other dynamic action. To determine the resulting void ratio e_2 an additional constitutive relation (cf. equ. 15) is needed, which is still lacking. Thus, the theory presented here is more or less kinematical and presupposes that e_2 is known.

5.2 Bubbles

Another mode of discontinuous consolidation consists in bubbles of very high water content (i.e. a slurry) which ascend through the granular medium and leave behind a compacted grain skeleton. The water squeezed out is contained in the bubble, see Figure 3. Such ascending bubbles appear in loose grain-water mixtures which have been recognized as unstable [4].

Considering the grain skeleton, the upper boundary of the bubble is a rarefaction wave and the lower boundary is a condensation wave. Within the bubble, the sand is moving downwards with the velocity v_s and the water is moving upwards with the velocity v_w. The speeds of the upper (A) and lower (B) discontinuities can be determined from the jump relation for the mass conservation:

$$U_A = v_s \frac{1 - n_2}{n_1 - n_2} \quad (11)$$

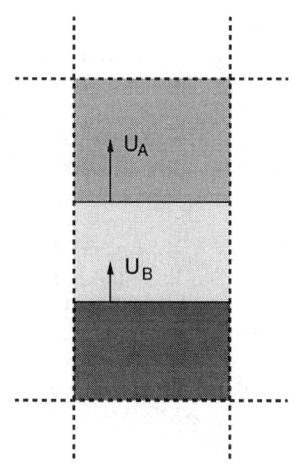

Figure 3: Slurry bubble moving upwards

$$U_B = v_s \frac{1 - n_2}{n_3 - n_2} \quad (12)$$

Herein, n_1, n_2 and n_3 are the porosities above, within and below the bubble, respectively. Of course, n_1 is the initial and n_3 is the final porosity.

The ratio of the velocities of water and sand can be found to

$$|v_s/v_w| = \frac{n_2}{1 - n_2} \quad (13)$$

The motion of the water relatively to the grains within the bubble is governed by a workrate balance,

$$\gamma_s(1 - n_2)v_s - \gamma_w n_2 v_w - \kappa(v_w - v_s)^2 = 0 \quad (14)$$

In order to determine n_2 and n_3 however, we need *additional constitutive relations.* Such relations could e.g. read

$$[n] = c[p] U \quad (15)$$

At time, we are planning a research project in order to experimentally check such constitutive assumptions.

REFERENCES

[1] Z.J. BAZANT: *Theory of vibration compaction of saturated granular medium*, Proceed. of the Intern. Symp. on Wave Propagation and Dynamic Properties of Earth Materials, Univ. of New Mexico, 1967.

[2] D.S. DRUMHELLER: *A theory for dynamic compaction of wet porous solids*, Int. J. Solids Structures, Vol. **23**, No. 2, (1987), pp. 211-237.

[3] P.L. IVANOV: *Compaction of non-cohesive soils by explosions,* Civil Engineering Publishing Company, Leningrad 1967 (in russian).

[4] R. JACKSON:*Hydrodynamic Stability of Fluid-Particle Systems* in: Fluidization, Second Edition, Academic Press 1985, pp. 47-72, (ISBN 0-12-205552-7).

[5] D. KOLYMBAS: *Dynamic compaction of saturated granular media,* Mech. Res. Comm. 9(6), (1982), pp. 351-358.

[6] H.K. KYTÖMAA: *Propagation and structure of solidification waves in concentrated suspensions,* Mechanics of Materials **9** (1990), pp. 205-215.

[7] L.W. MORLAND, A. SAWICKI: *A mixture model for the compaction of saturated sand,* Mechanics of Materials 2 (1983), pp. 203-216.

[8] W.K. NOWACKI, B. RANIECKI: *Theoretical analysis of dynamic compacting of soil arround a spherical source of explosion,* Arch. Mech. **39**, 4 (1987), pp. 359-384.

[9] R.F. SCOTT: *Soil properties from centrifuge liquefaction tests,* Mechanics of Materials **5** (1986), pp. 199-205.

[10] R.F. SCOTT: *Solidification and consolidation of a liquefied sand column,* Soils and Foundations, Vol. **26**, No. 4 (1986), pp. 23-31.

[11] D.R. SCOTT: *Magma Solitons,* Geophysical Research Letters, **11**, 11 pp. 1161-1164 (1984)

[12] J.A. WITEHEAD: *A laboratory demonstration of solitons using a vertical watery conduit in syrup,* Am. J. Phys. **55**,11 pp. 998-1003 (1987)

[13] P.OLSON, U.CHRISTENSEN: *Solitary wave propagation in a fluid conduit within a viscous matrix,* Journal of Geophysical Research **91**,B6 pp. 6367-6374 (1986)

[14] D. McKENZIE: *The generation and compaction of partially molten rock,* Journal of Petrology **25**,3 pp. 713-765 (1984)

[15] F.M. RICHTER, D. McKENZIE: *Dynamical models for melt segregation from a deformable matrix,* Journal of Geology **92** pp. 729-740 (1984)

[16] V. BARCILON, F.M.RICHTER: *Nonlinear waves in compacting media,* J.Fluid Mech. **164** pp. 429-448 (1986)

[17] D.R. SCOTT, D.J. STEVENSON, J.A. WHITEHEAD Jr: *Observations of solitary waves in a viscously deformable pipe,* Nature **319** pp. 759-761 (1986)

[18] M.J. LIGHTHILL: *Viscosity effects in sound waves of finite amplitude,* Surveys in mechanics pp. 250-351 (1965)

[19] R. ABEYARATNE, J.K. KNOWLES: *Kinetic relations and the propagation of phase boundaries in solids,* Arch.Rational Mech. Anal. **114** pp. 119-154 (1991)

[20] R.I.NIGMATULIN: *Dynamics of multiphase medria* (in russian), Nauka, Moscow 1987

[21] S.C. COWIN, W.J. COMFORT III: *Gravity-induced Density Discontinuity Waves in Sand Columns,* Journal of Applied Mechanics **498**, Vol. 49, September 1982

[22] E. KURIBAYASHI, F. TATSUOKA: *History of Earthquake-induced soil liquefaction in Japan,* Bulletin of Public Works Research Institute, Vol.38, edited by the Ministry of Construction, Tokyo, 1977

Post-critical response of soils and shear band evolution
Comportement post-critique des sols et évolution des bandes de cisaillement

Z. Mróz & J. Maciejewski
Institute of Fundamental Technological Research, Polish Academy of Sciences, Warsaw, Poland

ABSTRACT: The limit analysis theory is usually referred to an initial state of plastic flow of rigid-plastic materials for a specified configuration of external boundaries and interfaces. For a continuing deformation process the initial configuration changes and the mode of flow evolves. For dilatant and softening behaviour of cohesive soils, the failure mode is usually composed of localised deformation within shear bands. Some of these bands are material planes, while remaining bands move with respect to the material in order to constitute a kinematically admissible failure mode. A characteristic effect of switching to new failure modes occurs at particular states of the deformation process, resulting in an oscillatory force-displacement response. Three cases are treated in detail, namely, horizontal wall penetration into a soil, wedge and punch penetration into a semiplane. The theoretically predicted pattern of shear bands is confirmed experimentally.

Dans la théorie de l'analyse limite, on considère en général un état initial d'écoulement plastique du milieu rigide-plastique pour une configuration spécifiée des frontières et des interfaces. Dans un processus de déformation ininterrompu, la configuration change et le mode de déformation évolue. Dans le comportement des sols cohérents dilatants et adoucissants, le mode de rupture est habituellement caractérisé par l'apparition de déformations localisées en bandes de cisaillement. Certaines de ces bandes sont des plans matériels, mais d'autres se déplacent par rapport au matériau de façon à constituer un mode de rupture cinématiquement admissible. Un effet caractéristique de passage à un nouveau mode de rupture se produit à certains stades du processus de déformation, conduisant à des oscillations dans la réponse en terme de force-déplacement. Trois cas sont étudiés en détail: la pénétration d'un mur dans un sol, d'une arête et d'un poinçon dans un demi-plan. La figure de localisation prévue théoriquement est confirmée par l'expérience.

1. INTRODUCTION

The limit analysis theory is based on the assumption of a rigid-perfectly plastic material response for which the kinematically admissible failure mode is assumed to develop instantaneously, Izbicki and Mróz [6]. Usually, the plastic flow occurs along the strong velocity discontinuity surfaces and within finite plastic zones with continuous velocity and strain fields. As the deformation progresses, the initial mode will evolve during the post-yield period, thus resulting in configuration changes associated with stable or unstable load-displacement response.

When the plastic deformation in cohesive soils is associated with softening and varying dilatancy, it can be expected, that the localised mode of flow within shear bands becomes predominant. The initial failure mechanism is modified, with some shear bands becoming material interfaces and the remaining bands moving with respect to material so that the failure mechanism is kinematically admissible. We shall use the terms *material shear bands* and *adjusting bands* in discussing the evolving failure mechanisms. The other important effect occurs, namely, *mode switching* at discrete set of states with new material and adjusting bands generated and continued until a consecutive switching point is reached. In this way, a periodic pattern of shear bands develops with a characteristic oscillatory load-displacement curve. This type of deformation mode, associated with generation of active shear bands at switching points will be discussed in detail.

For an elastic-plastic model, the plastic zones are developing progressively with continuous velocity and strain fields. The onset of localisation occurs when the tangent stiffness matrix ceases to satisfy the ellipticity condition. For the non-associated flow rule, the strain localisation occurs in the hardening regime, cf. Rudnicki and Rice [13], and the progressive localised deformation occurs within the band in the softening regime. There are several approaches in treating the boundary-value problems with non-homogeneous localised deformation, such as smeared shear band approach, Pietruszczak and Mróz [12], non-local theories, Bazant an Feng Bao Lin [1], concept of gradient limiters, Schreyer and Chen [14], or approach following from the Cosserat medium accounting for couple stresses and micro rotations, Muhlhaus and Vardoulakis [11]. However, only simple boundary-value problems were solved for elasto-plastic soil models and the problem of shear band evolution for a finite deformation process was not treated. On the other hand, the rigid-plastic model accounting for hardening and softening effects provides a simple way to treat progressive soil deformation using the concept of kinematically admissible failure modes, mode switching effects, and the expected upper bound on applied loads.

In the next section, the fundamental constitutive assumptions on deformation response within the shear band will be discussed and in Section 4 some particular boundary-value problems, such as rigid wedge, punch and wall penetration into the soil, will be treated. These examples are fundamental in understanding the machine tool-soil interaction problems, especially soil cutting and moving processes.

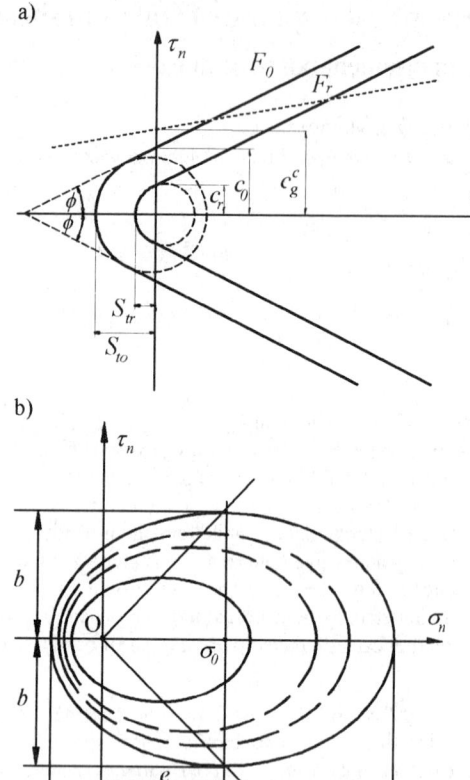

Fig. 1 Yield and critical state surfaces for soils, a) modified Coulomb condition, b) elliptical condition

2. CONSTITUTIVE MODELS FOR LOCALISED FLOW INTERFACES

The shear band is regarded as a thin interface between two rigid or plastic domains. The thickness d of the band (several times greater than the average grain size) is small as compared to a typical dimension of the problem. Denote by $\sigma_n, \tau_n, \sigma_t$ the stress within the band and by $\dot\varepsilon_n, \dot\gamma_n, \dot\varepsilon_t$ the strain rate components. Here (n, t) is the local Cartesian system with n normal to the band. The strain rate components are

$$\dot\gamma_n = \frac{[v_t]}{d}, \quad \dot\varepsilon_n = \frac{[v_n]}{d}, \quad \dot\varepsilon_t = 0 \tag{1}$$

The specific rate of dissipation per unit area of the interface equals:

$$D = (\sigma_n \dot\varepsilon_n + \tau_n \dot\gamma_n)d = \sigma_n[v_n] + \tau_n[v_t] \tag{2}$$

Consider first the Coulomb shear band, Fig. 1a.

During plastic deformation, the initial cohesion $c=c_0$ decreases and reaches its residual value $c=c_r$. We can write (assuming compressive stress and associated strains as positive)

$$F_1(\sigma_n, \tau_n, c) = \tau_n - \sigma_n \tan\varphi - c = 0 \tag{3}$$

The plastic potential is assumed in a form

$$G_1(\sigma_n, \tau_n, c_g^c) = \tau_n - \sigma_n \tan\varphi - c_g^c = 0 \tag{4}$$

where φ and ψ are the friction and dilatancy angles, and c_g^c denotes the apparent cohesion dependent on the actual normal stress σ_n or the shear stress τ_n at the interface. In particular, when the associated flow rule occurs, $\varphi=\psi$, then $c_g^c=c$. The plastic strain rate within the shear band for the Coulomb state equals:

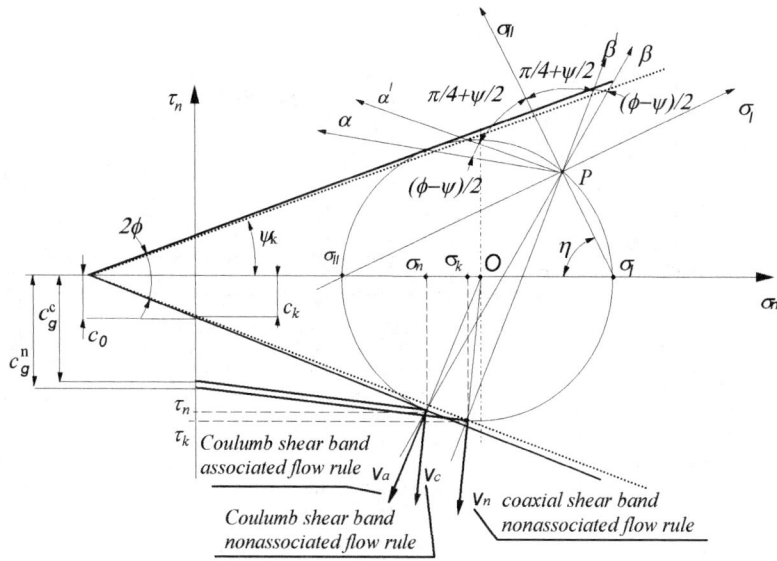

Fig. 2 Coulomb condition.

$$\dot{\gamma}_n = \frac{[v_t]}{d} = \dot{\lambda}\frac{\partial G_1}{\partial \tau_n} = \dot{\lambda}$$

$$\dot{\varepsilon}_n = \frac{[v_n]}{d} = \dot{\lambda}\frac{\partial G_1}{\partial \sigma_n} = -\dot{\lambda}\tan\psi \quad , \dot{\lambda} > 0 \quad (5)$$

$$\dot{\varepsilon}_t = 0$$

and the dissipation rate is

$$D = \dot{\lambda}\left[c + \sigma_n(\tan\varphi - \tan\psi)\right]d$$

$$= \left[\tau_n - (\tau_n - c)\frac{\tan\psi}{\tan\varphi}\right][v_t] = c_g^c[v_t] \quad (6)$$

In particular, when the associated flow rule occurs, $\phi = \psi$, from (6) it follows that

$$D = c[v_t] \quad (7)$$

and the apparent, stress dependent cohesion c_g^c becomes the material cohesion c.

Note that in the Coulomb shear band there is no coaxiality between stress and strain rate in the case of non-associated flow rule, $\phi \neq \psi$.

When the coaxiality of principal axes of stress and strain rate occurs, the shear band does not follow the Coulomb plane, but deviates from it, being inclined at the angle $\pm(\pi/4 + \psi/2)$ to the minor principal stress direction. Instead of (3), we now have

$$F^a(\sigma_n, \tau_n, c) = |\tau_n| - \sigma_n \tan\varphi_k - c_k = 0 \quad (8)$$

where

$$c_k = \frac{c\cos\psi\cos\varphi}{1 - \sin\psi\sin\varphi}$$

$$\tan\varphi_k = \frac{\cos\psi\sin\varphi}{1 - \sin\psi\sin\varphi} \quad (9)$$

an the dissipation rate is

$$D = \frac{1}{1 - \sin\psi\sin\varphi} \cdot$$

$$\left[c\cos\varphi\cos\psi - \sigma_n\frac{\sin\varphi - \sin\psi}{\cos\psi}\right][v_t] = \quad (10)$$

$$= c_g^n[v_t]$$

where c_g^n is the apparent stress dependent cohesion associated with the coaxial flow rule. Fig. 2 presents the Coulomb yield condition in the stress plane and the apparent yield locus (9) associated with the coaxial flow rule (dotted lines). The stress characteristics (α,β) and (α',β') associated with these two conditions are depicted in Fig. 2. Obviously, when $\phi = \psi$, there is $D = c[v_t]$ as predicted by (7).

For the limit tension regime, the condition $\dot{\varepsilon}_t = 0$ requires $[v_t]=0$ and σ_n follows the principal tensile strain. The dissipation rate now is

$$D = S_t \dot{\varepsilon}_n d = S_t \dot{\varepsilon}_1 d = S_t [v_n] \quad (11)$$

Figure 3 presents schematically three types of bands discussed in this section.

Let us now pass to density hardening model implying existence of the critical state. Consider a set of elliptical yield loci depending on varying material density and generated by similarity mapping with respect to the origin O, Fig. 1b. The yield locus has the form

$$F(\sigma_n, \tau_n, e, a) = (\sigma_n - e)^2 + \frac{\tau_n^2}{m^2} - a^2 = 0 \quad (12)$$

where m, a and e are the geometrical parameters namely a is the larger semiaxis, e the position of ellipse center and m is the ratio of semiaxes. The similarity coefficient α is specified as

$$\alpha = \frac{a-e}{a+e}, \quad \text{or} \quad e = \frac{1-\alpha}{1+\alpha} a = ka \quad (13)$$

Note that when $\alpha=0$, then $e=a$ and all ellipses pass through the origin, when $\alpha=1$, then $e=0$ and the ellipse center coincides with the origin O. The associated flow rule now provides the relations

$$\dot{\gamma}_n = \frac{[v_t]}{d} = \dot{\lambda} \frac{\partial F}{\partial \tau_n} = 2\dot{\lambda} \frac{\tau_n}{m^2}$$

$$\dot{\varepsilon}_n = \frac{[v_n]}{d} = \dot{\lambda} \frac{\partial F}{\partial \sigma_n} = 2\dot{\lambda}(\sigma_n - ka) \quad ,\dot{\lambda} > 0 \quad (14)$$

$$\dot{\varepsilon}_t = 0$$

where

$$\dot{\lambda} = \frac{1}{2a}\sqrt{[v_n]^2 + m^2[v_t]^2} \quad (15)$$

and the dissipation function has the form

$$D = a\left\{k[v_n] + \sqrt{[v_n]^2 + m^2[v_t]^2}\right\} \quad (16)$$

The present model provides a gradual transition to the critical state $\sigma_n = ka$, $\tau_n = mka$. The evaluation of the dilatancy angle along any stress path is now generated by (14), namely

$$\tan \psi = \frac{[v_n]}{[v_t]} = m^2 \frac{\sigma_n - ka}{\tau_n} = m^2 \frac{1 - ka/\sigma_n}{\tau_n / \sigma_n} \quad (17)$$

Assuming the shear band to be generated in the overconsolidated state, $\sigma_n \le e$, one can compute the en-

a) Coulomb shear band

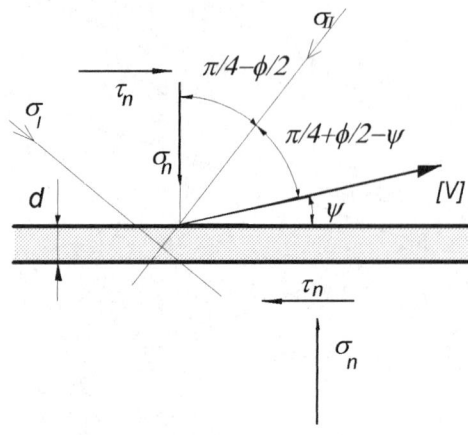

b) Coaxial shear band

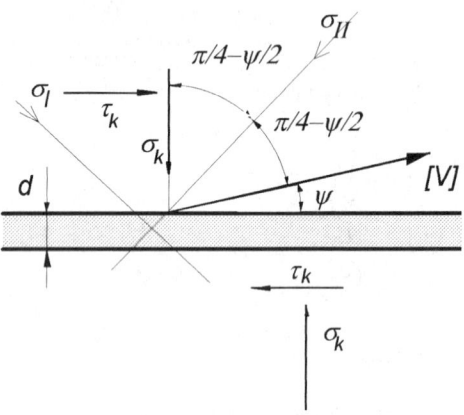

c) tensile band

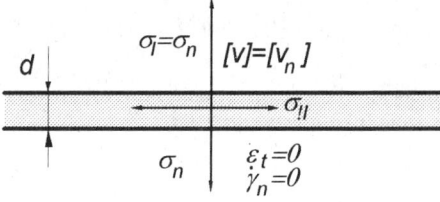

Fig. 3 Types of shear bands.

ergy dissipation required to achieve the critical state along a specified path. The density hardening could be specified by relation

$$\frac{a}{a_c} = \left(\frac{\rho - \rho_{min}}{\rho_{max} - \rho_{min}}\right)^n \quad (18)$$

where a_c denotes the maximal value of a reached at the maximal density ρ_{max}, and ρ_{min} is the minimal density. The dissipation rate at the critical state is obtained from (16)

$$D_c = am[v_t] \quad (19)$$

and depends on the actual material density at the critical state. For normally consolidated states, $\sigma_n \geq e$, the material exhibits hardening, hence, the localised zones do not occur. However, the concept of shear bands can be used in approximating the actual mode of flow by a set of rigid domains sliding with respect to each other along shear bands combined with compaction and hardening response.

3. INCREMENTAL BOUNDARY-VALUE PROBLEMS

The constitutive shear band models discussed in the preceding section can now be used in incremental analysis of boundary-value problems associated with progressive penetration of a rigid tool into the soil mass. The initial limit state and the associated failure mode is determined by the standard kinematic approach or in some cases by the complete solution. We may start from the work rate equation

$$\int \sigma_{ij} \dot{\varepsilon}_{ij}^k dA + \sum_i \left(\sigma_n [v_n^k] + \tau_n [v_t^k] \right) dl_i =$$
$$= \lambda_k \int T_i v_i^k dS \quad (20)$$

where the first and second terms represents the rate of dissipation in continuous and localised zones. In our analysis, the failure mode is approximated by a set of rigid blocks sliding with respect to each other along interface bands. Let us discuss the incremental procedure for the case of Coulomb yield condition associated with softening and dilatancy within bands, Fig. 1a. Assume the kinematic load multiplier λ_k has been found by minization with respect to geometric parameters. The consecutive step will require selection of material shear bands, and adjusting bands translating and rotating relative to the material. The respective block geometry will therefore be modified.

Fig 4a presents a moving band with normal velocity V_n and the material velocities v_1 and v_2 on both sides of the band. The density variation across the band can now be specified from the mass balance equation, namely

$$\rho_1 (v_{1n} - V_n) = \rho_2 (v_{2n} - V_n) \quad (21)$$

For the material band, the density variation is specified by assuming a finite band thickness d and using the relation

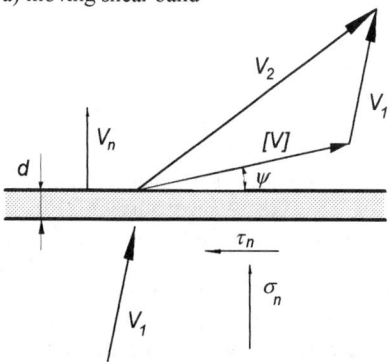

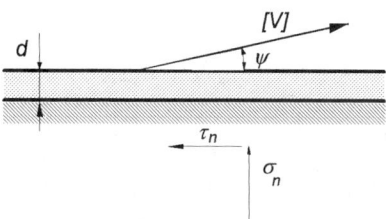

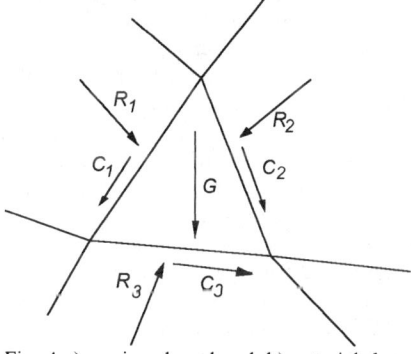

Fig. 4 a) moving shear band, b) material shear band, c) equilibrium of block system.

$$\dot{\rho} = -\rho \dot{\varepsilon}_n = -\rho \frac{[V_n]}{d} \quad (22)$$

Assume the linearly varying cohesion with the tangential displacement along the band, thus

$$c = c_0 - (c_0 - c_r) \frac{s}{s_0} , \quad for \quad s < s_0$$
$$c = c_r , \quad for \quad s > s_0 \quad (23)$$

where s_0 denotes the critical value of shear displacement inducing the residual state. Similarly, the dilatancy angle varies according to the relation

$$\psi = \psi_0\left(1 - \frac{s}{s_0}\right) \quad , \quad for \quad s \leq s_0 \qquad (24)$$
$$\psi = 0 \quad , \quad for \quad s > s_0$$

where the ψ_0 denotes the initial dilatancy angle.

Instead of the work balance equation (20), the equilibrium method is used in calculating the load factor at each configuration, cf. Mróz and Dresher [9]. In fact, assuming Coulomb interfaces between blocks, the tractions satisfy the Coulomb condition and the force diagram can be generated for each block, providing an assessment of the load factor, thus

$$\int \tilde{T}_1 dS_1 + \int \tilde{T}_2 dS_2 + \int \tilde{T}_3 dS_3 - \int \tilde{f} dV = 0 \qquad (25)$$

where S_1, S_2, S_3 are block side areas and $\tilde{T}_1(\sigma_{n1}, \tau_{n1}), \tilde{T}_2(\sigma_{n2}, \tau_{n2}), \tilde{T}_3(\sigma_{n3}, \tau_{n3})$ are the tractions satisfying the Coulomb condition or tool-soil friction condition, and the $\int \tilde{f} dV$ denotes the gravity force at the moving block. For the coaxial shear band, the respective material parameters specified by (9) are associated with the yield condition.

It should be noted that the equilibrium method provides the upper bound to the failure load when the interfaces between elements are the Coulomb bands. On the other hand, when the coaxial shear band are assumed, then the equilibrium method provides an upper to a the limit load associated with a fictitious Coulomb material for which the cohesion and the friction angle are specified by (9).

Hence, the limit load assessment obtained for non-associated flow rule could be lower than that for the Coulomb material with a non-associated and non-coaxial flow rule. However, the differences in limit load predictions for coaxial and non- coaxial flow rules are usually small.

After specifying the limit load factor and hodograph at each step, a new configuration is generated and the analysis repeated. It is also verified whether mode switching may occur. In the class of virtual failure modes, such mode is looked for to satisfy the conditions

$$\lambda > \lambda_k \quad or \quad \lambda = \lambda_k \quad , \frac{d\lambda}{ds} > \frac{d\lambda_k}{ds} \qquad (26)$$

where λ and λ_k are limit load factors for the actual and virtual modes and $d\lambda/ds$ is the load factor derivative with respect to the progressive tool penetration s. The relations (26) provide the mode switching condition. On the other hand, we have the following conditions for mode continuation

$$\lambda < \lambda_k \quad or \quad \lambda = \lambda_k \quad , \frac{d\lambda}{ds} < \frac{d\lambda_k}{ds} \qquad (27)$$

The incremental analysis algorithm is now summarised in the diagram:

Analysis algorithm

```
┌─────────────────────────────┐
│    Initial configuration    │
└──────────────┬──────────────┘
               ▼
┌─────────────────────────────┐
│ Tool incremental displacement │◄──┐
└──────────────┬──────────────┘   │
               ▼                   │
┌─────────────────────────────┐   │
│ Kinematically admissible mechanism │──┐
└──────────────┬──────────────┘   │  │
               ▼                   │  │
┌─────────────────────────────┐   │  │
│       Velocity field        │   │  │
└──────────────┬──────────────┘   │  │
┌─────────────────────────────┐   │  │
│      Block equilibrium      │   │  │
└──────────────┬──────────────┘   │  │
               ▼                   │  │
┌─────────────────────────────┐   │  │
│  Updating of soil parameters │   │  │
└──────────────┬──────────────┘   │  │
               ▼                   │  │
┌─────────────────────────────┐   │  │
│ Updating of soil configuration │  │  │
└──────────────┬──────────────┘   │  │
               ▼                   │  │
┌─────────────────────────────┐   │  │
│      Verification of         │   │  │
│      new failure mode        │   │  │
└────────┬──────────────┬─────┘   │  │
       Yes             No          │  │
         └──────────────┴──────────┘  │
                        └─────────────┘
```

4. EXAMPLES

In this section we shall consider three examples of incremental analysis associated with large deformation of soils.

4.1. Wedge indentation into a half-space

This classical problem of the theory of plasticity was treated by numerous researchers. Static and kinematic solution of wedge indentation into a weightless soil obeying the associated flow rule was obtained by Shield [15]. A kinematically admissible solution for this problem was discussed by Drescher and Michałowski [4] accounting for gravity forces and considering both associated and non-associated flow rules. The experimental study of wedge indentation into a granular material by Drescher et.al. [5] exhibited departure of stress and strain rate characteristics and good description was obtained by using the non-associated flow rule with vanishing dilatancy effect. The indentation force increased monotically with the depth of penetration. On the other hand, the experi-

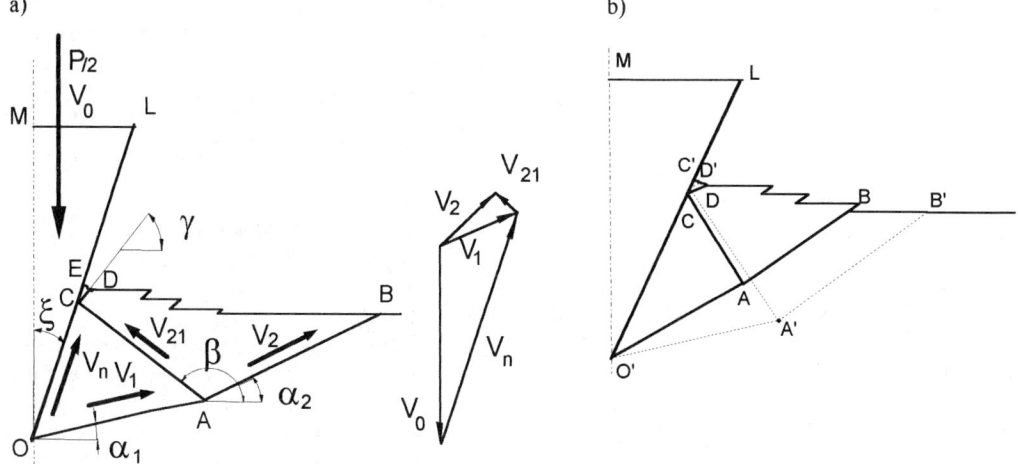

Fig. 5. a) Kinematically admissible mechanism and velocity hodograph, b) the mode switching configuration.

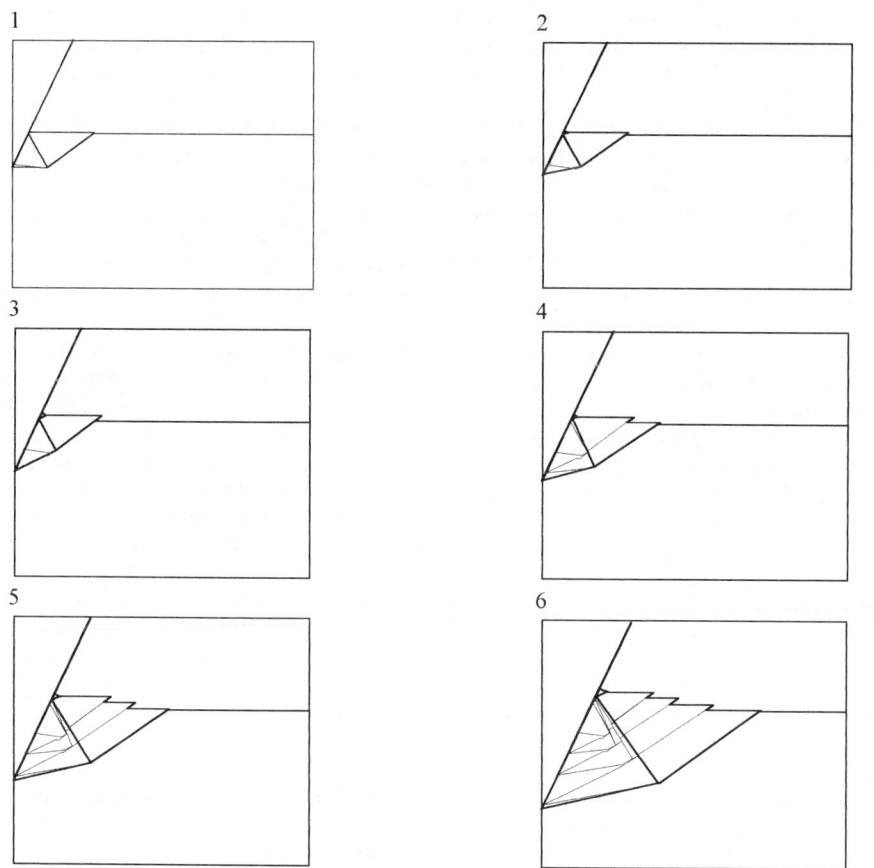

Fig. 6 The consecutive failure modes during the wedge penetration.

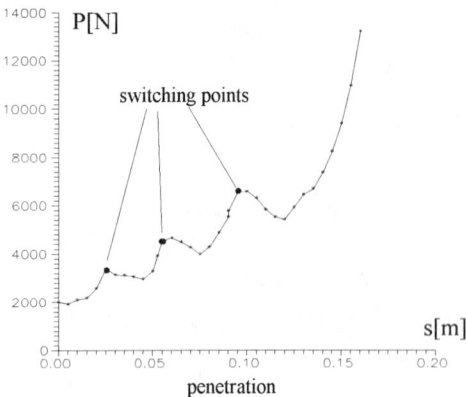

Fig. 7. Force-penetration diagram

mental study of wedge penetration into a dense sand by Butterfield and Andrawes [2] exhibited oscillatory effect on the load-displacement curve and localised zones of deformation moving with the material.

Consider the failure mode shown in Fig. 5 For the advanced flow, three rigid zones OAC, ABCD and CED are separated by shear bands with strong velocity discontinuities. The optimal failure mode is obtained by assuming the shear band AB and CA to be material whereas OA is an adjusting band moving with respect to the material . The shear band CD follows from the localised interaction of domain ABCD with the wedge flank. The velocity hodograph is shown in Fig . 5.

As the shear bands AB and AC are material, the cohesion in these bands takes the residual value c_r after an initial softening process. Similarly, the dilatancy angle tends to zero and velocity discontinuities follow the lines AB and AC. On the other hand, the shear band OA connecting moving wedge vertex O and the point A, moves toward undisturbed material, so the initial cohesion value c_0 should be ascribed to this band. The force method described in the previous section is applied at each step with the configuration update following the hodograph.

When the wedge vertex moves to the position O' a new failure mode composed of material bands C'A', A'B' and C'D', and the adjusting band A'O' is compared with the actual mode OABCDE, Fig. 4b. When P=P' and dP/ds > dP'/ds, the switching occurs to the new mode and the previously active bands become passive moving bodily with the material. Here P and P' are the calculated loads for both modes. When the new mode becomes active, first the load decreases because of softening along the material bands. Next due to configuration changes, the load starts to increase until the new consecutive switching point is reached. The material parameters were assumed as follows:

c_0 = 20000 [N/m^2]
S_{t0} = 5000 [N/m^2]
c_r = 5000 [N/m^2]
S_{tr} = 1250 [N/m^2]
ϕ = 23 [deg]
γ = 18000 [N/m^3]
δ = 15 [deg]
s_0 = 0.01 [m]
2ξ = 60 [deg]

where δ denotes the friction angle between the soil and wedge flank, s_0 is the softening parameter corresponding to the displacement of transition to the residual state, γ denotes the specific soil weight, 2ξ is the wedge opening angle, and c_0, S_{t0}, c_r, S_{tr} denote initial and residual values of cohesion and tensile strength.

Figure 6 presents the consecutive failure modes during the wedge penetration. Figure 7 shows the force-displacement diagram for the wedge .The mode switching points are marked on this diagram. Its character is very close to that observed by Butterfield and Andrawes[2], who observed also some material shear bands.

Figures 8 presents preliminary observation of the authors made in their tests .The wedge was indented into a dense sand mixed with a glycerine oil and cement. The initial cohesion was similar to that assumed in the calculation. A set of material shear bands is clearly seen near the surface. On the other hand, the adjusting bands OA are not seen as they move with respect to the material, thus resulting in the diffuse strain distribution. Figures 8a and 8b present the wedge at the penetration depths 0.15 m and 0.19 m. The first picture shows clearly the material shear bands propagating to the surface. Due to wedge penetration, the mode switching occurs and in Fig. 8b the consecutive material shear band develops with a new failure mode occurring for a finite penetration increment. There is no full symmetry of the failure process: in fact the consecutive failure modes on both sides of the wedge may develop at different instants. This is due to soil inhomogeneity and also to elastic wedge compliance.

For a layered soil the evolution of failure modes may follow different pattern. Fig 8 c,d presents such modes for a soil composed of distinct horizontal layers consolidated separately. It is seen that for a sufficiently large penetration depth the material shear band interacts with the interface between two layers inducting progressive *delamination* and *separation* of the upper layer from the bottom layer.

a)

b)

c)

d)

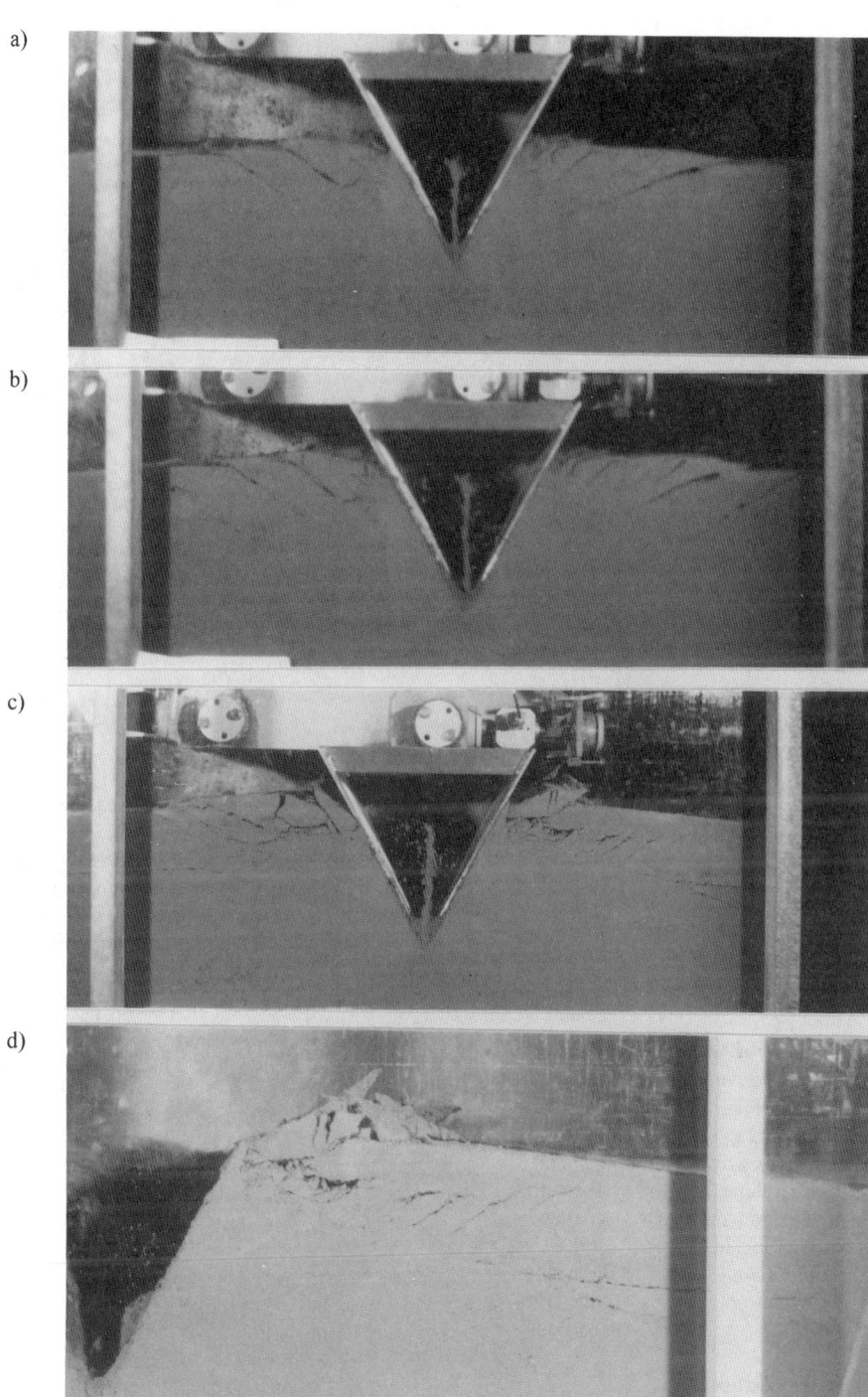

Fig. 8 Failure mode during wedge penetration

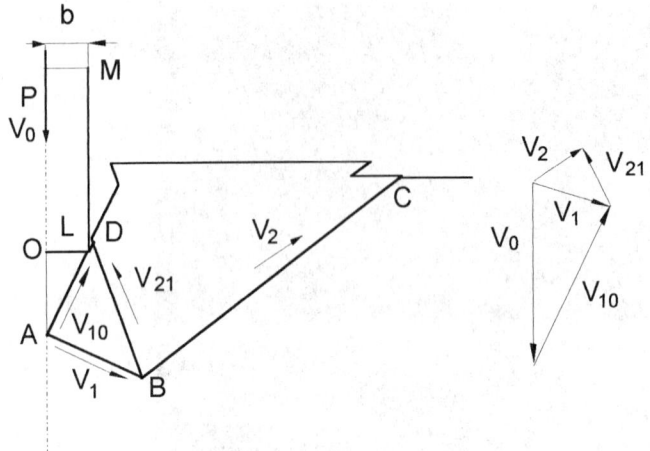

Fig. 9. Kinematically admissible mechanism and velocity holograph

The progressive delamination is inducted by the tension mode of failure propagation along the interface and subsequent flexure of the separated portion. The limit load associated with this mode is much lower as that associated with the material shear band propagating to the surface. A more detailed study of this mode will be presented in a separate paper.

A similar active band creation and passive band motion was studied by Michałowski [13] is his analysis of material flow in a converging hopper.

4.2. Punch indentation

Consider a rigid punch of width $2b$ penetrating into the semispace of a cohesive and ponderable soil of the specific gravity γ. In Figure 9 the failure mode is presented and the hodograph is constructed. The rigid domain OAD moves downward with the punch, the domain ABD moves along AB with respect to the undisturbed soil, the domain DBC is separated from the undisturbed soil by the material band BC. The shear band AB moves with respect to the material, hence the initial cohesion value is ascribed to AB.

The selected material parameters are:

$$c_o = 20000 \ [N/m^2]$$
$$S_{to} = 5000 \ [N/m^2]$$
$$c_r = 5000 \ [N/m^2]$$
$$S_{tr} = 1250 \ [N/m^2]$$
$$\phi = 23 \ [deg]$$
$$\gamma = 18000 \ [N/m^3]$$
$$\delta = 15 \ [deg]$$
$$s_o = 0.01 \ [m]$$
$$2b = 0.1 \ [m]$$

where the notation is the same as in the first example.

Figure 11 presents the evolution of consecutive failure modes and Fig. 10 shows the corresponding punch force-displacement diagram. The mode switching points are marked in this figure as points of local maxima on the diagram.

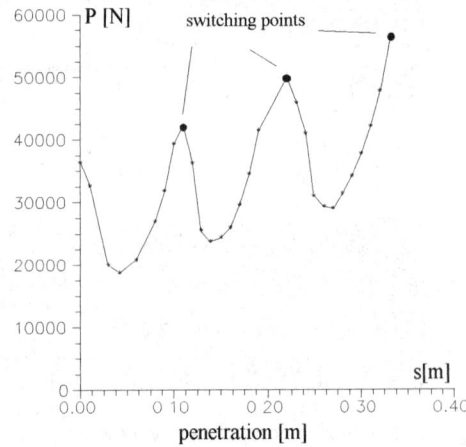

Fig. 10. Load-penetration diagram.

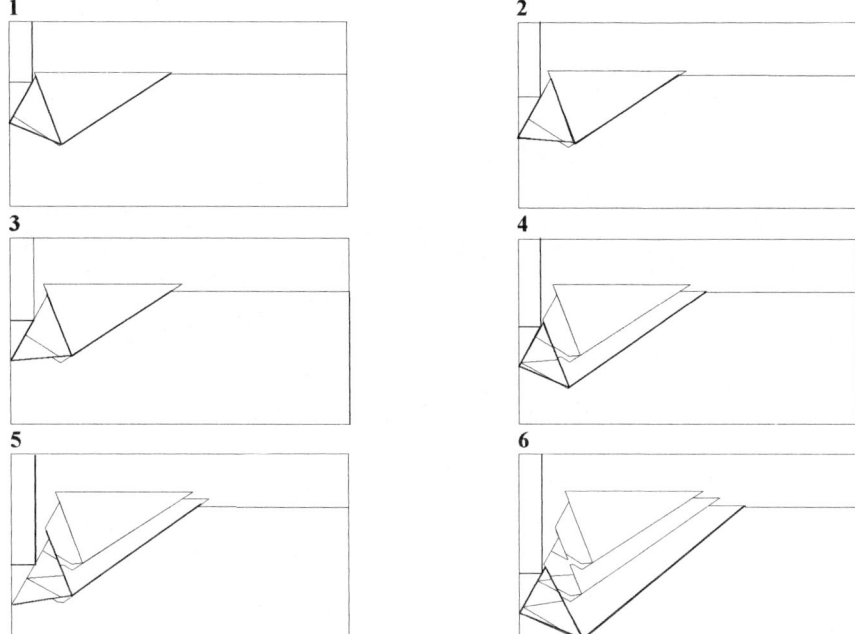

Fig. 11. The consecutive failure modes during the punch penetration

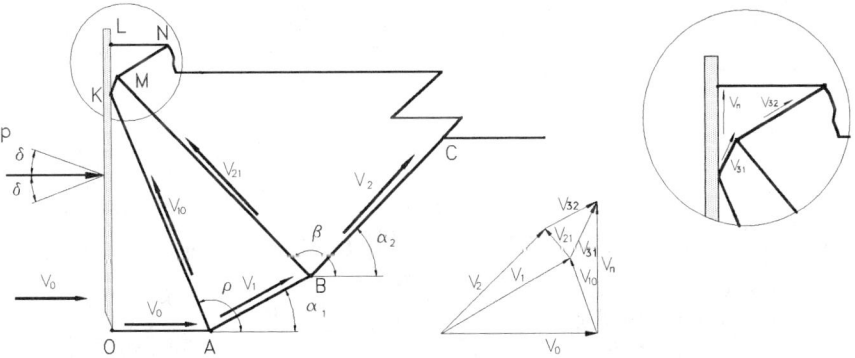

Fig. 12. Kinematically admissible mechanism and velocity holograph

4.3. Penetration of a vertical wall into the soil

Figure 12 presents the kinematically admissible failure mode composed of four rigid blocks OAK, AKMB, BMNC and KMNL. The block OAK moves horizontally with the plate, the block AKMN moves with the velocity v_1 with respect to soil and the block BCKM moves with the velocity v_2. The bands BC, BM, OA, AK are material with the residual cohesion c_r reached after initial softening period. On the other hand, the adjusting band AB moves with respect to the material. The bands KM and MN constitute the localised failure mode due to interaction of the soil with the plate. The material parameter were selected as follows:

c_o = 15000 [N/m²]
S_{to} = 7500 [N/m²]
c_r = 5000 [N/m²]
S_{tr} = 2500 [N/m²]
ϕ = 23.4 [deg]
δ = 15 [deg]
γ = 18000 [N/m³]
s_o = 0.01 [m]
h = 0.18 [m]

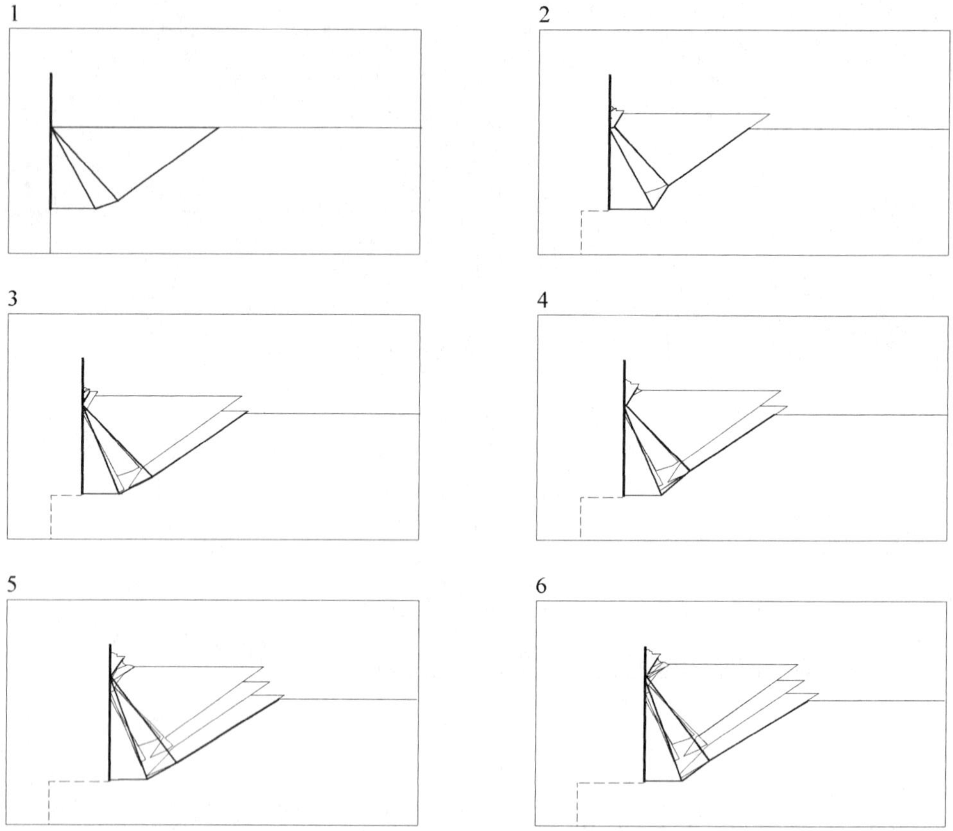

Fig. 13. Computer simulation of the rigid wall movement.

where h denotes the depth of penetration with respect to the soil level.

Fig 13 presents the evolving failure modes during consecutive stages of wall penetration. Figure 14 presents the plate force-displacement diagram. Similarly as previously, mode switching points occur at the load maxima of diagram.

The experiments carried out with the cohesive sand confirm this solution. In Figure 15 a-c, the characteristic set of blocks separated by material shear bands is seen after wall penetration into the soil. The oscillatory character of horizontal and vertical force variation is also observed in tests, Fig. 15d. The force (in N) is plotted against horizontal wall displacement (in units equal 0.154 mm).

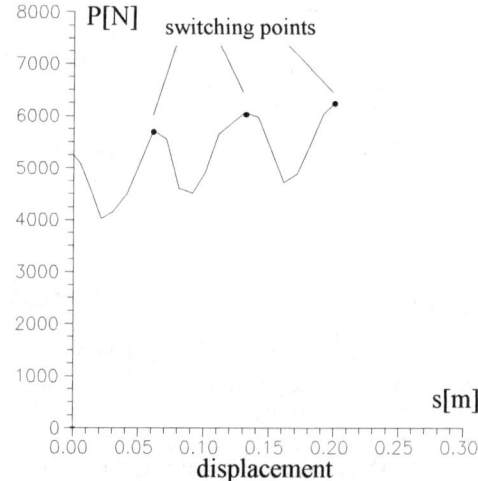

Fig. 14. Penetration load versus rigid wall displacement.

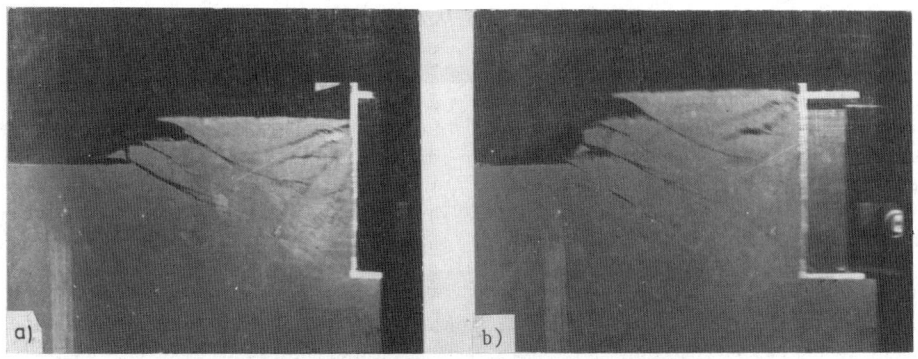

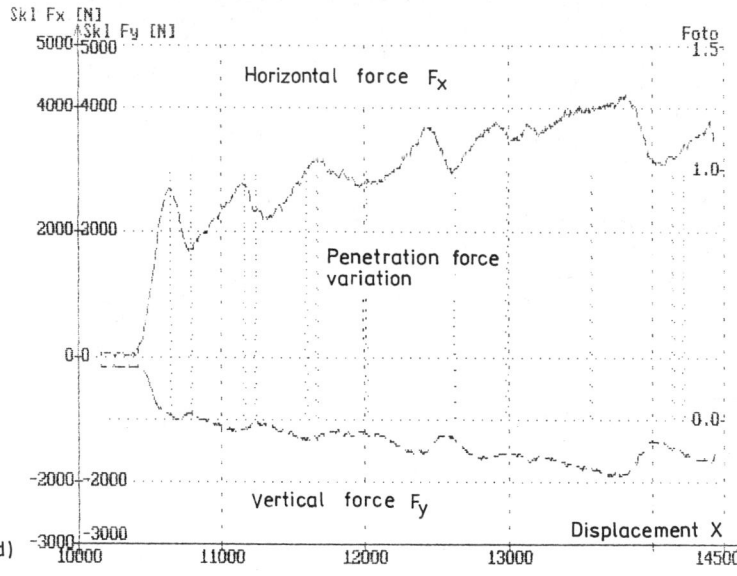

Fig. 15. Failure modes during the wall movement, through the depth h=0.1m ; a) the deformation stage with 2 shear bands, (s=0.04 m); b) the subsequent stage with 3 shear bands, (s=0.07m); c)the ultimate stage with 5 shear bands, (s=0.25m); d) horizontal and vertical force (in N) diagram versus wall displacement (in units equal to 0.154 mm)

5. CONCLUDING REMARKS

The present analysis exhibits a characteristic feature of solution for softening materials, namely creation of failure modes with material and moving shear bands, mode switching effect and oscillatory force-displacement diagram. The experimental observations confirm this evolution of failure modes resulting in creation of a set of blocks separated by material shear bands. The rigid-plastic softening response was assumed in the analysis. For elastic plastic model a similar feature should be exhibited, however, with the gradual progression of shear bands through the material and progressive creation of consecutive modes and new switching points. Such boundary-value problems will constitute a challenging task for finite element solutions.

Acknowledgement

The tests were carried out jointly with Prof. W.Trąmpczyński, dr. A.Jarzębowski, J.Cendrowicz and D.Szyba. The authors wish to express thanks for their generous assistance and discussions.

REFERENCES

1. Bazant, Z.P. & Feng-Bao Lin 1988. Non-local yield limit degradation. Int J. Num. Math Eng. 26: 1805-1823.
2. Butterfield, R. & K.Z. Andrawes 1972. An investigation of a plane strain continuous penetration problem. *Geotechnique 22* : 597-617.
3. Desrues, J. J. Lanier & P. Stutz 1985. Localisation of the deformation in test on sand sample. Eng. Fract. Mech vol 21: 909-921.
4. Drescher, A. & R.L. Michałowski 1984. Density variation in pseudo-steady plastic flow of granular media . *Geotechnique 34*: 1-10.
5. Drescher, A. K. Kwaszczyńska & Z. Mróz 1967. Statics and kinematics of granular medium in the case of wedge indentation. Arch. Mech. Stos. 19 : 99-113.
6. Izbicki, R. & Z. Mróz 1976. Methods of Limit Analysis in Soil and Rock Mechanics. (in Polish) . PWN
7. Michałowski, R.L. 1990. Strain localisation and periodic fluctuations in granular flow processes from hoppers. *Geotechnique 40* : 389-403.
8. Mróz, Z. 1985. Current trends and new directions in mechanics of geomaterials. in "Mechanics of Geomaterials". Ed Z.P.Bazant. J. Wiley. Sons.: 539-566.
9. Mróz, Z. & A. Drescher 1969. Limit plasticity approch to some cases of flow bulk solids. J. Eng. for Industry. Trans ASME. 91: 357-364.
10. Mróz, Z. & Cz. Szymański 1978. Non-associated flow rules in description of flow of granular materials.CISM- Springer Verlog. ed W. Olszak.
11. Muhlhlaus, H.B. & I. Vardoulakis 1987. The thickness of shear bands in granular materials. *Geotechnique* 37:271- 283.
12. Pietruszczak, S. & Z. Mróz 1981. Finite element analysis of deformation of strain- softening materials. Int. J. Num. Math Eng. .17. 327-334.
13. Rudnicki, J.W. & J.R. Rice 1975. Conditions for the localisations of deformation in pressure sensitive dilatant materials. J. Mech. Phys. Solids. 23:371-394.
14. Schreyer, H.L. & Z. Chen 1986. One dimensional softening with localisation. J. Appl. Mech. 53: 791-797.
15. Shield, R.T. 1953. Mixed boundary value-problem in soil mechanics. Q. Appl. Maths. 11: 61-75.
16. Trąmpczyński, W. & J. Maciejewski 1991. On the kinematically admissible solutions for soil-tool interaction description in the case of heavy machine working process.. Proc V-th Europ. Conf. Int. Soc. Terrain Vehicle Systems. Budapest.

2 Localisation and bifurcation conditions
  Localisation et critères de bifurcation

Existence and uniqueness for B.V. problems involving CLoE model

Existence et unicité dans des problèmes aux limites avec le modèle CLoE – Études des solutions avec bandes de cisaillement

D.Caillerie & R.Chambon
Laboratoire Sols, Solides, Structures (3S), UJF – INPG – CNRS, Grenoble, France

ABSTRACT : In this paper recent studies about mathematical problems coming from the use of constitutive equations having the same mathematical main features as CLoE model are investigated. In order to get some firm results we add some restricting assumptions. So the following results are not fully applicable but they are a first step to well understand mathematical and numerical results obtained for boundary value problems with CLoE models.
First, we prove an existence and uniqueness theorem for B.V. problems involving CLoE model in the considered simplifified framework, this theorem holds true for parameters of the model meeting an inequality. Then we study the problem of shear band localization and we prove that such a localization may appear only when the previous inequality is not fulfilled.

Ce papier présente des développements mathématiques récents pour des problèmes faisant intervenir des lois de comportement de caractéristiques principales analogues à celles de la loi CLoE. Des hypothèses restrictives ont été nécessaires pour établir des résultats rigoureux, par conséquent, ces premiers résultats ne sont pas complètement généraux mais ils permettent de mieux comprendre les analyses mathématique et numérique des problèmes aux limites où intervient la loi.

1- INTRODUCTION

We do not attempt here to detail the CLoE model. Such a study may be seen for instance in Chambon et. al. (1994). For the use of CLoE model in this paper, only the mathematical form is needed. For this purpose papers such as the ones of Chambon (1989a 1989b) are available.

The following results can be applied to the Kolymbas' (1987) model and the other models of this family called by their authors "hypoplastic" models (Kolymbas et. al. 1993), as they can be written in the same mathematical form, reminiscent of the hypoelasticity.

For simplicity we only study here the rate (or incremetal) problem. This means that the evolution problem is not investigated. For the same reason we work within the small strain assumption. This means that we do the confusion between the current and the initial configurations.

Let us finally give the principles of our notation. A component of a tensor (or a vector) is denoted by the name of the tensor (or the vector) accompanied with tensorial indices. All tensorial indices are in lower position as there is no need in the following of a distinction between covariant and contravariant components. Other indices have other meaning. The summation convention with respect to repeated tensorial indices has been adopted.

2 - EXISTENCE AND UNIQUENESS THEOREM

2.1 - *The Problem*

Let Ω be a bounded domain of $\mathbb{R}^3$ and Γ the boundary of Ω (Figure 1). So the tensorial incices belong to $\{1, 2, 3\}$. The space variable is denoted x_j. It is easy to simplify the analysis for a two-dimensional problem.

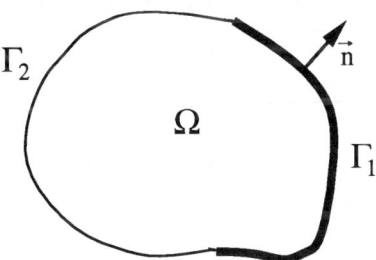

Figure 1 : The boundary value problem

We study the following boundary value problem, called problem P in the following, for the incremental unknowns $\dot{u}$ (velocity) and $\dot{\sigma}$ (stress-rate) which meet:

the balance equation:
$$\partial_j \dot{\sigma}_{ij} + \dot{f}_i = 0 \text{ in } \Omega \qquad (2.1.1)$$

∂_j denotes the partial derivative with respect to x_j. $\dot{f}_i$ is the given rate of body forces,

the constitutive equation:
$$\dot{\sigma}_{ij} = a_{ijkh} \, \varepsilon_{kh}(\dot{u}) + b_{ij} \, |\varepsilon(\dot{u})| \text{ in } \Omega \qquad (2.1.2)$$
$\varepsilon_{kh}(\dot{u})$ is the stain rate define by:
$$\varepsilon_{kh}(\dot{u}) = \frac{1}{2}(\partial_k \dot{u}_h + \partial_h \dot{u}_k)$$
and $|\varepsilon|$ denotes the norm of a second order tensor:
$|\varepsilon| = [\varepsilon_{ij} \, \varepsilon_{ij}]^{1/2}$
the boundary conditions:
$$\dot{\sigma}_{ij} n_j = \dot{F}_i \text{ on } \Gamma_1 \qquad (2.1.3)$$
$$\dot{u}_i = \dot{U}_i \text{ on } \Gamma_2 \qquad (2.1.4)$$
$\dot{F}_i$, and $\dot{U}_i$ are respectively the given rate of surface forces acting on Γ_1 and the given velocity on Γ_2.

Tensors a and b are assumed to meet the following minor symmetry conditions:
$a_{ijkh} = a_{ijhk} \qquad a_{ijkh} = a_{jikh}$
$b_{ij} = b_{ji}$

Remark.
i- Within our assumptions (small strain) there is no difference between the various stress rates.
ii- The coefficients of the constitutive equation a_{ijkh} and b_{ij} may depend on the space variable x_j and on the stress tensor components σ_{ij} which are data for this incremental problem, then the coefficients a_{ijkh} and b_{ij} depend only on x_j.
iii- Due to the term $b_{ij} |\varepsilon(\dot{u})|$ in the constitutive equation (2.1.2), this incremental problem is clearly non-linear.

2.2 - Assumptions

The coefficients a_{ijkh} and b_{ij} are assumed to meet the following relations:
i- The b_{ij}'s are bounded on Ω.
So we can define M by:
$$M = \sup_{x \in \Omega} |b| \qquad (2.2.1)$$
ii- The a_{ijkh}'s are bounded and coercive on Ω, so:
$\exists \, \alpha > 0$ such as:
$\forall \, \theta = (\theta_{ij})$ symmetric tensor:
$$\alpha \, \theta_{ij}\theta_{ij} \leq a_{ijhk}\theta_{kh}\theta_{ij} \text{ in } \Omega. \qquad (2.2.2)$$

2.3 - Theorem

The existence and uniqueness result for this problem is then:

Theorem
If $M/\alpha < 1$, the previous problem P has one and only one solution (in a sense which is precised in the following).

Remark.
Under the assumption: $M/\alpha < 1$, the constitutive equation satisfies almost everywhere in Ω the positive definiteness of the rate of second order work (Drucker 1950, Hill 1958), that is:
$\dot{\sigma}_{ij}\varepsilon_{ij}(\dot{u}) \geq 0$.

Indeed, from equation (2.1.2), we have:
$\dot{\sigma}_{ij}\varepsilon_{ij}(\dot{u}) = a_{ijkl}\varepsilon_{kl}(\dot{u})\varepsilon_{ij}(\dot{u}) + |\varepsilon(\dot{u})| \, b_{ij}\varepsilon_{ij}(\dot{u})$
Equation (2.2.1) implies that:
$b_{ij}\varepsilon_{ij}(\dot{u}) \leq M |\varepsilon(\dot{u})|$
then:
$\dot{\sigma}_{ij}\varepsilon_{ij}(\dot{u}) \geq \alpha \, \varepsilon_{ij}(\dot{u})\varepsilon_{ij}(\dot{u}) - M |\varepsilon(\dot{u})|^2 = (\alpha - M) |\varepsilon(\dot{u})|^2$

2.4 - Proof of the Theorem

The proof of this result is carried out for the weak formulation of this problem and not for the strong formulation (problem P) but, provided that the equations of P are considered in a weak sense (the sense of distributions), classical results of functional analysis prove that the problem P and its weak formulation are equivalent.

2.4.1 - Weak formulation of the problem
The weak formulation of this problem is got in a classical way: first, we reckon the virtual work relation for this incremental problem:

$$-\int_\Omega \dot{\sigma}_{ij} \, \varepsilon_{ij}(v) \, dx + \int_\Omega \dot{f}_i \, v_i \, dx + \int_{\Gamma_1} \dot{F}_i \, v_i \, dx = 0$$

for any v such that $v_i = 0$ on Γ_2 (i = 1, 2, 3).
This relation and the constitutive equation yields the weak formulation of the problem, which is:

Find $\dot{u} \in V_{ad}$ such that:
$$\forall \, v \in V \quad a(\dot{u},v) + b(\dot{u},v) = L(v) \qquad (2.4.1.1)$$

where V_{ad} and V are the spaces:
$V_{ad} = \{v = (v_1,v_2,v_3) \in [H^1(\Omega)]^3, \text{ s.a. } v = \dot{U} \text{ on } \Gamma_2\}$

$V = \{v = (v_1,v_2,v_3) \in [H^1(\Omega)]^3 \text{ s.a. } v = 0 \text{ on } \Gamma_2\}$

and

$$a(\dot{u},v) = \int_\Omega a_{ijkh} \, \varepsilon_{kh}(\dot{u}) \, \varepsilon_{ij}(v) \, dx$$

$$b(\dot{u},v) = \int_\Omega b_{ij} |\varepsilon(\dot{u})| \, \varepsilon_{ij}(v) \, dx$$

$$L(v) = \int_\Omega \dot{f}_i \, v_i \, dx + \int_{\partial\Omega} \dot{F}_i \, v_i \, dx$$

It is clear that, $a(\dot{u},v)$ is bilinear, $b(\dot{u},v)$ is linear with respect to v but not to $\dot{u}$, and that $L(v)$ is linear in v.

It is clear that $\dot{U}$ has to be the value on Γ_2 of an element of $[H^1(\Omega)]^3$, then it has to meet certain smoothness condition which will be assumed. Consequently, without going too much into details, there exists at least one element $\tilde{V}$ of $[H^1(\Omega)]^3$ such as $\tilde{V} = \dot{U}$ on Γ_2 and, by the change of unknown $\dot{u} = V + \tilde{u}$, it may be proved that the problem P is equivalent to same one with the data $\dot{U} = 0$.
Without loss of generality, we shall then prove the existence and uniqueness theorem for the problem P corresponding to the boundary condition :

$\dot{u} = 0$ on Γ_2
The weak formulation of which is :

Find $\dot{u} \in V$ such that :
$\forall\, v \in V \quad a(\dot{u},v) + b(\dot{u},v) = L(v)$ \hfill (2.4.1.2)

2.4.2 - Results of Functional Analysis
We sum up now some classical results of functional analysis :

V is an Hilbert space for the norm :

$$\|v\|_V = \left[\int_\Omega v_i \, v_i \, dx + \int_\Omega \frac{\partial v_i}{\partial x_j} \frac{\partial v_i}{\partial x_j} \, dx \right]^{1/2}$$

the corres-ponding scalar product in V is denoted $(u,v)_V$.

For any $\dot{u}$ in V, then $\varepsilon(\dot{u})$ is in $[L^2(\Omega)]^6$ and, according to the assumptions done for the b_{ij} (equation 2.2.1), $b(\dot{u},v)$ is continuous with respect to v on V, indeed :

$$|b(\dot{u},v)| \leq \int_\Omega \left|b_{ij} \, |\varepsilon(\dot{u})| \, \varepsilon_{ij}(v)\right| dx$$

$$\int_\Omega \left|b_{ij} \, |\varepsilon(\dot{u})| \, \varepsilon_{ij}(v)\right| dx \leq \int_\Omega M |\varepsilon(\dot{u})| \, |\varepsilon(v)| \, dx$$

$$\int_\Omega M |\varepsilon(\dot{u})| \, |\varepsilon(v)| \, dx \leq M \, \|\varepsilon(\dot{u})\|_{[L^2(\Omega)]^6} \, \|\varepsilon(v)\|_{[L^2(\Omega)]^6}$$

$|b(\dot{u},v)| \leq M \, \|\varepsilon(\dot{u})\|_{[L^2(\Omega)]^6} \, \|\varepsilon(v)\|_{[L^2(\Omega)]^6}$
and then :
$|b(\dot{u},v)| \leq M \, \|\dot{u}\|_V \, \|v\|_V$ \hfill (2.4.2.1)

According to the assumptions done for the a_{ijkh}, the bilinear form $a(\dot{u},v)$ is continuous and coercive on V. This means that :

$\alpha \, \|v\|^2 \leq a(v,v) \quad \forall\, v \in V$ \hfill (2.4.2.2)

We shall assume in the following that $\dot{f}_i$ and $\dot{F}_i$ are square-integrable respectively on Ω and on Γ_1 ($\dot{f} \in [L^2(\Omega)]^3$ and $\dot{F} \in [L^2(\Gamma_1)]^3$) which yields that $L(v)$ is continuous on V. These two assumptions are not really restrictive, indeed the space of square integrable functions is large and contains all the usual functions.

2.4.3 - Two Classical Theorems of Functional Analysis
To study the weak formulation of the problem, we shall use two classical theorems, the fixed point theorem and the Lax-Milgram theorem (see for instance Dautray R. and Lions J.L. 1988), let us recall these two theorems.

Fixed Point Theorem (in a Banach Space).
Let E be a Banach space (that is, a normed complete space) and S be an application from E to E such that :

$\forall\, u,v \in E \quad \|S(u) - S(v)\| \leq k \, \|u - v\|$ with $k < 1$
then the fixed point problem :
$u = S(u)$
has one and only one solution.

Lax-Milgram Theorem
Let H be a Hilbert space, $a(u,v)$ a bilinear continuous and coercive form on H and $\Lambda(v)$ a linear and continuous form on H then the problem :

Find $u \in H$ such that
$\forall\, v \in H \quad a(u,v) = \Lambda(v)$
has one and only one solution.

2.4.4 - Proof of the Existence and Uniqueness Theorem
Now, let us study the following problem :

For a given u' in V, find $u \in V$ such that
$\forall\, v \in V \quad a(u,v) = -b(u',v) + L(v)$
As seen before, $a(u,v)$ is bilinear, continuous and coercive on V, $b(u,v)$ and $L(v)$ are linear and continuous with respect to v, then the Lax-Milgram theorem may be applied and the problem has one unique solution u.

For a fixed form L, the solution u depends on u', this defines an application S from V to V such that :
u = S(u')

It is then clear that, if a solution $\dot{u}$ of the weak problem expressed by equation (2.4.1.1) exists, it is solution of the fixed point problem :
$\dot{u} = S(\dot{u})$
Then, in order to use the fixed-point theorem, we have to study the application S.

Lemma
$\forall\ w_1$ and $w_2 \in V$,
$$\|S(w_1) - S(w_2)\|_V \leq \frac{M}{\alpha} \|w_1 - w_2\|_V$$

Proof of the lemma.
$S(w_1)$ and $S(w_2)$ are defined by :
$\forall\ v \in V\ \ a(S(w_1),v) = -b(w_1,v) + L(v)$
$\forall\ v \in V\ \ a(S(w_2),v) = -b(w_2,v) + L(v)$
then by difference :
$\forall\ v \in V\ \ a(S(w_1) - S(w_2),v) = b(w_2,v) - b(w_1,v)$

Now,
$$b(w_2,v) - b(w_1,v) = \int_\Omega b_{ij}\,(|\varepsilon(w_2)| - |\varepsilon(w_1)|)\,\varepsilon_{ij}(v)\,dx$$

then using equation (2.4.2.1)
$$|b(w_2,v) - b(w_1,v)| \leq M \int_\Omega |\varepsilon(w_2 - w_1)| |\varepsilon(v)| dx$$
$$\leq M\,\|w_2 - w_1\|\,\|v\|$$

Substituting $v = S(w_1) - S(w_2)$ in (2.4.2.2) we get :
$\alpha\,(\|S(w_1) - S(w_2)\|_V)^2$
$\leq M\,\|w_1 - w_2\|_V\,\|S(w_1) - S(w_2)\|_V$
which ends the proof.

3 - SHEAR BAND ANALYSIS

3.1 - *The Problem*

In this second part of the paper, we study the possibility for a problem to present localized shear bands. We consider the problem of an initially homogeneous solid strained up to the current state. It is then submitted to a load rate on the straight loading path and we search the condition for which a shear band can exist.

This analysis can be seen as a condition of non uniqueness for a boundary value problem. It is necessary to see that first a bifurcation mode is assumed (shear band) and second that some boundary conditions are not necessarily met (Chambon et. al. 1984). We have to keep in mind these two restrictions.

In the following the analysis is a slightly different from the one of Chambon et. al. (1994), as here we use additional assumptions (Cf § 2.2) about a_{ijkh} and b_{ij} (the tensors which define the constitutive equation CLoE (2.1.2)) and as we assume a small strain-problem.

In this part, the vectors and tensors are denoted either by their components g_i, either by a bold character **g**.
Let **n** be the normal to the shear band, W^0 the inside of the shear band W^1 the outside of the shear band, $\dot{\sigma}^0$ the stress rate in W^0, $\dot{\sigma}^1$ the stress rate in W^1, $\dot{\varepsilon}^0$ the strain rate in W^0, and $\dot{\varepsilon}^1$ the strain rate in W^1.

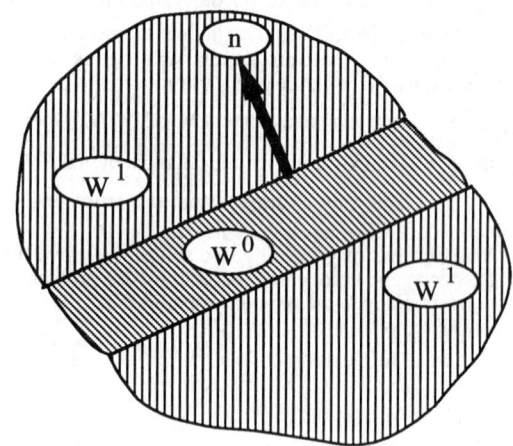

Figure 2 The shear band problem

The shear band problem can be written : a shear band is possible if a normal **n** and a vector **g** can be find such that :

$$\varepsilon^1 = \varepsilon^0 + (\mathbf{g} \otimes \mathbf{n})_s \tag{3.1.1}$$

where $(\mathbf{g} \otimes \mathbf{n})_s$ is the symetric part of the tensorial product $\mathbf{g} \otimes \mathbf{n}$, the previous equation can be called the kinematical condition.

$$\dot{\sigma}^0 \mathbf{n} = \dot{\sigma}^1 \mathbf{n}\ \text{ or }\ \dot{\sigma}^0_{ij} n_j = \dot{\sigma}^1_{ij} n_j \tag{3.1.2}$$
which can be called the statical condition (which is a balance equation)

$$\dot{\sigma}^0_{ij} = a_{ijkl}\dot{\varepsilon}^0_{kl} + b_{ij}\,|\dot{\varepsilon}^0|$$
$$\dot{\sigma}^1_{ij} = a_{ijkl}\dot{\varepsilon}^1_{kl} + b_{ij}\,|\dot{\varepsilon}^1| \tag{3.1.3}$$
the constitutive equation for W^0 and W^1.

Equation (3.1.2) yields :
$$a_{ijkl}\dot{\varepsilon}^0_{kl} n_j + b_{ij} n_j\,|\dot{\varepsilon}^0| = a_{ijkl}\dot{\varepsilon}^1_{kl} n_j + b_{ij} n_j\,|\dot{\varepsilon}^1| \tag{3.1.4}$$
this equation is clearly positively homogeneous of degree 1, then without loss of generality we may set $|\dot{\varepsilon}^0| = 1$ and the shear band problem (SBP) is then : $\dot{\varepsilon}^0$

being given such that $|\dot{\varepsilon}^0| = 1$, find two vectors **g**, **n** ($|\mathbf{n}| = 1$) and a real number r such that :

$$a_{ijkl} g_k n_l n_j + r b_{ij} n_j = 0$$
$$|\dot{\varepsilon}^0 + (\mathbf{g} \otimes \mathbf{n})_s| = 1 + r \quad (3.1.5)$$

3.2 - *Solution of the Problem*

First, we are going to set the problem under another form, to do this we need the following lemma :

Lemma
Let **n** be a normed vector, then the linear operator **L**, the matrix components of which are L_{ij} defined by :
$$L_{ij} = a_{ikjl} n_k n_l$$
is invertible and moreover for any vector **A** (the components of which are denoted A_i) we have :

$$\alpha |(\mathbf{A} \otimes \mathbf{n})_s|^2 = \frac{\alpha}{2}[|\mathbf{A}|^2 + (\mathbf{A}.\mathbf{n})^2]$$

$$\leq L_{ij} A_j A_i \quad (3.2.1)$$

Proof :
For any **A** we have (equation 2.2.2) :
$L_{ij} A_j A_i = a_{ikjl} A_i A_j n_k n_l \geq \alpha (A_i n_j)(A_i n_j)$
This proves the invertibility of the linear operator **L**.
For the second part of the lemma, we have :
$\frac{1}{2}(A_i n_j + A_j n_i)^2 = \frac{1}{2}[(A_i n_i)(A_j n_j) + A_i A_i]$
$= \frac{1}{2}[|\mathbf{A}|^2 + (\mathbf{A}.\mathbf{n})^2]$
Moreover:
$(A_i n_i)(A_j n_j) \leq A_i A_i$
So
$(A_i n_j)(A_i n_j) = A_i A_i \geq \frac{1}{2}[|\mathbf{A}|^2 + (\mathbf{A}.\mathbf{n})^2]$
which ends the proof.

Theorem
Now we proove that :
SBP (Cf 3.1) is equivalent to SBP' which is :
Find a normed vector **n** such that :
$|(\mathbf{G} \otimes \mathbf{n})_s| \geq 1$
where **G** is defined by :
$$G_i = - L_{ij}^{-1} b_{jk} n_k \quad (3.2.2)$$

Proof :
Let **n** be a normed vector, as **L** is invertible,
$G_i = - L_{ij}^{-1} b_{jk} n_k$ is well defined and SBP is clearly equivalent to find **n** and r such that
$|\dot{\varepsilon}^0 + r(\mathbf{G} \otimes \mathbf{n})_s| = 1 + r$
Considering r as the unknown, this equation is equivalent to
$r \geq -1$ and :
(3.2.3)
$r[r(|(\mathbf{G} \otimes \mathbf{n})_s|^2 - 1) + 2(\dot{\varepsilon}^0 : (\mathbf{G} \otimes \mathbf{n})_s - 1)] = 0$
Where **a:b** denotes the scalar product of two tensors which is, written with components : $\mathbf{a:b} = a_{ij} b_{ij}$
r = 0 is always a solution of this equation, it corresponds the to trivial solution $\dot{\varepsilon}^0 = \dot{\varepsilon}^1$.
We assume that $|(\mathbf{G} \otimes \mathbf{n})_s| \neq 1$.
So there is a non trivial solution of the equation (3.2.3) :

$$r = -2 \frac{(\dot{\varepsilon}^0 : (\mathbf{G} \otimes \mathbf{n})_s - 1)}{|(\mathbf{G} \otimes \mathbf{n})_s|^2 - 1}$$

it has to be greater or equal to -1, now :

$$1 + r = \frac{|(\mathbf{G} \otimes \mathbf{n})_s|^2 - 2(\dot{\varepsilon}^0 : (\mathbf{G} \otimes \mathbf{n})_s) + 1}{|(\mathbf{G} \otimes \mathbf{n})_s|^2 - 1}$$

$$= \frac{|(\mathbf{G} \otimes \mathbf{n})_s - \dot{\varepsilon}^0|^2}{|(\mathbf{G} \otimes \mathbf{n})_s|^2 - 1}$$

then $\{1 + r \geq 0\} \Leftrightarrow \{|(\mathbf{G} \otimes \mathbf{n})_s| > 1\}$
which ends the proof.

3.3 - *A necessary condition for localization*

We prove here that : $M/\alpha \geq 1$ is a necessary condition for the previous localisation condition.
Let us assume that : $M/\alpha < 1$
Equation (3.2.2) implies :
$L_{ik} G_k = - b_{ij} n_j$
so
$G_i L_{ik} G_k = - b_{ij} n_j G_i = - b_{ij} \left(\frac{G_i n_j + G_j n_i}{2}\right)$ as b_{ij} is symmetric (Cf § 2.1).
Equation (3.2.1) implies :
$G_i L_{ik} G_k \geq \alpha |(\mathbf{G} \otimes \mathbf{n})_s|^2$
and the first assumption in § 2.2 implies :
$b_{ij} \left(\frac{G_i n_j + G_j n_i}{2}\right)| \leq M |(\mathbf{G} \otimes \mathbf{n})_s|$
Finally :
$\alpha |(\mathbf{G} \otimes \mathbf{n})_s|^2 |b_{ij} \leq M |(\mathbf{G} \otimes \mathbf{n})_s|$
and then :
$\alpha |(\mathbf{G} \otimes \mathbf{n})_s|^2 |b_{ij} \leq M |(\mathbf{G} \otimes \mathbf{n})_s| \leq \frac{M}{\alpha} < 1$
So if $M/\alpha < 1$ localization is impossible.

4 - CONCLUSION

Finally the same condition $M/\alpha < 1$ implies existence and uniquemess of the rate boundary value problem and impossibilty of existence of a shear band in a classical shear band analysis. This result is not obvious as we know that the first problem is a global one and conversely the shear band analysis is only a local analysis as some boundary conditions are ignored and the initial state is assumed to be homogeneous. From this point of view our results are interesting.

If we consider now only the existence and uniqueness result, let us emphasize that :

This result is available for a complete problem. This is somewhat new. Most of such existence and uniqueness results are only available for incremental linear problems. Obviously some of these results, especially the ones concerning classical elastoplasticity with one loading criterion, can be extended to the complete problem.

This results is constructive, it gives us an effective algorithm which can be used in finite element codes.

THANKS

Part of the present research was developed within the ALERT network (CHM Project n° CHRX-CT93-0217).

REFERENCES

Drucker D. C. 1950 "Some implication of work hardening and ideal plasticity" Q. Appl. Math. 7 pp. 411-418

Chambon R. Desrures J. 1984 "Quelques remarques sur le problème de la localisation en bande de cisaillement" Mech. Res. Com. 11 pp. 145-153

Chambon R. 1989a "Une classe de loi de comportement incrémentalement non linéaire pour les sols non visqueux résolution de quelques problèmes de cohérence" C. R. Acad Sci. Paris 308 pp. 1571-1576

Chambon R. 1989b "Base théorique d'une loi de comportement incrémentale consistante pour les sols Rapport Interne I.M.G. Grenoble

Chambon R. Desrues J. 1994 "Shear modulus identification versus experimental localization data" this volume

Hill R. 1958 3A general theory of uniqueness and stability in elastic-plastic solids J. Mech. Phys. Solids 6 pp 236-249

Kolymbas D. 1987 "A novel constitutive law for soils" 2nd Int. Conf. on Const. Laws for Engineering Methods Tucson

Kolymbas D. Wu W. 1993 "Introduction to Hypoplasticity" Modern Approaches to Plasticity D. Kolymbas ed. Elsevier Science Publisher B.V.

Dautray R. and Lions J.L. 1988 "Analyse mathématique et calcul numérique pour les sciences et les techniques, vol. 4 Méthodes variationnelles" Massson Paris. (or the translation in Springer-Verlag).

Bifurcation conditions within hypoplastic constitutive theory

Conditions de bifurcation dans le contexte de la théorie hypoplastique

Zbigniew Sikora
Geotechnical Department, Technical University of Gdańsk, Poland

Sławomir Rybicki
Applied Mathematics Department, Technical University of Gdańsk, Poland

ABSTRACT: Based on some known theorems of the bifurcation theory the importance of a linearization process of the hypoplastic constitutive functions is presented. The linearization process should fullfil some conditions, which guarantee an isomorphic equivalence between solution families of the nonlinear function and the adequate linearized one. Based on Conley theorem and other topological theories one can define the Conley-Index, which can be calculated at a reference time point of the initial value problem. Some numerical criteria for the bifurcation are presented.

Sur la base de résultats connus de la théorie de la bifurcation, on discute l'importance de la linéarisation de lois de comportement hypoplastiques. La linéarisation doit remplir certaines conditions, qui garantissent une équivalence isomorphe entre les familles de solutions de la loi non linéaire et la loi linéarisée. Le théorème de Conley et d'autre théories topologiques permettent de définir l'indice de Conley, qui peut être calculé à un instant de référence du problème de Cauchy. On présente quelques critères numériques de bifurcation.

1 INTRODUCTION

In the recent years the constitutive theory of hypoplasticity is developed as an alternative approach to elastoplasticity in describing mechanical behaviour of granular materials, such as soils. There are some difficulties in applying elastoplasticity to geomaterials, e.g. decomposition of deformation into elastic and plastic parts and the transition between elastic and plastic deformation. Hypoplasticity for the granular materials has been proposed by Kolymbas (1987). Some concepts pertinent to elastoplasticity, e.g. yield surface, plastic potential, decomposition of the deformation into elastic and plastic parts, were abandoned to be used in formating the constitutive equations. The theoretical framework of hypoplastic theory is outlined following the recent work by Kolymbas (1987) and others Wu and Sikora (1990), Sikora (1992) and Wu (1993).

Some interesting realistic problems have been already solved based on correct defined boundary value problems (BVPs) within hypoplasticity (Sikora 1992). It is of interest, especially from physical and also numerical point of view, that the defined BVP is *well-posed*. A theoretical approach to check whether a BVP is well-posed or not is to check three following conditions: existance, uniqueness and stability of the solution. In contrary case the BVP is *ill-posed*. It is well known that the solution of problems, which are ill-posed has no practical value in the majority of cases.
The aim of the present paper is to discuss the uniqueness problem or in other words to analyse a possibility of a bifurcation the solution.

Every problem in applications contains several physical parameters which may vary over certain specified sets. Thus, it is important to understand the qualitative behaviour of the system as the parameters vary. A good design for a system will always be such that the qualitative behaviour does not change when the parameters are varied a small amount the value for which the original design was made. However, the behaviour may change when the system is subjected to large variations in the parameters. A change in the qualitative properties could mean a change in stability of the original system and thus the system must assume a state different from the original design. The value of the parameters where this change takes place are called **bifurcation values** or **bifurcation points**. *Knowledge of the bifurcation points is* **absolutely necessary** *for the complete understanding of the* system. Our objective is to present a method to check whether the system reaches a bifurcation value.

In general, two distinct aspects of bifurcation theory can be discussed – static and dynamic. **Static bifurcation theory** is concerned with the changes that occur in the structure of the set of zeros of a function as parameters in the function are varied. If the function is a gradient, then variational techniques play an important role and can be employed even for global problems. If the function is not a gradient or if more detailed information is desired, the general theory is usually local. At the same time, the theory is constructive and valid when several independent parameters appear in the function. In differential equations, the equilibrium solutions are the zeros of the vector field. Therefore, methods in static bifurcation theory are directly applicable.

Dynamic bifurcation theory is concerned with the changes that occur in the structure of the limit sets of solutions of differential equations as parameters in the vector field are varied. For example, in addition to discussing the way that the set of zeros of the vector field (the equilibrium solutions) change through the static theory, the stability properties of these solutions must be considered. In fact, there is an intimate relationship between changes of stability and bifurcation. The dynamics in a differential equation can also introduce other types of bifurcations; for example, periodic orbits, homoclinic orbits, invariant tori. This introduces several difficulties which require rather advanced topics from differential equations for their resolution.

In the following we formulate and utilize some already known theorems for a necessary and suficient conditions of a bifurcation point. After the introductory discussion, we define the basic problem and give a precise definition of a bifurcation point.

The first author defined a suficient condition for a bifurcation point of an initial value problem with a hypoplastic constitutive equation (see Sikora 1992, page 52) utilizing the Hadamard theory of wave propagation. This bifurcation condition defines a stress and streching tensors which allow a posibility of the onset of a shear band. Taking into account some conditions proposed by Hadamard (1903) a sufficient condition for bifurcation was derived (see Sikora 1992, formula (4.24)). This bifurcation condition plays an important role especially from numerical point of view because its matrix form can be checked at each Gauss point of the discretized volume and can be checked at every time reference point during the calculation of a initial-boundary value problem. However, the mentioned condition posesses a disadvantage, which consists of a description of the so called *jump parameter*. This parameter depends on the solution, so the bifurcation process could be discussed only as a non-trivial nonlinear optimization problem.

It is well known that bifurcation is of importance in solving of initial-boundary value problems. Therefore, it is of considerable interest for a given constitutive equation to check whether bifurcation of the solution curve is possible and in which direction is continued. However these questions are not trivial there has been ample research concerning stability criteria in the mathematical and mechanical literature (Chow and Hale 1982, Ciarlet 1978, Antman 1977, Keller and Langford 1972 and many others). However, little has been done to apply these criteria to a concrete family of constitutive equations. Owing to the great complexity of the most constitutive equations, the bifurcation condition can be hardly applied analytically. In the present paper we utilize some known theorems about bifurcation and show the possibility to use them in numerical calculations.

2 THE HYPOPLASTIC EQUATIONS

In this section, we describe some possible hypoplastic constitutive equations for sand, which were formulated based on the representation theorems for isotropic tensor-valued functions. For some other details the reader is refered to W. Wu (1993). We start with the general constitutive equation for hypoplasticity in the form

$$\dot{\mathbf{T}} = \overset{\circ}{\mathbf{T}} + \mathbf{W}\mathbf{T} - \mathbf{T}\mathbf{W} \quad (1)$$

$$\overset{\circ}{\mathbf{T}} = h(\mathbf{T}, \mathbf{D}) = \mathbf{L}(\mathbf{T}, \mathbf{D}) + \mathbf{N}(\mathbf{T})\|\mathbf{D}\| \quad (2)$$

or in another form as

$$\overset{\circ}{\mathbf{T}} = \mathbf{H}(e, \mathbf{T}, \mathbf{D}) \quad (3)$$

$$\overset{\circ}{\mathbf{T}} = a_1 f_b f_e \left[\mathbf{L}(\mathbf{T}, \mathbf{D}) + f_d f_a \mathbf{N}(\mathbf{T}, \mathbf{D}) \right] \quad (4)$$

where a_1 and f_a, f_b, f_e, f_d are material factors standing for the so called argotropy, barotropy and pyknotropy properties respectively. $\mathbf{T}$, $\mathbf{D}$ and $\mathbf{W}$ are Cauchy stress tensor, rate of deformation (stretching) and spin tensor respectively. e stands for a void ratio. The symbols $(\cdot)$ and $(\overset{\circ}{\cdot})$ indicate the time derivation and Jaumann time derivative respectively. $\mathbf{H}$ has a certain mathematical representation, so that all granular materials thus defined are assumed to be equal in quality (see for details Gudehus 1993). Using notions proposed by Kolymbas (1991), the above mentioned material properties have a following explanation:

- *argotropy*: the amount of **H** increases with the velocity gradient in a weakly sense, i.e. if $\lambda \gg 1$ then $\|\mathbf{H}(e,\mathbf{T},\lambda\mathbf{D})\| > \lambda\|\mathbf{H}(e,\mathbf{T},\mathbf{D})\|$,

- *barotropy*: the amount of **H** increases with mean pressure weaker than linear,

- *pyknotropy*: the constitutive function **H** depends explicitly on density, i.e. void ratio e.

For easiness we consider only the (1-2) case.
In order to describe irreversible deformation, equation (2) should be nonlinear in relation to **D**. The equation (2) shows this fact supposing that the **h** function is a sum of two tensor-valued functions **L** and **N**. **L** denotes bilinear tensor-valued operator and **N** is a tensor-valued function of the stress tensor **T**. $\|\cdot\|$ stands for the Euclidean norm of the stretching tensor and is defined by $\|\mathbf{D}\| = \sqrt{\operatorname{tr}(\mathbf{D}^2)}$. Since the functions **L** and **N** are isotropic their concrete forms can be obtained by invoking the representation theorems for isotropic tensor–valued functions.

Remark 1 *Because of $\sqrt{\operatorname{tr}(\mathbf{D}^2)}$, the constitutive equation (2) is incrementally nonlinear in **D**. Equation (2) defines a class of incrementally nonlinear constitutive equations.*

It was shown for the first time by Kolymbas (1987) with a special version, that equation (2) can describe many of the features pertinent to granular materials.
For the better understanding of the problem researched as well as the formulation of the hypoplastic constitutive equations we present the concept of a response envelope introduced by Gudehus (1979). Comparison with the laboratory tests provide only limited information under rather restricted paths. The response envelope has been proved to be a very efficient tool to study the general behaviour of a given constitutive equation.

For the graphical presentation we limited our considerations to the case of axisymmetry. The discussion will be performed in Rendulic plane, see Fig. 1, which is well-known in soil mechanics. A response envelope means in mathematical words an image of the mapping defined by (1-2) in the stress space **T**. Since the constitutive function **h** is homogeneous of degree one in **D** we can parametrize **D** as follows

$$\mathbf{D}(\alpha) = \begin{bmatrix} \sin(\alpha) & 0 & 0 \\ 0 & \frac{\cos(\alpha)}{\sqrt{2}} & 0 \\ 0 & 0 & \frac{\cos(\alpha)}{\sqrt{2}} \end{bmatrix} \longmapsto \quad (5)$$

$$\longmapsto \quad \forall_{\alpha \in <0,2\pi)} \quad \|\mathbf{D}(\alpha)\| = 1$$

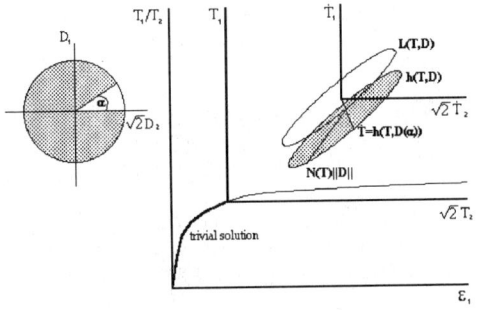

Fig. 1: Unit rate response and response envelope

obtaining the unit rate response, see also the parametrization in the fourth section, (29). If we now transformate the equations (1-2) at the stress level $\mathbf{T}_0 = (T_1^0, T_2^0, T_3^0)$ for all $\alpha \in <0, 2\pi)$ we obtain the image of the constitutive function **h** in the stress space as shown in Fig. 1. It is straightforward to prove that the response envelope of homogeneous hypoelastic constitutive equation (i.e. in case of function **L** as well) is always ellipse with the initial stress $\mathbf{T}_0$ laying in the center of the ellipse. If we look at the equation (1-2) we will easily see, taking into account (5), that the response envelope for the function **h** will be ellipse too, but parallel translated by quantity $\mathbf{N}(\mathbf{T})$, which is independent of α.

In the sequel we discuss a possible version of (2). Let us start with **L**. According to our assumptions that **L** is a bilinear tensor-valued function, we conclude based on the representation theorem for isotropic functions, that the form of the bilinear function **L** is a linear combination of the following five terms

$$(\mathbf{TD} + \mathbf{DT}), \quad \operatorname{tr}(\mathbf{T})\mathbf{D}, \quad \operatorname{tr}(\mathbf{D})\mathbf{T}, \quad \operatorname{tr}(\mathbf{TD})\mathbf{1}, \quad \operatorname{tr}(\mathbf{T})\operatorname{tr}(\mathbf{D})\mathbf{1} \quad (6)$$

The representation theorem provides only some possibilities for choice. Constitutive equation using terms in equation (6) does not need to be complete. We now turn to consider nonlinear part of the constitutive equation (2). It is assumed that **N** is a linear or nonlinear tensor-valued function of **T**. The Euclidean norm of the strain rate, i.e. $\|\mathbf{D}\| = \sqrt{\operatorname{tr}(\mathbf{D}^2)}$, fulfils the mentioned requirements in relating to response envelope and therefore should be used for the nonlinear part of the constitutive function (2). For the nonlinear tensor-valued function **N** we begin with its simplest form assuming that $\mathbf{N}(\mathbf{T})$ is linear in **T** and therefore the function **N** can be represented by linear combination of

$$\mathbf{T} \quad \text{and} \quad \operatorname{tr}(\mathbf{T}). \quad (7)$$

Within the above frame v. Wolffersdorff (1990) found a suitable constitutive equation in the following form

$$\overset{\circ}{\mathbf{T}} = C_1(\mathbf{TD}+\mathbf{DT})/2+ \\ C_2(\text{tr}(\mathbf{T})\mathbf{D}+\text{tr}(\mathbf{D})\mathbf{T})+ \\ (C_3\mathbf{T}+C_4\text{tr}(\mathbf{T})\mathbf{1})\|\mathbf{D}\| \quad (8)$$

where C_1, C_2, C_3, C_4 are material constants calibrated on the triaxial experiment results for the Karlsruhe sand. The parameters are equal to

$$C_1 = -96.56, \quad C_2 = -34.48, \\ C_3 = -153.1, \quad C_4 = 47.58 \quad (9)$$

A next example of a hypoplastic function, which is linear in $\mathbf{T}$, in all constitutive terms is a following function

$$\overset{\circ}{\mathbf{T}} = C_1(\mathbf{TD}+\mathbf{DT})/2 + C_2(\text{tr}\mathbf{T})\mathbf{D}+ \\ C_3\mathbf{T}\|\mathbf{D}\| + C_4(\text{tr}\mathbf{T})\|\mathbf{D}\|\mathbf{1} \quad (10)$$

with material parameters

$$C_1 = 66.265, \quad C_2 = 44.578, \\ C_3 = 149.676, \quad C_4 = -49.892 \quad (11)$$

calibrated to the triaxial experiment which was conducted with a dense specimen of Karlsruhe medium sand under confining pressure of 100 kPa. A detailed description of the calibration procedure can be found in Kolymbas (1987). It is worthwhile to point out that despite its simple mathematical structure involving solely four material constants many of the salient features pertinent to granular materials, such as nonlinearity, limit state and dilatancy, are well captured by the constitutive equations (8) or (10).

We go a step further and assume that $\mathbf{N}$ is nonlinear in stress $\mathbf{T}$ then the representation for this function is more complicated.
Among others a constitutive requirement due to objectivity can be obtained, i.e. we claim

$$\mathbf{h}(\mathbf{QTQ}^T, \mathbf{QTQ}^T) = \mathbf{Qh}(\mathbf{T},\mathbf{D})\mathbf{Q}^T \quad (12)$$

where $\mathbf{Q}$ is an orthogonal tensor. The requirement of the objectivity is satisfied if the function $\mathbf{h}$ is chosen according to the representation theorems for isotropic tensor-valued functions.
The general form of the function $\mathbf{N}$ can be determined by the following linear combination

$$\mathbf{N}(\mathbf{T}) = \alpha_1\mathbf{1} + \alpha_2\mathbf{T} + \alpha_3\mathbf{T}^2 \quad (13)$$

where α_i are functions of the three invariants of $\mathbf{T}$, i.e.

$$\alpha_i = \alpha_i\left(\text{tr}(\mathbf{T}), \text{tr}(\mathbf{T}^2), \text{tr}(\mathbf{T}^3)\right). \quad (14)$$

Kolymbas proposed a constitutive equation, with nonlinear tensor-valued function $\mathbf{N}$, in the following shape (Kolymbas 1981)

$$\overset{\circ}{\mathbf{T}} = C_1(\mathbf{TD}+\mathbf{DT})/2 + C_2\text{tr}(\mathbf{TD})\mathbf{1}+ \\ C_3\mathbf{T}\|\mathbf{D}\| + C_4\mathbf{T}^2\|\mathbf{D}\|/\text{tr}(\mathbf{T}) \quad (15)$$

where C_i for $i = 1,\ldots,4$ are material parameters calibrated on results of triaxial experiment with Karlsruhe sand and they are equal to

$$C_1 = -200.0, \quad C_2 = -47.9, \\ C_3 = 36.4, \quad C_4 = -252.7 \quad (16)$$

Please pay attention that the fourth term of the constitutive function (15) is not a polynomial function more but we assume that it can be suficiently approximated in form of high order polynomial function. This statement is of importance taking in to account the applicability the theorems presented in the next section.
Note additionaly, that the calibration of the constitutive law in sense of (1), to obtain values of the material parameters, is a strictly mathematical problem of nonlinear optimization which is based on some standard experimental data like triaxial or biaxial deformation. For some details the reader is refered to Sikora (1992).

3 BIFURCATION THEOREMS

In this paragraph we present three definitions of bifurcation points. We formulate four theorems, which give necessary and sufficient conditions for the existence of bifurcation points of solutions of ordinary differential equations.
Let us consider the so called double-initial value problem in local sense in the following form

$$\dot{\mathbf{T}} = \mathbf{h}(\mathbf{T},\mathbf{D}) = \mathbf{h}(\mathbf{T},\dot{\mathbf{E}}) \\ \mathbf{T}(t_0) = \mathbf{T}_0 \\ \mathbf{E}(t_0) = \mathbf{E}_0 \quad (17)$$

where $\mathbf{E}$ denotes a deformation tensor. Starting the loading process from a certain deformation state, say $\mathbf{E}_0$, we impose some other conditions to the deformation $\mathbf{E}(t)$ defined by boundary conditions of a boundary value problem. Thus we can state that the coordinates of the stretching tensor $\mathbf{D}$ belong to the finite set of parameters, say λ, to be varied. In this sense we can rewrite the initial value problem (17) in another form and utilize for bifurcation analysis.

Consider a parameterized family of differential constitutive equations as a sum of the linear part, $\mathbf{A}(\lambda)\mathbf{T}$, and the polynomial one, $\phi(\mathbf{T},\lambda)$, which has the order greater than one in relation to **T**, i.e.

$$\dot{\mathbf{T}} = \mathbf{h}(\mathbf{T}, \mathbf{D}) = \mathbf{h}(\mathbf{T}, \lambda) = \mathbf{A}(\lambda)\mathbf{T} + \phi(\mathbf{T}, \lambda) \quad (18)$$

where

$$\mathbf{h} : R^n \times R^k \to R^n \\ \mathbf{h}(\mathbf{0}, \lambda) = \mathbf{0} \quad (19)$$

is a C^1 map. Note that in case of a non-homogeneous initial value problem, like (17), we obtain the respectively homogeneous initial value problem throught the substitution $\bar{\mathbf{T}}(t) = \mathbf{T}(t) - \mathbf{T}_0$, thus the Fréchet derivative, see Appendix, of the constitutive function for stationary solution $\mathbf{T} = \mathbf{0}$ defines

$$\mathbf{A}(\lambda) = D_{\mathbf{T}}\mathbf{h}(\mathbf{0}, \lambda). \quad (20)$$

For simplicity the order of the problem can be limited to $k \leq n \leq 3$.

For the precise definition of the bifurcation point we take into account the Conley index formulated in the following theorem.

Theorem 1 (Conley) *(Conley 1978, Smoller 1983)*
Fix λ_0 such that $\sigma(\mathbf{A}(\lambda_0)) \cap \sqrt{-1} \cdot R = \emptyset$, where $\sigma(\mathbf{A})$ denotes the spectrum of $\mathbf{A}$. Under this assumption we can say that the origin is an isolated invariant set in the sense of Conley and its Conley index is a sphere whose dimension $\nu(\lambda_0)$ is equal to the sum of algebraic multiplicities of eigenvalues of $\mathbf{A}(\lambda_0)$ with positive real part.

Based on the above theorem the following definition is of importance for the solution of (17).

Definition 1 *A point $(\mathbf{0}, \lambda_0) \in R^n \times R^k$ is said to be a **bifurcation point** of bounded solutions of the equation (18), provided that for any open neighbourhood U of $(\mathbf{0}, \lambda_0)$ there is solution of (18) different from the solution of the form (19), i.e. $(\mathbf{0}, \lambda)$ and included in U.*

Let

$$\Lambda = \{\lambda \in R^k : \sigma(\mathbf{A}(\lambda)) \cap \sqrt{-1} \cdot R = \emptyset\}. \quad (21)$$

Computing the Conley index, i.e. the number of those eigenvalues which the real part is positive, as in Theorem 1, we can prove the following bifurcation theorem.

Theorem 2 (Conley) *(Conley 1978, Smoller 1983)*
Fix $\lambda_0, \lambda_1 \in \Lambda$ and assume that $\nu(\lambda_0) \neq \nu(\lambda_1)$. Then the interval $\{\mathbf{0}\} \times [\lambda_0, \lambda_1]$ contains a bifurcation of bounded solutions of the family

$$\dot{\mathbf{T}} = \mathbf{h}(\mathbf{T}, \lambda). \quad (22)$$

Notice that if we assume that $det(A(\lambda))$ is different from zero then we know that there is no bifurcation points of stationary solutions of our system. In other words nontstationary bounded solutions bifurcate from the interval $\{\mathbf{0}\} \times [\lambda_1, \lambda_2]$.
From now on we will consider sufficient conditions for the existence of bifurcation points of stationary solutions of the family (22).

Definition 2 *A point $(\mathbf{0}, \lambda_0) \in R^n \times R^k$ is said to be a bifurcation point of stationary solutions of the equation (22), provided that for any open neighbourhood U of $(\mathbf{0}, \lambda_0)$ there is a stationary solution of (22) different from the solution of the form $(\mathbf{0}, \lambda)$ and included in U.*

Let us denote by $\Lambda_+, \Lambda_- \subset R^k$ the following sets

$$\Lambda_+ = \{\lambda \in R^k : det(\mathbf{A}(\lambda)) > 0\} \\ \Lambda_- = \{\lambda \in R^k : det(\mathbf{A}(\lambda)) < 0\} \quad (23)$$

Theorem 3 (Krasnosielski) *(Chow and Hale 1982, Krasnosielski 1956, Deimling 1985)*
If $\lambda_0 \in \Lambda_+$ and $\lambda_1 \in \Lambda_-$, then the interval $\{\mathbf{0}\} \times [\lambda_0, \lambda_1]$ contains a bifurcation point of stationary solutions of the family (22).

The above theorem is a useful tool in order to show the existence of bifurcation points of stationary solutions of autonomous differential equations. But very often it is of our interest to research the local topological structure of the flow of dynamical system. Therefore we consider another definition of a bifurcation point, which is not equivalent to the previous one, and formulate a sufficient condition for the existence of a bifurcation point int the sence of this definition.

Definition 3 *We say that two differential equations $\dot{\mathbf{T}} = \mathbf{h}(\mathbf{T}, \lambda_1)$ and $\dot{\mathbf{T}} = \mathbf{h}(\mathbf{T}, \lambda_2)$ are equivalent if there is a homeomorphism $\mathbf{F} : R^n \to R^n$ which takes orbits of the first equation on the orbits of the second equation.*

Finally we present another definition of a bifurcation point, which is very convenient from the numerical point of view.

Definition 4 *We say that a point $(0, \lambda_0)$ is said to be a bifurcation point of the solutions of the family (22) if for any neighbourhood U of $(0, \lambda_0)$ there is $\lambda_1 \in U$ such that families $\dot{\mathbf{T}} = \mathbf{h}(\mathbf{T}, \lambda_0)$ and $\dot{\mathbf{T}} = \mathbf{h}(\mathbf{T}, \lambda_1)$ are not equivalent.*

From the local version of the Hartman–Grobmann theorem (Hartman 1964, Szlenk 1982) we obtain the following theorem.

Theorem 4 *Fix $\lambda_0, \lambda_1 \in \Lambda$, (21), and assume that $\nu(\lambda_0) \neq \nu(\lambda_1)$. Then the interval $\{0\} \times [\lambda_0, \lambda_1]$ contains a bifurcation of solutions of the family (22).*

Unfortunately in this case we know nothing about the existence of bifurcation points of stationary solutions.

4 APPLICATIONS

In this section we apply the above theorems to some differential equations of hypoplastic constitutive theory for granular materials. We examine in the previous presented constitutive equations (8), (10) and (15) in axisymetric conditions. In the sequel we examine the existence of bifurcation points for the versions of the hypoplastic constitutive equations which are linear in stress $\mathbf{T}$. Consider a family of ordinary differential equations the form

$$\dot{\mathbf{T}} = \mathbf{h}(\mathbf{T}, \lambda) = \mathbf{A}(\lambda) \cdot \mathbf{T} + \phi(\mathbf{T}, \lambda), \quad (24)$$

where $\mathbf{h} : R^3 \times R^3 \to R^3$ and

$$\mathbf{D} \sim \lambda. \quad (25)$$

4.1 Example no. 1

We examine the equation (8) with the appropriate material parameters (9). The tensor $\mathbf{A}(\mathbf{D})$ for the constitutive function (8) dependent on principle strain and stress obtains the following form

$$\mathbf{A}(\mathbf{D}) = \begin{bmatrix} a_{11} & a_{12} & a_{13} \\ a_{21} & a_{22} & a_{23} \\ a_{31} & a_{32} & a_{33} \end{bmatrix} \quad (26)$$

where

$$\begin{aligned} a_{ii} &= C_1 D_i + C_2(D_i + \mathrm{tr}\mathbf{D}) + \\ & \quad (C_3 + C_4)\|\mathbf{D}\| \\ a_{ij} &= C_2 D_i + C_4 \|\mathbf{D}\|, \quad \text{for } i \neq j \end{aligned} \quad (27)$$

and $C_{i(i=1,...,4)}$ as given in (9).
In order to research the set of bifurcation points of (24) we restrict our consideration to a one-

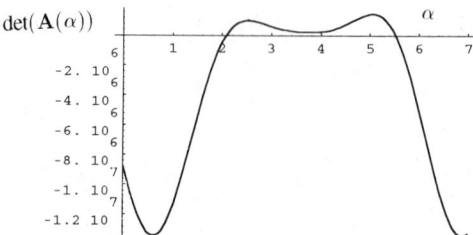

Fig. 2: Graph of the determinant for the eqn. (8)

dimensional path on a sphere S^2, see also (5)

$$\alpha : [0, 2\pi] \to S^2 \quad (28)$$

of parameters of the form

$$\alpha(t) = \left(\sin(t), \frac{1}{\sqrt{2}} \cos(t), \frac{1}{\sqrt{2}} \cos(t) \right) \quad (29)$$

Note that the sphere is defined as

$$S^2 = \{(D_1, D_2, D_3) = (\alpha_1, \alpha_2, \alpha_3) : \\ \alpha_1^2 + \alpha_2^2 + \alpha_3^2 = 1\} \quad (30)$$

We take into account a map ψ of the form

$$\lambda \to \det(\mathbf{A}(\alpha(\lambda))). \quad (31)$$

It is easy to draw a graph of this map using for example a computer program MATHEMATICA. Using this program one can compute eigenvalues of matrices $\mathbf{A}(\alpha(\lambda))$, see fig. 2. In other words using computer programs one can verify assumptions of theorems of the third section.
The map ψ changes sign at exactly two points of the interval $[0, 2\pi]$. Moreover, we know that for any $\lambda \in [0, 2\pi]$ we have $\sigma(\mathbf{A}(\alpha(\lambda))) \cap \sqrt{-1} \cdot R = \emptyset$. Applying Theorems 1 and 2 we prove that there are exactly two bifurcation points of bounded solutions of the family (24) included in the interval $[0, 2\pi]$.
Using Theorem 3 we can prove much more. We are able to prove that there are exactly two bifurcation points of stationary solutions of the family included in the interval $[0, 2\pi]$.
Now let us apply Theorem 4 to the linearization of the equation (24) at the origin. The map ψ changes sign at two points. It means that the number of algebraic multiplicities of negative eigenvalues of $\mathbf{A}(\alpha(\lambda))$ changes at least two point of the path α. Eigenvalues of the matrices $\mathbf{A}(\lambda(\alpha))$ are real. Therefore, applying Theorem 4 we prove the existence of bifurcation points of solutions of the family

$$\dot{\mathbf{T}} = \mathbf{h}(\mathbf{T}, \mathbf{D}) = \mathbf{A}(\mathbf{D}) \cdot \mathbf{T} \quad (32)$$

in the sense of Definition 4.

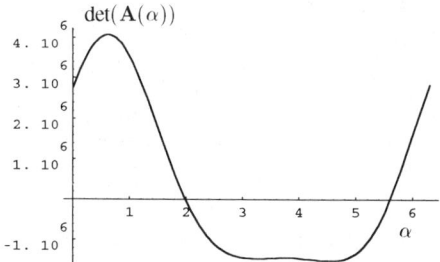

Fig. 3: Graph of the determinant for the eqn. (10)

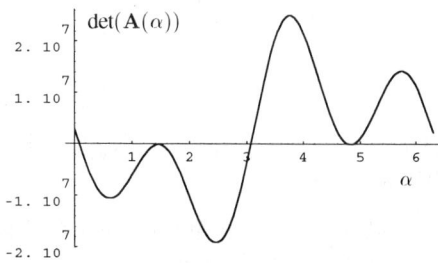

Fig. 4: Graph of the determinant for the eqn. (15)

4.2 Example no. 2

Let us consider the family of ordinary differential equations in the form of (10) with material parameters defined in (11). The coordinates of the tensor $\mathbf{A}(\mathbf{D})$ refering to (24) for the equation (10) has the following form

$$a_{ii} = (C_1 + C_2)D_i + (C_3 + C_4)\|\mathbf{D}\| \\ a_{ij} = C_2 D_i + C_4 \|\mathbf{D}\|, \quad \text{for} \quad i \neq j \qquad (33)$$

where $C_{i(i=1,\ldots,4)}$ the material parameters as written in (11).
In the Fig. 3 we can see bifurcation points for the case of (10) based on determinant of (33). Note that the constitutive function (10) here examined like the previous examined function (8) is linear in $\mathbf{T}$, so all four terms constitute the matrix (33). Like in the previous case we obtain exactly two bifurcation points.

4.3 Example no. 3

In this example we examine the constitutive function (15) consisting of three terms which are tensor-valued functions linear in $\mathbf{T}$ and the fourth term is non-linear in $\mathbf{T}$. Thus, the form of the tensor $\mathbf{A}(\mathbf{D})$ is determined only by the first three terms. Note, that for the use of the formulated theorems we need to assume that the fourth term can be suficiently approximated by a polynomial tensor-valued function to fulfil the requirement (19). In this example one can emphasize the influence of the nonlinearities in equation (15). Thus we consider constitutive equation (15) with its linearization in the form (24) which becomes

$$a_{ii} = (C_1 + C_2)D_i + C_3\|\mathbf{D}\| \\ a_{ij} = C_2 D_j, \quad \text{for} \quad i \neq j \qquad (34)$$

where $C_{i(i=1,\ldots,4)}$ as defined in (16).
In order to research the set of bifurcation points of (24) using the equation (15) we restrict our consideration, like in the previous cases, to a one-dimensional path (28). In this case we check again the map of the graph (31), see Fig. 4. The map ψ changes sign at exactly six points of the interval $[0, 2\pi]$.
Applying Theorems 1, 2 we prove that there are exactly six bifurcation points of bounded solutions of the family (24) included in the interval $[0, 2\pi]$.
Now let us apply Theorem 4 to the linearization of the equation (15) at the origin. The map ψ changes sign at six points. It means that the number of algebraic multiplicities of negative eigenvalues of $\mathbf{A}(\alpha(\lambda))$ changes at least six points of the path α. Eigenvalues of the matrices $\mathbf{A}(\alpha(\lambda))$ are real. Therefore, applying Theorem 4 we prove the existence of bifurcation points of solutions of the family (32) in the sense of Definition 4.
In the above examples we showed the existence of different kinds of bifurcation points of solutions of ordinary differential equations. This analysis was considered for the so called *element test* but the same analysis could be repeted for more complicated cases of a differential problem. The homogeneity of the constitutive tensor-valued function with respect to the strain rate allows, as mentioned, a parametrization of the strain rate tensor as shown in (5). If we look at the last diagrams we can find some zeros of the respectively determinants with respect to some parameters λ. Each parameter λ, which is called a bifurcation point, defines a direction of the unit strain rate tensor. The bifurcation point localizes a possible discontinuity plane which could be understood as a boundary of a shear-band at a material point. An onset of a shear-banding is possible if and only if the mentioned bifurcated strain rate direction with respect to the particulary boundary conditions of the BVP belongs to the set of strain rate directions determining the deformation process in question.
However the problem of linearization stands of importance. An answer of the linearization problem gives the Hartman-Grobman Theorem (Chow and Hale 1982, p. 109-110). This theorem defines some conditions which guarantees that the flows generated by the nonlinear differential equation, e.g. (24),

and linear one, e.g. (32), are identically, in relation to an existing and unique homeomorphism, in the whole considered space.

5 NUMBER OF BRANCHES OF ZEROS ANALYTIC FUNCTIONS

In this short section we would like briefly to show how to compute the number of curves of zeros of analytic maps $\mathbf{h} : (R^n, \mathbf{0}) \to (R^{n-1}, \mathbf{0})$ in terms of local topological indices of maps defined explicitly in terms of the map $\mathbf{h}$.
Let $\mathbf{h} = (\mathbf{h}_1, \ldots, \mathbf{h}_{n-1}) : (B^n, \mathbf{0}) \to (R^{n-1}, \mathbf{0})$ be an analytic map, where B^n is a small ball centered at the origin. Let us denote $X = \mathbf{h}^{-1}(\mathbf{0}) \cap B^n$. Assume that $\mathbf{0} \in R^n$ is an isolated singular point in X, i.e. $\mathbf{0} \in R^n$ is an isolated point in

$$\{\mathbf{x} \in X : rank[D\mathbf{h}(\mathbf{x})] < n-1\}. \quad (35)$$

If B^n is small enough the set $X - \{\mathbf{0}\}$ is void or a finite disjoint union of analytic curves. Let $G : (R^n, \mathbf{0}) \to (R, \mathbf{0})$ be an analytic map such that $\mathbf{0} \in R^n$ is isolated in $\{\mathbf{x} \in X : G(\mathbf{x}) = 0\}$. It is well known fact that if B^n is small enough then G has a constant sign on each connected component of $X - \{\mathbf{0}\}$. Let

1. $b(\mathbf{h}) = $ the number of branches of $X - \{\mathbf{0}\}$,

2. $b_+(G, \mathbf{h}) = $ the number of branches of $X - \{\mathbf{0}\}$ on which G is positive,

3. $b_-(G, \mathbf{h}) = $ the number of branches of $X - \{\mathbf{0}\}$ on which G is negative.

Of course $b(\mathbf{h}) = b_-(G, \mathbf{h}) + b_+(G, \mathbf{h})$.
Let

$$\Delta = \frac{\partial(G, \mathbf{h}_1, \ldots, \mathbf{h}_{n-1})}{\partial(\mathbf{x}_1, \ldots, \mathbf{x}_n)} \quad (36)$$

be the Jacobian of a map

$$(G, \mathbf{h}_1, \ldots, \mathbf{h}_{n-1}) : B^n \to R^n \quad (37)$$

and let

$$H = (\Delta, \mathbf{h}_1, \ldots, \mathbf{h}_{n-1}) : (B^n, \mathbf{0}) \to (R^n, \mathbf{0}) \quad (38)$$

Theorem 5 *The origin is isolated in $H^{-1}(\mathbf{0})$ and*

$$b_+(G, \mathbf{h}) - b_-(G, \mathbf{h}) = 2 \cdot deg(H). \quad (39)$$

Notice that the above theorem allows us to compute the number of branches of zeros of $\mathbf{h}$ on which the map G is positive and negative.
It is worth to point out that if $\mathbf{h}$ and G are polynomial maps then the map H is a polynomial. There is a computer program written by A. Lęcki from University of Gdańsk, which computes the local topological degree for polynomial maps. This program is based on Eisenbud & Levine algorithm (see for this Eisenbud 1977 and Szafraniec 1988). Using this program one can compute the number of branches of zeros of polynomial maps and localize these branches.

6 CONCLUSIONS

The above presented analysis allows to state the following conclusions.

1. (The necessary condition)
   The bifurcation point λ must belong to the following set Z:

$$Z = \{\lambda \in R^3 : \det(\mathbf{A}(\lambda)) = 0\} \quad (40)$$

2. (The sufficient condition)
   If $\lambda_0 \in Z$, see(40), and the determinant of the tensor $\mathbf{A}(\lambda_0)$ of the linearized form of constitutive equation changes its sign at the point λ_0, then the point λ_0 is a bifurcation point, see Theorem 3.

3. (The neutral condition)
   If $\lambda_0 \in Z$, see(40), and the determinant of the tensor $\mathbf{A}(\lambda_0)$ of the linearized form of constitutive equation is equal to zero at the point λ_0, then we can not say whether the point is or not a bifurcation point.

4. The role of nonlinearities in the hypoplastic constitutive equations is fundamental. The linearization process has a great influence of the number of the bifurcation points and their types.

From the above analysis it can be seen that this intricate phenomenon of bifurcation can be well explained in the realm of hypoplasticity using the known theorems of bifurcation theory.

APPENDIX – Fréchet derivative

We consider two Banach spaces $\mathcal{B}_1$ and $\mathcal{B}_2$ (they may be one and the same) and a continuous operator $\mathbf{h}$ mapping $\mathcal{B}_1$ into $\mathcal{B}_2$.

Definition 5 *The operator $\mathbf{h}$ has a Fréchet derivative $\mathbf{A}(x_0)$ in the point x_0 in $\mathcal{B}_1$, if $\mathbf{A}(x_0)$ is a linear, bounded operator with the property*

$$\mathbf{h}(x_0 + v) - \mathbf{h}(x_0) = \mathbf{A}(x_0)v + \mathbf{R}(x_0, v) \quad (41)$$

where the operator **R** satisfies

$$\|\mathbf{R}(v)\|_{\mathcal{B}_2} = \mathbf{O}(\|v\|_{\mathcal{B}_1}) \quad (42)$$

when $\|v\|_{\mathcal{B}_1} \to 0$.

The notation $\mathbf{O}(\|v\|_{\mathcal{B}_1})$ means that

$$\mathbf{O}(\|v\|_{\mathcal{B}_1}) \stackrel{def}{=} \lim_{\|v\|_{\mathcal{B}_1} \to 0} \frac{\|\mathbf{R}(v)\|_{\mathcal{B}_2}}{\|v\|_{\mathcal{B}_1}} = 0. \quad (43)$$

h is called continuously differentiable on an open set $\Omega \subset \mathcal{B}_1$, if

- $\mathbf{A}(x_0)$ is continuous map from Ω to the Banach space $\mathcal{A}(\mathcal{B}_1, \mathcal{B}_2)$ (the set of linear operators mapping $\mathcal{B}_1$ into $\mathcal{B}_2$), and

- **R** is continuous and satisfies

$$\|\mathbf{R}(x_0, v)\|_{\mathcal{B}_2} = \mathbf{O}(\|v\|_{\mathcal{B}_1}) \quad (44)$$

uniformly for all x_0 in closed subsets of Ω.

ACKNOWLEDGMENTS

The authors are indebted to the Polish Scientific Research Committee for the financial support of this project. A part of this project is worked out within grant No. 7 7205 9203.

REFERENCES

Antman, S.S. 1970. Biffurcation problems for nonlinear elastic structures. Academic Press, In Applications of bifurcation theory. Rabinowitz (ed).

Chow, S.N. & J.K. Hale 1982. Methods of bifurcation theory. Springer Verlag. New York.

Ciarlet, Ph.G. 1978. The finite element methods for elliptic problems. North-Holland. Amsterdam, New York, Oxford.

Conley, C. 1978. Isolated invariant sets aand the morse index. In Conference Board of the Mathematical Sciences. Rhode Island. AMS, Providence.

Deimling, K. 1985. Nonlinear functional analysis. Springer Verlag. Berlin.

Eisenbud, D. & H. Levine. 1977. An algebraic formula for the degree of a C^∞-map. Ann. of Math. 106:19-44

Gudehus, G. 1979. A comparison of some constitutive laws for soils under radially symmetric loading and unloading. In Third International Conference on Numerical Methods in Geomechanics. W.Wittke (ed). 1309-1323

Gudehus, G. 1993. A comprehensive equation of state for granular materials. Soils and Foundations, Jap. Soc. Soil Mech., submitted for publication.

Hadamard, J. 1903. Leçons sur la propagation des ondes et les équations l'hydrodynamique. Librairie Scientifique. A.Herrmann (ed). Paris. Ch.6

Hartman, P. 1964. Ordinary differential equations. John Wiley and Sons. New York.

Keller, H.B. & W.F. Langford. 1972. Iterations, perturbations and multiplicities for nonlinear bifurcation problems. Arch. Rational Mech. Anal. 48:83-108

Kolymbas, D. 1981. Bifurcation analysis for sand sample with non-linear constitutive equation. Ing. Arch. 50:131-140

Kolymbas, D. 1987. A novel constitutive law for soils. In 2nd Int. Conf. on Constitutive Laws for Engn. Materials. 319-323. Tucson

Kolymbas, D. 1991. An outline of hypoplasticity. Archive of Applied Mechanics. 61:143-151

Krasnosielski, M.A. 1956. Metody topologiczne w teorii nieliniowych równań całkowych, współczesne problemy matematyki. Nauka. Moscow. In polish

Sikora, Z. 1992. Hypoplastic flow of granular materials – A numerical approach. Veröffentlichungen des Institutes für Bodenmechanik und Felsmechanik der Universität Fridericiana in Karlsruhe. Heft 123. Technical University of Karlsruhe.

Smoller, J. 1983. Schock waves and reactiondiffusion equations. Springer Verlag. New York.

Szafraniec, Z. 1988. On the number of branches of an 1-dimensional semianalytic set. Kodai Math. Journ. 11:78-85

Szlenk, W. 1982. Wstęp do teorii gładkich układów dymanicznych. PWN. Warsaw. In polish

v. Wolffersdorff, P. 1990. Technical University of Karlsruhe. Germany. Private communication.

Wu, W. & Z.Sikora. 1990. Localized bifurcation in hypoplasticity. Int. J. Engng. Sci. 29:195:201

Wu, W. 1993. Ein mathematisches Modell der konstitutiven Beziehungen für granulare Materialien. PhD thesis, IBF, Technical University of Karlsruhe, Heft 127.

A dynamical interpretation of flutter instability
Une interprétation dynamique de l'instabilité de flottement

D. Bigoni
Istituto di Scienza delle Costruzioni, University of Bologna, Italy

J. R. Willis
School of Mathematical Sciences, University of Bath, UK

ABSTRACT: The differential equations governing the development of a small disturbance in material initially stressed to the flutter condition are shown to admit no solution, except for special sets of initial conditions. A viscous regularization, which modifies the dynamic but preserves the static response of the material, is introduced, which permits the solution of the incremental initial value problem for all initial data. The results provide the first physical interpretation of the flutter instability proposed by Rice (1976). In particular, flutter corresponds to an oscillating motion of the material particles, which blows up with time. This behaviour is similar to what is observed in structural systems subjected to follower loads.

On montre que les équations différentielles qui gouvernent le développement d'une petite perturbation dans un matériau soumis à un état de contrainte qui vérifie la condition de flottement n'admettent pas de solution, sauf pour des conditions initiales particulières. Une régularisation viscoplastique est introduite, qui modifie la réponse dynamique mais préserve la réponse statique; ceci permet d'obtenir une solution du problème de Cauchy incrémental pour toutes les données initiales. Le résultat fournit la première interprétation physique de l'instabilité de flottement proposée par Rice (1976). En particulier, le flottement correspond à un mouvement oscillatoire des particules, avec une solution explosive dans le temps. Ce comportement est similaire à celui qu'on observe dans les structures soumises à des charges suiveuses.

1 INTRODUCTION

Flutter instabilities occurring in structural elements subjected to follower loads have been known from the early works of Nikolai (1928), Pflüger (1950), Beck (1952) and Ziegler (1953, 1956) and have been thoroughly studied in succeeding years (Bolotin 1963, Leipholz 1964, Herrmann and Jong 1965, Como 1966, Augusti 1966, Nemat-Nasser and Herrmann 1966a, 1966b, Prasad and Herrmann 1969, Dubey and Leipholz 1975, Alliney and Tralli 1984, Laudiero et al. 1991). In the case of a structural system, flutter instability consists of a vibrational motion of increasing amplitude, when adjacent configurations of static equilibrium for the system are absent. This circumstance occurs when the eigenvalue problem governing the vibration frequencies of the system admits complex eigenvalues. In particular, the condition for the onset of flutter is given by the coalescence of two eigenvalues. For continuous media, Rice (1976) considered the eigenvalues of the acoustic tensor in the context of studying localization. He termed the situation where two real eigenvalues coalesce and then move into the complex plane, that of flutter instability, by analogy with terminology for structures. In contrast to the mechanics of structures, the flutter instability so defined for continuous media is not understood in dynamical terms. In fact, all work to date has concerned evaluations of the constitutive parameters for the occurrence of flutter (Loret et al. 1990, Loret and Harireche 1991, An and Schaeffer 1992, Loret 1992, Bigoni and Zaccaria 1992, 1994, Bigoni 1994).

A brief review of the papers on flutter in continuous media reveals that flutter is much more frequent than one might expect. In fact, flutter was detected in the case of mixture theories of plasticity (Loret and Harireche 1991) and for finite theories of plasticity in the presence of hypoelasticity with asymmetric constitutive law (An and Schaeffer 1992, Bigoni 1994). Moreover, in the usual theories of infinitesimal elastoplasticity, the *onset* of flutter, i.e. the coalescence of two eigenvalues of the acoustic tensor, is always possible, even for associative flow-law. Therefore, a generic perturbation can always induce flutter. For instance, in the case of flow laws obeying deviatoric associativity, a perturbation in the direction of the plastic flow, non-coaxial with the yield function gradient, is sufficient to yield flutter (Loret 1992).

All of the works quoted above refer to the algebraic condition of occurrence of flutter in a continuous medium, without any exploration of the physical meaning of the criterion. The purpose of the present work is to show with a simple example how flutter instability may on one hand be related to the integrability of differential equations governing the dynamic modion of a body, and on the other hand to a particular type of dynamic instability. In the example that we will consider, it is important to observe that flutter instability occurs even if the constitutive operator is positive definite, and therefore second order work positiveness and strong ellipticity are verified. This circumstance suggests that the classical definitions of material stability should be extended to cover the possibility of flutter.

The problem which is addressed in this work corresponds to a medium which is stable (i.e. has acoustic tensor with real and positive eigenvalues) up to some level of stress but, at some critical level, two eigenvalues coalesce and thereafter turn into the complex plane. It is envisaged that the body is at rest, in (unstable) equilibrium, in a state of uniform stress just greater than that associated with coalescence. For the sake of this first investigation, a particular constitutive equation has been selected, which corresponds to a non-symmetric linear constitutive operator, with two complex conjugate eigenvalues. The acoustic tensor, corresponding to the particular constitutive law, has two complex conjugate eigenvalues too. It should be noted that the assumed constitutive law is linear but it may correspond to the loading branch of an elastoplastic constitutive operator. For this problem, it will be shown that the dynamic equations of motion do not possess solutions for arbitrary initial conditions — that is, that the incremental dynamic initial value problem is ill-posed. Therefore, a viscous regularization is introduced by assuming a particular viscous response of the material under shear. The response of the medium to quasi-static deformations is unaltered. The solution shows that, after loading by a small impulse, two symmetric waves travel in opposite directions in the body. There is a wave front across which the displacements suffer a finite jump. Contrary to the well-known elastic symmetric solution (see, e.g. Graff 1975), the displacements are not constant behind the wave front. Therefore, the material particles experience an oscillation in time, after the passage of the wave front. The oscillation grows exponentially with time, remaining finite for every finite value of time. Therefore, we propose to interpret this behaviour, which is similar to the motion experienced by structural elements under follower loads, as a physical consequence of flutter. It should be mentioned that the conclusions presented herein have some similarities with those of Sandler and Rubin (1987). However, the instabilities discussed in that work related to the loading/unloading behaviour, whereas here it is demonstrated that even incremental loading presents difficulties associated with ill-posedness, unless the governing equations are modified.

2 NON-EXISTENCE OF POST-FLUTTER SOLUTIONS

We consider a medium, uniformly stressed into the flutter regime. Its plane-strain response to small perturbations which generate disturbances depending on x_1 and t only is described by the constitutive relation:

$$\begin{bmatrix} \sigma_{11} \\ \sigma_{12} \end{bmatrix} = \begin{bmatrix} 1 & \varepsilon \\ -\varepsilon & 1 \end{bmatrix} \begin{bmatrix} u_1 \\ u_2 \end{bmatrix}_{,1}, \qquad (2.1)$$

where σ_{ij} and u_i are the stress and displacement components, respectively. ε is a (positive) scalar parameter which provides the asymmetry of the constitutive operator. It should be noted that the constitutive operator has two complex conjugate eigenvalues, both with positive real part. The equations of motion of a wave in such a material are (ρ is the mass density, t the time variable and x the space variable, which coincides with the direction 1):

$$\frac{1}{\rho}\begin{bmatrix} 1 & \varepsilon \\ -\varepsilon & 1 \end{bmatrix}\begin{bmatrix} u_1 \\ u_2 \end{bmatrix}_{,xx} = \begin{bmatrix} u_1 \\ u_2 \end{bmatrix}_{,tt}. \qquad (2.2)$$

Therefore, the constitutive operator (2.1) coincides with the acoustic tensor. The equations of motion (2.2) are strongly elliptic, and thus no characteristic solutions are possible.

Now solutions are sought in the following form:

$$\mathbf{u} = \mathbf{f}[x + c(1+i\varepsilon)^{1/2} t], \qquad (2.3)$$

and therefore

$$\mathbf{u}_{,xx} = \mathbf{f}'' \quad \text{and} \quad \mathbf{u}_{,tt} = c^2(1+i\varepsilon)\mathbf{f}''. \qquad (2.4)$$

It can be easily verified that

$$\mathbf{f} = \begin{bmatrix} 1 \\ i \end{bmatrix} g[x + c(1+i\varepsilon)^{1/2} t], \qquad (2.5)$$

is a solution of (2.2) for any function g. Another solution is

$$\mathbf{f} = \begin{bmatrix} 1 \\ i \end{bmatrix} h[x - c(1+i\varepsilon)^{1/2}t], \qquad (2.6)$$

and the complex conjugate functions are solutions as well. Therefore the general solution is:

$$\mathbf{u} = \mathcal{R}e \begin{bmatrix} 1 \\ i \end{bmatrix} \times$$

$$\times \left\{ g[x + c(1+i\varepsilon)^{1/2}t] + \overline{h}[x - c(1+i\varepsilon)^{1/2}t] \right\}. \qquad (2.7)$$

Let us analyze the initial value problem

$$\mathbf{u}(x,0) = 0, \ \mathbf{u}_{,t}(x,0) = \mathbf{v}(x). \qquad (2.8)$$

The initial conditions (2.8) imply

$$g(x) = -\overline{h}(x), \qquad (2.9)$$

$$\mathbf{u}_{,t}(x,0) = +2\mathcal{R}e \begin{bmatrix} 1 \\ i \end{bmatrix} \left\{ c(1+i\varepsilon)^{1/2} g'(x) \right\}, \qquad (2.10)$$

i.e.

$$\mathbf{u}_{,t}(x,0) =$$

$$+ \begin{bmatrix} c[(1+i\varepsilon)^{1/2}g'(x) + (1-i\varepsilon)^{1/2}\overline{g}'(x)] \\ ci[(1+i\varepsilon)^{1/2}g'(x) - (1-i\varepsilon)^{1/2}\overline{g}'(x)] \end{bmatrix}.$$

$$(2.11)$$

If, in particular,

$$\mathbf{u}_{,t}(x,0) = \begin{bmatrix} v_1(x) \\ 0 \end{bmatrix}, \qquad (2.12)$$

one obtains

$$(1-i\varepsilon)^{1/2}\overline{g}'(x) = (1+i\varepsilon)^{1/2}g'(x), \qquad (2.13)$$

$$v_1(x,0) = +2c(1+i\varepsilon)^{1/2}g'(x), \qquad (2.14)$$

which are mutually inconsistent unless v_1 is an analytic function of x, real when x is real. This is not the case, for example when

$$v(x) = \mathbf{V}\delta(x). \qquad (2.15)$$

To verify this claim, let us assume $\rho = 1$, for simplicity, and re-write the differential problem (2.2), with initial conditions (2.8), with $v(x)$ given by (2.15),

$$\mathbf{C}\mathbf{u}_{,xx} + \mathbf{V}\delta(x)\delta(t) = \mathbf{u}_{,tt}, \qquad (2.16)$$

where C is the constitutive tensor defined in (2.1). Taking the Fourier transform on the time variable, one gets (a superscript ^ denotes the Fourier transform of a function, a superscript − denotes the complex conjugate of a number):

$$\mathbf{C}\hat{\mathbf{u}}_{,xx} + \omega^2 \hat{\mathbf{u}} + \mathbf{V}\delta(x) = 0, \qquad (2.17)$$

where

$$\hat{u}(x,\omega) = \int_{-\infty}^{\infty} u(x,t)e^{i\omega t}dt. \qquad (2.18)$$

By introducing the change of variables

$$\hat{\mathbf{u}} = \begin{bmatrix} -i & i \\ 1 & 1 \end{bmatrix} \hat{\mathbf{w}}, \qquad (2.19)$$

the differential problem (2.17) becomes

$$\begin{cases} \hat{w}_{1,xx} + \Omega_+^2 \hat{w}_1 + \frac{1}{2}\{i,1\}\{V\}\frac{i}{i-\varepsilon}\delta(x) = 0 \\ \hat{w}_{2,xx} + \Omega_-^2 \hat{w}_2 + \frac{1}{2}\{-i,1\}\{V\}\frac{i}{i+\varepsilon}\delta(x) = 0 \end{cases},$$

$$(2.20)$$

where $\{V\}$ represents the column vector corresponding to $\mathbf{V}$ and

$$\Omega_+^2 = \omega^2 \frac{i}{i-\varepsilon} \quad \text{and} \quad \Omega_-^2 = \omega^2 \frac{i}{i+\varepsilon}. \qquad (2.21)$$

Solutions of (2.20) are selected in the form:

$$\begin{cases} \hat{w}_1 = \frac{i}{2\Omega_+}\frac{1}{2}\{i,1\}\{V\}\frac{i}{i-\varepsilon}e^{i\Omega_+|x|} \\ \hat{w}_2 = \frac{i}{2\Omega_-}\frac{1}{2}\{-i,1\}\{V\}\frac{i}{i+\varepsilon}e^{i\Omega_-|x|} \end{cases}. \qquad (2.22)$$

The exact choices of the branches for Ω_+, Ω_- are imposed by the requirement of causality, which implies that $\hat{w}_1, \hat{w}_2$ must be analytic in the lower half of the complex ω-plane. Thus, $\Omega_+ = \omega(1+i\varepsilon)^{-1/2}$, $\Omega_- = \omega(1-i\varepsilon)^{-1/2}$ and the square roots have positive real parts. It follows that

$$\begin{cases} u_1 = \{\mathcal{R}e(I_1), \mathcal{I}m(I_1)\}\{V\} \\ u_2 = \{-\mathcal{I}m(I_1), \mathcal{R}e(I_1)\}\{V\} \end{cases}, \qquad (2.23)$$

where

$$I_1(x,t) =$$

$$\frac{i}{8\pi}\{i,1\}\{V\}\left[\frac{1}{1+i\varepsilon}\right]^{1/2} \int_{-\infty}^{\infty} \frac{1}{\omega} e^{i\omega|x|[\frac{1}{1+i\varepsilon}]^{1/2}} e^{-i\omega t} d\omega.$$

$$(2.24)$$

It is easy now to show that I_1 diverges. In fact,

$$\frac{\partial I_1}{\partial t} = -i \left[\frac{1}{i|x|\left[\frac{1}{1+i\varepsilon}\right]^{1/2} - t} e^{i\omega|x|[\frac{1}{1+i\varepsilon}]^{1/2}} e^{-i\omega t} \right]_{-\infty}^{\infty},$$

$$(2.25)$$

in which

$$\left[\frac{1}{1+i\varepsilon}\right]^{1/2} = \left[\frac{1-i\varepsilon}{1+\varepsilon^2}\right]^{1/2} \text{ has negative imaginary part,}$$

and therefore (2.25) grows when $\omega \to \infty$.

In conclusion of this Section, we stress that we are unable to solve the dynamic problem (2.2) with general initial impulse conditions (2.8). This fact shows clearly that *in the case of flutter the equations of motion governing small perturbations may have no solution.* The implications of this finding for numerical methods are evident.

3 VISCOUS REGULARIZATION AND SOLUTION OF INITIAL VALUE PROBLEM UNDER FLUTTER CONDITIONS

In this Section the problem posed in the previous Section is resolved by introducing a viscous regularization. To this purpose, the following viscous-non-symmetric constitutive equation is assumed for the body:

$$\begin{bmatrix} \sigma_{11} \\ \sigma_{12} \end{bmatrix} = \begin{bmatrix} \int_{-\infty}^{\infty} dt' \delta(t-t') & \frac{\varepsilon}{\tau} \int_{-\infty}^{\infty} dt' H(t-t') e^{-(t-t')/\tau} \\ -\frac{\varepsilon}{\tau} \int_{-\infty}^{\infty} dt' H(t-t') e^{-(t-t')/\tau} & \int_{-\infty}^{\infty} dt' \delta(t-t') \end{bmatrix} \times \begin{bmatrix} u_{1,1}(t') \\ u_{2,1}(t') \end{bmatrix},$$

(3.1)

where $H(\cdot)$ is the Heaviside step function, t is the time, ε and τ are material parameters, σ_{ij} are the stress and u_i the displacement components. It is important to note that the scalar parameter ε is related to the non-symmetry of the constitutive equation, whereas τ is related to the viscosity of the material when subjected to shear. By introducing the constitutive matrix $[C]$ and the column vectors $\{\sigma\}$ and $\{u\}$, equation (3.1) may be written as

$$\{\sigma\} = [C]\{u\}_{,x}$$ (3.2)

The problem to be considered is the motion of an infinite body, governed by the constitutive equation (3.1), with density $\rho = 1$, and loaded by an impulse load occurring at $t = 0$ over all points of the plane $x = 0$. For this problem, the equation of the motion can be written as

$$[C]\{u\}_{,xx} + \{V\}\delta(x)\delta(t) = \{u\}_{,tt},$$ (3.3)

where $\{V\}$ is the vector which specifies the initial conditions. By taking the Fourier transform of (3.3) on the time variable and using the convolution theorem, one obtains the following differential problem:

$$\begin{bmatrix} 1 & \frac{\varepsilon}{1-i\omega\tau} \\ -\frac{\varepsilon}{1-i\omega\tau} & 1 \end{bmatrix} \{\hat{u}\}_{,xx} + \omega^2\{\hat{u}\} + \{V\}\delta(x) = 0.$$

(3.4)

By introducing the following change of variables:

$$\{\hat{u}\} = \begin{bmatrix} -i & i \\ 1 & 1 \end{bmatrix} \{\hat{w}\},$$ (3.5)

the differential problem (3.4) becomes:

$$\begin{cases} \hat{w}_{1,xx} + \Omega_+^2 \hat{w}_1 + \frac{1}{2}\{i,1\}\{V\}\frac{i+\omega\tau}{i+\omega\tau-\varepsilon}\delta(x) = 0 \\ \hat{w}_{2,xx} + \Omega_-^2 \hat{w}_2 + \frac{1}{2}\{-i,1\}\{V\}\frac{i+\omega\tau}{i+\omega\tau+\varepsilon}\delta(x) = 0 \end{cases}$$

(3.6)

where

$$\Omega_+^2 = \omega^2 \frac{i+\omega\tau}{i+\omega\tau-\varepsilon} \quad \text{and} \quad \Omega_-^2 = \omega^2 \frac{i+\omega\tau}{i+\omega\tau+\varepsilon}.$$ (3.7)

Solutions of (3.6) are selected in the form:

$$\begin{cases} \hat{w}_1 = \frac{i}{2\Omega_+}\frac{1}{2}\{i,1\}\{V\}\frac{i+\omega\tau}{i+\omega\tau-\varepsilon} e^{i\Omega_+|x|} \\ \hat{w}_2 = \frac{i}{2\Omega_-}\frac{1}{2}\{-i,1\}\{V\}\frac{i+\omega\tau}{i+\omega\tau+\varepsilon} e^{i\Omega_-|x|} \end{cases}.$$ (3.8)

The use of condition (3.5), together with the causality argument given earlier, yields the antitransform:

$$\begin{cases} u_1 = \{\mathcal{R}e(I_1), \mathcal{I}m(I_1)\}\{V\} \\ u_2 = \{-\mathcal{I}m(I_1), \mathcal{R}e(I_1)\}\{V\} \end{cases},$$ (3.9)

where

$$I_1(x,t) = \frac{1}{2\pi}\int_{-\infty}^{\infty} \frac{i}{2\Omega_+}\frac{i+\omega\tau}{i+\omega\tau-\varepsilon} e^{i\Omega_+|x|} e^{-i\omega t} d\omega.$$

(3.10)

The integral (3.10) can be rewritten in the form:

$$I_1(x,t) = \frac{i}{4\pi}\int_{-\infty}^{\infty} \frac{1}{\omega}\left[\frac{i+\omega\tau}{i+\omega\tau-\varepsilon}\right]^{\frac{1}{2}} e^{i\Omega_+|x|} e^{-i\omega t} d\omega.$$

(3.11)

The integral I_1 has a simple pole at $\omega = 0$ and two branch points at $\omega = (\varepsilon-i)/\tau$ and $\omega = -i/\tau$ (see Fig.1). When $\omega \to \infty$, $\Omega_+ \sim \omega$; therefore, for $t-|x| < 0$ closing the contour in the upper half plane gives zero. Thus, a wave front does exist, corresponding to $|x|/t = 1$. For $t-|x| > 0$, closing the contour in the lower half plane yields:

$$I_1 = -2\pi i \mathcal{R}es(I_1, 0)$$

$$-\frac{i}{4\pi}\oint \frac{1}{\omega}\left[\frac{i+\omega\tau}{i+\omega\tau-\varepsilon}\right]^{\frac{1}{2}} e^{i\Omega_+|x|} e^{-i\omega t} d\omega,$$

(3.12)

where the contour integral is to be evaluated on any closed contour enclosing the branch cut, and

$$\mathcal{R}es(I_1, 0) = \frac{i}{4\pi}\left[\frac{i}{i-\varepsilon}\right]^{\frac{1}{2}}.$$ (3.13)

The following coordinate transformation can be performed in the contour integral:

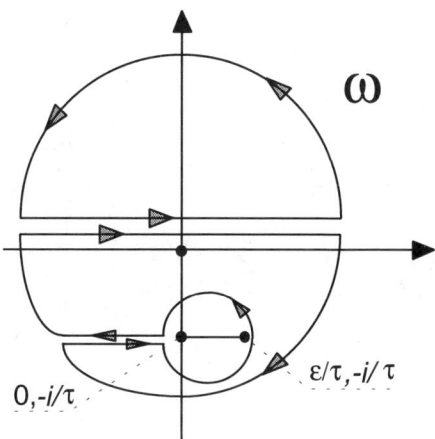

Fig 1. The complex ω-plane and the contours employed in the evaluation of I_1.

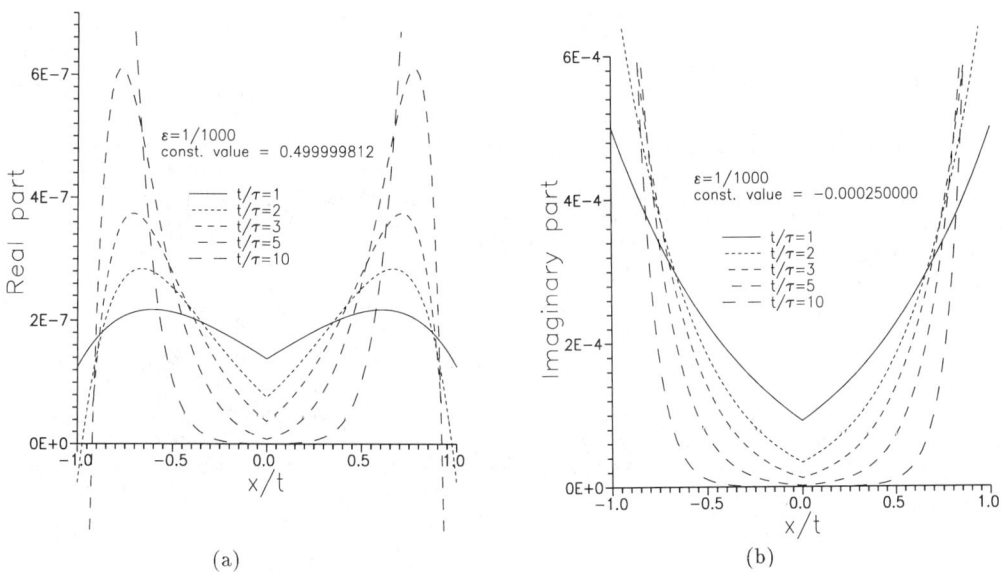

(a)   (b)

Fig 2. The variation of the real part (a) and the imaginary part (b) of the integral I_1, plotted against x/t for various t/τ, when $\varepsilon = 1/1000$.

$$s = \frac{i + \omega\tau}{\varepsilon}, \qquad (3.14)$$

therefore obtaining:

$$I_1 = \frac{1}{2}\left[\frac{i}{i-\varepsilon}\right]^{\frac{1}{2}} - \frac{i\varepsilon}{4\pi} e^{-t/\tau}$$

$$\oint \frac{1}{\varepsilon s - i}\left[\frac{s}{s-1}\right]^{\frac{1}{2}} e^{-i\varepsilon s t/\tau}\, e^{\frac{|x|}{t}\frac{t}{\tau}(1+i\varepsilon s)\sqrt{s/(s-1)}}\, ds\,. \qquad (3.15)$$

The last integral may be evaluated numerically, e.g. on the circle $s(\vartheta) = c + re^{i\vartheta}$, where $\vartheta \in [-\pi, \pi]$.

It is worth noting that the value of the integral I_1 is given by the sum of a contribution independent of t, x and τ (which will be called "constant" in the figures) and of a contribution dependent on t/τ and $|x|/t$. Numerical values of the integral in equation (3.15) are reported in Figs. 2(a,b), 3(a,b), 4(a,b), for $0 \leq |x|/t \leq 1$ and different values of t/τ (1,2,3,5,10). Values of $\varepsilon = 1/10, 1/100, 1/1000$ are explored. In figures

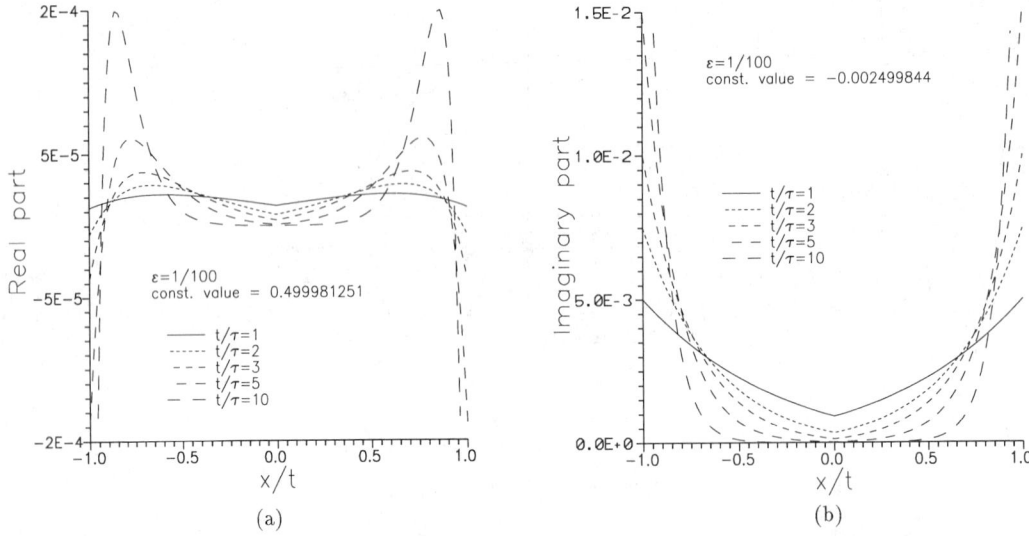

Fig 3. As for Fig. 2, except that $\varepsilon = 1/100$.

"a" the real part of the contour integral appearing in (3.15) is reported, whereas figures "b" refer to the imaginary component. From such figures it can be seen that all curves approach a well-defined value on the wave front $|x|/t = 1$. Moreover, the qualitative trends of the curves are identical for different values of ε. Finally, from figures "a", it appears clearly that the effect of flutter consists, according to this model, of an oscillating motion of the particles after the passage of the wave front. The oscillation blows up when $t/\tau \to \infty$ but remains finite for every finite value of t/τ. The same effect can be appreciated, perhaps more directly, from Fig. 5(a,b), where the real and imaginary components are reported of the contour integral appearing in (3.15), for $\varepsilon = 1/100, \tau = 100$, and for three different particle positions ($x = 10, 100, 200$).

4 CONCLUSIONS

In order to appreciate the effect of flutter, a simple problem of dynamic motion of a continuous medium has been analyzed, namely, the propagation of a small disturbance in a space of material. A first constitutive law was analyzed, for which the acoustic tensor and the constitutive tensor have two complex conjugate eigenvalues with positive real parts. In this case the dynamic equation of motion cannot be solved. For the same case, a viscous regularization has been introduced and the problem solved. The main difference with the same wave problem in the case of a (symmetric) elastic material is that, after the passage of the wave front, the material particles suffer an oscillation which blows up when time increases. However, the oscillation remains finite for every finite value of the ratio time/viscosity parameter. It is worth noting that the considered material is stable in the sense of second order work positiveness (and thus it is strongly elliptic also), but is unstable in the sense of the algebraic condition of flutter (which is calculated for quasistatic disturbances). Of course, the physical relevance of the analysis presented here depends on the credibility of the proposed modification to the constitutive law. The relaxation time τ that was introduced will depend upon micromechanical processes and is likely to be small - perhaps of the order of grain size divided by wave speed c. The blow-up shown in the figures will therefore occur rapidly, and nonlinear terms neglected in this analysis will become important. It appears, nevertheless, that admission of time-dependent influences of microstructural events will be essential if the evolution of disturbances from the "flutter" state are to be resolved.

The model presented here was selected for the purpose of illustration only; the development and analysis of more realistic models, respecting specific microstructural features, is a subject of on-going study.

ACKNOWLEDGEMENTS

D. Bigoni would like to acknowledge the financial support of both the Italian Ministry of University and Scientific and Technological Research (M.U.R.S.T.) and the Italian National Council of Research (C.N.R.- Contr. 91.02914.CT07).

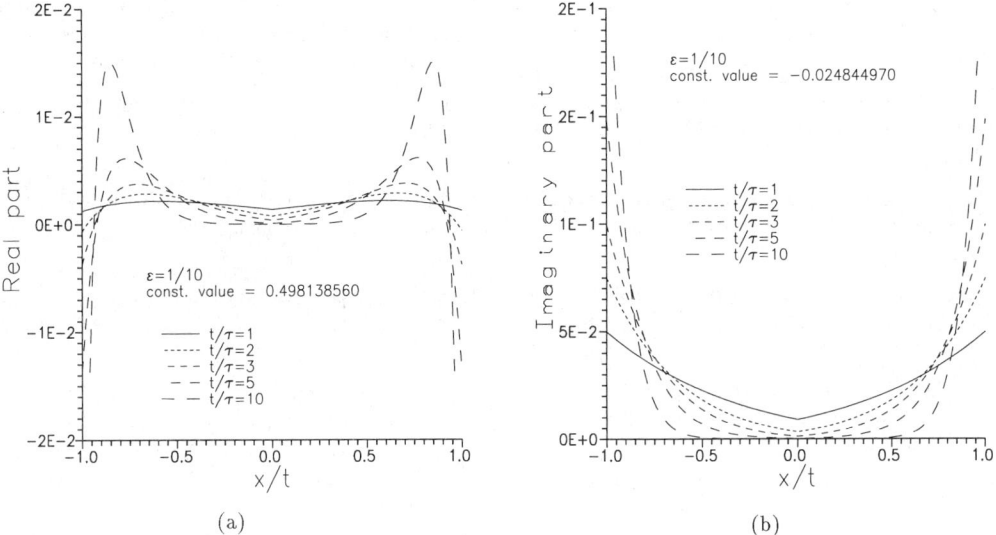

Fig 4. As for Fig. 2, except that $\varepsilon = 1/10$.

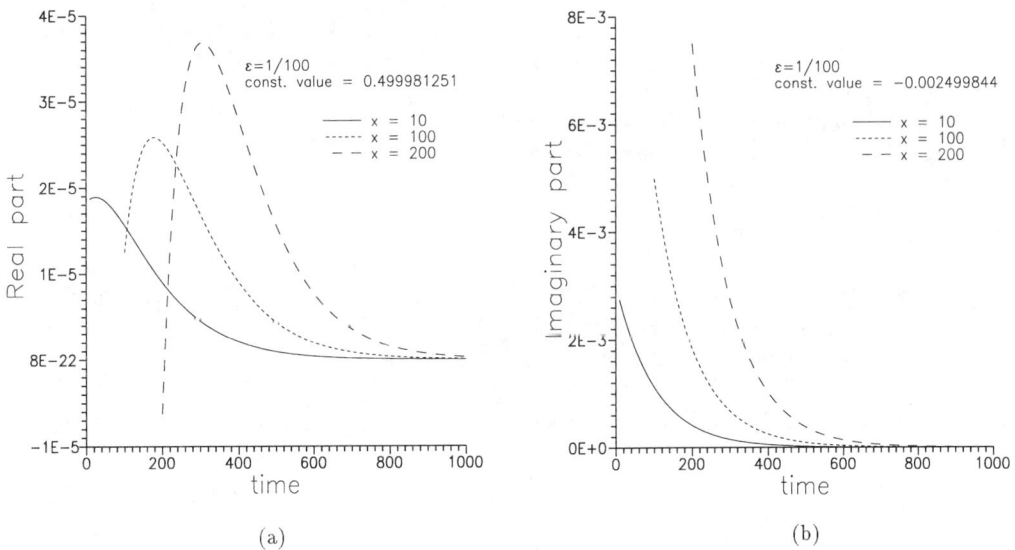

Fig 5. Plots of the variation of the real part (a) and the imaginary part (b) of the integral I_1 against t, for three selected positions x, showing growth of the disturbance immediately behind the wave-front, and subsequent decay.

REFERENCES

Alliney, S. & Tralli, A. 1984. Extended variational formulations and F. E. models for non-linear beams under nonconservative loading. *Comput. Meth. Appl. Mech. Engng.* 46: 177.

An. L. & Schaeffer, D. 1990 The flutter instability in granular flow. *J. Mech. Phys. Solids.* 40: 683.

Augusti, G. 1966 Su di un' asta inelastica compressa da forza trascinata. *Giornale del Genio Civile*, (in Italian). 103.

Beck, M. 1952. Die Knicklast des einseitig eigenspannten, tangential gedrückten Stabes. *ZAMP*, 3: 225.

Bigoni, D. 1994. On flutter instability in elastoplastic constitutive models. Submitted.

Bigoni, D. & Zaccaria, D. 1992. Stability in Mandel sense for elastiplastic solids at finite strain. *XI International Congress AIMETA*, Trento, Sept-Oct 1992. 35.

Bigoni, D. & Zaccaria, D. 1994. On eigenvalues of the acoustic tensor in elastoplasticity. *Eur. J. Mech., A/Solids* (in the press).

Bolotin, V. V. 1963. Nonconservative problems of the theory of elastic stability. New York: Pergamon Press.

Como, M. 1966. Lateral buckling of a cantilever subjected to a transverse follower force. *Int. J. Solids Struct.* 2: 515.

Dubey, R. N.& Leipholz, H. H. E. 1975. On variational methods for nonconservative problems. *Mech. Res. Commun.* 2: 55.

Herrmann, G. & Jong, I.-C. 1965. On the destabilizing effect of damping in nonconservative elastic system. *J. Appl. Mech.* 32: 592.

Graff, K. F. 1975. Wave motion in elastic solids. Oxford: Clarendon Press.

Laudiero, F., Savoia, M.& Zaccaria, D. 1991. The influence of shear deformations on the stability of thin-walled beams under non-conservative loading. *Int. J. Solids Struct.* 27: 1351. itemLeipholz, H. H. E. 1964. Über den Einfluss der Dämpfung bei nichkonservatiken Stabiliitätsprobleme elastischer Stabe. *Ing. Arch.* 33: 308.

Loret, B. 1992. Does deviation from deviatoric associativity lead to the onset of flutter instability? *J. Mech. Phys. Solids* 40: 1363.

Loret, B. & Harireche, O. 1991. Acceleration waves, flutter instabilities and stationary discontinuities in inelastic porous media. *J. Mech. Phys. Solids* 39: 569.

Loret, B., Prevost, J. H. & Harireche, O. 1990. Loss of hyperbolicity in elastic-plastic solids with deviatoric associativity. *Eur. J. Mech., A/Solids* 9: 225.

Nemat-Nasser, S. & Herrmann, G. 1966a. Torsional stability of cantilever bars subjected to nonconservative loading. *J. Appl. Mech.* 33: 102.

Nemat-Nasser, S. & Herrmann, G. 1966b. Adjoint systems in nonconservative problems of elastic stability. *AIAA J.* 4: 2221.

Nikolai, E. L. 1928. On the stability of the rectilinear form of equilibrium of a bar in compression and torsion. *Izv. Leningr. Politechn. in-ta.* 31.

Pflüger, A. 1950. Stabilitätsprobleme der Elastostatik. Berlin: Springer.

Prasad, S. N. & Herrmann, G. 1972. Adjoint variational methods in conservative stability problems. *Int. J. Solids. Struct.* 8: 29.

Rice, J. R. 1976. The localization of plastic deformation. *Theoretical and Applied Mechanics*, Koiter W. T., Ed. p. 207. Amsterdam: North-Holland.

Sandler, I. S. & Rubin, D. 1987. The consequences of non-associated plasticity in dynamic problems. In *Constitutive Laws for Engineering Materials*, Desai, C. S. et al., Eds. p 345. Amsterdam: Elsevier.

Ziegler, H. 1953. Linear elastic stability. *ZAMP*. 4: 89.

Ziegler, H. 1956. On the concept of elastic stability. In *Advances in Applied Mechanics*, IV: 351. New York: Academic Press.

Stability problems related to static liquefaction of loose sand
Problèmes de stabilité en relation avec la liquéfaction statique dans le sable lâche

Claudio di Prisco & Roberto Nova
Milan University of Technology (Politecnico), Italy

ABSTRACT: The condition for the occurence of a flow slide in an infinite subaqueous slope is first given. Such a condition is next compared to other conditions which may lead to unstable soil behaviour. It is shown that depending of the criterion adopted, a slope may be considered to be unstable for different values of the inclination. Spontaneous collapse according to the criterion given in this paper in undrained conditions may occur when the slope inclination is much less than the friction angle but it is larger than the value corresponding to the loss of positive definiteness of the stiffness matrix.

On présente d'abord la condition de rupture par écoulement d'une pente infinie submergée. Cette condition est ensuite comparée à d'autres conditions qui peuvent conduire à un comportement instable des sols. On montre que suivant le critère envisagé, une pente peut être considérée comme instable pour différentes valeur de son inclinaison. La rupture spontanée en conditions non drainées peut se produire alors que l'inclinaison est bien plus petite que l'angle de frottement interne, mais cependant plus grande que la valeur correspondant à la perte du caractère défini positif de la matrice de rigidité.

1. INTRODUCTION

The occurence of flow slides in loose saturated sands has been an intriguing research topic since long (Hazen (1920), Terzaghi (1950), Casagrande (1975)). Experimental evidence is in fact that even flat subaqueous slopes may collapse under the triggering effect of apparently minor causes, such as small earthquakes or rapid tide level variation. Such a phenomenon cannot be explained by a classical limit equilibrium approach, however.

Liquefaction of sand was therefore evocated as the phenomenon responsible for such an ackward result. Recently Sladen et al. (1985) and Lade (1991) proposed two criteria to evaluate the stability of a slope in undrained conditions, that although based on conceptually different premises, identify the loss of stability with the level at which the stress deviator reaches a maximum in an undrained triaxial compression test. According to such criteria even very flat slopes are unstable in the at rest conditions. One may wonder how relatively steep subaqueous slopes of sandy soil could exist in nature.

Here a different approach will be followed. The condition for the occurence of a flow slide in an infinite slope will be first investigated. By means of a convenient elastoplastic strain hardening model such a condition will be compared to other instability criteria. It will be shown that spontaneous collapse in undrained conditions takes place when the inclination is much lower than the friction angle, but it is larger than the values corresponding to the Lade stability criterion and to the Hill (1958) sufficient condition for stability (loss of positive definiteness of the stiffness matrix).

Relatively steep subaqueous sandy slopes may therefore exist in nature, but it is shown that a perturbation of limited intensity (but not infinitesimal) may give rise to a flow slide even for flat slopes.

2 SPONTANEOUS COLLAPSE OF SUBMERGED INFINITE SLOPES

Consider an element of an infinitely long planar submerged slope, such as that shown in Fig. 1. Water is assumed to be in a steady state. Therefore the resultant of the water pressures on the edges of the element is vertical and counterbalances the weight of

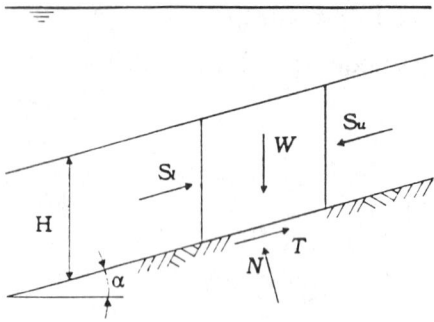

Fig. 1 a) Forces acting on an element of an infinite submerged slope.

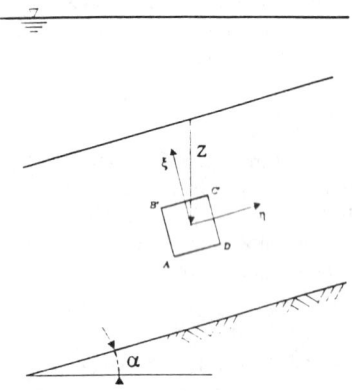

Fig. 2 New reference frame for the element of Fig. 1

an element of equal volume entirely filled with water. Since the slope is infinitely long and no distinction can be made between the upper and the lower vertical edges of the element, the resultant of the effective stresses on the vertical edges must balance out. The effective normal and shear stresses at the base of the element are then:

$$\sigma' = \gamma' z \cos^2 \alpha \qquad (1)$$

$$\tau = \gamma' z \cos\alpha \sin\alpha \qquad (2)$$

where γ' is the buoyant unit weight of the soil, α is the slope and z is the depth at which the stress state is calculated

It is readily apparent from the Mohr circle of stresses that the resultant of the shear force and the normal force acting on the vertical edges is directed along the slope. However, the intensity of such force cannot be determined by means of equilibrium equations only.

Traditionally, the analysis of the stability of the slope, in the classical Soil Mechanics sense of the word, is performed by comparing the stress acting on the potential sliding surface and the available strength. Since such surface is necessarily a plane parallel to the slope, because of the symmetry of the problem, it turns out that the slope is stable insofar the inclination angle α is less than the friction angleϕ', relative to plane strain conditions.

Such a result is in contradiction with the experimental evidence that even very flat slopes, with an inclination of less than 5°, may fail under perturbations small in size but rapid enough to prevent water drainage. To explain why collapse may occur even when α is much lower than ϕ' an alternative approach is necessary.

Consider therefore the element of Fig. 2, where ξ and η are the axes orthogonal and parallel to the slope, while the axis χ is normal to the plane of the figure. Note that the state of stress is the same as that of the element of Fig. 1, although, the stress vectors are different on differently inclined edges. Such an element is representative of the behaviour of the entire slope. In fact the problem can be treated as one dimensional because of the aforementioned symmetry and because seepage can be neglected. Undrained conditions are assumed to be valid because only rapid perturbation of the equilibrium state are considered.

Plane strain conditions imply $\dot{\varepsilon}_\chi = \dot{\gamma}_{\chi\xi} = \dot{\gamma}_{\chi\eta} = 0$, and since the slope is infinitely long $\dot{\varepsilon}_\eta = 0$. Undrained conditions imply then $\dot{\varepsilon}_\xi = 0$, so that the only non zero component of the strain rate vector is $\dot{\gamma}_{\xi\eta}$. Stress rates are related to strain rates for a soil element by D_{ijhk}, the stiffness tensor of the element. For simple symmetry conditions, linked to the uniformity of the material in the χ direction, it is readily apparent that the axis χ is not only principal axis of strain rates but also of stress rates.

If the stiffness tensor is ordered into a matrix form, D, the incremental constitutive relationship becomes:

$$\begin{Bmatrix} \dot{\sigma}'_\xi \\ \dot{\sigma}'_\eta \\ \dot{\sigma}'_\chi \\ \dot{\tau}_{\xi\eta} \\ \dot{\tau}_{\chi\xi} \\ \dot{\tau}_{\chi\eta} \end{Bmatrix} = \begin{bmatrix} D_{11} & D_{12} & D_{13} & D_{14} & D_{15} & D_{16} \\ D_{21} & D_{22} & D_{23} & D_{24} & D_{25} & D_{26} \\ D_{31} & D_{32} & D_{33} & D_{34} & D_{35} & D_{36} \\ D_{41} & D_{42} & D_{43} & D_{44} & D_{45} & D_{46} \\ D_{51} & D_{52} & D_{53} & D_{54} & D_{55} & D_{56} \\ D_{61} & D_{62} & D_{63} & D_{64} & D_{65} & D_{66} \end{bmatrix} \begin{Bmatrix} \dot{\varepsilon}_\xi \\ \dot{\varepsilon}_\eta \\ \dot{\varepsilon}_\chi \\ \dot{\gamma}_{\xi\eta} \\ \dot{\gamma}_{\xi\chi} \\ \dot{\gamma}_{\chi\eta} \end{Bmatrix} \qquad (3)$$

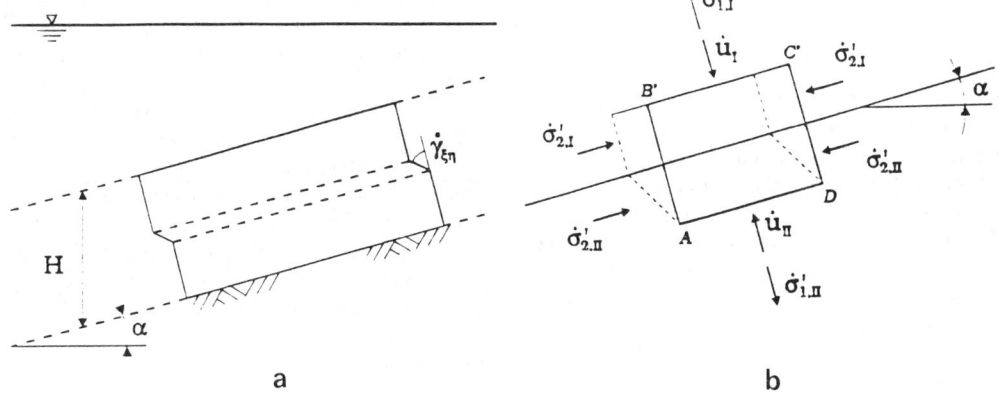

Fig. 3 a) Possible collapse mechanisms when shear stiffness D44 is nil; b) state of stress rate across the band.

and for the particular problem at hand application of the undrained and plane strain conditions, reduces Eq(3) to the simpler form:

$$\begin{Bmatrix} \dot{\sigma}'_\xi \\ \dot{\sigma}'_\eta \\ \dot{\sigma}'_\chi \\ \dot{\tau}_{\xi\eta} \\ \dot{\tau}_{\chi\xi} \\ \dot{\tau}_{\chi\eta} \end{Bmatrix} = \begin{Bmatrix} D_{14} \\ D_{24} \\ D_{34} \\ D_{44} \\ 0 \\ 0 \end{Bmatrix} \dot{\gamma}_{\xi\eta} \qquad (4)$$

If at a certain depth z:

$$D_{44} = 0 \qquad (5)$$

a bifurcation in the form of a shear band may occur. An arbitrary value of $\dot{\gamma}_{\xi\eta}$ in the layer at depth z is compatible with zero strain elsewhere, while equilibrium across the plane parallel to the slope can be satisfied with no change of the external loading. The first of Eqs. 4 gives in fact the effective stress rate normal to the discontinuity, which can be compensated by an instantaneous variation of the pore pressure in the layer in order to keep zero the total stress rate:

$$\dot{u} = -\dot{\sigma}'_\xi = -D_{14}\dot{\gamma}_{\xi\eta} \qquad (6)$$

Since the value of $\dot{\gamma}_{\xi\eta}$ is not specified and can be very large, when condition (5) is met, the slope may spontaneously collapse, with the mechanism shown in Fig. 3a. Equation (5) gives therefore the condition for loss of stability of the slope. Such a static condition of instability is analogous to the latent instability of a beam under compressive load in which the overall stiffness (elastic and geometric) is zero, see e.g. Ziegler (1968). The state of stress rate across the band is shown in Fig. 3b.

In order that such a mechanism could occur it is sufficient that only the layer at depth z behaves in undrained conditions, the other layers, in particular those close to the free surface, being free to drain.

The locus for which D_{44} is nil will be called undrained stability locus. Since for virgin soils stiffness varies roughly linearly with σ', in the plane σ'-τ, the stability locus is approximately a straight line passing through the origin.

This undrained stability locus is obtained by analysing a bifurcated mechanism but it is analogous to the stability line concept proposed by Lade (1991) for undrained triaxial compression tests. In fact when D_{44} is nil, the second order work, defined as:

$$d^2W = 1/2\,\dot{\varepsilon}_{ij}\dot{\sigma}'_{ij} = 1/2\,D_{44}\dot{\gamma}_{\xi\eta}\dot{\gamma}_{\xi\eta} \qquad (7)$$

associated to this particular strain rate tensor is also nil.

Moreover the second order work can be written as:

$$d^2W = 1/2\,\dot{e}_{ij}\dot{s}_{ij} + 1/2\,\dot{v}\dot{p}', \qquad (8)$$

where $\dot{s}_{ij}$ is the deviator stress rate tensor, $\dot{e}_{ij}$ is the deviator strain rate tensor, $\dot{v}$ is the volumetric strain rate and $\dot{p}'$ is the effective mean pressure rate.

If $\dot{\varepsilon}_{ij}$ fulfills the constant volume condition, the second term of Eq.(8) is nil. Consequently the second order work is nil if and only if the deviator stress rate tensor $\dot{s}_{ij}$ is normal to the deviator strain rate tensor $\dot{e}_{ij}$.

The same result is obtained if $\dot{s}_{ij}$ is nil; in triaxial conditions this means $\dot{q} = 0$, which coincides with the stability line concept proposed by Lade, which is determined as the locus of the peaks of the stress deviator in undrained triaxial compression tests.

The stability locus according to Lade criterion can be formally obtained in the following way. If the stress-strain relationship is written as:

$$\dot{\varepsilon}_{ij} = C_{ijhk}\dot{\sigma}_{hk} \qquad (9)$$

C_{ijhk} being the compliance tensor, i.e. the inverse of D_{ijhk}, and we impose the condition $\dot{v} = 0$, taking account that at peak the stress rate has zero deviatoric component, we get:

$$\dot{\varepsilon}_{ij}\delta_{ij} = \delta_{ij}C_{ijhk}\dot{\sigma}_{hk} = \delta_{ij}C_{ijhk}\delta_{hk}\dot{p}' = 0, \qquad (10)$$

where $\dot{p}'$ is a non-zero scalar. Thus

$$\delta_{ij}C_{ijhk}\delta_{hk} = 0 \qquad (11)$$

turns out to be the mathematical condition for the determination of the stability line, which is different from Eq. (5)

Note further that the application of the Lade's stability line to the case considered here would be incorrect, because the kinematics of failure are quite different in axisymmetric conditions and in an infinitely long slope.

It is worth noting that the stability locus concept should not be confused with the static stability condition introduced by Hill (1958). The Hill sufficient condition for stability is in fact:

$$d^2W_{ij} > 0 \qquad \forall \dot{\varepsilon}_{ij}, \qquad (12)$$

i.e. the second order work must be positive for every strain rate tensor and not only for the strain rate tensor which describes the particular strain path followed. It can be easily shown that Eq(12) is violated first when the matrix D becomes positive semidefinite. This occurs when the determinant of the symmetric part of D, D_s, is zero:

$$\det D_s = 0 \qquad (13)$$

Condition (5) is also different from the usual condition for the occurrence of a shear band, the nullity of the determinant of the acoustic tensor (Rudnicki and Rice (1975)). The difference is in that,

in order to obtain the latter, equilibrium across the plane of discontinuity is imposed on effective stresses, i.e. either the soil is dry or the pore pressure is the same inside and outside the band. To obtain condition (5), instead, equilibrium was imposed on total stresses, so that pore pressures turn out to be different at the two sides of the band.

The usual shear band analysis can be performed via the acoustic tensor; the kinematic conditions allow only one plane of discontinuity to occur: the plane defined by a normal vector directed as the axis ξ. Therefore the acoustic tensor is then:

$$A_{jh} = n_i D_{ijhk} n_k \qquad (14)$$

where $n_i = (1 \quad 0 \quad 0)^t$.

In matrix form the acoustic tensor associated to the plane of normal ξ, is given by:

$$A_{ij} = \begin{bmatrix} D_{11} & D_{14} & D_{15} \\ D_{41} & D_{44} & D_{45} \\ D_{51} & D_{54} & D_{55} \end{bmatrix} \qquad (15)$$

and spontaneous collapse will occur therefore when:

$$\det A_{ij} = 0 \qquad (16)$$

Since however for the symmetry of the problem and the plane strain condition imposed, the off-diagonal terms in the third row are nil, the condition for the occurrence of a shear band in drained conditions is simply:

$$D_{11}D_{44} - D_{14}D_{41} = 0 \qquad (17)$$

3. INSTABILITY CONDITIONS FOR STRAINHARDENING ELASTOPLASTICITY

Strainhardening elastoplasticity theory is commonly used to reproduce and to interpret the mechanical behaviour of geomaterials. In order to model the mechanical behaviour of sand, and particularly the static liquefaction phenomena it is necessary that the plastic potential and the loading function do not coincide: the flow rule must be non associated (Nova (1991)).

In this case the instability phenomena previously described can appear in the hardening regime, i.e. when the hardening modulus H is still positive.

Maier and Hueckel (1979) proved that, in order to fulfill the Hill stability condition, Eq. (13), the following inequality must be obeyed:

$$H > H_c/2 + 1/2\sqrt{(n^t D^e n)(m^t D^e m)} \quad (18)$$

where n and m are vectors normal to the yield surface and the plastic potential, respectively:

$$n \equiv \partial f / \partial \sigma' \quad (19)$$

$$m \equiv \partial g / \partial \sigma' \quad (20)$$

D^e is the elastic stiffness matrix and

$$H_c \equiv -n^t D^e m. \quad (21)$$

In the following the value of H which fulfills Eq.18 with the equal sign and corresponds to the loss of positive definiteness of the stiffness matrix will be called H_H.

The stiffness tensor can be put in the following form (Maier and Hueckel (1979)):

$$D = D^e - \frac{UM^t}{(H - H_c)} \quad (22)$$

where

$$U \equiv D^e m \quad (23)$$

$$M \equiv D^e n \quad (24)$$

Condition (5) can be written as:

$$D^e_{44} - \frac{U_4 M_4}{H - H_c} = 0, \quad (25)$$

and then the value of H which fulfills Eq.(5), H^*, is given by:

$$H^* = H_c + U_4 M_4 / D^e_{44}. \quad (26)$$

Condition (11) can be put in explicit form:

$$\delta^t C^e \delta + n^t \delta \, m^t \delta \, / H = 0 \quad (27)$$

where $\delta = (1\ 1\ 1\ 0\ 0\ 0)^t$ is the vector equivalent of the Kronecker δ. Therefore, the value of H which fulfills Eq(27), H_L, is given by:

$$H_L = (n^t \delta \, m^t \delta) K \quad (28)$$

K being the elastic bulk modulus.

Finally the value of H at the onset of a shear band in drained conditions, Eq.(17), by using Eq. (22), can be written as:

$$H_A = H_c + b/a \quad (29)$$

where:

$$a = D^e_{11} D^e_{44} - D^e_{14} D^e_{41}$$

$$b = D^e_{11} U_4 M_4 + D^e_{44} U_1 M_1 - \quad (30)$$

$$D^e_{14} U_4 M_1 - D^e_{41} U_1 M_4$$

4. SPECIALIZATION FOR A PARTICULAR CONSTITUTIVE MODEL

In order to make a comparison between the different instability criteria presented so far, we shall use here an elastoplastic constitutive model, characterized by an anisotropic hardening law (di Prisco and Nova (1992), di Prisco et al. (1993)). Such a model is an extension of the model called "Sinfonietta classica" (Nova (1988)), used for the Cleveland Workshop (Saada and Bianchini (1988)). At variance with the latter, the former model is also able to reproduce correctly the behaviour of sand even in non-monotonic tests such as the circular test in the deviatoric plane, or in tests in which the loading direction varies abruptly.

For the sake of brevity, the presentation of the model will not be repeated here. A thoroughful description is given in di Prisco et al. (1993) and in di Prisco (1993). It will be enough to recall here that the shape of the yield function is similar to that experimentally determined by Nova and Wood (1978) in the triaxial plane and by Lade (1977) in the deviatoric plane. Hardening, which depends on both volumetric and deviatoric strains, as in the Nova and Wood (1979) model, is such that the current yield surface widens, rotates and changes shape along with plastic straining. The flow rule is non-associated. Elastic behaviour is described by a non linear law (Lade and Nelson (1987)).

A series of triaxial test were conducted on Hostun RF sand at a relatively density of 30%. The calibration of the fourteen parameters (three elastic, friction and dilatancy angles at failure in compression and extension, plastic isotropic compressibility, initial size of the yield locus, initial and limit shape parameters of the yield function and two evolution parameters of the yield function and plastic potential) was done on the basis of the results obtained in one isotropic loading-unloading test, one drained and one undrained triaxial compression tests. The results of the calibration tests together with the results obtained in similar tests at a different confining pressure are

a

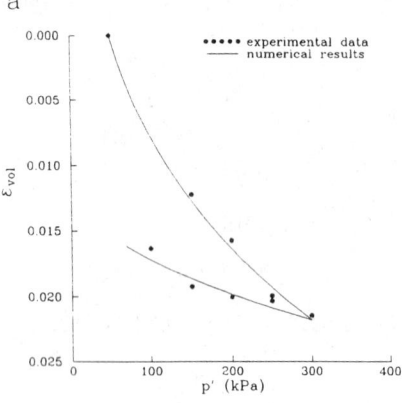

b

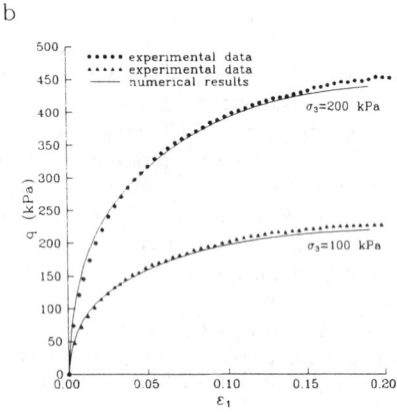

c

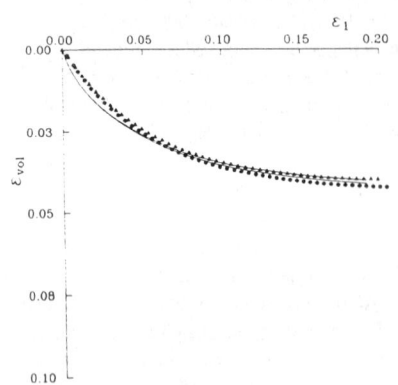

d

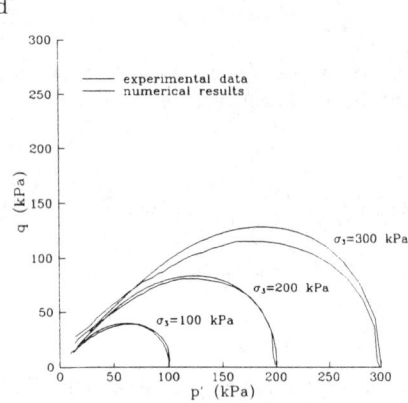

e

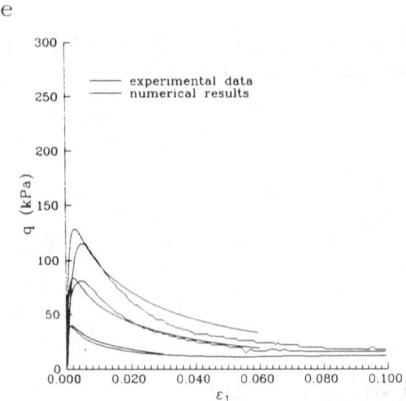

Fig. 4 Comparison between experimental and numerical results for loose Hostun RF sand: a) isotropic loading unloading; b) drained triaxial compression, deviatoric stress strain relationship; c) drained triaxial compression, volumetric strain; d) undrained triaxial compression, effective stress path; e) undrained triaxial compression, deviatoric stress strain relationship; .

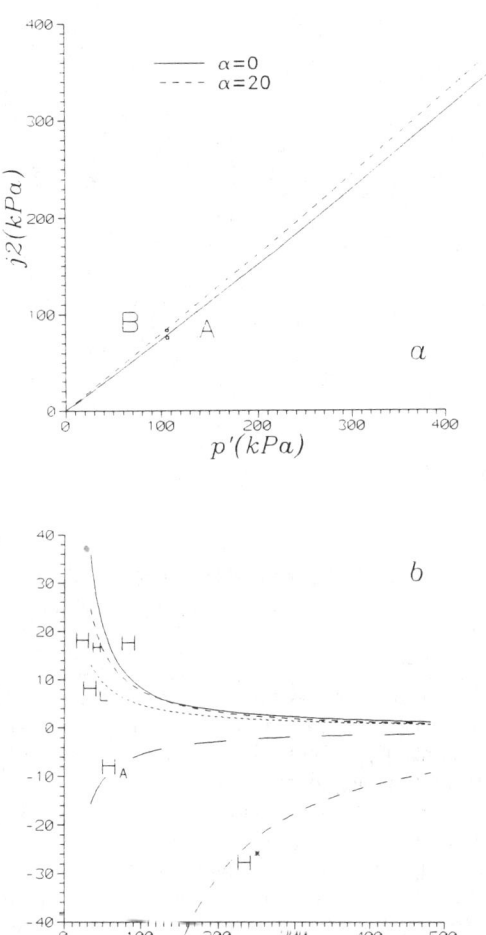
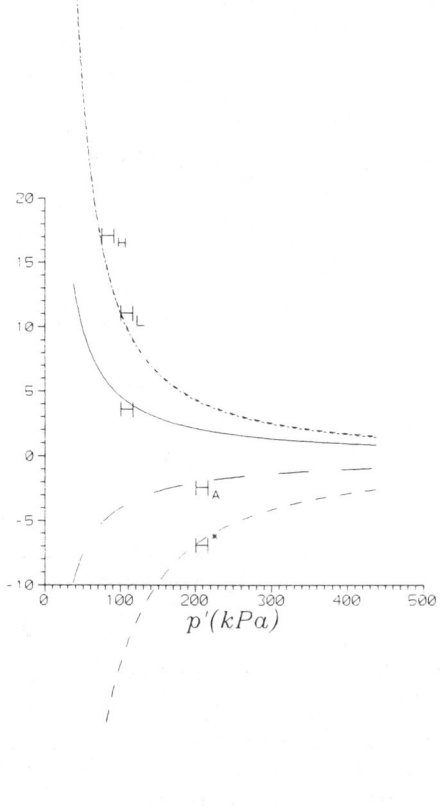

Fig. 5 a) Stress paths during the process of sedimentation for α=0° and α=20°, b)evolution of the moduli H during the process of sedimentation with α=0°, c) evolution of the moduli H during the process of sedimentation with α=20°.

Tab.1 Constitutive parameters

B_p	C_p	t_p	$\hat{\vartheta}_C$	$\hat{\vartheta}_E$	ξ_C	ξ_E
.0049	18.	10	.25	.039	-.25	-.04
P_0	β_{f0}	β_f	RF	α_c	B_0	γ_c
1.	1.1	.5	.28	.25	1028.	3.7

shown in Fig. 4. The values of the parameters retrieved from the calibration are given in Table 1.

In the paper by di Prisco et al. (1993) it is shown that the model is able to reproduce the observed behaviour of sand even in tests more complex than triaxial compression. In particular, tests in which the Lode angle of stress varies either continuously (circular testing in the deviatoric plane) or abruptly (loading-unloading -reloading at a different Lode angle) can be accurately reproduced.

This model will be used henceforth as a predicting tool for test paths which are difficult to perform experimentally. In particular in order to determine the onset of failure of a submerged slope, it would be necessary to perform a simple shear test starting from an initial condition difficult to be realized in the laboratory, since the ratio between shear and normal stress at the base of the apparatus should be imposed a priori, as it will be clear in the following section. A numerical conceptual test will be therefore performed in lieu of an actual experiment.

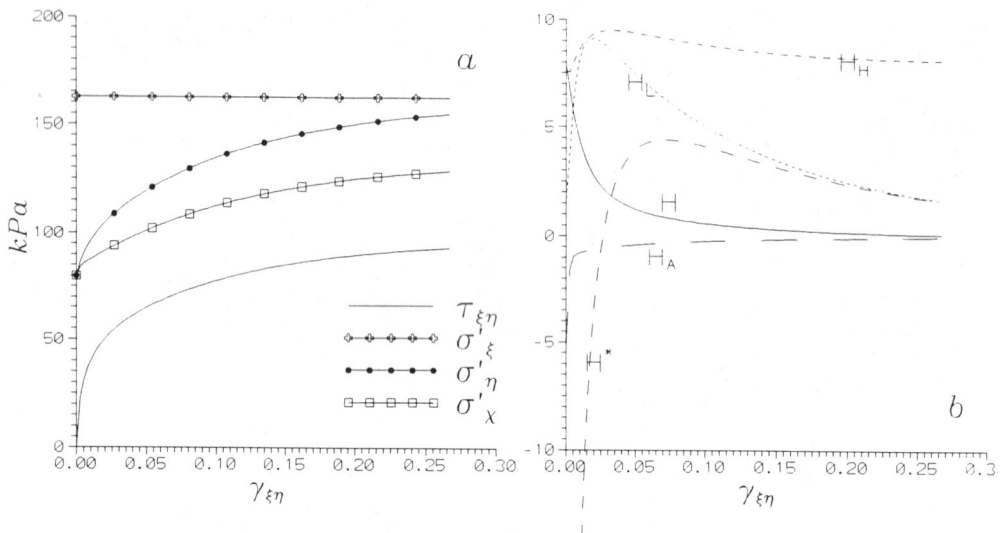

Fig.6 Numerical results for a drained shear stress path of an element of a K_0 consolidated ($\alpha=0°$) layer a) evolution of stresses, b) evolution of moduli H

5 SIMULATION OF NON-TRIVIAL SIMPLE SHEAR TESTS

Consider the slope shown in Fig.1. If we assume that the slope was formed by deposition of a sand layer on a preexisting rigid substratum inclined of an angle α to the horizontal, then symmetry and plane strain conditions impose that, during the deposition of the sand layer, $\dot{\varepsilon}_\xi > 0, \dot{\varepsilon}_\eta = 0, \dot{\varepsilon}_\chi = \dot{\gamma}_{\chi\eta} = \dot{\gamma}_{\chi\xi} = 0$ and in order to satisfy equilibrium:

$$\dot{\tau}_{\xi\eta} = \dot{\sigma}'_\xi \tan\alpha \tag{31}$$

In Fig.5a, the stress path in the plane p'-j$_2$, for different values of α during the process of sedimentation is shown; j$_2$ is the square root of the second invariant of the stress deviator:

$$j_2 = \sqrt{s_{ij} s_{ij}} \tag{32}$$

The evolution of the hardening modulus H and of the moduli H_H, H^*, H_L and H_A, previously introduced, during K_0 sedimentation ($\alpha = 0°$) is depicted in Fig.5b.

For a given instabilty condition, the material is stable insofar H is greater than the value corresponding to it. The onset of instability takes place when the hardening modulus is equal to the value appropriate for that condition, i.e. when the evolution lines for H and the instability moduli cross each other.

We can observe that the hardening modulus H is larger than any of the moduli associated to the onset of a certain type of instability. Therefore the material is stable according to every condition.

We can note however that H is very close to H_H which means that the instability condition in the sense of the loss of positive definiteness of the stiffness matrix is near to be met. In the sense of Hill a horizontal soil layer in the at rest condition is stable but in a condition for which a small perturbation may induce instability.

Similarly in Fig.5c the evolution of H, H_H, H^*, H_A and H_L is shown for the sedimentation process on a slope with an inclination of 20°. It is interesting to note that in this case H is less than H_H and H_L but it is greater than H^* and H_A. This implies that even in the at rest condition the slope is unstable according to Hill and Lade criteria, although it is not for the other two criteria.

Note however that, although unstable in a general sense, the slope can not slide, the boundary conditions of the problem restricting the kinematic condition of collapse.

If we now decide to impose a drained shear path, characterized by $\dot{\gamma}_{\xi\eta} > 0$ and $\dot{\sigma}'_\xi = 0$, starting from point A of Fig.5a, the evolution of stresses is shown in Fig.6a, while the moduli H, H_H, H^*, H_L and H_A vary as depicted in Fig.6b. The hardening modulus H very soon becomes smaller than H_H and H_L. At a later stage the H line crosses the H^* line, and only

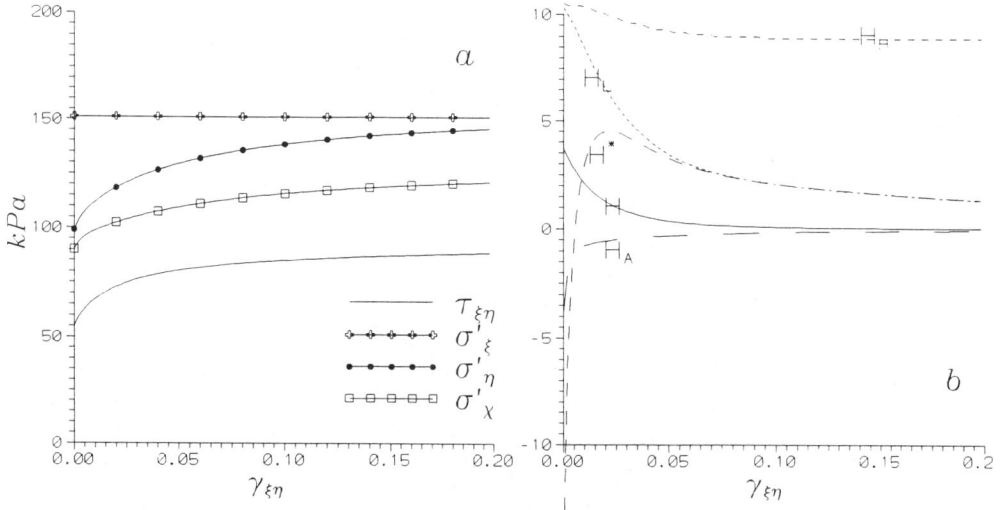

Fig.7 Numerical results for a drained shear stress path of an element in a layer deposited with an inclination α =20° a) evolution of stresses, b)evolution of moduli H

asymptotically joins the H_A line. That means that, for a loose sand, in drained conditions, along the stress path of simple shear, the shear band condition, in small deformations, can be achieved only asymptotically.

The undrained stability condition is fulfilled when the shear strain is approximately 0.03. Assume to stop the drained stress path at this stage and to give a rapid perturbation (not allowing drainage). The slope will then manifest its instability. This means that the slope could become unstable under rapid perturbation (not allowing drainage) at this stage.

Numerical results relative to a similar conceptual experiment ($\dot{\gamma}_{\xi\eta}>0. and.\dot{\sigma}'_\xi=0$) but with higher inclination (α=20°) are shown in Fig.7a. In Fig.7b the evolution of H, H_A H_L H_H and H^* is illustrated. In this case the H value is less than the critical Hill and Lade values since the beginning of the shearing phase. A shear strain smaller than $\gamma=0.01$ is necessary to reach the instability criterion for undrained conditions.

If we impose an undrained shear programme, ($\dot{\gamma}_{\xi\eta}>0. and.\dot{\varepsilon}_\xi=0$), starting from the at rest condition we obtain the results shown in Fig. 8 and 9 for the two inclinations considered.

In Fig.8a it is shown the stress-strain behaviour for the case α=0°. A peak is associated to every component of the effective stress tensor, the peak of the τ-γ curve appearing when $\gamma_{\xi\eta} \cong 0.007$. In Fig.8b it is worth noting that the undrained instability condition (5) is fulfilled before the shear band condition (17). In fact the latter is fulfilled after the peak of the τ-γ curve.

Note that a comparison between the undrained stability condition and the condition for shear band occurrence is relevant, since we may conceive to stop the loading path at any point and to give either a drained or an undrained perturbation to check the overall stability of the slope.

Similar results are presented in Fig.9 for α=20°. As expected for higher slope inclination a much smaller straining (γ=0.001) is necessary to cause collapse.

It can be shown with a similar analysis that spontaneous collapse in undrained conditions occurs only for α = 24.5°. The existence of submerged sandy slopes with inclinations of the same order of magnitude is then justified. Note however that such a collapse value is much lower than that corresponding to the friction angle for the material considered, which, in plane strain conditions, is approximately ϕ' = 36°. In this condition the mobilized angle of friction is 32°. According to Hill and Lade criteria the onset of instability occurs for α=2° and α=8°, respectively.

If the shear stress increment necessary to cause collapse, according to the criterion of Eq(5), is normalized with respect to a convenient measure of the state of stress at the base of the layer, we may see, Fig. 10, that the influence of depth is negligible and that, as a first approximation, a single curve may describe the variation of the normalized shear stress

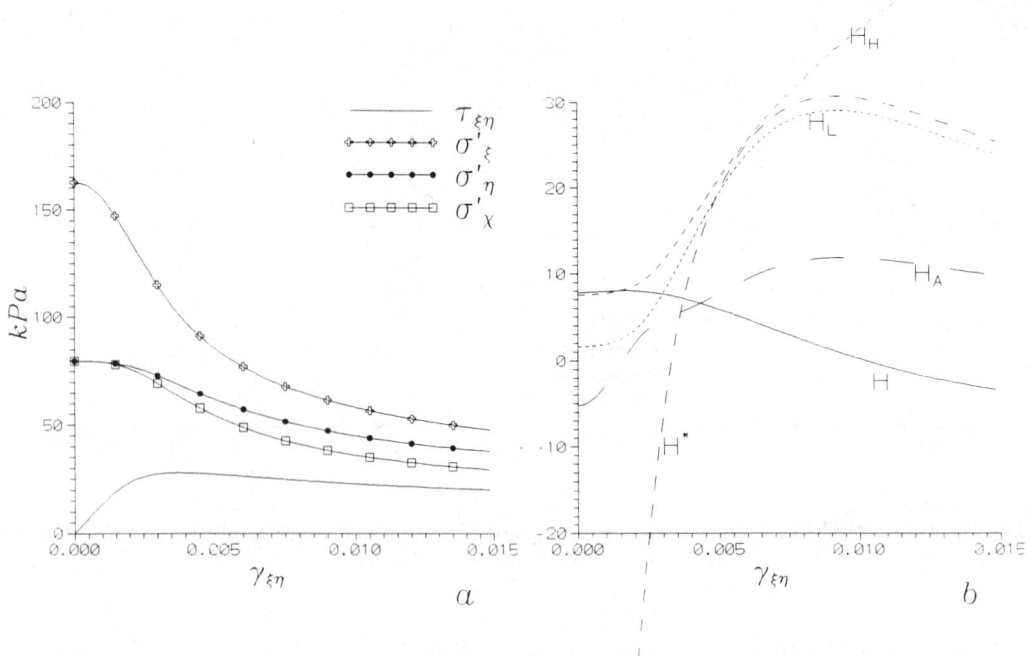

Fig.8 Numerical results for an undrained shear stress path of an element of a K_0 consolidated ($\alpha=0°$) layer: a) evolution of stresses, b) evolution of moduli H

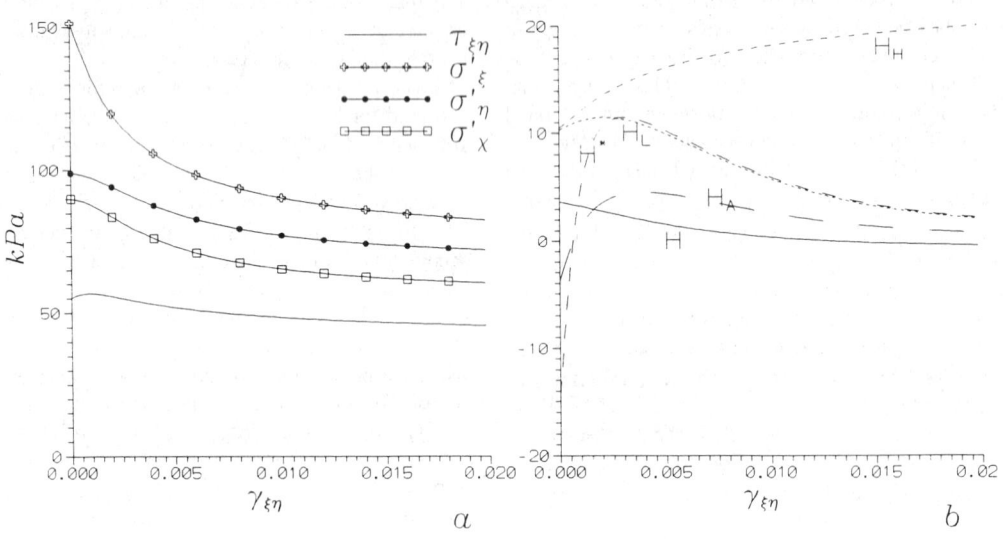

Fig.9 Numerical results for an undrained shear stress path of an element of a layer deposited at an inclination $\alpha=20°$: a) evolution of stresses, b) evolution of moduli H

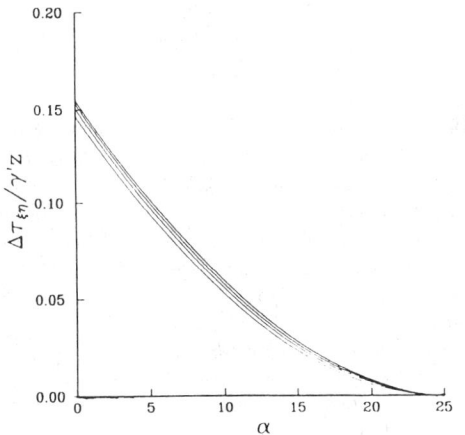

Fig. 10 Normalised shear stress increment versus slope inclination for different depths z

at peak with slope inclination.

From the practical viewpoint such a result is convenient since the curve of Fig. 10 can be used to predict the effect of an environmental attack on the stability of the slope. As an example, we may note that for a slope with an inclination as small as 10°, a shear stress increment equal approximately to the 5% of the vertical stress is enough to cause collapse. Such a limited disturbance is likely to occur in small earthquakes or because of sea waves of moderate intensity. This result gives therefore a possible explanation of the experimental evidence that even very flat slopes may collapse under apparently minor disturbances.

6 CONCLUSIONS

In this study the problem of the stability of submerged slopes was tackled from various viewpoints. A theoretical criterion was first established to determine the conditions for the occurrence of spontaneous collapse of a sand slope under an infinitesimal perturbation, which is rapid enough to prevent drainage. It was found that, under the boundary conditions imposed by symmetry and no volume change, this occurs when the shear stiffness is zero in at least one layer at a certain depth z. It has been shown that a shear band may occur at this stage. The total normal stress rates are continuous across the band, but, at variance with ordinary shear band analysis, rates of effective stresses, as well as those of pore pressures, are not.

Other instability criteria were then recalled and the critical expressions of the hardening moduli associated to such criterion were derived for a generic elastoplastic strainhardening model. A particular model was then employed to reproduce numerically the behaviour of Hostun R.F. sand observed in experiments. It was shown that the elastoplastic law considered can describe correctly the experimental evidence.

It was then assumed that the model could be reliably used as a predicting tool for the slope stability problem. The in situ state of stress was numerically reproduced for different slope inclinations. It was shown that according to Hill and Lade criteria even sedimentation process may become unstable for small inclinations, while it is not for the criterion presented in this paper.

The conditions for collapse were determined by increasing the shear strains in undrained conditions. It was found that the normalised incremental shear stress necessary to cause collapse decreases with slope inclination and that, for a virgin soil, spontaneous collapse occurs when the slope is inclined of about 24.5°, for the sand considered. The mobilised friction angle corresponding to this situation is 32°, while the limit friction angle in drained plane strain conditions is of about 36°. This is the value of the inclination for which drained collapse could occur spontaneously.

The existence in nature of relatively steep submerged slopes is therefore possible. However, a small perturbation in undrained conditions may trigger a catastrophic collapse. For instance a relatively small earthquake characterized by an acceleration of about .05 g is sufficient to cause the failure of a 10° slope made of the material considered in the paper. Clearly, the higher the inclination the lower the perturbation necessary to produce failure. The occurrence of flow slides even for very flat slopes can therefore be explained, provided the occurrence of a sufficient external excitation is considered.

7 ACKNOWLEDGEMENT

This research was conducted in the framework of Project 2, Localisation phenomena in Geomaterials, of the ALERT Geomaterials Programme, funded by the EU (Human Capital and Mobility). Financial support from Italian CNR and MURST is also gratefully acknowledged.

8 REFERENCES

Casagrande A. 1975. Liquefaction and cyclic deformation of sands, a critical review. *Proc. 5th Pan. American Conference, Soil Mechanics and Foundations Eng.*, Buenos Aires, 5: 79-133

di Prisco C., Nova R. 1992. Isotropic versus kinematic hardening in the mathematical modelling of soil behaviour. *Proc. Int. Conf. Computational Plasticity*, D.R.J. Owen et al. eds., Barcelona, 827-838, Swansea: Pineridge Press

di Prisco C., Nova R., Lanier J. 1993. A mixed isotropic kinematic hardening constitutive law for sand. in D. Kolymbas ed *Modern approaches to plasticity*, ,83-124, Rotterdam: Balkema

di Prisco C. 1993. *Studio sperimentale e modellazione matematica del comportamento anisotropo delle sabbie*. Tesi di dottorato in Ingegneria Geotecnica, Politecnico di Milano

Hazen A. 1920. Hydraulic fill dams. *ASCE trans.* 83: 1713-1745

Hill R. 1958. A general theory of uniqueness and stability in elastic-plastic solids. *J. Mech. Phys. Solids* 6: 236-249

Lade P.V. 1977. Elasto-plastic stress-strain theory for cohesionless soil with curved yield surfaces. *Int. J. Solids Structures*, .13: 1019-1035

Lade P.V. 1991. Non associated flow and instability of slopes. *Proc. VII IACMAG* Cairns: 487-492 Rotterdam: Balkema

Lade P.V., Nelson R.B. 1987. Modelling the elastic behaviour of granular materials. *Int. J. Num. Anal. Meth. Geomech.* 2: 521-542

Maier G., Hueckel T. 1979. Nonassociated and coupled flow rules of elastoplasticity for rock-like materials. *Int. J. Rock Mech. Min. Sci. & Geomech. Abstr.*, 16: 77-92

Nova R. 1988. Sinfonietta classica: an exercise on classical soil modelling. in Saada A. and Bianchini G. eds., *Constitutive Equations for Granular non-cohesive soils*, 501-520 Cleveland, 22-24 July 1987, Balkema,

Nova R. 1991. A note on sand liquefaction and soil stability. *Proc. Third Int. Conf. Contitutive Laws for Engineering Materials: Theory and Applications*, January 7-12, 1991 Tucson, Arizona, USA, 153-156

Nova R., Wood D.M. 1978. An experimental programme to define the yield function for sand. *Soils and Foundations* 18, 4, 77-86

Nova R., Wood D.M. 1979. A constitutive model for sand in triaxial compression. *Int. J. Num. Anal. Meth. Geomech.* 3, 3, 255-278

Rudnicki J.W., Rice J.R. 1975. Conditions for the localisation of deformation in pressure sensitive dilatant materials. *J. Mech. Phys. Solids*, 23, 371-394

Saada A., Bianchini G. 1988. *Proc. Int. Workshop on Constitutive Equations for Granular non-cohesive Soils*, Cleveland, 22-24 July 1987, Balkema

Sladen J.A., D'Hollander R.D., Krahn J. 1985. The liquefaction of sands, a collapse surface approach. *Can.Geotech. J.* 22, 564-578

Terzaghi K. 1950. *Mechanisms of landslides: application of Geology to Engineering practice*. Berkeley volume, Geol. Soc. of America, 181-194

Ziegler H. 1968. *Principles of structural stability*. Blaisdell Publ. Company.

3 Localisation and constitutive modelling
   Localisation et lois de comportement

Stability and uniqueness in geomaterials constitutive modelling
Stabilité et unicité dans la modélisation rhéologique des géomatériaux

F. Darve
Laboratoire Sols, Solides, Structures, Institut de Mécanique de Grenoble, ALERT Géomatériaux, France

ABSTRACT : Loss of stability and loss of uniqueness are not equivalent notions. In the case of geomaterials (soils, rocks, concretes) particularly, where the plastic strains are not only strongly non-associated but the behaviour is moreover rather incrementally non-linear, these questions need to be more investigated in detail. After reviewing these concepts and establishing a sufficient condition of stability and another of uniqueness, we consider particular classes of material and loading : loose sands under isochoric conditions. The main features of an incrementally non-linear constitutive relation are briefly recalled. The application of this model to loose Hostun sand under isochoric loading allows to exhibit stress-strain states with loss of stability first and then loss of uniqueness. Approached paths with multiple "sharp bends" are considered in the analysis of the condition of loss of uniqueness.

D'un point de vue général, la perte d'unicité et la perte de stabilité ne sont pas des notions équivalentes. Dans le cas des géomatériaux (sols, bétons, roches) en particulier, pour lesquels non seulement les déformations plastiques sont fortement non associées mais aussi le comportement est notablement non linéaire incrémentalement, cette question doit être examinée en détail. Après une revue de ces concepts, et l'établissement d'une condition suffisante de stabilité et d'une autre d'unicité, on considère une classe particulière de matériaux et un chargement particulier: les sables lâches en condition non-drainées. Les caractéristiques principales d'une loi de comportement incrémentalement non-linéaire sont rappelés brièvement. L'application de ce modèle au sable d'Hostun lâche en chargement non-drainé permet d'exhiber les états de contrainte-déformation correspondant aux premières pertes de stabilité et d'unicité. Dans l'analyse des conditions de perte d'unicité, on considère des chemins décomposés en multiples " marches d'escalier ".

1 - INTRODUCTION

From a constitutive point of view a possible definition of instability could state : after a given strain history, a stress-strain state is unstable if at least exists one particular "small" loading which induces "large" responses.

This definition of mechanical instabilities implies that all stress states on the plastic limit surface are unstable since any "small" incremental stress directed outside the plastic surface (as for example particular dead loads) will induce "large" strains. Another class of clear physical instabilities is generally met when, during a certain loading history, a component of the stress or strain tensor goes through an extremum value (maximum or minimum). At such a stress-strain state it is generally possible to find a certain stress or strain controlled loading for which "large" responses would be obtained, showing an unstable behaviour after our definition. Of course for another type of control (usually that may be for example a strain controlled path) for the "same" loading path any particular instability could not be detected.

The question of instability may also be investigated by considering the sign of the work of second order. This tool, if not based on thermodynamics principles, is interesting to be utilized and we will analyse it before applying to the case of isochoric loading on loose sands.

The sign of the work of second order has been considered by Hill (1958, 1959) by assuming that for any infinitesimal displacement the increase in internal energy shall exceed the work of the external forces. The consequences of this statement in elasticity and hypoelasticity are clear. For incrementally non-linear constitutive relations we will utilize the sign of the work of second order as a tool to investigate the stability locally, what means from a constitutive point of view.

Our sufficient analytical condition of stability is thus formulated as follows :

$$\forall \, \underline{d\sigma}, \, \underline{d\varepsilon} : d^2 W = \underline{d\sigma} \cdot \underline{d\varepsilon} > 0 \qquad (1)$$

where $d\underline{\sigma}$ and $d\underline{\varepsilon}$ are linked by the "incremental" constitutive relation :

$$d\underline{\varepsilon} = \underline{F}(d\underline{\sigma}) \quad (2)$$

where $\underline{F}$ does not depend on time increment dt for rate-independent materials.

Incremental stress $d\underline{\sigma}$ and incremental strain $d\underline{\varepsilon}$ are classically defined as :

$d\underline{\sigma} = \overset{\triangledown}{\underline{\sigma}}\, dt$, with $\overset{\triangledown}{\underline{\sigma}}$ an objective derivative of Cauchy stress tensor,

$d\underline{\varepsilon} = \underline{D}\, dt$, with $\underline{D}$ the strain rate, symmetric part of the deformation rate.

In other words, if there is at least one particular incremental loading for which the work of second order is negative or null, the corresponding stress-strain state (for the considered strain history) is characterized as potentially unstable. We will see further that sufficient analytical condition (1) is linked, in the case of loose sands under isochoric conditions, to mechanical instabilities.

We are going to study the question of uniqueness also by restricting to only a constitutive point of view what means to restrict the study to spatially local considerations. From this point of view uniqueness is coinciding with invertibility of the incremental constitutive relation or with its biunivocity.

In order to prove the local invertibility we have to use the theorem of implicit functions of several variables : if the Jacobian of the non-linear transformation (here from $\mathbb{R}^6$ to $\mathbb{R}^6$) is not null, it is possible to locally invert the transformation. Local invertibility means means that, in six-dimensional $d\underline{\varepsilon}$ space, for each incremental strain direction there is a certain hyper-cone (which surrounds this direction) where $d\underline{\sigma}$ is defined in a unique manner from $d\underline{\varepsilon}$. Let us note that this condition is not necessary : the Jacobian can be null and the transformation remain invertible. As the Jacobian of the transformation is exactly the determinant of the gradient-matrix of function $\underline{F}$ (as defined by relation (2)), a sufficient condition of local invertibility is then given by :

$$\det\left(\frac{\partial \underline{F}}{\partial(d\underline{\sigma})}\right) \neq 0 \quad (3)$$

In this paper we consider only rate-independent (non-viscous) materials. For this class of materials their mechanical behaviour is independent of the rate of loading, thus it comes :

$$\forall \lambda \in \mathbb{R}^+ : \lambda\, d\underline{\varepsilon} \equiv \underline{F}(\lambda d\underline{\sigma})$$

Therefore $\underline{F}$ is an homogeneous function of degree one and Euler's identity can be applied to $\underline{F}$:

$$\underline{F}(d\underline{\sigma}) \equiv \frac{\partial \underline{F}}{\partial(d\underline{\sigma})}\, d\underline{\sigma}$$

This essential identity proves the existence of a tangent constitutive tensor $\underline{M}$:

$$\underline{M} = \frac{\partial \underline{F}}{\partial(d\underline{\sigma})}$$

which is moreover an homogeneous function of degree zero with respect to $d\underline{\sigma}$. Thus $\underline{M}$ depends only on the direction of $d\underline{\sigma}$, which can be characterized by unit vector $\underline{u}$:

$$\underline{u} = \frac{d\underline{\sigma}}{\|d\underline{\sigma}\|}, \text{ with } \|d\underline{\sigma}\| = \sqrt{d\sigma_{ij}.d\sigma_{ij}}$$

Therefore constitutive relation (2) is equivalent to (4) :

$$d\underline{\varepsilon} = \underline{M}(\underline{u})\, d\underline{\sigma} \quad (4)$$

The sufficient condition of local uniqueness is then given by :

$$\forall \underline{u} : \det \underline{M} \neq 0 \quad (5)$$

Further we will try to exhibit a condition of global uniqueness and to apply it to the case of isochoric loading on loose sands.

Local and global uniquenesses are different notions. The local uniqueness is equivalent to the reciprocal invertibility inside given hypercones which surround the considered incremental strain or incremental stress directions. The global uniqueness corresponds to a general invertibility or a complete biunivocity between the incremental strain and incremental stress spaces.

2 - CONSTITUTIVE MODELLING OF GEOMATERIALS

We recall here briefly the main features of the constitutive models which are utilized in this paper. Relation (4) constitutes the general formulation of rate-independent constitutive models. If we write it in the six-dimensional stress-strain spaces, we obtain :

$$d\varepsilon_\alpha = M_{\alpha\beta}(u_\gamma)\, d\sigma_\beta \quad (\alpha, \beta, \gamma = 1, \ldots 6) \quad (4)$$

In order to describe the variations of 36 functions $M_{\alpha\beta}$ with u_γ, we can utilize polynomial series developments as approached expressions :

$$M_{\alpha\beta}(u_\gamma) = M^1_{\alpha\beta} + M^2_{\alpha\beta\gamma} u_\gamma + M^3_{\alpha\beta\gamma\delta} u_\gamma u_\delta + \ldots \quad (6)$$

From (4), (6) and the definition of $\underline{u}$, it comes :

$$d\varepsilon_\alpha = M^1_{\alpha\beta}\, d\sigma_\beta + \frac{1}{\|d\underline{\sigma}\|} M^2_{\alpha\beta\gamma}\, d\sigma_\beta d\sigma_\gamma + \ldots \quad (7)$$

The first term of relation (7) represents the general expression of all hypoelastic models and the first two terms the incrementally non-linear constitutive models of second order.

In this paper we consider as applications only loading paths in fixed stress-strain principal axes. In such cases, relation (7) takes the following simple form (for more details see Darve (1990), Darve and Dendani (1989) :

$$\begin{bmatrix} d\varepsilon_1 \\ d\varepsilon_2 \\ d\varepsilon_3 \end{bmatrix} = \frac{1}{2}(\underset{\sim}{N}^+ + \underset{\sim}{N}^-) \begin{bmatrix} d\sigma_1 \\ d\sigma_2 \\ d\sigma_3 \end{bmatrix}$$
$$+ \frac{1}{2\|d\underset{\sim}{\sigma}\|}(\underset{\sim}{N}^+ - \underset{\sim}{N}^-) \begin{bmatrix} (d\sigma_1)^2 \\ (d\sigma_2)^2 \\ (d\sigma_3)^2 \end{bmatrix} \quad (8)$$

where :

$$\|d\underset{\sim}{\sigma}\| = \sqrt{(d\sigma_1)^2 + (d\sigma_2)^2 + (d\sigma_3)^2}$$

and $\underset{\sim}{N}^+$, $\underset{\sim}{N}^-$ are two 3x3 matrices which depend on state variables and discrete memory parameters. $\underset{\sim}{N}^+$ and $\underset{\sim}{N}^-$ are expressed as follows (Darve, 1990a) :

$$\underset{\sim}{N}^+ = \begin{bmatrix} \frac{1}{E_1^+} & -\frac{V_2^{1+}}{E_2^+} & -\frac{V_3^{1+}}{E_3^+} \\ -\frac{V_1^{2+}}{E_1^+} & \frac{1}{E_2^+} & -\frac{V_3^{2+}}{E_3^+} \\ -\frac{V_1^{3+}}{E_1^+} & -\frac{V_2^{3+}}{E_2^+} & \frac{1}{E_3^+} \end{bmatrix} \quad (9)$$

where E_i and V_i^j are tangent moduli and tangent Poisson's ratios along "generalized triaxial paths" in compression (indices "+") or in extension (indices "-").

$$\begin{cases} E_i = \left(\frac{\partial f}{\partial \varepsilon_i}\right)_{\sigma_j, \sigma_k} \\ V_i^j = -\left(\frac{\partial g}{\partial c_i}\right)_{\sigma_j, \sigma_k} \; ; \; V_i^k = -\left(\frac{\partial h}{\partial \varepsilon_i}\right)_{\sigma_j, \sigma_k} \end{cases}$$

and f, g and h are defined by :

$$\begin{cases} \sigma_i = f(\varepsilon_i, \sigma_j, \sigma_k) \\ \varepsilon_j = g(\varepsilon_i, \sigma_j, \sigma_k) \\ \varepsilon_k = h(\varepsilon_i, \sigma_j, \sigma_k) \end{cases}$$

for these "generalized triaxial paths" such that the two lateral stresses σ_j, σ_k are kept constant in fixed principal axes. Relation (8) can be also interpreted as a non-linear interpolation between the six basic paths defined by :

$$d\underset{\sim}{\sigma} = \begin{bmatrix} \pm 1 \\ 0 \\ 0 \end{bmatrix}, \begin{bmatrix} 0 \\ \pm 1 \\ 0 \end{bmatrix}, \begin{bmatrix} 0 \\ 0 \\ \pm 1 \end{bmatrix};$$

and corresponding to the generalized triaxial paths. If we consider a linear interpolation between these six paths, we obtain the "octo-linear" model whose expression is given by :

$$\begin{bmatrix} d\varepsilon_1 \\ d\varepsilon_2 \\ d\varepsilon_3 \end{bmatrix} = \frac{1}{2}(\underset{\sim}{N}^+ + \underset{\sim}{N}^-) \begin{bmatrix} d\sigma_1 \\ d\sigma_2 \\ d\sigma_3 \end{bmatrix}$$
$$+ \frac{1}{2}(\underset{\sim}{N}^+ - \underset{\sim}{N}^-) \begin{bmatrix} |d\sigma_1| \\ |d\sigma_2| \\ |d\sigma_3| \end{bmatrix} \quad (10)$$

where $|d\sigma_i|$ means the absolute value of $d\sigma_i$.

Two asymptotic cases are interesting to be considered :

• the hypo-elastic case for which $\underset{\sim}{N}^+$ and $\underset{\sim}{N}^-$ are identical :

(8) and (10) are then degenerating into the same expression :

$$d\underset{\sim}{\varepsilon} = \underset{\sim}{N} \, d\underset{\sim}{\sigma}$$

where $\underset{\sim}{N}$ is exactly the tensor of the orthotropic non-linear elasticity.

• the perfectly plastic case :

For the octo-linear model the plasticity criterion is given by : $\det \underset{\sim}{N}^{-1} = 0$, that is to say :
$$E_1 \, E_2 \, E_3 = 0$$

and a generalized flow rule comes from :
$$\underset{\sim}{N}^{-1} \, d\underset{\sim}{\varepsilon} = \underset{\sim}{0} \text{ with } \det \underset{\sim}{N}^{-1} = 0$$

3 - STABILITY CONDITIONS

In order to study the stability we will base our analysis on the sufficient condition (1) :

$$\forall \, d\underset{\sim}{\sigma}, d\underset{\sim}{\varepsilon} : d\underset{\sim}{\sigma} \cdot d\underset{\sim}{\varepsilon} > 0$$

Utilizing general form (4) of rate-independent constitutive relations and decomposing the constitutive tensor into its symmetric part $\underset{\sim}{M}^s$ and antisymmetric part $\underset{\sim}{M}^A$, it comes :

$$d\underset{\sim}{\sigma} \cdot d\underset{\sim}{\varepsilon} = d\underset{\sim}{\sigma} \, \underset{\sim}{M} \, d\underset{\sim}{\sigma} = d\underset{\sim}{\sigma} \, \underset{\sim}{M}^s \, d\underset{\sim}{\sigma} + d\underset{\sim}{\sigma} \, \underset{\sim}{M}^A \, d\underset{\sim}{\sigma} = d\underset{\sim}{\sigma} \, \underset{\sim}{M}^s \, d\underset{\sim}{\sigma}$$

(1) implies therefore :

$$\forall \, \underset{\sim}{u} : \det \underset{\sim}{M}^s(\underset{\sim}{u}) > 0 \quad (11)$$

Now if we assume that, at the beginning of the loading, the behaviour is "close" to elasticity or at least not violating the thermodynamic principles for elasticity (where the eigenvalues of $\underset{\sim}{M}^s$ are strictly positive) and that these eigenvalues of $\underset{\sim}{M}^s$ are varying continuously with the loading, any eigenvalue can not become negative during a loading without annulling the determinant of $\underset{\sim}{M}^s$.

Thus inequality (11) implies inequality (1). (11) and (1) are then equivalent. Therefore a sufficient condition of stability is given by :

$$\forall \, \underset{\sim}{u} : \det \underset{\sim}{M}^s(\underset{\sim}{u}) > 0 \quad (12)$$

This demonstration is strictly valid only for incrementally piece-wise linear constitutive models, what means for the octo-linear model here. But as, after Euler's identity, in the general case of incrementally non-linear constitutive relations, $\underset{\sim}{M}$ is both the gradient matrix of $\underset{\sim}{F}$ and a directional

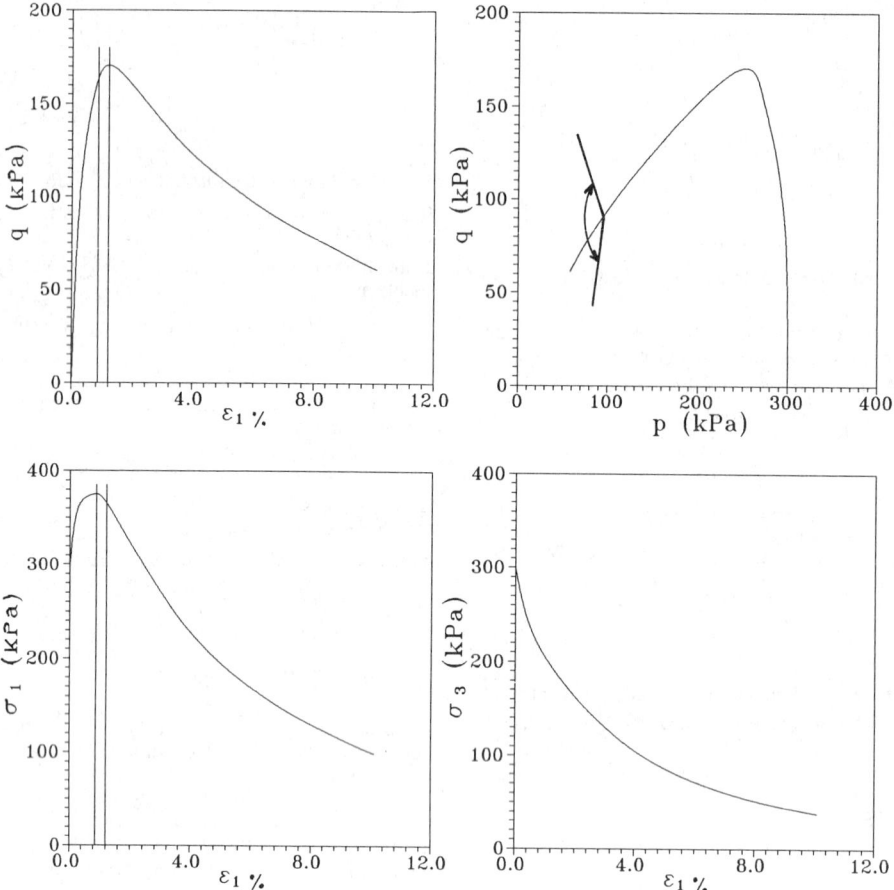

Fig. 1 : Typical numerical results from the incremental non-linear constitutive relation in modelling undrained loose Hostun sand. The maximum of σ_1 appears before the maximum of q. The cone of unstable directions is plotted in (p - q) plane (Meghachou, 1993)

linearization of $\underline{F}$, inequality (12) is also valid in the non-linear case.

Now let us consider the particular case of isochoric loading on loose sands as an application of this general analysis.

Fig. 1 presents typical results from the incremental non-linear model (relation (8)) when applied to axisymmetric isochoric loading on loose Hostun sand. These results, while coming from a numerical analysis, are completely representative of the classical experimental behaviour. We notice the maximum of q ($q = \sigma_1 - \sigma_3$) in (p - q) plane and also the maximum of the axial stress σ_1 (all the stresses are here intergranular, effective stresses). This last maximum appears before the maximum of q, since at the maximum of q :

dq = 0

Thus :

$d\sigma_1 = d\sigma_3$

As σ_3 is monotonously decreasing, $d\sigma_3$ is always negative, what means that at the maximum of q, σ_1 is already decreasing. This point can be verified on the results of fig. 1.

Both these maxima of q and of σ_1 are unstable states after the classical definition of instability : it exists at least particular "small" loading which induces a "large" response.

In this case this particular loading is defined by :

$\begin{cases} d\sigma_1 = \text{constant positive or } dq = \text{constant positive} \\ d\varepsilon_2 = d\varepsilon_3 = - 0.5 \, d\varepsilon_1 \end{cases}$

Following this loading, when the maxima of σ_1 and q are reached, these stress-strain states will exhibit instabilities clearly appearing with "dead loads" : a "small" continuation of the loading at these maxima

induces "large" strains as experimentally shown.
The question which must be asked now is: are these physical instabilities detected by criterion (1) or equivalently (12)?
Let us express the work of second order for an axisymmetric isochoric loading path:

$$d^2W = d\sigma_1 d\varepsilon_1 + 2d\sigma_3 d\varepsilon_3$$

With: $d\varepsilon_1 + 2d\varepsilon_3 = 0$, it comes:

$$d^2W = dq\, d\varepsilon_1 \quad (13)$$

Thus at and from the maximum of q the sufficient condition of stability is no more valid.

At the maximum of σ_1 the work of second order is positive for an undrained loading but is certainly negative for other incremental directions. In fact let us consider the octo-linear model (relation (10)). This model is incrementally piece-wise linear with 8 different determinations for the constitutive tensor $\underset{\sim}{M}$, associated with 8 "tensorial zones" (Darve and Labanieh, 1982 and Darve, 1990b). These 8 determinations are obtained by considering all possible combinations of the first columns of $\underset{\sim}{N}^+$ and $\underset{\sim}{N}^-$ for the first column of $\underset{\sim}{M}$, the second columns of $\underset{\sim}{N}^+$ and $\underset{\sim}{N}^-$ for the second of $\underset{\sim}{M}$, the third of $\underset{\sim}{N}^+$ and $\underset{\sim}{N}^-$ for the third of $\underset{\sim}{M}$.

Here in the case of axisymmetric isochoric loading with $d\sigma_1$ and $d\sigma_3$ having negative values the octo-linear constitutive relation takes the following form:

$$\begin{bmatrix} d\varepsilon_1 \\ d\varepsilon_3 \\ d\varepsilon_3 \end{bmatrix} = \begin{bmatrix} \dfrac{1}{E_1^-} & -\dfrac{v_3^{1-}}{E_3^-} & -\dfrac{v_3^{1-}}{E_3^-} \\ -\dfrac{v_1^{3-}}{E_1^-} & \dfrac{1}{E_3^-} & -\dfrac{v_3^{3-}}{E_3^-} \\ -\dfrac{v_1^{3-}}{E_1^-} & -\dfrac{v_3^{3-}}{E_3^-} & \dfrac{1}{E_3^-} \end{bmatrix} \begin{bmatrix} d\sigma_1 \\ d\sigma_3 \\ d\sigma_3 \end{bmatrix} \quad (14)$$

The determinant of $\underset{\sim}{M}$ keeps the same sign (negative here) in a given tensorial zone defined here by:

$$\begin{cases} d\sigma_1 < 0 \\ d\sigma_3 < 0 \end{cases}$$

The cone of unstable directions is then given by negative $d\sigma_1$ and $d\sigma_3$, what is equivalent in (p - q) plane to the conditions:

$$\begin{cases} \dfrac{dq}{dp} > -\dfrac{3}{2} \\ \dfrac{dq}{dp} < 3 \end{cases} \quad (15)$$

This cone has been plotted on fig. 1. We see that the undrained stress path enters exactly inside the cone in (p - q) plane, when the maximum of σ_1 is reached, what means a little bit before the maximum of q on the figure.
This result is analytically obvious since the limit of the cone correspond to: $d\sigma_1 = 0$, condition precisely verified at the maximum of σ_1.

It is clear that the cone of instability could appear still before, when the determinant of the symmetric part of the constitutive tensor (14) becomes negative. It would imply that the undrained path could be potentially unstable still before the maximum of σ_1, in the sense that there would be, from stress-strain states on the undrained loading, other directions which are unstable. Here we have shown that the undrained stress response enters into the cone of unstable directions, after the octo-linear model, exactly at the maximum of σ_1, what corresponds to clear physical instabilities.

4 - UNIQUENESS CONDITIONS

4.1 - Sufficient condition of uniqueness

We have seen in the introduction that the theorem of implicit functions allows to exhibit a sufficient condition of local uniqueness (by implying the possibility of a local inversion of the constitutive function $\underset{\sim}{F}$ inside a certain hypercone):

$$\forall \underline{u} : \det \underset{\sim}{M} \neq 0 \quad (5)$$

This condition (5) does not coincide at all generally with a global condition of uniqueness.
We must add now the other property verified by $\underset{\sim}{F}$, that is its homogeneity of degree one. The following theorem can be demonstrated:
Let $\underset{\sim}{F}$ be a function from $\mathbb{R}^n$ to $\mathbb{R}^n$, of C^1 class, homogeneous of degree p and such that:

$$\forall \underline{U} \in \mathbb{R}^n : d\underset{\sim}{F}(\underline{U})$$

is an isomorphism, then $\underset{\sim}{F}$ is surjective.
Therefore, for the considered constitutive relation, as soon as condition (5) is verified, $\underset{\sim}{F}$ is surjective, what means that:

$$\forall \underline{d\varepsilon}\, \exists \underline{d\sigma} : \underline{d\varepsilon} = \underset{\sim}{F}(\underline{d\sigma})$$

In order to ensure the global uniqueness, we need to demonstrate the biunivocity of $\underset{\sim}{F}$, what is equivalent to its bijectivity or equivalently its surjectivity and its injectivity.
Unfortunately, generally this property of injectivity is not verified by any homogeneous function with a never null Jacobian.
What can be demonstrated is the following property: Let Σ be an hypersurface surrounding the origin of the six-dimensional $\underline{d\sigma}$ space and $\underset{\sim}{F}$ be an homogeneous function, of C^1 class, such that $\underline{dF}$ is an isomorphism. Then, if any tangent plane to Σ does not contain the origin of $\underline{d\sigma}$ space, the same property is verified by the tangent planes to $\underset{\sim}{F}(\Sigma)$.
The demonstration is essentially based on the fact that

$$\frac{\partial \underline{F}}{\partial (d\underline{\sigma})} \underline{u}$$

is proportional to $d\underline{\varepsilon}$, what is obvious after Euler's identity for homogeneous functions.

If we except some eventual peculiar forms for $\underline{F}(\underline{\Sigma})$ which are difficult to conceive, the other possibility to meet a loss of injectivity (since the previous property eliminates all the possible cases of singularities by bends) is to obtain, for one going round the origin of $d\underline{\sigma}$, space, more than one going round the origin of $d\underline{\varepsilon}$ space (as simple bidimensional example).
A simple example of such a loss of injectivity in the case of an homogeneous function of two variables, whose Jacobian is always strictly positive, is given by :

$$e^{i\theta} \rightarrow e^{2i\theta}$$

In fact such a situation is not met by our models because of their interpolating nature. The octo-linear constitutive relation (equation (10)) as the non-linear one (equation (8)) can be considered as interpolations between the basic paths (which are these "generalized triaxial paths" defined previously in paragraph 2) : the octo-linear model corresponds to a linear interpolation while the non-linear one is a non-linear of second order interpolation.

Now, let us consider the eight determinants of the eight constitutive tensors which are associated to the eight tensorial zones of the octo-linear model (as defined in paragraph 3). If these determinants are all strictly positive or all strictly negative, then the transformation of the eight tensorial zones by $\underline{F}$ are tangent hypercones, which do not intersect each other (since all the trihedrons are direct or respectively all indirect). The limits of these hypercones are identical with the non-linear model, since the used basic paths for the interpolation are the same.

As initially for infinitesimal strains (for example in propagating waves) the behaviour must be close to elasticity and as in elasticity the constitutive determinant is strictly positive, thus we must transform condition (5) into the following :

$$\forall \underline{u} : \det \underline{M} > 0 \qquad (16)$$

In summary :
• the non-nullity of the Jacobian ensures the local uniqueness after the theorem for implicit junctions,
• the global uniqueness needs moreover surjectivity and injectivity of $\underline{F}$,
• the homogeneity of $\underline{F}$ ensures its surjectivity by a general theorem as soon as the Jacobian is never null,
• the injectivity is not generally verified. But bends are impossible after the theorem on the tangent planes and the overlaps are also excluded by the chosen type of interpolation relations, if the Jacobian is strictly positive.
In conclusion, condition (16) is a sufficient condition of uniqueness.

4.2 - Discussion

This condition has to be compared with the sufficient condition of stability (11) :

$$\begin{cases} \text{sufficient condition of stability :} \\ \forall \underline{u} : \det \underline{M}^s(u) > 0 \end{cases} \quad (11)$$

$$\begin{cases} \text{sufficient condition of uniqueness :} \\ \forall \underline{u} : \det \underline{M}(u) > 0 \end{cases} \quad (16)$$

Several conclusions can be issued from this comparison :
- for associated elasto-plastic constitutive relations, where the constitutive tensor is symmetric, stability and uniqueness are coinciding notions. That is no more the case in non-associated elasto-plasticity and a fortiori for incrementally non linear models.
- since these notions are not generally confounded, they must be studied strictly inside the plastic limit criterion.
- for a definite positive matrix the determinant of its symmetric part is always lower than the determinant of the matrix itself (R. Nova, 1989). Thus the loss of stability will generally precede the loss of uniqueness with respect to a given parameter monotonically increasing with the loading.
The last general comment is illustrated by the case of incremental incompressibility. Such a state is defined by :

$$\forall d\underline{\sigma} : d\varepsilon_v = d\varepsilon_1 + d\varepsilon_2 + d\varepsilon_3 \equiv 0 \qquad (17)$$

For a piece-wise linear constitutive relation, condition (17) implies that the three first lines of the constitutive tensor are linearly dependent, thus the determinants of all the constitutive tensors for this stress-strain state must be null. Condition (17) must be fulfilled also inside one given tensorial zone only. In such a case only the constitutive determinant associated to this tensorial zone will be null.
In the non-linear case, as the constitutive gradient matrix represents at the same time a directional linearization for the constitutive relation, the previous result can be generalized and the Jacobian (that means the constitutive determinant) will be null for all the stress directions or at least for a finite range of directions. We will see later on such an example.

4.3 - Application

Fig. 2 shows typical results obtained from the incrementally non-linear model (relation (8)) by modelling axisymmetric undrained loading on loose Hostun sand. The computation is stopped when the nullity of the Jacobian is met.
In order to analyse this loss of uniqueness, two tools are available : the octo-linear model which allows analytical considerations because of its very simple form and the non-linear model with only possible numerical analyses.
Let us consider first the octolinear constitutive relation. In the considered case the determinant is equal to :

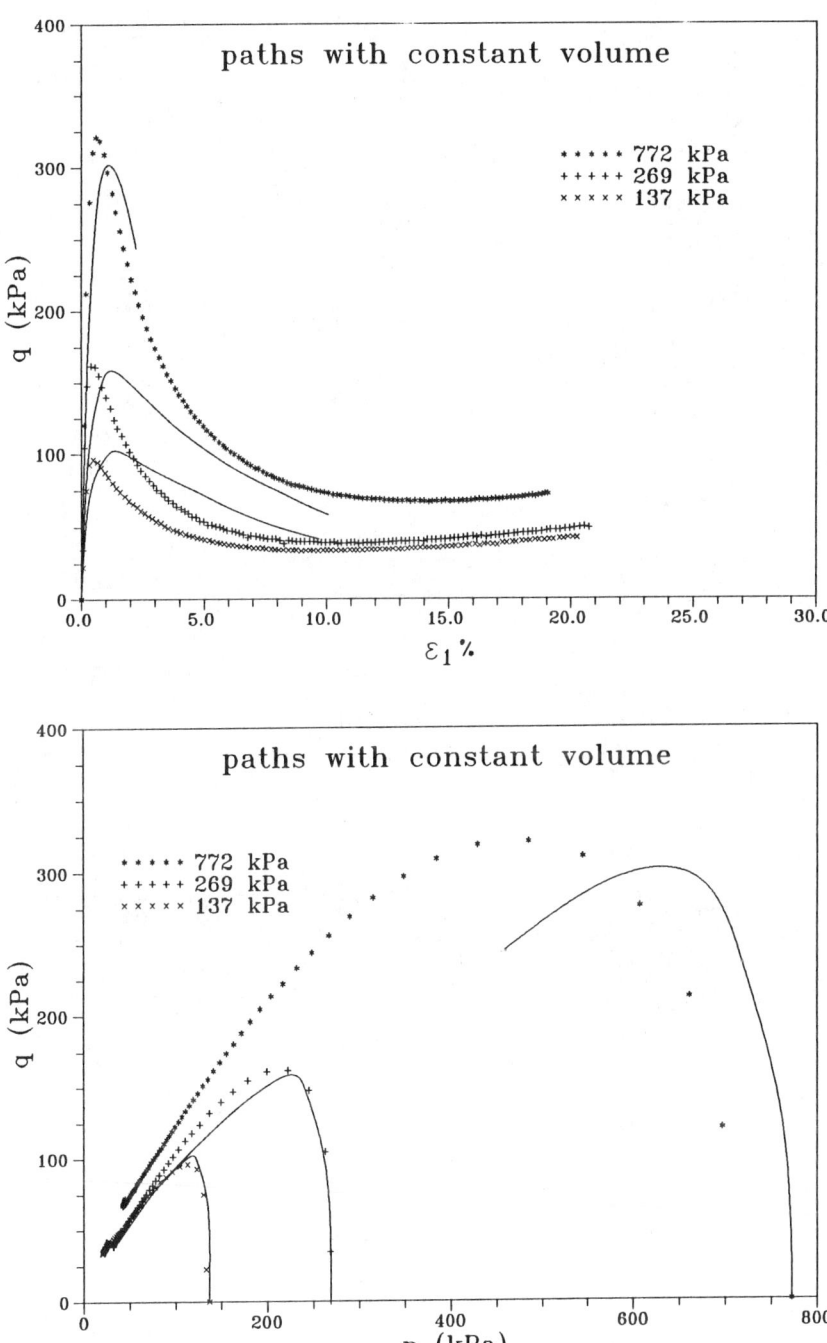

Fig. 2 : Modelling of undrained loading paths on loose Hostun sand by the incremental non-linear constitutive relation (continuous lines). The computations stop when a nullification of the jacobian is detected (Meghachou, 1993)

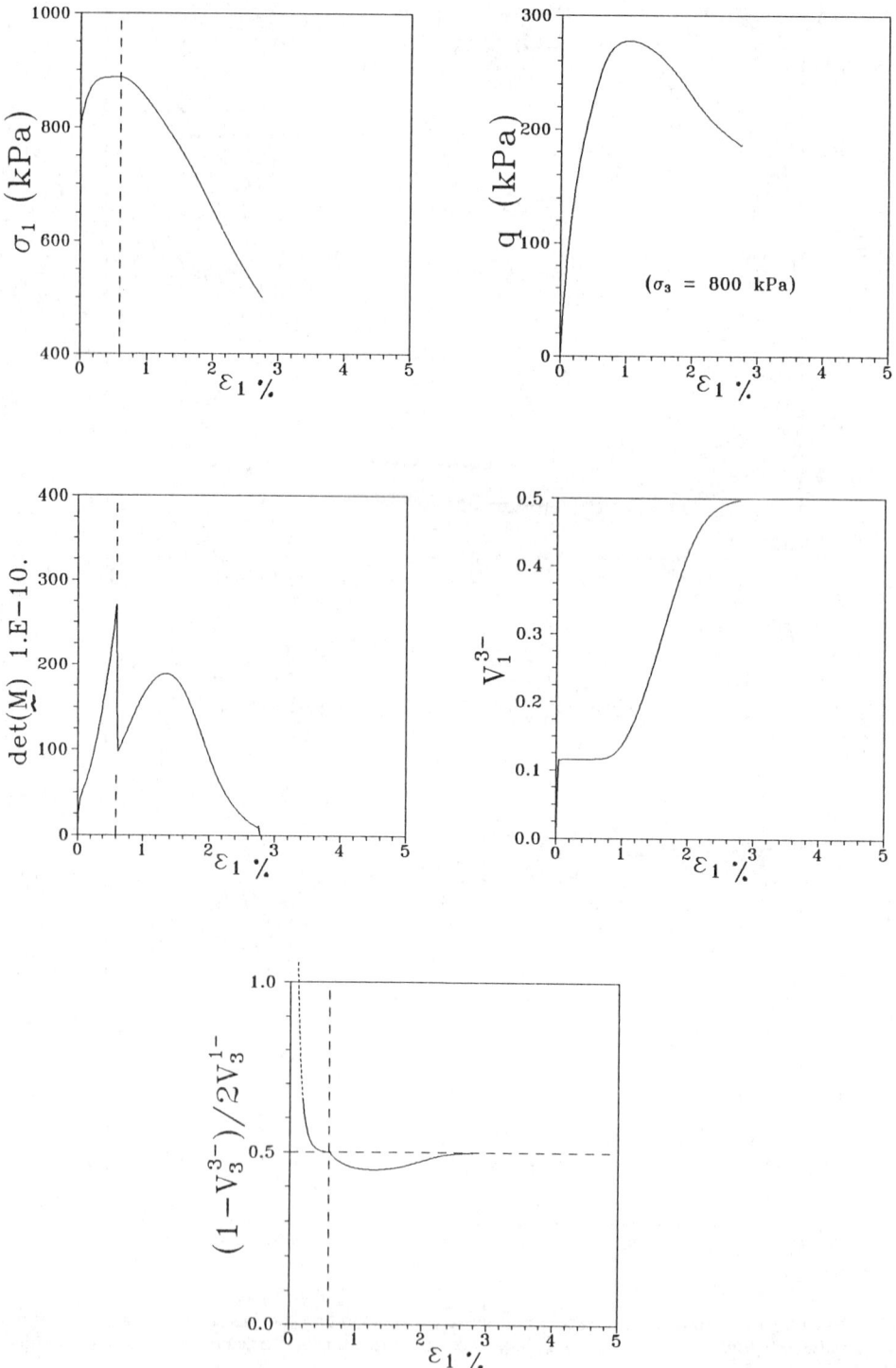

Fig. 3 : Variations of constitutive functions for the octo-linear model for undrained loading path on loose Hostun sand : analysis of the nullification of the constitutive determinant (Meghachou, 1993).

$\det \underline{M} = (1 + V_3^{3-})(1 - V_3^{3-} - 2V_1^{3-}V_3^{1-}) / E_1^-(E_3^-)^2$ (18)

This determinant is null if both the following conditions are fulfilled :

$$\begin{cases} 1 - 2V_1^{3-} = 0 \\ 1 - V_3^{3-} - V_3^{1-} = 0 \end{cases} \quad (19)$$

These conditions can be verified for loose sands, in the framework of our models essentially because they are lightly dilatant then contractant in drained triaxial extensions.

Fig. 3 shows the variations of V_1^{3-} and of $(1 - V_3^{3-})/2V_3^{1-}$ and allows to prove that relations (19) are verified.

Now if we compute the volume variations, we find :

$$d\varepsilon_v = d\varepsilon_1 + 2d\varepsilon_3 = \frac{1 - 2V_1^{3-}}{E_1^-} d\sigma_1 + 2\frac{1 - V_3^{3-} - V_3^{1-}}{E_3^-} d\sigma_3 \quad (20)$$

Thus relations (19) imply :

$\forall d\sigma_1 < 0, \forall d\sigma_3 < 0 : d\varepsilon_v \equiv 0$

Therefore an incremental incompressibility condition is met at this stress-strain state of loss of uniqueness (at least inside the tensorial zone which corresponds to negative $d\sigma_1$ and $d\sigma_3$). The incremental stresses $d\sigma_1$ and $d\sigma_1$ for isochoric loading conditions are equal to :

$$\begin{cases} d\sigma_1 = E_1^- \frac{1 - V_3^{3-} - V_3^{1-}}{1 - V_3^{3-} - 2V_1^{3-}V_3^{1-}} d\varepsilon_1 \\ d\sigma_3 = E_3^- \frac{V_1^{3-} - 0.5}{1 - V_3^{3-} - 2V_1^{3-}V_3^{1-}} d\varepsilon_1 \end{cases} \quad (21)$$

(19) and (21) show that $d\sigma_1$ and $d\sigma_3$ are both undetermined at this stress-strain state. This state is a bifurcation point.

Such a conclusion could explain classical experimental results, which show a very brutal change in the direction of the stress path in (p - q) plane after q peak for loose sands under undrained triaxial loading.

Same basic analyses can be developed from the incrementally non-linear model. We have shown (Darve and Chau, 1987) that the strain response envelopes (in the sense defined by Gudehus, 1979) are progressively degenerating along an undrained path.

Let us recall here briefly the principle of the construction of these diagrams. At a given stress-strain state in the stress bissector plane we consider all the incremental stress vectors, having the same intensity but directed in all the directions. In the

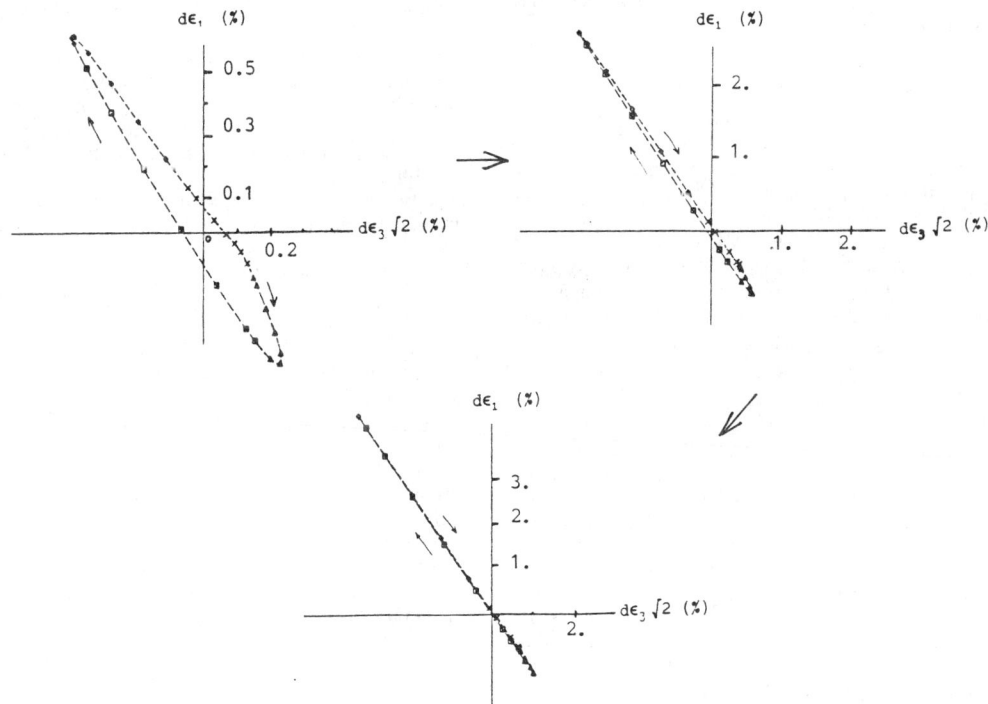

Fig. 4 : Progressively degenerating response-envelopes until the state of loss of uniqueness, where the response-envelope is exactly coinciding with the line of isochoric condition (Darve and Chau, 1987)

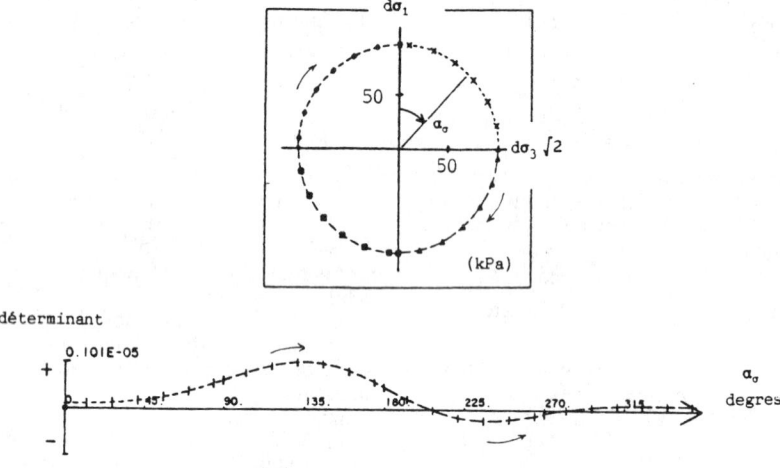

Fig. 5 : Variations of the constitutive determinant with the direction of the incremental stress vector. The range of negative values correspond to the cone of non-uniqueness directions (Darve and Chau, 1987)

associated strain bissector plane the extremities of the strain response vectors form a diagram, called "the strain response envelope".
Fig. 4 shows these degenerating diagrams, which at the stress-strain state of loss of uniqueness become a straight line. The slope of this straight line is exactly :

$$\frac{\Delta\varepsilon_1}{\sqrt{2}\Delta\varepsilon_3} = -\sqrt{2}$$

which corresponds to the isochoric condition. This result is completely in agreement with the previous analysis : at an incrementally incompressible state, as defined by condition (17), the strain response envelope must verify :

$$d\varepsilon_1 + 2\,d\varepsilon_3 = 0$$

Fig. 5 presents the variations of the constitutive determinant at this state of loss of uniqueness, as defined just before. We see that it is null (in fact negative) for $d\sigma_1$ and $d\sigma_3$ negative.

After the sufficient condition of uniqueness, as defined by inequality (16), it is sufficient to find at least one stress direction for which the Jacobian is negative or null, in order to no more fulfill the condition. But if we consider a given stress path, it will enter into the cone of non-uniqueness directions at a given stress state generally different from the stress state for which the first non-uniqueness direction has appeared. In other words between these two stress states the various states could develop a loss of uniqueness but the considered path did not meet this possibility. We can say also that, if the sufficient condition (16) is essentially not directionally dependent, the nullity of the constitutive determinant itself is strongly directionally dependent.

For the octo-linear model the cone of non-uniqueness directions is the same as the cone of unstable directions. As the stress path in undrained loading lies inside this cone from σ_1 peak, as soon as the cone of non-uniqueness directions appears the undrained path will verify the condition of loss of uniqueness.
An illustration of the fact that the nullity of the constitutive determinant is directionally dependent and also stress-strain history dependent is now presented by considering paths with "sharp bends" which verify in an approximate manner the isochoric condition.
The two types of decompositions which have been chosen are presented on fig. 6. They are defined by :

I $\begin{cases} q/p = \text{constant and } d\varepsilon_1 < 0 \text{ (dilatant behaviour)} \\ p = \text{constant and } d\varepsilon_1 > 0 \text{ (contractant behaviour)} \end{cases}$

II $\begin{cases} q/p = \text{constant and } d\varepsilon_1 < 0 \text{ (dilatant behaviour)} \\ \sigma_3 = \text{constant and } d\varepsilon_1 > 0 \text{ (contractant behaviour)} \end{cases}$

At all the ends of both the two parts, in case I as in case II, the isochoric condition is verified as it appears on fig. 7. Finally the computation stops because of an artefact : both the parts become contractant and the isochoric condition cannot be fulfilled.
The verification of the objective character of the method was necessary and this aspect is illustrated on fig. 8, where four different lengths of the decomposed paths are compared :

$$\sqrt{(\Delta q)^2 + (\Delta p)^2} = 8, 4, 2, 1 \text{ kPa}$$

Two conclusions emerge from fig. 8 :
• when the length of the paths is decreasing, the

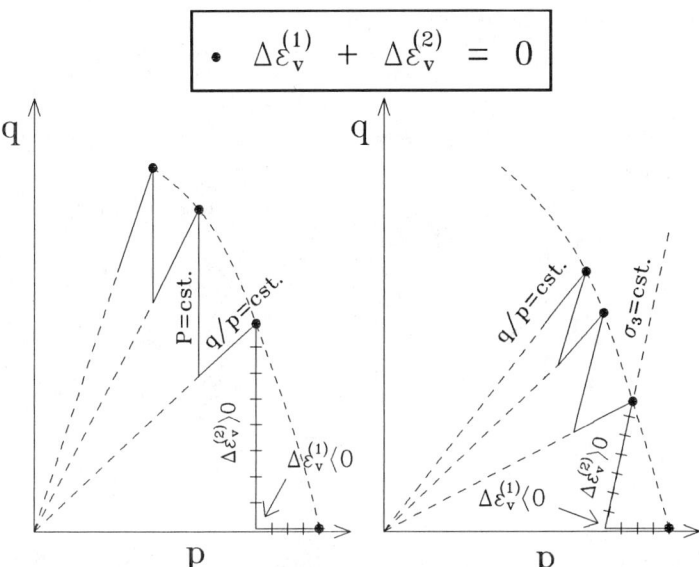

Fig. 6 : The two considered decompositions of loading path which have been utilized to simulate the isochoric condition in order to analyse the loss of uniqueness in undrained loading (Meghachou, 1993)

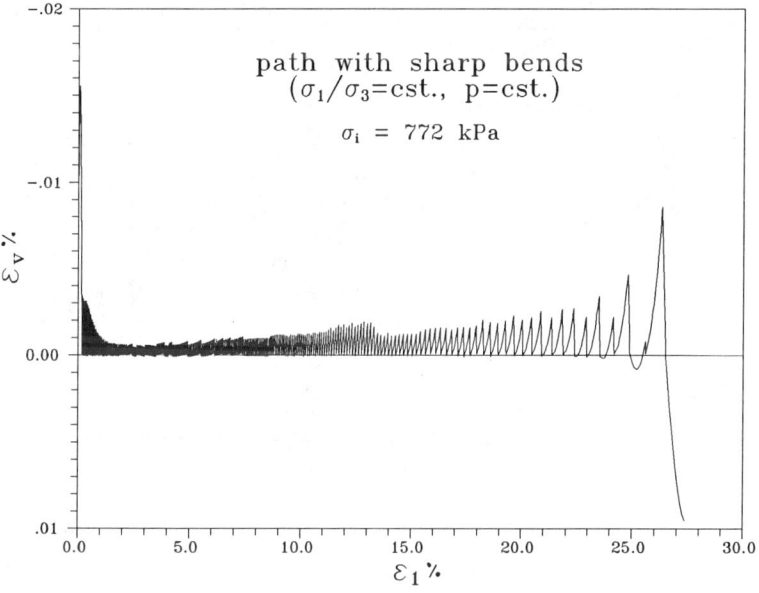

Fig. 7 : Validation of the simulation of the isochoric condition in the case of decomposed paths. The computation stops when the isochoric condition can not be more fulfilled because of an artefact (Meghachou, 1993)

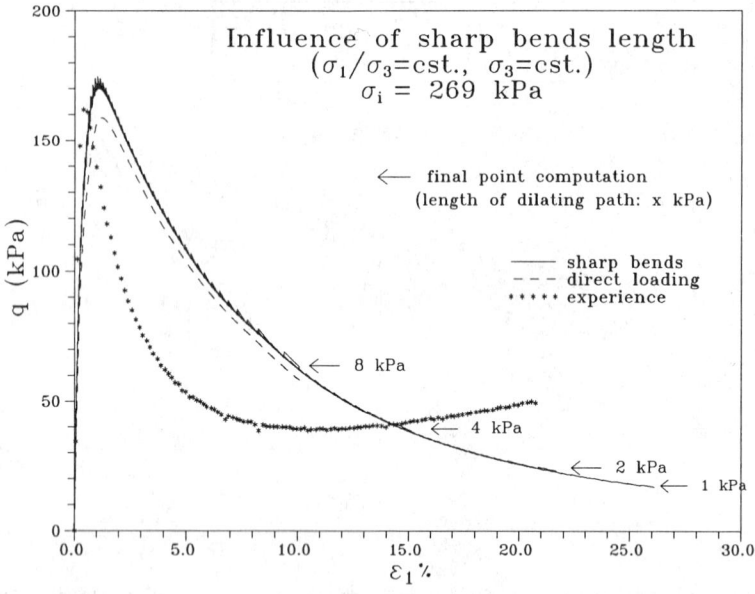

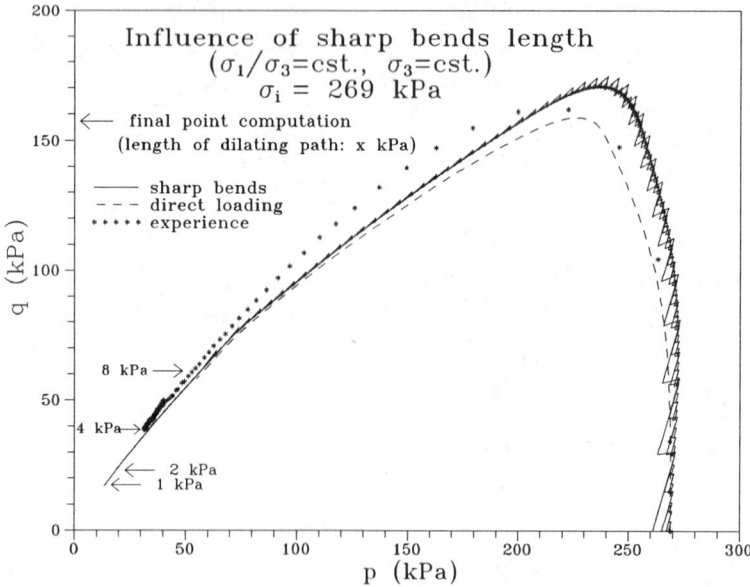

Fig. 8 : Validation of the objective character of the method : when the length of sharp bends decreases, the response diagrams converge. Four cases are considered :
$\sqrt{(\Delta p)^2 + (\Delta q)^2}$ = 8 ; 4 ; 2 ; 1 kPa (Meghachou, 1993)

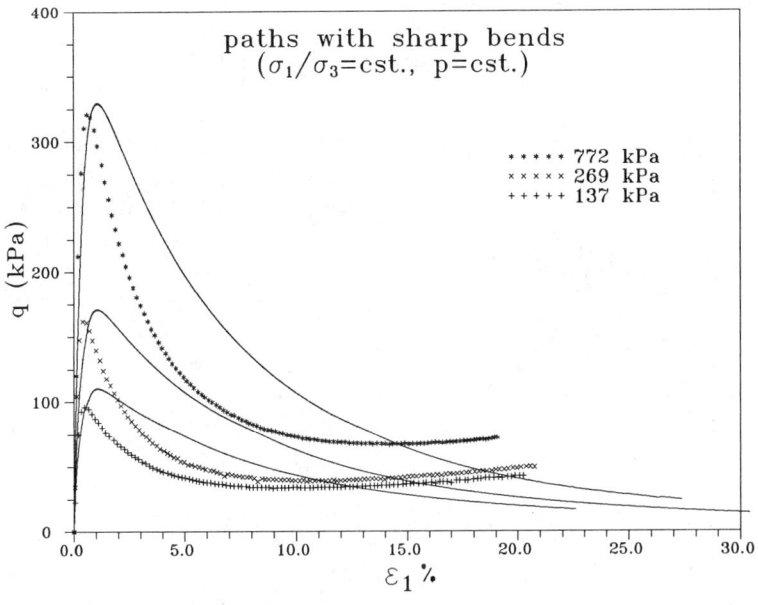

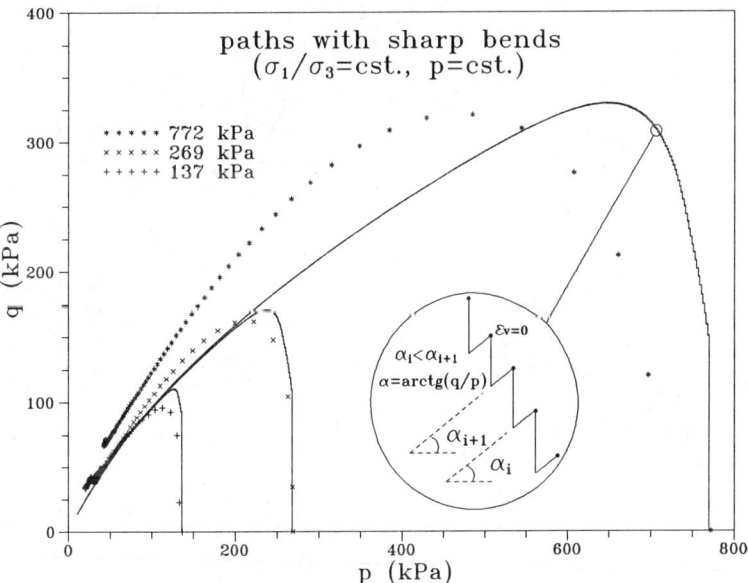

Fig. 9 : Simulation of undrained loading by decomposed paths of type I. The nullification of the constitutive determinant is no more met for the current incremental stress directions and for this particular stress-strain history (Darve, Flavigny and Meghachou, 1992)

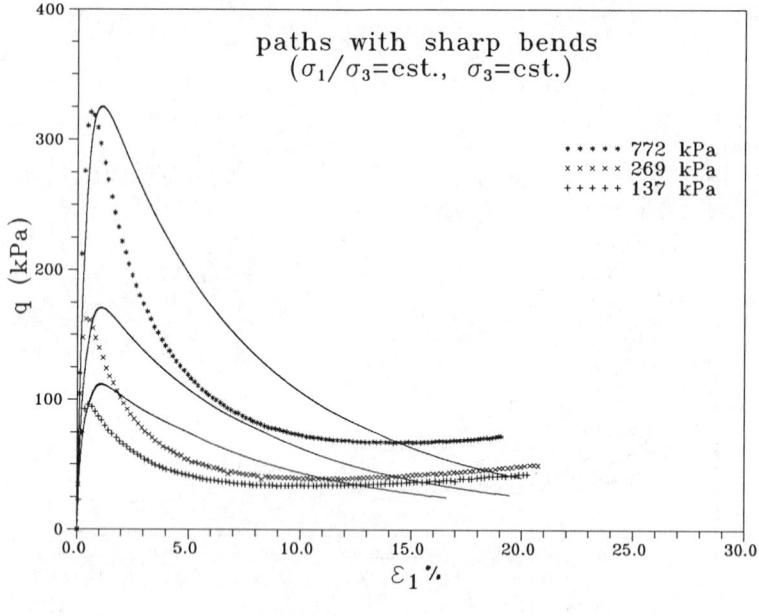
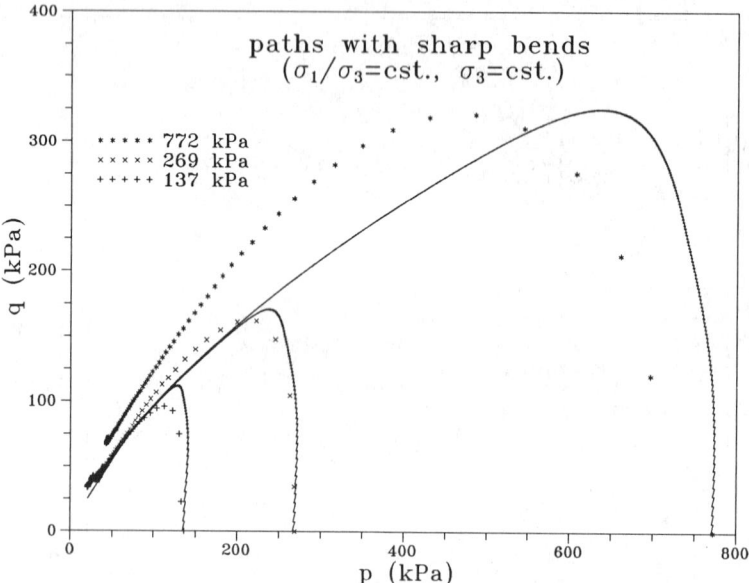

Fig. 10 : Simulation of undrained loading by decomposed paths of type II. The nullification of the constitutive determinant is no more met for the current incremental stress directions and for this particular stress-strain history (Darve, Flavigny and Meghachou, 1992)

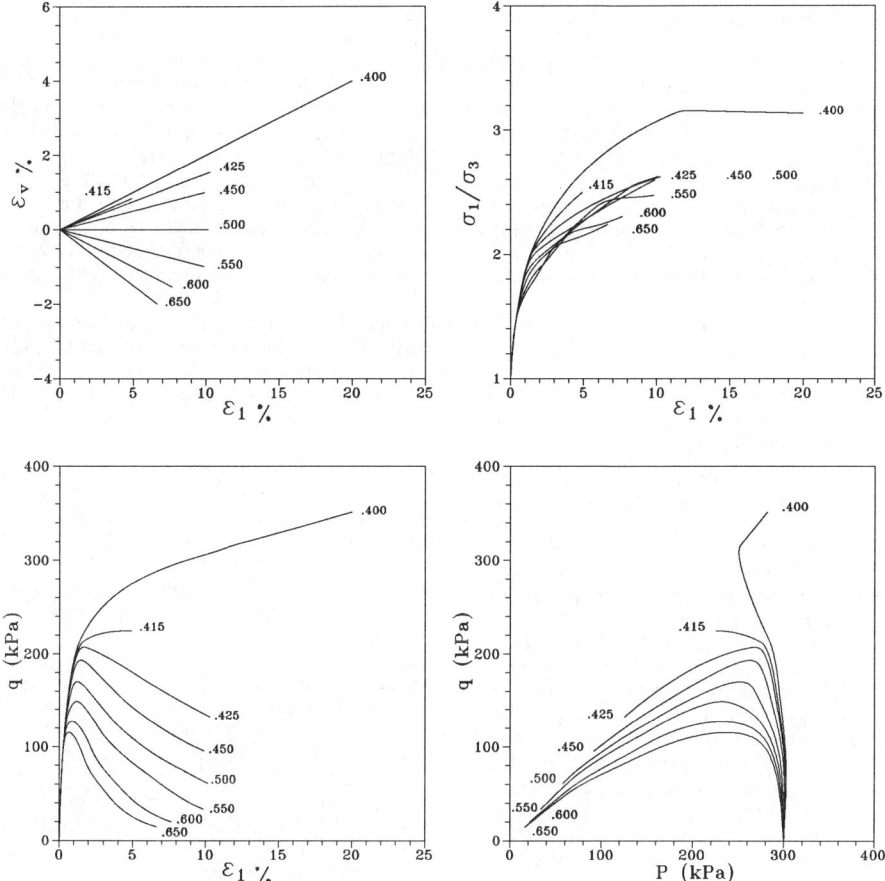

Fig. 11 : Modelling of rectilinear proportional strain paths by the incremental non-linear constitutive relation applied to Hostun loose sand :

$$d\varepsilon_3 / d\varepsilon_1 = -R = \text{constant}$$

The computations stop when the nullification of the jacobian is detected, except for R = 0.4 (Meghachou, 1993)

response diagrams converge to given unique curves, showing the objective character of these responses. Let us recall that this convergency is not a priori obvious since, when the length of paths decreases, the number of bends increases about proportionally and since moreover the model is incrementally non-linear,

• the converged responses are not coinciding with the diagrams obtained by a direct integration of the constitutive model along an undrained path. This point illustrates the fact that, for an incrementally non-linear constitutive relation, the "principle of superposition for incremental loading" is never strictly valid. This aspect has been studied more in details in Darve, Flavigny, Meghachou (1994).

Now fig. 9 and 10 present the response diagrams obtained with the incrementally non-linear model from the loading paths defined by conditions I and II. As we see it on these figures, the point of nullity of the constitutive determinant is no more met, because of the directions of the incremental stress vector which are different from the previous ones inducing the nullity and also because of the different stress-strain history. This result does not imply that the loss of uniqueness is suppressed : simply for the new kind of considered loading paths the nullity of the Jacobian is not met even if these stress-strain states lie inside a domain of loss of uniqueness.

Finally fig. 11 shows that the undrained path must be not considered as a particular case with respect to this discussion about the losses of stability and uniqueness. The loading paths are axisymmetric strain proportional paths defined by :

$$\begin{cases} d\varepsilon_1 = \text{constant} \\ d\varepsilon_3/d\varepsilon_1 = -R = \text{constant} \\ d\varepsilon_2 = d\varepsilon_3 \end{cases}$$

The values of R are given on the figure. For R = 0.5 we have the usual undrained loading. The ends of the response diagrams indicate the losses of uniqueness by nullity of the constitutive determinant. All the paths exhibit such a loss of uniqueness except for R = 0.4, which corresponds to a contractant path. In this last case we have not verified that the Jacobian is not null in any direction (a priori it is certainly null for a certain range of directions) : in any case the incremental stress directions which are exhibited for R = 0.4 do not belong to this range, as it is rather obvious after the results in (p - q) plane on fig. 11.

5 - CONCLUSIONS

In order to study stability and uniqueness in geomaterials constitutive modelling, we have exhibited as tools two sufficient conditions which are generally applicable as well in the case of incrementally non-linear relations.

As soon as the material is non-associated, these two conditions are distinct. Loss of stability generally precedes loss of uniqueness and both can appear strictly inside the plastic limit surface.

By applying these conditions to the case of loose sands under isochoric loading, we have shown that, when the sufficient condition of stability is no more fulfilled, clear physical instabilities can be detected (i.e. a small increase of certain dead loads induces a general brutal failure of the sample) and that the stress-strain states of loss of uniqueness correspond to a transition through a point of incremental incompressibility, which is basically a bifurcation point. This bifurcation could be linked to experimental results, showing a brutal change in the stress path direction in (p - q) plane for loose sands under undrained triaxial loading.

Finally we have illustrated on examples the fact that, if the sufficient condition of loss of uniqueness is essentially not directionally dependent, the nullity of the constitutive Jacobian is on the contrary strongly directionally and stress-strain history dependent.

REFERENCES

Hill, R., 1958. A general theory of uniqueness and stability in elastic-plastic solids. J. of the Mech. and Phys. of Solids, 6 : 236-249

Hill, R., 1959. Some basic principles in the mechanics of solids without a natural time. J. of the Mech. and Phys. of Solids, 7 : 209-225.

Darve, F., 1990a. Incrementally non-linear constitutive relationships. In : Geomaterials Constitutive Equations and Modelling. F. Darve (Ed.), pp 213-238. London, Chapman and Hall.

Darve, F., Dendani, H., 1989. An incrementally non-linear constitutive relation and its predictions. In : Constitutive Equations for Granular Non-cohesive Soils. A. Saada and G. Bianchini (eds.) pp 237-354, Rotterdam, Balkema.

Darve, F., Labanieh, S., 1982. Incremental constitutive law for sands and clays. Simulation of monotonic and cyclic tests, Int. J. for Num. and Anal. Meth. in Geomechanics, 6 : 243-275.

Darve, F., 1990b. The expression of rheological laws in incremental form and the main classes of constitutive equations. In : Geomaterials Constitutive Equations and Modelling. F. Darve (ed.) pp 123-147, London, Chapman and Hall.

Nova, R., 1989. Liquefaction stability bifurcations of soils via strainhardening plasticity. Workshop on localisation and bifurcation of granular bodies, Gdansk.

Darve, F., Chau, B., 1987. Constitutive instabilities in incrementally non-linear modelling. In : Constitutive Laws for Engineering Materials. Theory and Applications. C.S. Desai and G.H. Gallagher (eds.), pp 301-310, Elsevier Science Publ. Co.

Gudehus, G., 1979. A comparison for some constitutive laws for soils under radially symmetric loading and unloading. 3rd Int. Conf. on Num. Meth. in Geomechanics. Wittke (ed.), pp 1309-1323, Rotterdam, Balkema.

Meghachou, M., 1993. Stabilité des sables lâches. Essais et Modélisations. Thèse de Doctorat, Grenoble.

Darve, F., Flavigny, E., Meghachou, M., 1994. Yield surfaces and principle of superposition revisited by incrementally non-linear constitutive relations. Int. J. of Plasticity, to appear.

Darve, F., Flavigny, E., Meghachou, M., 1992. Numerical modelling of undrained behaviour of very loose sands by loading paths with sharp bends. In : Numerical Models in Geomechanics. G.N. Pande and S. Pietruszczak (eds.), pp 85-94, Rotterdam, Balkema.

A new effective non-local strain-measure for softening plasticity
Une nouvelle mesure non-locale de déformation pour la plasticité avec adoucissement

P.A. Vermeer
University of Stuttgart, Germany

R.B.J. Brinkgreve
Delft University of Technology, Netherlands

ABSTRACT: A brief review of softening regularization methods is given, but the attention is focused on the non-local method for treating softening plasticity. Restriction is made to the "one-dimensional" problem of necking in a tension bar. It is shown that the existing non-local model does not provide a full regularization and a modified non-local model is introduced. In contrast to the existing model, the modified model allows for a true analytical solution of the necking problem. As an alternative to this theory, we briefly review the strain-gradient theory. In fact, both the modified non-local model and the strain-gradient model allow for a fairly similar analytical solution. Numerical results show full regularization, i.e. no mesh-dependency and outstanding convergence of equilibrium iterations. The extension of the 1D-theory is not presented, but some 2D-results are shown for shear-banding.

Après une brève revue des méthodes de régularisation, l'accent est mis sur la méthode non-locale pour le traitement de la plasticité avec adoucissement. On restreint le sujet au problème unidimensionnel du rétrécissement localisé dans une barre en traction. On montre que le modèle non-local existant ne fournit pas une régularisation complète, et on propose un modèle modifié. Au contraire du modèle existant, le modèle modifié donne une vraie solution analytique du problème du rétrécissement. En tant que solution alternative, on passe brièvement en revue le modèle à gradient de déformation. On constate que le modèle non-local modifié et le modèle à gradient conduisent à des solutions remarquablement proches. Les résultats montrent une régularisation complète, c'est à dire pas de dépendance vis-à-vis du maillage et convergence remarquable des itérations d'équilibre. Bien qu'on ne présente pas l'extension de la théorie 1D, quelques résultats 2D sont montrés concernant les bandes de cisaillement.

1 INTRODUCTION

Rock, concrete and dense soils exhibit material softening under many conditions of loading. Here a material is said to undergo material softening when the second-order work is negative, i.e. $\dot{\sigma}_{ij} \dot{\varepsilon}_{ij} < 0$ in terms of rates of stress and strain. For rock and concrete it is a logical phenomenon as the micro-structural response to stress involves fissuring, and thus weakening of the material. For densely packed sand, softening beyond peak strength cannot be explained by cracking, but rather from dilation due to particle rearrangement. For dense clays, both fissuring and dilation would seem to be the cause of material softening.

In some soil tests, softening is quite severe, as one observes both material softening and geometrical softening due to bulging, necking or shear-banding. Indeed, material softening tends to induce such non-homogeneous deformations and thus geometrical softening. For general experimental evidence, the reader is referred to Read & Hegemier (1984). Note that we use the expression geometrical softening rather than structural softening as for instance used by Schreyer & Chen (1986), because we consider fissuring and cracking as material softening. On deriving constitutive models from experimental data, material softening needs to be isolated from geometrical softening as otherwise such models would obtain an overdose of material softening.

Classical (local) constitutive models involving material softening cannot be used straightforwardly in numerical simulations because of mesh-sensitivity. Near failure, all deformation tends to concentrate in the narrowest shear-bands that can be resolved by the mesh, i.e. a shear-band thickness of about one element. Hence, upon mesh refinement shear-bands become narrower and computed load-displacement curves will change considerably. From an engineering point of view mesh dependency is solved on using the softening scaling technique as proposed by Pietruszczak & Mróz (1981), but an influence of mesh alignment remains. A more rigorous solution is to use adaptive remeshing in the zone of localization down to an element size of about the physical shear-band thickness; see among others Pastor et al. (1991). However, this method was developed rather to model a correct shear-band thickness than to solve the numerical problems related with softening.

Modern computational procedures have proved to be quite successful in solving boundary value problems for non-softening plasticity. Indeed, fast and accurate procedures have been designed to find the succession of deformation fields leading to perfectly plastic failure. In our experience computational efforts increase considerably when changing from perfect plasticity to softening plasticity. For low degrees of softening, iterative solution procedures will converge relatively fast, but the convergence usually deteriorates when increasing the rate of softening. For some problems, iterative solution procedures fail to give any solution at all because of divergence. Indeed, numerical difficulties may often be attributed to inappropriate numerical discretisation schemes or to dubious procedures in implementing the actual model described, but this does not apply to the observed difficulties. When changing from hardening plasticity or perfect plasticity to softening plasticity, well-posedness of the rate boundary value problem is lost and the equations cease to be a proper description of the physics. In numerical simulations this loss of well-posedness becomes not only manifest through an extreme mesh sensitivity, but also through non-converging iterative procedures. Scaling techniques or adaptive remeshing as mentioned above may be a remedy for mesh-sensitivity, but the convergence-problem remains. Rigorous regularization techniques as considered by Sluys (1992) and de Borst et al. (1992) are needed.

In this paper we will review some existing regularization techniques and express preferences for the non-local model and the strain-gradient model. As the latter is not easy to use in existing finite element codes, we concentrate on the non-local model. It will be shown that the existing non-local model needs to be modified in order to ensure full regularization. The final aim is to solve shear-banding problems in rock-like materials. However, in most chapters restriction is made to 1D-theories. This is done to obtain insight and analytical solutions.

2 REVIEW OF PREVIOUS WORK

Classical models of softening plasticity suffer from the drawback that any width of the localization zone is theoretically admissible, whereas experiments show a distinct relationship with a micro-structural length scale. For example, in sands one observes shear-bands with a well-defined thickness of about ten times the average grain size; see among others Desrues (1984) and Mühlhaus & Vardoulakis (1987). Classical models need to be extended in such a manner that they render a distinct thickness of shear-bands, rather than any value that happens to be favoured by the particular mesh being used in a numerical simulation. Such extensions of classical models are called regularizations.

One way of regularization is to include some viscosity, but this is only effective for dynamic boundary value problems in which inertia forces play an active role, as indicated by Sandler & Wright (1984) and Sluys (1992). Another way of regularization is to use the Cosserat continuum model which includes rotations and couple stresses, as illustrated by Vardoulakis (1989) and Mühlhaus (1986). On reviewing regularizations Sluys (1992) also considers the option of using a Cosserat continuum model, but he obtains by far the best results for another model, namely the strain-gradient model. For this reason we are inclined to use the latter, but since it can be viewed as an approximation of fully non-local models we will concentrate on this more general model.

In general, a fully non-local model is based on average stresses, σ_{ij}^*, and average strains, ε_{ij}^*, where

$$\sigma_{ij}^* = \frac{1}{A} \int \int \int w(x_n') \, \sigma_{ij}(x_n + x_n') \, dx_1' \, dx_2' \, dx_3' \quad (2.1)$$

$$\varepsilon_{ij}^* = \frac{1}{A} \int \int \int w(x_n') \, \varepsilon_{ij}(x_n + x_n') \, dx_1' \, dx_2' \, dx_3' \quad (2.2)$$

x_n' is a local coordinate with $n = 1, 2$ or 3. The symbol w denotes a weighting function, usually taken as the error function, and

$$A = \int \int \int w(x_n') \, dx_1' \, dx_2' \, dx_3' \quad (2.3)$$

A fully non-local model is obtained by introducing a relationship between the average stresses and average strains. The relationship between average stress and local stress complies with the relationship between macroscopic stress and microscopic stress for granular bodies. Thornton (1979), Christoffersen et al. (1981) and others have given definitions of the form:

$$\sigma_{ij}^* = \frac{1}{V_r} \int_{V_r} \sigma_{ij} \, dV \quad (2.4)$$

where V_r is a so-called representative volume; a small sphere around the material point considered. Hence, the macroscopic stress is an average of the more rapidly varying microscopic stress. Equation (2.1) simply reduces to Equation (2.4) when assuming a block-type function $w(x'_n)$ with $w = 1$ in the interior of V_r and $w = 0$ outside the representative sphere. Similarly, ε_{ij}^* and ε_{ij} can be conceived as macroscopic strain and microscopic strain respectively. It is basically because of this clear physical background that we prefer the non-local theory.

In most applications the non-local or strain-gradient formulation is restricted to a specific part of the constitutive relation. Vardoulakis & Aifantis (1991) include a gradient term in the flow rule and Sluys (1992) extends the yield condition with a gradient term. In fact, these models can also be defined by an extended formulation of a particular strain measure in the softening rule, for example a non-local hardening / softening parameter ε^*:

$$\varepsilon^* = \int\int\int w(x_n') \kappa(x_n + x_n') \, dx_1' \, dx_2' \, dx_3'$$

where $\kappa = \sqrt{\varepsilon_{ij}^p \varepsilon_{ij}^p}$ (2.5)

Hence, all non-local aspects are concentrated in one parameter, whereas general stresses and strains remain local. This enables a near-local treatment of the constitutive relation.

On using the Taylor expansion of κ around $x_n = 0$, it can be derived that

$$\varepsilon^* = \kappa + B\nabla^2\kappa + \text{higher-order derivatives} \quad (2.6)$$

where B is a constant that depends on the weighting function w. On omitting the higher-order derivatives, this non-local model reduces to the strain-gradient model, as previously illustrated by Mühlhaus & Aifantis (1991).

At first sight the strain-gradient model would appear to be highly attractive. However, the model is difficult to implement in existing finite element codes. Sluys (1992) solved the problem by introducing a plastic strain measure as an additional unknown, similar to the displacement components. This gives a significant expansion of the degrees-of-freedom in a finite element analysis. We tried to circumvent the use of an enlarged matrix equation by a direct computation of the gradient of κ, but we did not succeed in getting a sufficiently smooth distribution as needed for the calculation of ε^*. Hence, it would seem that a gradient regularization requires complex extensions of existing finite element codes. For this reason we did not proceed with the strain-gradient method.

3 EXISTING NON-LOCAL PLASTICITY MODEL

For the sake of simplicity, the existing non-local theory of plasticity is applied to a one-dimensional tension bar of infinite length ($-\infty < x < \infty$). The relations for the one-dimensional elastoplastic behaviour are given below:

$$\dot{\varepsilon} = \dot{\varepsilon}^e + \dot{\varepsilon}^p \quad \text{(additivity of strain rates)} \quad (3.1)$$

$$\dot{\sigma} = E\dot{\varepsilon}^e = E(\dot{\varepsilon} - \dot{\varepsilon}^p) \quad \text{(stress-strain relation)} \quad (3.2)$$

$$f = \sigma - \sigma_t \quad \text{(yield function)} \quad (3.3)$$

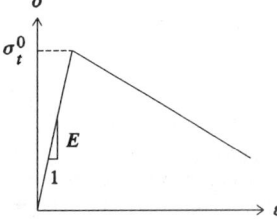

Figure 1. One-dimensional stress-strain relation

E is Young's modulus of elasticity and σ_t is the ultimate tensile strength. A simple linear non-local softening rule is adopted:

$$\dot{\sigma}_t = -h\,\dot{\varepsilon}^* \quad (3.4)$$

In this relation h is the softening modulus. The softening is controlled by an auxiliary non-local strain rate $\dot{\varepsilon}^*$. In accordance with Bazant et al. (1987), the non-local strain rate is defined as:

$$\dot{\varepsilon}^*(x) = \int_{-\infty}^{\infty} w(r)\,\dot{\varepsilon}^p(x+r)\,dr \quad (3.5)$$

In this formulation $w(r)$ is a weighting function that depends on the distance r from point x. In order to preserve the total amount of plastic strain, the integral of w over the domain of r must be equal to 1. A well-suited weighting function is the error function:

$$w(r) = \frac{1}{l\sqrt{\pi}} e^{-\left(\frac{r}{l}\right)^2} \qquad \int_{-\infty}^{\infty} w(r)\,dr = 1 \quad (3.6)$$

The parameter l is a characteristic length which is related to the width of the localization zone. The aim of the non-local model is to obtain a well-posed boundary value problem, to ensure mesh-

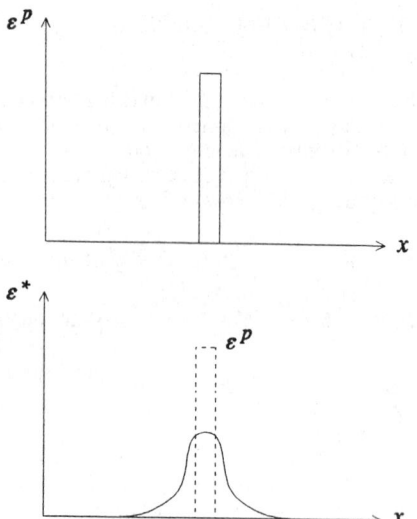

Figure 2. Distribution of strain in a bar with an imperfection
a. Plastic strain at the onset of plasticity
b. Non-local strain that follows from ε^p

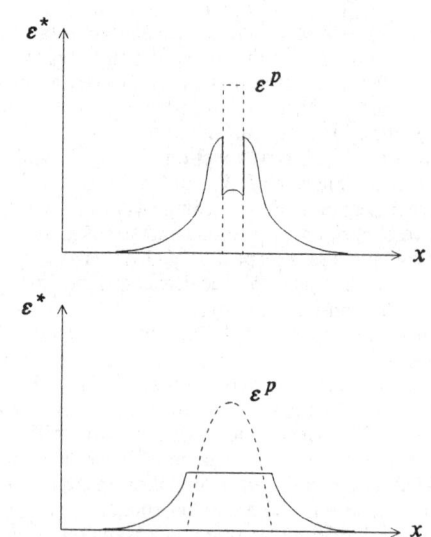

Figure 3. Distribution of new non-local strain for $\alpha > 1$
a. Initial distribution
b. Final distribution

independency and to promote convergence of numerical procedures.

However, in our experience one-dimensional calculations using the existing non-local model in this form resulted in localization in a single point. Hence, the results were not objective with respect to the schematization. Moreover, the convergence during the iteration process was very bad. Two-dimensional calculations gave better results, but full mesh-independence was not obtained. A qualitative explanation will be given below for the fact that the current non-local formulation is not effective as a regularization method for 1D applications.

Consider again a one-dimensional softening tension bar with a small imperfection where the tensile strength is slightly lower than in the rest of the bar. Hence, we consider a bar with a short section where σ_t^0 is somewhat smaller than in the rest of the bar. Due to this fact at the onset of plasticity, the plastic strain is concentrated in this weak zone as shown in Figure 2a. Note that the precise shape of the plastic strain is of no importance, but we consider the simple "block" function in Figure 2a for convenience. The non-local strain ε^* is a redistribution of the plastic strain, which always results in a smoother curve. For the distribution of the plastic strain as considered in Figure 2a the non-local strain looks like a Gaussian curve as visualised in Figure 2b. The exact shape is influenced by the choice of the weighting function. In general, the maximum will be in the centre of the imperfection. According to the softening rule, which says that the softening depends on the non-local strain ε^*, the material will soften mostly in the centre point. As a consequence, the plastic strain is more and more concentrated in the centre point and cannot spread in order to become a fully developed neck. If the non-local strain were such that the maximum extended to the edges of the plastic zone, the material would soften in a wider range and the plastic strain would spread to become a real neck. These considerations led to the formulation of a modified non-local plasticity model, which is presented in the next section.

4 MODIFIED NON-LOCAL PLASTICITY MODEL

In this modified non-local formulation a new non-local strain rate is introduced which consists of a local and a purely non-local part. It will be shown that this new formulation is indeed effective and does not suffer from the problems that occur in the existing non-local formulation.

For the case of the one-dimensional tension bar as introduced in the previous section the new non-local strain rate can be written as:

$$\dot{\varepsilon}^*(x) = (1-\alpha)\dot{\varepsilon}^p(x) + \alpha \int_{-\infty}^{\infty} w(r)\dot{\varepsilon}^p(x+r)\,dr \quad (4.1)$$

where α is a new parameter, the value of which can be chosen more-or-less arbitrarily. Note that for $\alpha = 0$ the classical (local) theory is obtained. For $\alpha = 1$ the above expression is equal to the non-local formulation as given in equation (3.5). Hence, based on the considerations at the end of the previous section, the current formulation cannot be effective for $\alpha = 1$. As α is increased, one makes sure that zones of low plastic strains, ε^p, get higher non-local strains, ε^*. At the same time zones of high plastic strains get lower non-local strains upon an increasing α.

To illustrate the effect of α we consider again the imperfection of Figure 2a. The error function as given in equation (3.6) is taken as the weighting function. For $\alpha = 1$, the block-type distribution of ε^p results in the Gaussian-type distribution of Figure 2b for ε^*. The distribution of the non-local strain for $\alpha > 1$ is plotted in Figure 3a. It can be seen that ε^* has a maximum at the edges of the plastic zone. As a result, at these points the highest amount of softening occurs. This induces a converging redistribution of plastic strains in such a way that the non-local strain will finally be constant over a particular width, as plotted in Figure 3b. Hence, an effective non-local regularization is obtained for $\alpha > 1$. Note that $\alpha < 1$ cannot be effective, so that this case is not taken into consideration at all.

5 ANALYTICAL SOLUTION FOR MODIFIED NON-LOCAL MODEL

In this section it will be shown that the modified non-local model allows for an analytical solution. It is important to note that the analytical solution given here is directly based on the non-local formulation and not on a differential approximation as done by Bazant and Zubelewicz (1988). The analytical solution confirms the observation that for the existing non-local formulation (or $\alpha = 1$ in the modified non-local formulation) the localization zone reduces to a point.

Consider now necking of the tension bar in which the plastic strain rate is formulated as:

$$\dot{\varepsilon}^p(x) = -\frac{\dot{\sigma}}{h}(1 + \cos x') \quad (5.1)$$

where $x' = \dfrac{2\pi x}{L}$

with $-\infty < x < \infty$ and L being the width of the localization zone. In reality a neck will occur very locally for $-L/2 \le x \le L/2$ and outside this zone the plastic strain vanishes. However, for convenience, we will consider periodic necking with equation (5.1) valid for any value of x. It will be shown that this periodic necking is an analytical solution of the modified non-local model.

The expression for the plastic strain rate as given in equation (5.1) substituted into equation (4.1) with the weighting function of equation (3.6) gives:

$$\dot{\varepsilon}^*(x) = -\frac{\dot{\sigma}}{h}\Big[(1-\alpha)(1+\cos x') + \quad (5.2)$$

$$+ \frac{\alpha}{l\sqrt{\pi}} \int_{-\infty}^{\infty} e^{-\left(\frac{r}{l}\right)^2}(1 + \cos(x'+r'))\,dr\Big]$$

where $r' = 2\pi r / L$. In the integral the cosine can be split into two parts using the standard goniometric rule $\cos(x'+r') = \cos x' \cos r' - \sin x' \sin r'$. Hence, equation (5.2) becomes:

$$\dot{\varepsilon}^* = -\frac{\dot{\sigma}}{h}\Big[1-\alpha+(1-\alpha)\cos x' + \frac{\alpha}{l\sqrt{\pi}}\int_{-\infty}^{\infty} e^{-\left(\frac{r}{l}\right)^2} dr +$$

$$+ \frac{\alpha}{l\sqrt{\pi}}\cos x' \int_{-\infty}^{\infty} e^{-\left(\frac{r}{l}\right)^2}\cos r'\,dr +$$

$$- \frac{\alpha}{l\sqrt{\pi}}\sin x' \int_{-\infty}^{\infty} e^{-\left(\frac{r}{l}\right)^2}\sin r'\,dr \Big] \quad (5.3)$$

The integrals in the above expression have a particular form. According to Gröbner & Hofreiter (1961) the integrals can be evaluated as:

$$\int_{-\infty}^{\infty} e^{-\left(\frac{r}{l}\right)^2} dr = l\sqrt{\pi} \quad (5.4a)$$

$$\int_{-\infty}^{\infty} e^{-\left(\frac{r}{l}\right)^2}\cos r'\,dr = l\sqrt{\pi}\,e^{-\left(\frac{\pi l}{L}\right)^2} \quad (5.4b)$$

$$\int_{-\infty}^{\infty} e^{-\left(\frac{r}{l}\right)^2}\sin r'\,dr = 0 \quad (5.4c)$$

Substitution of equations (5.4) into equation (5.3) gives:

$$\dot{\varepsilon}^*(x) = -\frac{\dot{\sigma}}{h}\Big[1+(1-\alpha)\cos x' + \alpha \cos x'\,e^{-\left(\frac{\pi l}{L}\right)^2}\Big] \quad (5.5)$$

In order to obtain a stable localization zone, the

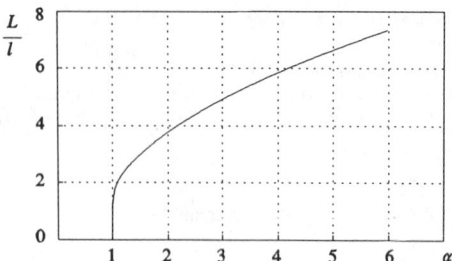

Figure 4. Width of the localization zone as a function of α

amount of softening should be constant and maximal within the range $-L/2 \leq x \leq L/2$. This is obtained if the non-local strain rate ε^* is constant within the localization zone, i.e. independent from x, which means that the derivative towards x should be zero:

$$\frac{d\dot{\varepsilon}^*(x)}{dx} = -\frac{\dot{\sigma}}{h}\left[-(1-\alpha)\frac{2\pi}{L}\sin x' - \alpha\frac{2\pi}{L}\sin x' e^{-\left(\frac{\pi l}{L}\right)^2}\right]$$
$$= 0 \qquad (5.6)$$

Elaboration of equation (5.6) yields an expression for the width of the localization zone, L, which appears to be a function of α:

$$L^2 = \frac{\pi^2 l^2}{\ln(\alpha) - \ln(\alpha-1)} \qquad (5.7)$$

The relation between L and α is plotted in Figure 4. From this relation it can be seen that α must be greater than 1 in order to have a well-defined finite width. In the limiting case of α approaching unity, the width of the localization zone reduces to zero. In the limiting case of α going to infinity, the width of the localization zone also goes to infinity.

$$\lim_{\alpha \downarrow 1} L(\alpha) = 0 \qquad \lim_{\alpha \to \infty} L(\alpha) = \infty \qquad (5.8)$$

Hence, the existing non-local theory ($\alpha = 1$) does not produce a solution to the necking problem. One needs a factor α as used in the modified theory presented here. For $\alpha > 1$ the theory incorporates a sound analytical solution for the necking problem.

Preliminary conclusion: the fact that the modified non-local theory allows for an analytical solution of the necking problem creates some confidence in this approach. It is surprising that the analytical solution is relatively independent of the particular value of the constant α being used. Indeed, on changing its value one can also change the value of the internal length parameter, l, in such a manner that the width of the neck remains constant. Many combinations of α and l will give the same neck, as can be seen from equation (5.7). At the same time such combinations theoretically give the same strain distribution within the neck. As a consequence, all these combinations of α and l also provide the same rate of elongation in a tension test. On using the analytical solution to evaluate the rate of elongation, one obtains for a bar with a single neck:

$$\dot{L}_{bar} = \int \dot{\varepsilon}\, dx = \int (\dot{\varepsilon}^e + \dot{\varepsilon}^p)\, dx = \dot{\sigma}(L_{bar}/E - L/h) \qquad (5.9)$$

6 DISCUSSION OF ANALYTICAL SOLUTION

In reality localization will only occur in a single neck and not in periodic necks as considered in the analytical solution. In the following it will be evaluated to what extent the analytical solution, as given in equation (5.1) and the corresponding width of the localization zone, as given in equation (5.7), apply to real non-periodic necking.

To this end we consider the relationship (4.1) between the non-local strain rate and the plastic strain rate. In fact, this relationship can be conceived as an averaging procedure. The averaging procedure is visualised in Figure 5. In this plot the large curves represent the periodic plastic strain rate, $\dot{\varepsilon}^p$, as defined by equation (5.1). The smaller curve is the weighting function, w, as defined by equation (3.6). The non-local strain is largely determined by the product between w and $\dot{\varepsilon}^p$. Let us first consider a point near the centre of the neck as illustrated in Figure 5a. Here the non-local strain of point A is determined by the shaded plastic strains. Plastic strains outside the shaded area have hardly any influence. Hence, for near-centre points, such as point A in Figure 5a, the periodicity of the analytical solution has hardly any influence.

For near-edge points the situation is different. Consider for example point B as indicated in Figure 5b. Again, the influence zone is indicated by shaded plastic strains. Point B is positioned in the same neck as point A, but not entirely determined by the plastic strains within this neck. Indeed, a smaller part of the shaded area is located in another neck on the left-hand side. Hence, near-edge points in single necks will differ from near-edge points in periodic necks. As a consequence, single necks will not exactly match the periodic necks considered here. It can be argued that single necks will be somewhat narrower than periodic necks, but differences will be smaller when larger

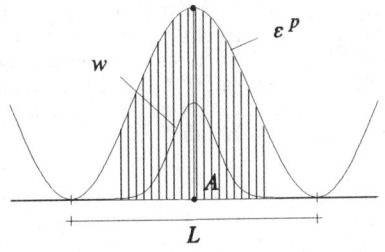

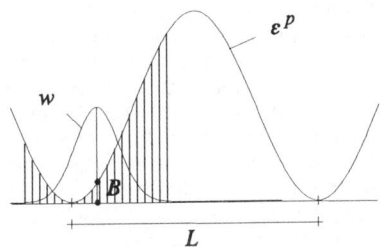

Figure 5. Plastic strain rate for periodic necking and weighting function, w, for computation of non-local strain rate, ε^*.
a. Weighting function for near-centre point
b. Weighting function for near-edge point.

values of α are considered. Indeed, for a given internal length, l, the width of necks increase as α is increased. As a consequence, the w-curves in Figure 5 become small with respect to the ε^p-curves.

7 COMPARISON WITH STRAIN-GRADIENT THEORY

The strain-gradient theory is another effective regularization technique to provide well-posedness in softening problems. According to the gradient theory the auxiliary strain rate ε^* can be formulated as:

$$\dot{\varepsilon}^*(x) = \dot{\varepsilon}^p(x) + \frac{l_g^2}{4} \frac{d^2 \dot{\varepsilon}^p(x)}{dx^2} \qquad (7.1)$$

l_g is an internal length parameter; the subscript g has been added in order to emphasise the difference with the non-local theory.

Similar to the modified non-local theory, the strain-gradient theory also allows for an analytical solution of the necking problem (de Borst & Mühlhaus, 1991). In fact, the analytical solution is identical to the solution as given in Section 4, and equation (5.1) applies to the strain-gradient theory as well. However, in this case the solution is also exact for a single neck. The width of the localization zone is found to be:

$$L_g = \pi \, l_g \qquad (7.2)$$

Considering the results as presented by de Borst & Mühlhaus (1991) and the intensive study by Sluys (1992), it can be concluded that the strain-gradient model is highly effective as a regularization technique. In fact, the strain-gradient theory would be very satisfying if its numerical implementation were not so awkward. Comparison of equation (7.2) with equation (5.7) shows that both theories give the same width of the localization zone when using:

$$l = l_g \sqrt{\ln\left(\frac{\alpha}{\alpha - 1}\right)} \qquad \text{or} \qquad (7.3)$$

$$\alpha = \frac{\exp(l^2 / l_g^2)}{\exp(l^2 / l_g^2) - 1}$$

As it remains a powerful regularization technique, we will compare it with the modified non-local theory. From equation (7.3) it follows that the internal length parameter, l, as used in the modified non-local model, is not equal to the internal length, l_g, as used in the strain-gradient model. Only for the special case of $\alpha \approx 1.58$ do both parameters coincide.

The strain-gradient theory might be conceived as an approximation of the non-local theory. Indeed, on expanding the non-local strain rate as defined by equations (3.5) and (3.6) in a Taylor series around $r = 0$, one obtains

$$\dot{\varepsilon}^*(x) = \frac{1}{l\sqrt{\pi}} \int_{-\infty}^{\infty} e^{-\left(\frac{r}{l}\right)^2} \dot{\varepsilon}^p(x) \, dr +$$

$$+ \frac{1}{l\sqrt{\pi}} \int_{-\infty}^{\infty} r \, e^{-\left(\frac{r}{l}\right)^2} \frac{d\dot{\varepsilon}^p(x)}{dx} \, dr +$$

$$+ \frac{1}{l\sqrt{\pi}} \int_{-\infty}^{\infty} \frac{r^2}{2} e^{-\left(\frac{r}{l}\right)^2} \frac{d^2 \dot{\varepsilon}^p(x)}{dx^2} \, dr +$$

$$+ \frac{1}{l\sqrt{\pi}} \int_{-\infty}^{\infty} \frac{r^3}{6} e^{-\left(\frac{r}{l}\right)^2} \frac{d^3 \dot{\varepsilon}^p(x)}{dx^3} \, dr + \ldots \qquad (7.4)$$

Every odd derivative term cancels in this Taylor expansion, and it follows that:

$$\dot{\varepsilon}^*(x) = \dot{\varepsilon}^p(x) + \frac{l^2}{4}\frac{d^2\dot{\varepsilon}^p(x)}{dx^2} + \text{higher-order derivatives} \quad (7.5)$$

The difference between equation (7.5) and equation (7.1) is just the higher order derivatives. In general, Taylor expansions give accurate approximations after a few terms. For this reason researchers are inclined to think that the higher-order derivatives in equation (7.5) are negligible. However, from the fact that the strain-gradient model is effective and the non-modified non-local model is not, it can already be concluded that the higher-order derivative terms in equation (7.5) have an important influence and that they are certainly not negligible.

It is concluded that the modified non-local theory and the strain-gradient theory provide very similar solutions to the necking problem, although internal length parameters play somewhat different roles. However, for the special choice of $ln(\alpha) - ln(\alpha-1) = 1$ ($\alpha \approx 1.58$), their internal length parameters are equivalent.

8 NUMERICAL IMPLEMENTATION

Finite element application of the modified non-local softening model involves an incremental approach where an iterative procedure is used to satisfy the equilibrium condition. To be sure about convergence, an elastic stiffness matrix is used in an accelerated initial stress procedure (Vermeer & Van Langen, 1989). We have great doubts about the use of a tangent stiffness matrix for the current non-local model. An exact formulation of the tangent matrix is virtually impossible. An approximated tangent matrix based on the assumption that the non-local strain rate can be approximated by the local plastic strain rate, yields a stiffness matrix that would produce much too soft a structural response. Performing equilibrium iterations with such a matrix tends to give convergence problems. An elastic stiffness matrix does definitely not suffer from these problems and has therefore been adopted.

Prescribed displacements are applied at the bar ends. The calculation procedure is similar to ordinary "local" finite element calculations that involve initial stress-type procedures. An essential subroutine is the computation of stress increments for given strain increments. Therefore this topic will be described in full detail.

Time integration of equations (3.2) and (3.4) respectively yields:

$$\Delta\sigma = E(\Delta\varepsilon - \Delta\varepsilon^p) \quad (8.1)$$

$$\Delta\sigma_t = -h\,\Delta\varepsilon^* \quad (8.2)$$

In the case of elastic behaviour, the plastic strain increment and the non-local strain increment are zero. For plastic behavior, the non-local strain increment $\Delta\varepsilon^*$ is obtained by time integration of equation (4.1). Substitution of the plastic strain increment $\Delta\varepsilon^p$ by $\Delta\varepsilon - \Delta\varepsilon^e$ gives:

$$\Delta\varepsilon^*(x) = (1-\alpha)\left(\Delta\varepsilon(x) - \Delta\varepsilon^e(x)\right) + \quad (8.3)$$
$$+ \frac{\alpha}{l\sqrt{\pi}}\int_{-\infty}^{\infty} e^{-\left(\frac{r}{l}\right)^2}\left(\Delta\varepsilon(x+r) - \Delta\varepsilon^e(x+r)\right)dr$$

In a one-dimensional application the elastic strain increment $\Delta\varepsilon^e$ can be directly calculated from the applied stress increment. Hence, the non-local strain increment is related with the total strain increment $\Delta\varepsilon$ rather than with the plastic strain increment $\Delta\varepsilon^p$. For a homogeneous bar the elastic strain increment is constant, which gives:

$$\frac{1}{l\sqrt{\pi}}\int_{-\infty}^{\infty} e^{-\left(\frac{r}{l}\right)^2}\Delta\varepsilon^e(x+r)\,dr = \Delta\varepsilon^e(x) \quad (8.4)$$

Hence, the non-local strain increment can be written as:

$$\Delta\varepsilon^*(x) = \Delta\varepsilon^p(x) - \alpha\Delta\varepsilon(x) + \frac{\alpha}{l\sqrt{\pi}}\int_{-\infty}^{\infty} e^{-\left(\frac{r}{l}\right)^2}\Delta\varepsilon(x+r)\,dr \quad (8.5)$$

Using this equation, one computes the non-local strain in a material point from a given distribution of physical strain. Hence, the entire strain distribution must be known before any non-local strain can be computed.

For a material point in plastic state which is subject to loading, the change of stress is such that the new state of stress complies with the yield condition. As a result, the value of the yield function is zero at the end of the time-step:

$$f = 0 \qquad \sigma = \sigma_t \quad (8.6)$$

where

$$\sigma = \sigma^0 + \Delta\sigma \qquad \sigma_t = \sigma_t^0 + \Delta\sigma_t \quad (8.7)$$

The superscript ^{0} indicates the stresses at the end of the previous time-step. Substitution of the equations (8.7) for the current stress and tensile strength into equation (8.6), with the help of equations (8.1), (8.2) and (8.5), yields an expression for the plastic strain increment:

$$\Delta\varepsilon^p = \frac{1}{E-h}\left[\sigma^0 - \sigma_t^0 (E-\alpha h)\Delta\varepsilon + \frac{\alpha h}{l\sqrt{\pi}}\int_{-\infty}^{\infty} e^{-\left(\frac{r}{l}\right)^2}\Delta\varepsilon(r)\,dr\right] \quad (8.8)$$

Note that this expression for the plastic strain increment is exact, even in the case of transition from elastic to plastic behaviour. In case the material point was already in plastic state at the end of the previous time-step, the term $\sigma^0 - \sigma_t^0$ in the above expression is cancelled.

Backsubstitution of the plastic strain increment into equation (8.1) gives the final stress increment. The calculation of the stress increment is summarized in Table 1.

Table 1. Calculation of stress increment for a given strain increment

1. $\Delta\sigma = E\, \Delta\varepsilon$

2. $f = \sigma^0 + \Delta\sigma - \sigma_t^0 - \alpha h\, \Delta\varepsilon + \dfrac{\alpha h}{l\sqrt{\pi}} \displaystyle\int_{-\infty}^{\infty} e^{-\left(\frac{r}{l}\right)^2} \Delta\varepsilon(r)\, dr$

3. if $f > 0$ then

4. $\Delta\varepsilon^p = \dfrac{f}{E - h}$

5. $\Delta\sigma = \Delta\sigma - E\, \Delta\varepsilon^p$

6. $\Delta\sigma_t = -h\, \Delta\varepsilon^p + \alpha h\, \Delta\varepsilon - \dfrac{\alpha h}{l\sqrt{\pi}} \displaystyle\int_{-\infty}^{\infty} e^{-\left(\frac{r}{l}\right)^2} \Delta\varepsilon(r)\, dr$

7. end if

8. $\sigma = \sigma^0 + \Delta\sigma$

9. $\sigma_t = \sigma_t^0 + \Delta\sigma_t$

9 NUMERICAL VERSUS ANALYTICAL SOLUTION

Again we consider the necking problem for a tension bar, but now the problem is solved numerically. One of the aims is to verify mesh-independence of the modified non-local theory. Hence the problem is to be solved for various meshes with different finesses. The second aim is to verify whether or not the analytical solution is approximately retrieved; both for the width of the neck and for the rate of elongation of the tension bar. To investigate this in detail, we will consider two combinations of α and l which should give the same neck width of $L = 31.42$ mm. The calculations are fully determined by the data:

Length of Bar $L_{bar} = 100$ mm
Youngs Modulus $E = 20 \cdot 10^6$ kPa

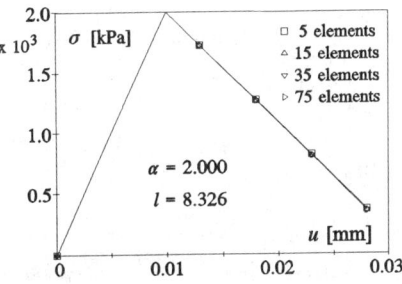

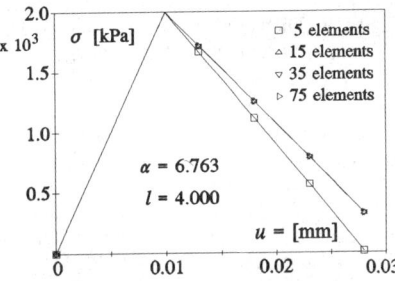

Figure 6. Stress-displacement curves of softening bar
  a. Relatively small α-value
  b. Relatively large α-value

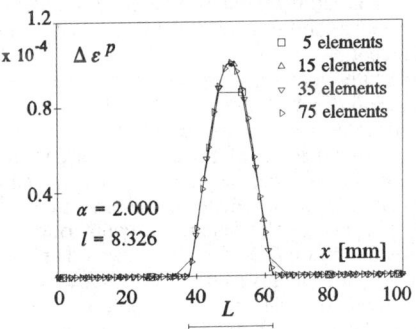

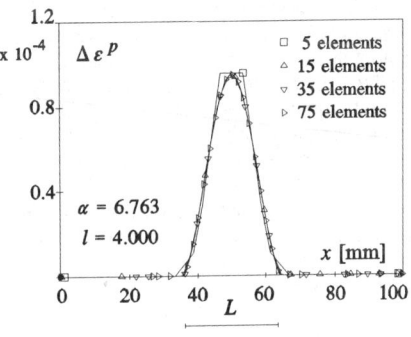

Figure 7. Plastic strain increments of softening bar
  a. Relatively small α-value
  b. Relatively large α-value

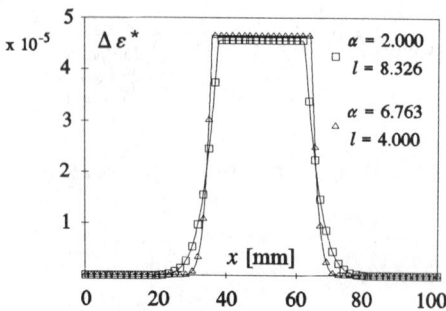

Figure 8. Distribution of non-local strain increment (75 elements)

Softening Modulus	$h = 2 \cdot 10^6$ kPa
Initial Strength	$\sigma_t^0 = 2 \cdot 10^3$ kPa
Amplification Factor	$\alpha = 2.000$ and 6.763
Internal Length	$l = 8.326$ and 4.000
Number of Elements	$N = 5, 15, 35$ and 75

High-order 5-noded line elements are used, but we could just as well have used low-order line elements. Calculations are done with four Gaussian integration points per 5-noded element. For all calculations a neck was triggered by means of a small perturbation; the tensile strength of the most central Gaussian integration point was reduced by 0.1 percent.

Computed load-displacement curves are shown in Figure 6. From Figure 6a it is observed that four different meshes give exactly the same curve. For the second series of calculations only the very coarse mesh produces deviating results, as can be seen from Figure 6b. Hence an excellent degree of mesh-independence is found. On comparing Figures 6a and 6b, it is also observed that both sets of α and l give virtually identical results, as expected from the analytical solution. Except for the coarsest mesh in Figure 6b, all calculations show a softening slope $\Delta\sigma/\Delta u$ of about $93 \cdot 10^3$ kPa/mm, as follows from the analytical solution.

Computed distributions of plastic strain increments are shown in Figure 7. These data demonstrate again mesh-independence. Very pronounced strain localizations are observed and the lengths of these necks can simply be measured. For the computations with $\alpha = 2.000$, we measure a length of about 25 mm (Figure 7a) and a length of about 29 mm is found when applying a large value of $\alpha = 6.763$ (Figure 7b). The last one approaches the analytical solution of $L = 31.42$ mm. The latter is in line with the argument in Section 6, that larger α-values will give closer approximations of the analytical solution.

Figure 8 shows distributions of the non-local strain measure ε^*. Here the analytical solution indicates a block function. The best approximation of this block function is again obtained for the higher value of the factor α, as expected from the discussion in Section 6.

10 EXTENSION TO 2D SHEAR-BANDING

In previous sections the necking problem is considered on the basis of a one-dimensional theory. Such an approach can obviously be criticized as the occurrence of necks is also influenced by geometric non-linearity. Indeed, in the middle of a neck, stresses will be much higher than outside the neck, as here the cross section of the bar is significantly reduced. In fact, the 1D-solution of the necking problem would have little value, if it did not suggest that the modified non-local theory could be successfully extended to a general three-dimensional theory.

In this study we will not present the general modified non-local theory as this topic merits another paper. However, the power of a general theory will be demonstrated by showing some preliminary numerical results. Rather than focusing the attention again on the necking problem, we consider the shear-banding problem in a sample of soil or rock. Although realistic shear-bands can be handeled, it is not intended here to model the shear-band realisticly, but to show the effectiveness of the modified non-local method in two-dimensional applications for an enlarged shear-band thickness. For the sake of convenience, pressure dependent frictional behaviour is not considered. Instead, the simple Von-Mises plasticity model is adopted. Hence, the strength of the material is related to cohesion alone. The material behaves perfectly elastic up to yield. Then cohesion decreases linearly as a function of deviatoric plastic strain. We assume that the cubical samples are tested in a biaxial apparatus to create a plane state of strain. Further data are:

Sample Height	$H_{sample} = 0.25$ m
Internal Length	$l = 0.02$ m
Amplification Factor	$\alpha = 2.0$
Tensile strength	$s = 20$ kPa
(cohesion $= 10$ kPa)	

Localization of strain is numerically triggered by reducing the tensile strength of one Gaussian integration point by 10 percent. It can be argued that the 1D-solution for the necking problem also applies to the present shear-banding problem. Hence, equation (5.7) would suggest a shear-band thickness of 0.075 m.

Numerical results as shown in Figure 10 give a shear-band thickness of about 0.06 m. Indeed, the thickness is smaller than the theoretical value, but this is in line with the one-dimensional results for

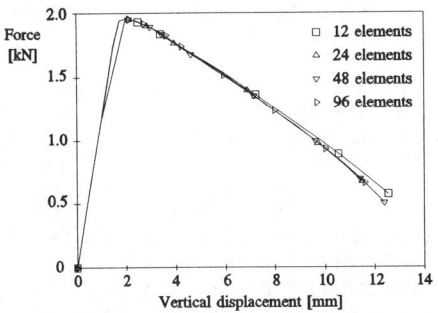

Figure 9. Load-displacement curves of biaxial test problem

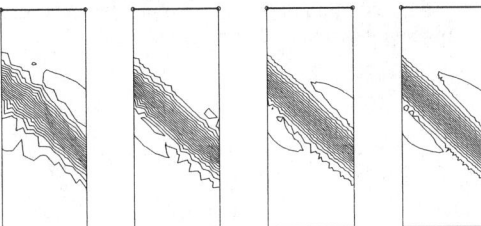

Figure 10. Contourlines of displacement increments
a. 12 elem. b. 24 elem. c. 48 elem. d. 96 elem.

$\alpha = 2.0$. Mesh-independence is obvious both from Figure 9 and Figure 10. On using a coarse mesh of only 12 triangular elements, the solution is already fairly accurate, as the results from the 96-element mesh are not so different. Here it should be noted that high-order 15-noded elements are used with full 12-point Gaussian integration. As a first identification a 15-noded triangle can be compared with a group of four 6-noded triangles, as this gives the same number of nodes and Gaussian stress points.

11 CONCLUSIONS

This paper reports on a pilot study on a novel non-local theory. Material softening is described by means of a plastic strain-dependent strain-softening parameter. The difference with a classical theory of hardening and softening is that we introduce an auxiliary strain measure, ε^*, that depends not only on the plastic strain in the material point considered, but also on the plastic strain around this point. The difference with the strain-gradient theory is that the auxiliary strain is a spatial average, involving evaluation of a strain integral rather than a strain-gradient. For this reason, the new non-local theory can be implemented relatively easily in existing finite element codes. The difference with other such non-local theories is as follows.

The proposed definition of the non-local strain-softening parameter, ε^*, ensures (analytical) solutions of localization problems such as necking in tension tests and shear-banding in compression tests. Numerical solutions are essentially mesh-independent as shown in this study. In contrast to classical softening models, equilibrium iterations within a loading step show fast convergence. The latter topic has hardly been mentioned in this paper, but we noticed a remarkable difference. Some shear-banding problems with high degrees of softening could not even be solved without the introduction of a non-local strain-softening parameter because of diverging equilibrium iterations. No such difficulties were found for the present model.

The new non-local theory is subject to further study as several questions remain. On studying the one-dimensional necking problem, we found that the amplification factor, α, needs to be larger than unity, but its precise value is found to be relatively unimportant. In fact, all numerical data are virtually controlled by the imposed width of the localization zone. The question is whether or not this also applies to general three-dimensional problems. Here the value of the factor α might, for instance, influence the orientation of shear-bands. Indeed, for the shear-bands reported here, no variations of α were considered. Moreover, we did not yet consider any frictional material with a non-associated flow rule for the rate of plastic strain. This will be the subject of another paper.

ACKNOWLEDGEMENT

This research was supported by the Technology Foundation (STW)

REFERENCES

Bazant Z.P., Lin F.B., Pijaudier-Cabot G. (1987), Yield limit degradation; Non-local continuum model with local strain. In: Proc. Int. Conf. Computational. Plasticity, Barcelona (eds. Owen, Hinton, Onate). pp. 1757-1780.

Bazant Z.P., Zubelewicz A. (1988), Strain softening bar and beam: Exact nonlocal solution. Int. J. Solids & Structures, Vol. 24, No. 7, pp. 659-673.

Christoffersen J., Mehrabadi M.M., Nemat-Nasser S. (1981), A micromechanical Description of Granular Material Behavior. J. Applied Mechanics, Vol. 48, No. 2, pp. 339-344.

de Borst R., Mühlhaus H.B. (1991), Computational strategies for gradient continuum models with a view to localization of deformation. Proc. 4 Int. Conf. on Nonlinear Eng. Comp. (eds. N. Bicanic, P. Marovic, D.R.J. Owen, V. Jovic, A. Mihanovic). Pineridge Press, Swansea, U.K., pp. 239-260.

de Borst R., Mühlhaus H.B., Pamin J., Sluys L.J. (1992), Computational modelling of localization of deformation. Proc. 3 Int. Conf. Computational Plasticity. (eds. D.R.J. Owen, E. Onate, E. Hinton). Pineridge Press, Swansea, U.K., pp. 483-508.

Desrues J. (1984), La localisation de la déformation dans les matériaux granulaires. Dissertation. Institut National Polytechnique de Grenoble.

Gröbner W., Hofreiter N. (1961), Integraltafel. Erster teil, Unbestimmte integrale. Zweiter teil, Bestimmte integrale. Springer-Verlag, Wien.

Mühlhaus H.B. (1986), Sherfugenanalyse bei granularem Material im Rahmen der Cosserat-Theorie. Ingenieur Archiv, Vol. 56, pp. 389-399.

Mühlhaus H.B., Aifantis E.C. (1991), A variational principle for gradient plasticity. Int. J. Solids & Structures, Vol. 28, pp. 845-857.

Mühlhaus H.B., De Borst R., Aifantis E.C. (1991), Constitutive models and numerical analyses for inelastic materials with microstructure. In: Proc. Seventh Conf. Int. Assoc. Comp. Meth. Adv. Geomech. (eds. G. Beer, J.R. Booker). Balkema, Rotterdam, pp. 337-386.

Mühlhaus H.B., Vardoulakis I. (1987), The thickness of shear bands in granular materials. Géotechnique 37, pp. 271-283.

Pastor M, Peraire J., Zienkiewicz O.C. (1991), Adaptive remeshing for shear band localization problems. Ingenieur Archiv, Vol. 61, pp. 30-39.

Pietruszczak S., Mróz Z. (1981), Finite element analysis of deformation of strain-softening materials. Int. J. Num. Meth. Eng. Vol. 17, pp. 327-334.

Read H.E., Hegemier G.A. (1984), Strain softening of rock, soil and concrete - A review article. Mechanics of Materials Vol. 3, pp. 271-294.

Rudnicki J.W., Rice J.R. (1975), Conditions for the localization of deformation in pressure-sensitive dilatant materials. J. Mech. Phys. Solids, Vol. 23, pp. 371-394.

Sandler I.S., Wright J.P. (1984), Summary of strain-softening. In: Theoretical foundation for large-scale computations of nonlinear material behavior (ed. S. Nemat-Nasser). Proc. of DARPA-NSF workshop, Evanston, Ill, Oct. 1983. Martinus Nijhoff Publishers, The Netherlands, pp. 285-296.

Schreyer H.L., Chen Z. (1986), One-dimensional softening with localization. J. Appl. Mech. 53, pp. 791-797.

Sluys L.J. (1992), Wave propagation, localization and dispersion in softening solids. Dissertation. Delft University of Technology.

Thornton C. (1979), The conditions for failure of a face-centered cubic array of uniform rigid spheres. Géotechnique 29, No. 4, pp. 441-459.

Vardoulakis I. (1989), Shear banding and liquefaction in granular materials on the basis of a Cosserat continuum theory. Ingenieur Archiv, Vol. 59, No. 2, pp. 106-114.

Vardoulakis I., Aifantis E.C. (1991), A gradient flow theory of plasticity for granular materials. Acta Mechanica 87, pp. 197-217.

Vermeer P.A., Van Langen H. (1989), Soil collapse computations with finite elements. Ingenieur-Archiv 59, pp. 221-236

Shear moduli identification versus experimental localisation data

Identification des modules des cisaillement en fonction de résultats expérimentaux

R.Chambon, J.Desrues & D.Tillard
Laboratoire Sols, Solides, Structures (3S), UJF – INPG – CNRS, Grenoble, France

ABSTRACT : In this paper, new aspects of the work about CLoE models are presented. The principles of this model and of a localization analysis for this model were presented in the second workshop held in Gdansk in 1989. Applications for sand were presented in the Workshop "Modern approaches to plasticity" held in Horton last year (Desrues et al. 1993). Here we want to focus on the idenfication of the shear moduli.

On présente quelques nouveaux aspects du développement des modèles de type CLoE. Les principes du modèle lui-même, ainsi que l'analyse de localisation avaient été présentés lors du second Workshop de la présente série à Gdansk en 1989. Des applications pour les sables ont été discutées au Workshop "Modern approach to plasticity" à Horton en 1993. Le présent article concerne l'identification des modules de cisaillement.

1 INTRODUCTION

When devising CLoE model our essential motivations were dictated by the objective of efficiency in finite element code numerical modeling, including problems involving strain localization. This objective implies reasonably accurate modeling of monotonous loading paths, a rigorous approach of ultimate stress-states (consistency), and a non restricted shear band analysis leading to an explicit bifurcation criterion.

Our model has been designed to model monotonic or quasi-monotonic loading paths (not cyclic loading paths). Our goal is to model geotechnical problems, from settlement calculations to bearing capacity problems. When approaching limit loads in geotechnical structures, shear band localization becomes an essential feature of the behaviour ; hence localized strain must be considered in the numerical modeling. One of the possible ways for that is to use a constitutive model able to predict the onset of shear banding on the basis of the bifurcation theory.

It is well known that for soils, loading behavior is rather different than unloading. Moreover micromechanical considerations (numerous potential slipping planes) as well as experimental results at the macroscopic level implie that classical elastoplasticity with only two linear zones is not always sufficient. From the point of view of shear banding, classical elastoplastic models (isotropic hardening) does not give good results as it is the elastic shear modulus which is involved.

One of the aims of Cloe model is to provide a model with some free parameters (which however do not influence the classical loading paths) the value of which being adjusted owing to experiments exhibiting strain localization. There is no physical reason for shear moduli (in a wide sense) to depend on the material behavior along loading paths with fixed axes so the free parameters chosen are : the shear moduli.

In order to reach this goal it is necessary to have a non-linear form for which a complete shear band analysis is tractable. A simple form is needed. As we want also that the model involve incremental non-linearity, a rather simple non-linear form was chosen. The simple mathematical form of CLoE allows both description of irreversibility, and complete and tractable shear band analysis with "free" shear moduli.

First use of such a model in a rather simple version can be seen in Desrues (1984) and Chambon et al. (1985). More information is available in Chambon (1989a, 1989b), Chambon et al. (1990) (1994), Charlier et al. (1991), Desrues et al. (1991) (1993), Hammad (1991).

In the following the basic features of the model : irreversiblity, flow rule (in a specific sense), failure surface, limit surface for CloE model are discussed first. The non-linear shear band analysis is presented in details. Then the assumption done for shear moduli is presented. Finally, one application is done for a shale. This application concerns a shale for a lower range of pressure for which it is over consolidated. In this case the classical CLoE model was extended to cohesive soils in a simple manner.

Let us finally give the principles of our notations: Vectors and tensors are denoted by boldface letters. A component is denoted by the name of the tensor (or vector) accompanied with tensorial indices. All tensorial indices are in lower position as there is no need in the following of a distinction between covariant and contravariant components. Other indices have no tensorial meaning. The summation convention with respect to repeated tensorial indices has been adopted.

2 CLOE MODEL FOR GEOMATERIALS : BASIC CONCEPTS

2.1 *Definitions*

Under rather general assumptions (especially no viscosity), a constitutive equation may be written :
$$\tilde{\sigma} = f(D)$$
where $\tilde{\sigma}$ is an objective stress-rate, D the rate of deformation, and f is a priori a non-linear tensor valued function depending on the state.

A simple form for non-linear tensor valued function f has been chosen. It is assumed that :
$$\tilde{\sigma} = A:D + b\|D\| = A:(D + B\|D\|)$$
or using components :
$$\tilde{\sigma}_{ij} = A_{ijkl}D_{kl} + b_{ij}\|D\| = A_{ijkl}(D_{kl} + B_{kl}\|D\|)$$
where :
$$\|D\|^2 = (D_{kl}D_{kl})$$

A is a fourth rank tensor, b and B are second rank tensors, all depending on the state which in current applications reduces to the stress tensor σ. Let us notice that it is possible to write the Kolymbas' (1987) model in this form.

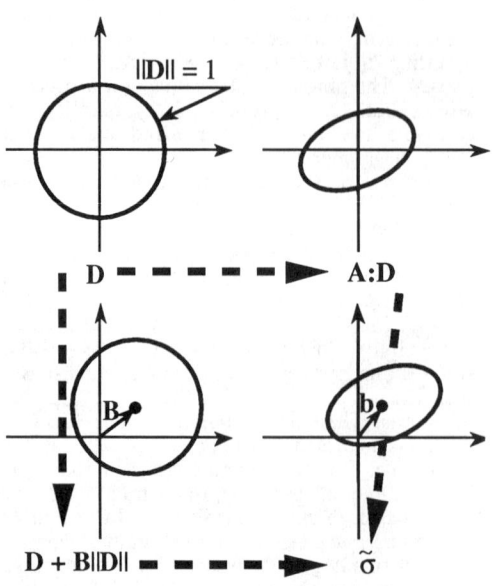

Figure 1. Basic principle of CLoE.

As we will use it extensively in the following, let us recall the principle of Gudehus' (1979) diagrams. This allowed us to get geometrical illustrations and to understand better the calculations done with CLoE model.

Due to the homogeneity of the degree one with respect to rate of deformation of the constitutive equation, the behavior may be represented by the stress increment responses to unit strain increments. Thus in the strain-rate space, the end of a unit strain-rate generates an hypersphere sketched in Figure 1 as a circle. The corresponding points in the stress-rate space lie thus on an hypersurface sketched in the same figure as an ellipse.

In order to transform the hypersphere into this hypersurface following a CLoE model two ways are possible:

1. first multiply D by A, which corresponds geometrically to the transformation of the hypersphere into an hyperellipsoïde, and then add b which corresponds geometrically to a translation of the hyperellipsoïde.

2. first add B to D, which corresponds geometrically to a translation of the hypersphere, and then multiply the result by A which geometrically produces the hyperellipsoïde.

Notice that D or d are linked to the irreversible nature of behavior of geomaterials.

2.2 *Inversibility and Failure Surface*

The problem of inversibility is important. If the model is not inversible, the same boundary-value problem can give two different homogeneous solutions.

The previous constitutive equation has an inverse if : for every stress-rate $\tilde{\sigma}$ it is possible to find D such that $\tilde{\sigma} = A:(D + B\|D\|)$. As the law is positively homogeneous, inversibility is then equivalent to the following : for every stress-rate direction $\tilde{S}$ ($\|\tilde{S}\|=1$), there is a unique $\lambda>0$ and a unique D for which $\|D\| = 1$ such that : $\tilde{\sigma} = \lambda\tilde{S}$ and $\tilde{\sigma} = A:(D + B\|D\|)$. D has to be the solution of : $\lambda\tilde{S} = A:(D + B\|D\|)$. So existence of A^{-1} is a necessary condition to insure inversibility. Assuming that this condition is met yields the following condition : equation

$$\lambda^2\|A^{-1}:\tilde{S}\|^2 - 2\lambda(A^{-1}:\tilde{S}):B + \|B\|^2 - 1 = 0 \quad (2.2.1)$$

must have one and only one positive solution. Hence it is necessary that :
$\|B\|^2 - 1 < 0$ or $\|B\| < 1$ (Figure 2). Consequently for $\|B\| = 1$, the law is no more a one to one correspondance (Figure 3). Hence equation $\|B\| = 1$ characterises the failure surface (we will see in the next section why it is justified to call it failure surface). So here we define the failure surface as the set of states for which $\|B\|$ is equal to one. As the stress is the only state parameter, this failure surface is isotropic.

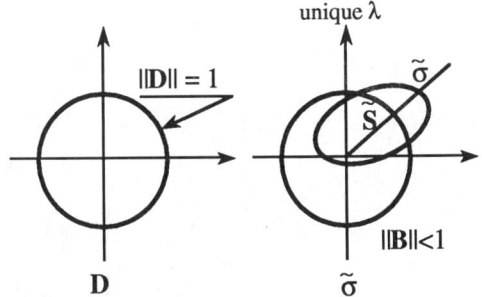

Figure 2. Inversibility : current stress-state.

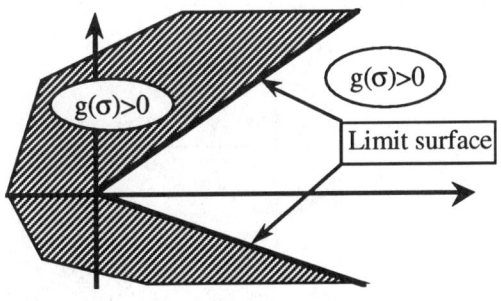

Figure 4. Limit surface.

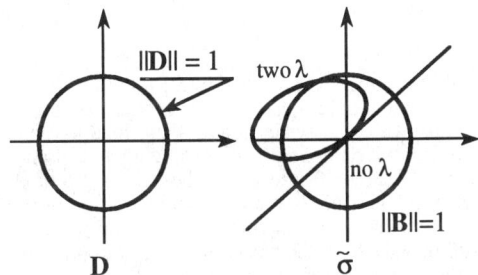

Figure 3. Inversibility : stress lying on the failure surface.

2.3 Flow rule

For stress-states corresponding to the failure surface, $\|B\| = 1$. Then if :

$D_{kl} = -\mu B_{kl}$,

for any $\mu > 0$, the stress-rate $\tilde{\sigma}$ corresponding to **D** is equal to zero. Physically this means that we have a steady stress-state for continuing flow with constant strain-rate. This is the reason for which we called failure surface the surface defined by : $\|B\| = 1$. Moreover for these stress-states -**B** can be interpreted as a flow rule.

2.4 Consistency Condition

A very common notion in soil mechanics is the notion called here limit surface. It is the surface, defined in stress space, splitting this space into two parts : a part of accessible and a part of non accessible stress-states (Figure 4). In the following for Cloe model we do the following assumption : the failure surface is identical with the limit surface.

It is common to assume for soils and rocks that plastic flow occurs when a specimen is submitted to a classical triaxial loading path. This means that with the previous assumption, the limit surface is reached in such loading paths. From an incremental point of view this assumption induces a constraint to be met

by the model : for stress-states lying on the failure surface no stress increment can be directed outside the failure surface. For example the case illustrated in Figure 5b corresponds to a model violating our assumption as some stress-rate can be directed outwards. On the contrary the case illustrated in Figure 5a meets our assumption.

For classical elastoplastic models this condition can be met by the convergence of the tangential plane of the current yield surface to the tangential plane of the failure surface.

For stress-states lying on the failure surface as $\|B\| = 1$, equation (2.2.1) written with components reads :

$\lambda^2 A^{-1}_{ijkl}\tilde{S}_{kl} A^{-1}_{ijmn}\tilde{S}_{mn} - 2\lambda A^{-1}_{ijkl}\tilde{S}_{kl}B_{ij} = 0$

So the possible stress-rates ($\lambda > 0$) must be such that : $A^{-1}_{ijkl}\tilde{S}_{kl}B_{ij} \geq 0$

Let $g(\sigma_{ij}) = 0$ be the failure (or limit) surface equation, $g(\sigma_{ij}) > 0$ defining the inaccessible zone (Figure 4), all the possible stress-rates have to be directed inwards (Figure 4). This condition reads :

$\dfrac{\partial g}{\partial \sigma_{ij}} \tilde{\sigma}_{ij} \leq 0.$

As $\tilde{\sigma} = \lambda \tilde{S}$ it is necessary that :

$A_{ijkl} \dfrac{\partial g}{\partial \sigma_{ij}} = -\beta B_{kl}$

with β being a positive scalar. More details are available in Chambon et al. (1994). Meeting this consistency condition insures a correct approach of the failure surface.

Otherwise if this condition is not met it is likely that for some stress-states lying outside the failure surface but inside the limit surface the constitutive equation is no more inversible. In his model, Kolymbas (1987) does not do the assumption corresponding to the consistency condition.

2.5 Interpolation Procedure

As the state reduces to the stress, if we use the

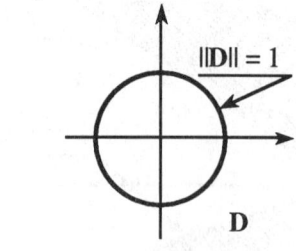

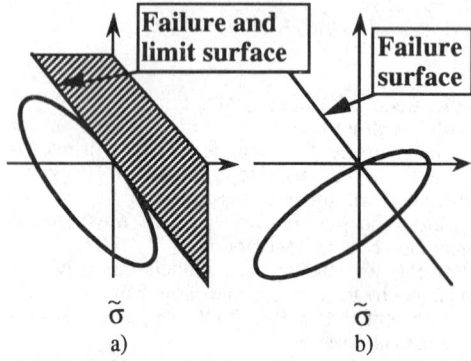

Figure 5. consistency condition.

components of the tensors **A** and **B** with respect to principal stress axes then the following consequences arise from the representation theorem for tensorial functions of tensors (Chambon 1989a, 1989b, Chambon et al. 1990 1994, Charlier et al. 1991, Desrues et al.1991, 1993). We have necessarily :

$$\mathbf{A} = \begin{pmatrix} a & f' & e' & 0 & 0 & 0 \\ f & b & d' & 0 & 0 & 0 \\ e & d & c & 0 & 0 & 0 \\ 0 & 0 & 0 & g & 0 & 0 \\ 0 & 0 & 0 & 0 & h & 0 \\ 0 & 0 & 0 & 0 & 0 & j \end{pmatrix}$$

$$\mathbf{B} = \begin{pmatrix} k & 0 & 0 \\ 0 & 1 & 0 \\ 0 & 0 & m \end{pmatrix}$$

If the stress-state is axisymmetrical ($\sigma_2=\sigma_3$) then :

$$\mathbf{A} = \begin{pmatrix} a & e' & e' & 0 & 0 & 0 \\ e & b & d & 0 & 0 & 0 \\ e & d & b & 0 & 0 & 0 \\ 0 & 0 & 0 & b-d & 0 & 0 \\ 0 & 0 & 0 & 0 & h & 0 \\ 0 & 0 & 0 & 0 & 0 & h \end{pmatrix}$$

$$\mathbf{B} = \begin{pmatrix} k & 0 & 0 \\ 0 & 1 & 0 \\ 0 & 0 & 1 \end{pmatrix}$$

and if the stress-state is isotropic we have :

$$\mathbf{A} = \begin{pmatrix} a & d & d & 0 & 0 & 0 \\ d & a & d & 0 & 0 & 0 \\ d & d & a & 0 & 0 & 0 \\ 0 & 0 & 0 & a-d & 0 & 0 \\ 0 & 0 & 0 & 0 & a-d & 0 \\ 0 & 0 & 0 & 0 & 0 & a-d \end{pmatrix}$$

$$\mathbf{B} = \begin{pmatrix} k & 0 & 0 \\ 0 & k & 0 \\ 0 & 0 & k \end{pmatrix}$$

Given an axisymmetric stress-state (triaxial compression or extension) the knowledge of the response of the material along some loading paths (classical triaxial loading and unloading and "pseudo isotropic" paths ; figure 6) supplies the values of {k, l, a, e, e'} and of (b + d). A simple assumption is done to deduce the behavior along "pseudo isotropic" loading paths from the behavior along isotropic path. Another simple assumption is done for (b - d) and thus all the components of tensors **A** and **B** except h are known for every axisymmetric stress-state.

In the present state of CLoE we assume that all components of tensors **A** and **B** are homogeneous of the degree one with respect to the mean stress p. Finally it is only necessary to know the behavior of the soil along one triaxial stress path and along the isotropic loading path. The details about this part are given in Desrues et al. (1993) and Chambon et al. (1994).

Let us do two comments. First the assumption of the homogeneity of the degree one with respect to the mean stress has implications which may be unrealistic if we use the model for rather different pressure. In the present state CLoe is usable only in the vicinity of a given mean pressure. This is illustrated in the applications discussed in § 4. Second the shear moduli {g, h, j} are free, as mentionned in our motivations, except for stress-states exhibiting some symmetry (isotropic and axisymmetric states).

Finally an interpolation (figure 7) is done between compression and extension. Given a stress-state, a surface homothetic to the failure surface, is used to define two image states in the deviatoric plane : one in axisymmetric triaxial compression and the other in axisymmetric triaxial extension. The interpolation uses the Lode angle and a scalar quantity $\bar{q}$ called deviator ratio. $\bar{q}$ is the ratio of the current deviator over the deviator at failure corresponding to the same Lode angle.

This interpolation takes into account the consistency condition on the failure surface and leaves free the shear moduli {g, h, j}. Let us notice

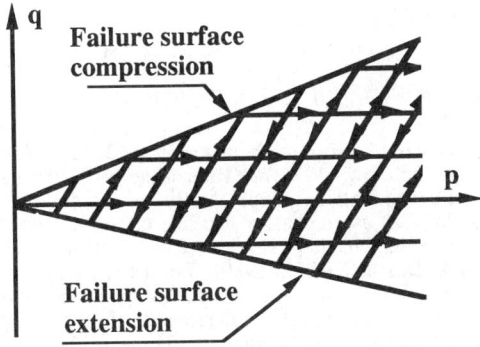

Figure 6. Principle of interpolation in the p q plane.

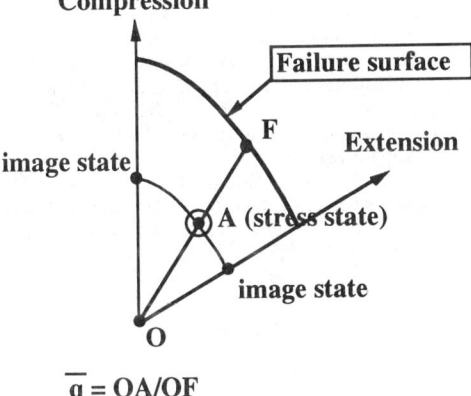

$\overline{q} = OA/OF$

Figure 7. Principle of interpolation in the deviatoric plane.

that {g, h, j} are not analogous to elastic shear moduli, as the term b‖D‖ of the CLoE model implies a coupling between shear and isotropic part of the model. The details of the interpolation can be seen in Chambon et al. (1993) and (1994).

For the following applications to cohesive soils we simply add to the actual pressure a constant value, necessary to obtain what we call equivalent pressure, like in the so called correspondance-state theorem (which states that the limit equilibrium conditions for problem involving a cohesive material with internal friction can be calculated using a non-cohesive equivalent material with the same friction angle and adding an isotropic stress equivalent to the cohesion. Entering experimental triaxial data, with some interpolation, it is possible to gain a good model for general true triaxial loading paths. However this interpolation does not give for every stress-state the shear moduli. These moduli are given through the following shear band analysis.

3 SHEAR BAND ANALYSIS

Firstly, it is necessary to emphasize that only shear bands are studied here. From a theoretical point of view this is clearly a strong restriction : we perform a mode-assumed bifurcation analysis, and not a general bifurcation study. On the other hand, failure in soil masses is in most cases characterized by intense strain localization, so the shear band mode is especially important.

A first attempt of a complete study of uniqueness of boundary value problems with Cloe model can be seen in Caillerie et al. (1994).

We consider the problem of an initially homogeneous solid strained up to the current state. It is then submitted to a load rate on the straight loading path. It is useful to recall first the basic equations used in a shear band analysis in order to well understand the following application for CLoE model. In the sequel we follow the Rice' (1976) presentation of the basic equations.

3.1 Basic Equations for Shear Band Analysis

1. Kinematical Condition

A solution of the resulting rate equilibrium problem corresponds to a homogeneous velocity gradient denoted as $\dot{F}^0$. Another solution involving a shear band, is considered. It is assumed that the velocity gradient is equal to $\dot{F}^0$ outside the shear band and equal to :

$$\dot{F}^1 = \dot{F}^0 + \vec{g} \otimes \vec{n}$$

inside . $\vec{n}$ is the normal vector to the band and $\vec{g}$ is some vector as we can see in Figure 8. Using components this yields :

$$\dot{F}^1_{ij} = \dot{F}^0_{ij} + \vec{g}_i \vec{n}_j$$

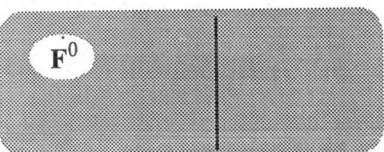

Homogeneous solution

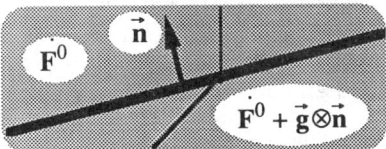

Solution involving a shear band

Figure 8. Shear Band Analysis.

2. Statical Condition

Let $\dot{\sigma}^1$ be the Cauchy stress-rate with respect to a fixed frame inside the shear band and $\dot{\sigma}^0$ outside. Along the boundaries of the band, equilibrium equations in a rate form can be written :

$$\dot{\sigma}^1.\vec{n} = \dot{\sigma}^0.\vec{n}$$

although $\dot{\sigma}^1$ and $\dot{\sigma}^0$ are not objective see e.g. Rudnicki et al. (1975) or Rice et al. (1980)). Using components we have :

$$\dot{\sigma}^1_{ij}\vec{n}_j = \dot{\sigma}^0_{ij}\vec{n}_j$$

It is necessary to notice here that from the point of view of a boundary value problem, all the requirements are met except the boundary conditions corresponding to the external boundary of the shear band (see Chambon et al. (1984) Chambon (1986)).

3. Bifurcation Condition

As seen in § 2.1., under rather general assumptions (especially no viscosity), the constitutive equations may be written :

$$\tilde{\sigma} = h(\dot{F})$$

where $\tilde{\sigma}$ is an objective stress-rate, $\dot{F}$ the velocity gradient and h a non-linear tensor-valued function depending on the state. The constitutive equations can be written alternatively :

$$\dot{\sigma} = l(\dot{F})$$

or using components :

$$\dot{\sigma}_{ij} = l_{ij}(\dot{F})$$

where $\dot{\sigma}$ is the material stress-rate with respect to a fixed frame and l another tensor valued function. Hence using the kinematical condition, the statical condition and the constitutive equations, shear band bifurcation is possible if there exist some unit normal vector $\vec{n}$ and a vector $\vec{g} \neq \vec{0}$ such that :

$$\{l_{ij}(\dot{F}^0 + \vec{g} \otimes \vec{n}) - l_{ij}(\dot{F}^0)\}\vec{n}_j = 0 \qquad (3.1.3.1)$$

Let us notice that the previous relation does not distinguish between the inside and the outside of the shear band as it can be written alternatively :

$$\{l_{ij}(\dot{F}^1 + \vec{g}' \otimes \vec{n}) - l_{ij}(\dot{F}^1)\}\vec{n}_j = 0$$

with $\vec{g}' = -\vec{g}$.

4. The Linear Case

If f is a linear function, then l is linear and we can define the fourth rank tensor L such that :

$$\dot{\sigma}_{ij} = L_{ijkl}\dot{F}_{kl}$$

replacing $\dot{\sigma}_{ij}$ in the statical condition and then $\dot{F}_{kl}$ in terms of F^0 and $\vec{g}$ we get th condition :

$$L_{ijkl}\vec{g}_k\vec{n}_l\vec{n}_j = 0$$

As this equation has to be satisfied for $\vec{g} \neq \vec{0}$ thus :

$$\det(L_{ijkl}\vec{n}_i\vec{n}_j) = 0 \qquad (3.1.4.1)$$

often called bifurcation criterion or localization condition (Rice (1976)) or Rice criterion. Tha same result was obtain independantly by Vardoulakis (1976). Let us emphasize that this criterion is only valid for incrementally linear models (Chambon et al. 1984, Desrues et al. 1989). For (incrementally) bilinear models such as classical elastoplastic models (isotropic hardening), extensions of the previous result can be used either directly (Rice et al. 1980) or using the linear comparison solid (Hill 1978 for instance) or extensions of this notion (Raniecki et al. 1981). For (incrementally) multilinear solids it is possible to use a generalization of the Rice criterion (Chambon 1986).

There is no general method to solve this problem with incrementally non-linear models. To our knowledge the first rigorous result (without any additional assumption) was given by Desrues (1984) and Chambon et al. (1985). As said in the introduction, the quoted analysis used a rather simple preliminary version of CLoE models for incompressible solids. So the following analysis (which can be seen in Chambon 1989b, and the result of which was published in Chambon et al. (1990)) is a generalization of the previous result.

3.2 *Shear Band Analysis for Cloe Model*

1. Analysis

As the difference between $\tilde{\sigma}$ and $\dot{\sigma}$ is linear with respect to $\dot{F}$, the constitutive equations can be written using $\dot{\sigma}$ in the following form :

$$\dot{\sigma} = M:D + b\|D\| \text{ or using components :}$$
$$\dot{\sigma}_{ij} = M_{ijkl}\dot{F}_{kl} + b_{ij}\|D\|$$

where M is another fourth rank tensor depending on the state. Knowing A it is rather easy to calculate M as :

$$M_{ijkl} = A_{ijkl} + 1/2\,(\sigma_{il}\,\delta_{jk} - \sigma_{ik}\,\delta_{jl} + \sigma_{jl}\,\delta_{ik} - \sigma_{jk}\,\delta_{il})$$

where δ is the identity tensor.
As the kinematical and statical conditions, and the constitutive equations are positively homogeneous of the first degree, $\|D^0\| = 1$ can be assumed (let us notice that 1 is the inverse of a time). Let r be defined by :

$$\|D^1\| = (1+r)\|D^0\| = 1+r$$

Hence the bifurcation condition becomes : there is a unit normal vector $\vec{n}$ and a vector $\vec{g} \neq \vec{0}$ such that :

$$M_{ijkl}\vec{g}_k\vec{n}_l\vec{n}_j + r\,b_{ij}\vec{n}_j = 0$$
$$\text{and} \qquad (3.2.1.1)$$
$$\|1/2(\dot{F}^0_{ij} + \vec{g}_i\vec{n}_j + \dot{F}^0_{ji} + \vec{g}_j\vec{n}_i)\| = 1+r$$

Given $\vec{n}$ the previous condition is a non-linear system of four simultaneous equations the unknown of which are the three components of $\vec{g}$ and r (the non-linearity is due to the fourth equation). Let us denote **P** a second rank tensor such that : $P_{ij} = M_{ikjl}\, \vec{n}_k \vec{n}_l$. **P** can be seen as the acoustic tensor associated with **M** and $\vec{n}$.

a) Assume that **P** has an inverse denoted P^{-1}. Hence : $\vec{g}_i = - r\, P^{-1}_{ij}\, b_{jk}\vec{n}_k$
and the non-linear equation reads :

$$\|1/2(F^0_{ij} - r\, P^{-1}_{il}\, b_{lk}\vec{n}_k\vec{n}_j + F^0_{ji} - r\, P^{-1}_{jl}\, b_{lk}\vec{n}_k\vec{n}_i)\| = 1+r$$

In other words, bifurcation is possible if there exists some $\vec{n}$ for which the following equation in the single unknown r has a solution different from 0 :

$$\|D^0_{ij} - r/2\, P^{-1}_{il}\, b_{lk}\vec{n}_k\vec{n}_j - r/2\, P^{-1}_{jl}\, b_{lk}\vec{n}_k\vec{n}_i\| = 1+r$$

The previous equation raised to the power two, gives a new form of the bifurcation criterion :
$(N^2 - 1)\, r^2 + 2(S - 1)\, r = 0$ and $r \geq -1$
with :
$N^2 = \|1/2(P^{-1}_{il}\, b_{lk}\vec{n}_k\vec{n}_j + P^{-1}_{jl}\, b_{lk}\vec{n}_k\vec{n}_i)\|$

and :
$S = -D^0_{ij}\, P^{-1}_{il}\, b_{lk}\vec{n}_k\vec{n}_j$
The only interesting ($r \neq 0$) solution for r is :
$r = 2(1-S)/(N^2-1)$.
Generally, at the beginning of a loading process, shear banding does not occur. The value of **b** and consequently the values of N^2 and S are close to zero and thus the value of $2(1-S)/(N^2-1)$ is close to -2. It is not greater than -1. As it is reasonable to assume that **b** and P^{-1} vary continuously with the state variables and consequently with the loading parameter, localization occurs when :
$2(1-S)/(N^2-1) = -1$ or $+\infty$.
This means either $N^2 - 2S + 1 = 0$, or $N^2 = 1$.
The first solution means $\|D^1\| = 0$ and consequently $D^1 = 0$ and as $r = -1$:
$D^0_{ij} = -1/2\, P^{-1}_{il}\, b_{lk}\vec{n}_k\vec{n}_j - 1/2\, P^{-1}_{jl}\, b_{lk}\vec{n}_k\vec{n}_i$
So $S = \|D^0\| = 1$. Finally the first solution reads : $N^2 = 1$ and is then the same as the second one. This proves that the two solutions lead to the same mathematical conditions and thus correspond to the same physical phenomena. First there is a symmetry between the inside and the outside of the shear band (cf section 3.1.3) in the equations involved in a shear band analysis. Secondly in the two solutions the ratio of the strain-rates norms inside and outside is infinite.
Finally if **P** is invertible, the bifurcation criterion corresponding to shear bands for the studied constitutive equations is :

$$\|1/2(P^{-1}_{il}\, b_{lk}\vec{n}_k\vec{n}_j + P^{-1}_{jl}\, b_{lk}\vec{n}_k\vec{n}_i)\| = 1 \quad (3.2.1.2)$$

b) **P** has no inverse. Some vector $\vec{G} \neq \vec{0}$ (the components of which are denoted $\vec{G}_i$) exists such that : $M_{ijkl}\vec{n}_k\vec{n}_l\vec{G}_j = 0$. Hence if $\vec{g}_j = \lambda\, \vec{G}_j$, the components of $\vec{g}$ and $r = 0$ are solution of system (3.2.1.1.) --and so bifurcation is possible-- if :
$\|D^0_{ij} + 1/2(\vec{G}_i\vec{n}_j + \vec{G}_j\vec{n}_i)\| = 1$.
This equation has always a solution $\lambda \neq 0$ (except if $D^0_{ij}\vec{G}_i\vec{n}_j = 0$. In this case, the other solution is also $\lambda = 0$).
So for this second case the bifurcation criterion reads :

$$\det(M_{ikjl}\, \vec{n}_k\vec{n}_l) = 0 \quad (3.2.1.3)$$

Let us sum up the previous results. The bifurcation condition is an equation for which the constitutive equations (i.e. **A** and **b**) are known, and the unknown is $\vec{n}$. If one of the equations : (3.2.1.2) and (3.2.1.3) have a solution, localization is possible. Along a loading process it is then possible to detect the lower value of the loading parameter for which at least one of the two criteria is met.

2. Discussion

Finally we found two conditions. The second one corresponds to the classical Rice condition for an acoustic tensor. However this condition is not written with a linear incremental stiffness tensor but with the tensor **A** of CLoE model which has not the same physical meaning. Our numerical experiments about CloE model applied to various soils and rocks show that this criterion is never met before the first condition.

As previously seen the first condition (equation 3.2.1.2.), the one which is in fact the first met implies that the ratio of the strain-rates norms inside and outside the band is infinite. Such solutions are also basic for incremental multilinear models (Chambon 1986). This is a result of the analysis and it gives us a result which was postulated by Kolymbas (1981) in his bifurcation analysis for a more complicated incrementally non-linear model. Physical interpretation of this result is the following : for our model, when localisation occurs, the whole strain of the specimen is concentrated in a part which can be seen as the inside of a shear band. In other words the outside part is like a rigid body. This result has also a theoretical consequence. As the onset of localisation corresponds to r infinite D^0 is negligible with respect to D^1. So $\dot{\sigma}^0$ is negligible with respect to $\dot{\sigma}^1$, and the statical condition becomes :
$\dot{\sigma}^1_{ij}\vec{n}_j = 0$
finally
$\dot{\sigma}^1_{ij}\vec{g}_j\vec{n}_j = 0 = 1/2(\vec{g}_i\vec{n}_j + \vec{g}_j\vec{n}_i)\,\dot{\sigma}^1_{ij} = 0$

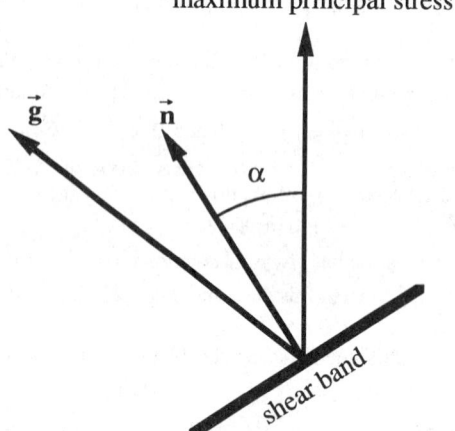

Figure 9. Definition of α.

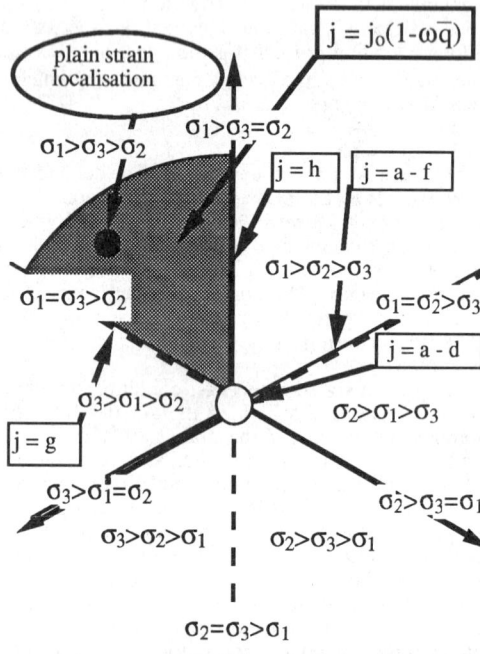

Deviatoric plane

Figure 10. Shear Moduli.

This means that under the small strain assumption (which means that : $\dot{\sigma} = \tilde{\sigma}$) localization implies loss of positive definiteness of the rate of second order work (Drucker 1950, Hill 1958).

4 APPLICATION

4.1 *Use of Localisation Criteria*

If it is assumed that $\vec{n}$ is in the plane of the two extreme principal stresses, this vector is characterised by a, the angle between the maximum principal stress direction and $\vec{n}$ (cf Figure 9). Let us define t by :

t = tg α

Equation (3.2.1.3) has a classical form (acoustic tensor). Equation (3.2.1.2) is a polynom of the fourth degree with respect to t^2, the coefficients of which depends on the components of **A** and **b**. The coefficients of the polynom were caculated by mean of a symbolic calculation package. Given these components it is easy to do a numerical calculation of the two localization criteria during any use of the model (i.e. in finite element codes or in a program of simulation of homogeneous loading paths).

4.2 *Formulation about Shear Moduli*

If it is assumed that $\vec{n}$ is in the plane of the two extreme principal stresses, it is easy to use the previous criteria. In section 2.5 we have seen that the knowledge of the behaviour along true triaxial loading paths is not sufficient to specify the shear moduli {g, h} and, except if the stress-state enjoies some symmetry. On the contrary {g, h, j} are involved in the localisation criteria, which therefore can be used for their identification as proposed in a different constitutive framework by Vardoulakis (1980).

More precisely only one of these moduli is involved. For example only j influences localisation if σ_2 is the intermediate stress.

We have summarized the state of affairs in figure 10 drawn in the deviatoric plane, for j. In the following we assume that :

$$j = j_0 (1 - \omega \bar{q}) \qquad (4.2.1)$$

where j_0 is the value of j for the isotropic stress-state corresponding to the current mean stress, $\bar{q}$ is defined in section 2.5 and ω the parameter to be identified.

Such an assumption gives us a model usable for plane calculation. For a true tridimensionnal use we have to do not only a permutation between g, h and j but we need also an assumption for j when σ_2 is not the intermediate stress.

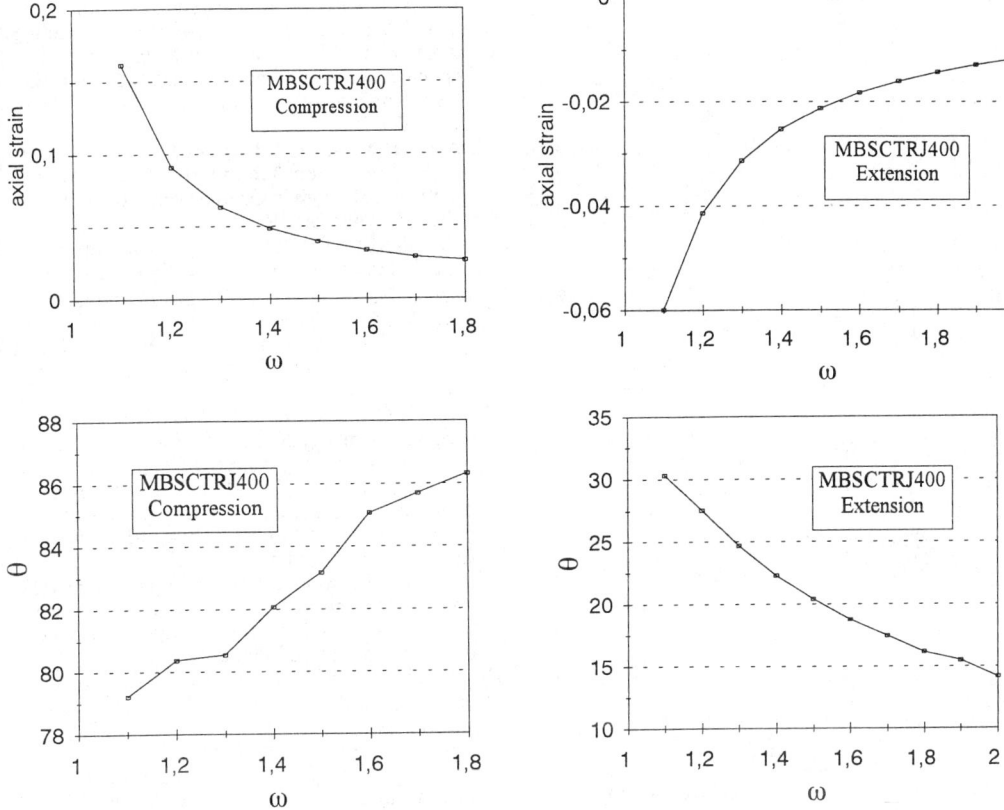

Figure 11. Beaucaire Marl : Localization for axisymetric compression loading paths.

Figure 12. Beaucaire Marl : Localization for axisymetric extension loading paths.

4.3 Application to Beaucaire Marl

Using our identification code "CLoE Id", we got the parameters of the Beaucaire Marl except ω (Tillard 1992). We perform then the following numerical tests : for classical (axisymmetric) compression loading (Fig 11), for classical (axisymmetrical) extension loading (Fig 12) and finally for plane strain loading (Fig 13), we simulate loading programs and we stop the calculation when the localization criterion is met. We do this for different values of ω.

Let us notice that equation (4.2.1) can give negative j value for large values of ω which means in the vicinity of the limit surface. However we tested the positiveness of the value of j up to the final state of the calculations. This means that up to localisation j is positive. For ω greater than 2 this was not possible. This is the reason why ω is limited to 2.

Qualitatively it may be seen that CLoE gives good results. It is more difficult to localize for axisymmetric states than for plane strain states. For the latter, localisation is possible whatever the value of ω.

Tillard (1992) performed plane strain experiments with this marl. The results give values of θ between 62° and 72°, and axial strain between 2% and 3%. Thus it is possible to assume that the value of ω is between 1. and 1.1.

These first results are quite encouraging for the use of CLoE model in soil and rocks mechanics.

ACKNOWLEDGMENT

Part of the present research was developped within the ALERT network (CHM Project N° CHRX-CT93-0217)

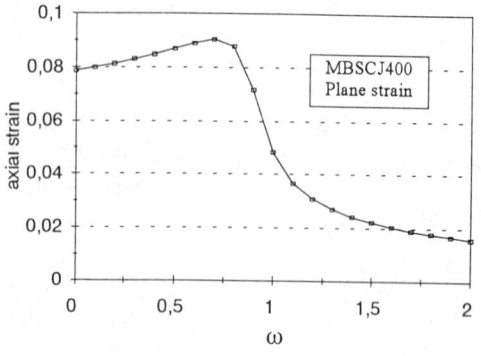

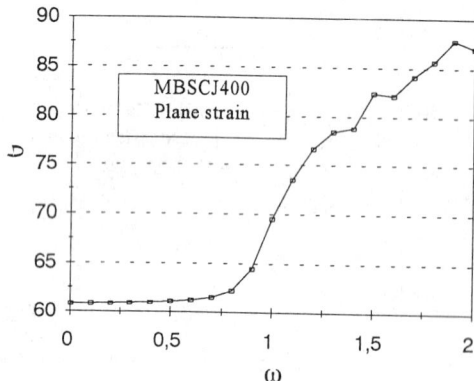

Figure 13. Beaucaire Marl : Localization for plane strain loading paths.

REFERENCES

Caillerie D. Chambon R. 1994 "Existence and uniqueness for boundary problem involving CLoE model" this volume.

Chambon R. Desrues J. 1984 "Quelques remarques sur le problème de la localisation en bande de cisaillement" Mech. Res. Com. 11 pp. 145-153

Chambon R. Desrues J. 1985 "Bifurcation par localisation et non linéarité incrémentale : un exemple heuristique d'analyse complete" in Plasticity Instability E.N.P.C. Paris pp. 101-119

Chambon R. 1986 "Bifurcation en bande de cisaillement, une approche avec des lois incrémentalement non linéaires" J. Méc Théor. Appl Vol 5 n°2 pp. 277-298

Chambon R. 1989a "Une classe de loi de comportement incrémentalement non linéaire pour les sols non visqueux résolution de quelques problèmes de cohérence" C. R. Acad Sci. Paris 308 pp. 1571-1576

Chambon R. 1989b "Base théorique d'une loi de comportement incrémentale consistante pour les sols Rapport Interne I.M.G. Grenoble

Chambon R. Charlier R. Desrues J. Hammad W. 1990 "A rate type constitutive law including explicit localisation : development and implementation in a finite element code" 2nd Eur. Spec. Conf. on Num. Meth. in Geotech. Engn. Santander

Chambon R. Charlier R. Desrues J. & Hammad W. 1994 " International Journal for Numerical and Analytical Methods in Geomechanics in print

Charlier R. Chambon R. Desrues J. Hammad W. 1991 "Shear bifurcation in soil modelling; a rate type constitutive law for explicit localisation analysis" 3rd Int. Conf. on Constitutive laws for Engn Materials Tucson

Desrues 1984 ""La localisation de la déformation dans les milieux granulaires" Thèse d'état Grenoble

Desrues J. Chambon R. 1989 "Shear band anlysis for granular materials : the question of incremental non-linearity Ingenieur Archiv 59 pp. 187-196

Desrues J. et al. 1991 "Soil Modelling with regard to consistency : Cloe, a new rate Type constitutive model" 3rd Int. Conf. on Constitutive laws for Engn Materials Tucson

Desrues J. Chambon R. 1993 "A new rate type constitutive model for geomaterials : Cloe" Modern Approaches to Plasticity D. Kolymbas ed. Elsevier Science Publisher B.V.

Drucker D. C. 1950 "Some implication of work hardening and ideal plasticity" Q. Appl. Math. 7 pp. 411-418

Gudehus G. 1979 "A comparison of some constitutive laws for soil under radially symmetric loading and unloading" Numerical methods in geomechanics W.Wittke ed. A.A. Balkema Rotterdam

Hammad W. 1991 "Localisation en bande de cisaillement dans les sables" Thèse de doctorat Grenoble

Hill R. 1958 "A general theory of uniqueness and stability in elastic-plastic solids J. Mech. Phys. Solids 6 pp 236-249

Hill R. 1978 "Aspect of invariance in solids" Advances in Applied Mechanics, Academic Press, New York, Vol 18 pp. 1-75

Kolymbas D. "Bifurcation anlysi for sand sample with non-linear constitutive equation" Ingenieur-Archiv 50 pp 131-1400

Kolymbas D. 1987 "A novel constitutive law for soils" Constitutive Law for Engineering Materials pp 319-326 C.S. Desai ed. Elsevier New York

Raniecki B. Bruhns O.T. 1981 "Bounds to bifurcation stresses in solids with non associated plastic flow rule at finite strain" J. Mech. Phys. Solids vol. 29 pp. 153-172

Rice J. 1976 "The localization of plastic deformation" International Congress of Theoritical and Applied Mechanics W.D. Koiter ed. North Holland Publishing Comp.

Rice J. Rudnicki J. 1980 "A note of some features of the theory of localization of deformation" Int. J. Solids Struct. vol 16 pp. 597-605

Rudnicki J. Rice J. 1975 "Conditions for the localization of deformation in pressure sensitive dilatant materials" J. Mech. Phys. Solids vol. 23 pp. 371-394

Tillard D. 1992 "Etude de la rupture dans les géomatériaux cophésifs. Application à la marne de Beaucaire." These Grenoble

Vardoulakis I. 1976 "Equlibrium theory of shear bands in plastic bodies" Mech. Res. Comm. vol 3 pp 209-214

Vardoulakis I. 1980 "Shear band inclination and shear modulus of sand in biaxial tests" International Journal for Numerical and Analytical Methods in Geomechanics vol 4 pp 103-119

Beyond invertibility surface in granular materials
Au-delà de la rupture dans les milieux granulaires

Wei Wu
Lahmeyer International Ltd, Frankfurt a. M., Germany

Andrzej Niemunis
Institute of Soil Mechanics and Rock Mechanics, Karlsruhe University, Germany

ABSTRACT: Usual constitutive equations for granular materials express stress rate as a function of strain rates. The inverse function exists only below a so-called yield (= invertibility or failure) surface. The notion 'failure' is avoided here in order to reserve it for boundary value problems only. Recent investigations on hypoplastic equations for granular materials show that the yield surface can be surpassed by some stress paths. This is contradictory to the conventional yield surface, since the stress states are allowed to move along but never across it. The accessible stress states, and the stability surface are also investigated. Theoretical findings about the accessible stress states are verified qualitatively by presenting results of triaxial tests on sand specimens.

Des résultats théoriques récents concernant les lois de comportement hypoplastiques pour les milieux granulaires montrent que la surface de rupture (failure surface), caractérisée par un taux de contrainte nul, peut être franchie par certains chemins de contrainte après la rupture. Ceci est en contradiction avec le concept conventionnel de rupture, car ordinairement les états de contrainte sont supposés pouvoir se déplacer sur la surface, mais pas la franchir. Dans cet article, les relations et les différences entre diverses lois de comportement sont discutées, du point de vue de la rupture. On explore les états accessibles, et la surface de stabilité. Les résultats théoriques obtenus concernant les états accessibles sont vérifiés qualitativement par comparaison avec des essais triaxiaux sur des échantillons de sable.

1 INTRODUCTION

We are concerned here with the following Ansatz for hypoplastic constitutive equations (Wu and Kolymbas 1990):

$$\overset{\circ}{\mathbf{T}} = \mathbf{L}(\mathbf{T}) : \mathbf{D} + \mathbf{N}(\mathbf{T})\|\mathbf{D}\|, \quad (1)$$

where $\mathbf{T}$ stands for the Cauchy stress, $\overset{\circ}{\mathbf{T}}$ the Jaumann stress rate and $\mathbf{D}$ the stretching. $\mathbf{L}(\mathbf{T}) : \mathbf{D}$ and $\mathbf{N}(\mathbf{T})$ in (1) are isotropic tensor valued functions. $\mathbf{L}$ is a fourth order tensor and the colon : denotes an inner product between two tensors, $\| \cdot \|$ stands for a norm and is definded by $\|\mathbf{D}\| = \sqrt{\mathbf{D} : \mathbf{D}}$. The following notation will be used: bold lower and upper case letters denote vectors and tensors. Granted that the behaviour to be described is rate independent constitutive equation (1) is necessarily positively homogeneous of the first degree in $\mathbf{D}$. We refer to Wu and Kolymbas (1990), Kolymbas (1991), Wu (1992) and Wu and Niemunis (1993) for details on the framework of the hypoplastic constitutive model. Though the analyses to be presented are not restricted to a specific version of the hypoplastic model, we will take the following constitutive equation as a heuristic example (Wu 1992)

$$\overset{\circ}{\mathbf{T}} = C_1(\mathrm{tr}\mathbf{T})\mathbf{D} + C_2\frac{(\mathbf{T}:\mathbf{D})\mathbf{T}}{\mathrm{tr}\mathbf{T}} + \left(C_3\frac{\mathbf{T}^2}{\mathrm{tr}\mathbf{T}} + C_4\frac{\mathbf{T}_d^{\,2}}{\mathrm{tr}\mathbf{T}}\right)\|\mathbf{D}\|, \quad (2)$$

where C_i ($i = 1, \cdots, 4$) are dimensionless constants. The deviatoric stress tensor in (2) is given by $\mathbf{T}_d = \mathbf{T} - 1/3(\mathrm{tr}\mathbf{T})\mathbf{1}$. The following constants have been found for dense Karlsruhe medium sand with an initial void ratio of about 0.5: $C_1 = -106.5$, $C_2 = -801.5$, $C_3 = -797.1$, $C_4 = 1077.7$ (Wu and Bauer 1994).

The hypoplastic constitutive equations have been employed successfully to describe various aspects of the behaviour of granular materials. In particular, remarkable progress was achieved by Wu and Bauer (1993) by integrating the critical state into (1). The constitutive model by Wu and Bauer covers a broad range of stress level and density and accounts for both initial deformation and fully developed flow. The present paper continues our previous investigation on hypoplasticity.

Our presentation opens in Section 2 with the classification of constitutive models. The significance of the classification will become apparent in the discussion about yield and invertibility in Section 3. The finding that the yield surface can be surpassed for specific stress paths gives rise to the investigation on accessible stress states and their bound in Section 4. Section 4 is accomplished with the experimental verification of the fact that the inveriibility surface can be surpassed. Finally, the hypoplastic model is investigated with respect to the second order work density in Section 5.

2 CLASSIFICATION OF CONSTITUTIVE MODELS

In order to show the position of hypoplasticity in the midst of constitutive models, the following classification and terminology are proposed along the line of work by Hill (1959). Consider an element under uniform stress and subjected to a closed strain circuit. We proceed to classify the constitutive models according to the recovery of stress at the end of the strain circuit and whether the constitutive equation admits a potential.

1. Hyperelasticity: A constitutive model is said to be hyperelastic if the stress is recovered on all strain circuits. The term hyperelasticity was due to Truesdell (1955). The stress rate and strain rate are related through an elastic potential. Owing to the existence of the elastic potential there exists a unique relation between stress and strain. The behaviour described by a hyperelastic constitutive equation is path independent. Note that the stress strain relation need not be linear.

2. Hypoelasticity: The stress rate and strain rate are related through a tensorial function

$$\overset{\circ}{\mathbf{T}} = \mathsf{L}(\mathbf{T}) : \mathbf{D}, \qquad (3)$$

which need not be derivable from a potential. Hypoelasticity was introduced by Truesdell (1955). The behaviour described by a hypoelastic constitutive equation is in general path dependent. The stress is recoverable only on special strain circuits, e.g. with zero enclosed area. This can be easily seen from the fact that $\overset{\circ}{\mathbf{T}}$ and $-\overset{\circ}{\mathbf{T}}$ are obtained by setting $\mathbf{D}$ and $-\mathbf{D}$ into (3). Hypoelastic constitutive equations are incrementally linear. Applications of hypoelastic constitutive models to granular materials has been made among others by Stutz (1973), by Romano (1974) and by Davis and Mullenger (1978).

3. Hyperplasticity: By hyperplasticity we mean elastoplastic constitutive models that admit plastic potentials. Usually, two or more equations, representing different branches of the stress-strain behaviour for loading and unloading, are employed in hyperplastic models. Switch functions are needed to distinguish between the different branches. The stress at the end of the strain circuit is recovered for the one branch and is not recovered for the other. Hyperplastic constitutive equations are sometimes called bilinear or multi-linear models in the sense that the relation between the stress rate and strain rate is linear within distinct sectors in the stress or strain space.

4. Hypoplasticity: The stress is not recovered on any strain circuit. In particular, no plastic potential is used in the hypoplastic model and the relation between the stress rate and strain rate is incrementally nonlinear in the sense that no matter how small the strain rate may be the behaviour is always path dependent.

3 YIELD AND INVERTIBILITY CONDITION

In the present paper we will restrict our attention to the homogeneous case with uniform stress and strain. The subject of concern is the plastic flow predicted by constitutive models rather than the plastic flow observed in experiments. A material element is said to yield if it cannot sustain further increase of stress. In other words if at a given stress $\mathbf{T}$ further straining $\mathbf{D} \neq \mathbf{0}$ results in no variation in stress

$$\overset{\circ}{\mathbf{T}} = \mathbf{0}. \qquad (4)$$

The stretching $\mathbf{D}$ satisfying (4) characterizes the flow rule. The phenomena of yield is described by the pair $(\mathbf{T}, \mathbf{D})$. On the yield surface there exists at least one $\mathbf{D}$ such that the stress rate vanishes. Before proceeding further it is useful to introduce the so-called response envelope (Gudehus 1979). The response envelope has been proved to be an efficient instrument to study the behaviour of the constitutive equations. The

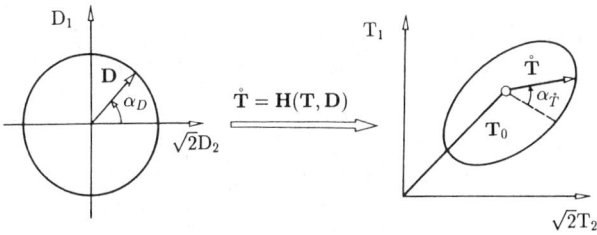

Figure 1: Response envelope on the $T_1 - T_2$ plane.

response envelope for a given stress state $\mathbf{T}_0$, called *initial stress*, is the surface spanned by all stress rates $\mathring{\mathbf{T}}$ which result from the corresponding strain rate $\mathbf{D}$ with the same magnitude $\|\mathbf{D}\|$. The response envelopes can be presented either in the stress rate space or in the stress space by adding the stress increment to the initial stress $\mathbf{T} = \mathbf{T}_0 + \mathring{\mathbf{T}}\Delta t$. To visualise the results graphically, we consider the triaxial stress state. In this case, the response envelopes can be represented on the planes $T_1 - T_2$ and $D_1 - D_2$ with the indices 1 and 2 for the principal stresses, see Figure 1. The following properties of the response envelope are useful in the following discussion. For details we refer to Gudehus (1979) and Wu and Kolymbas (1990).

- The distance from a point on the response envelope to the point of the initial stress characterizes the tangential stiffness for a given direction of stretching (directional stiffness). As a consequence, the response envelope passes through the initial stress at the yield surface.

- The response envelope of the hyper- and hypoelastic constitutive equations is an ellipse with the initial stress in the centre of the ellipse.

- The response envelope of the hypoplastic constitutive equation (1) is an ellipse shifted with reference to the initial stress.

In passing, the response envelopes need not be convex since such a restriction follows neither from the requirement of uniqueness nor from the consideration of any work inequality. The only requirement on the shape of the response envelopes is that the region enclosed by the response envelope should be simply connected (Royis 1989, Wu and Kolymbas 1990). It remains to point out that the above properties of the response envelopes can be easily generalized to the multi-dimensional case. In this case the response envelopes of the hyper- and hypoelastic constitutive equations are ellipsoids in lieu of ellipses.

3.1 Hypoelasticity

The hypoelastic flow was first investigated by Truesdell (1956). However, it was not until Tokuoka (1971) that the general yield criterion for the hypoelastic constitutive equations was laid down. Consider hypoelastic constitutive equation (3). For the vanishing stress rate we have

$$\mathring{\mathbf{T}} = \mathsf{L} : \mathbf{D} = 0. \tag{5}$$

From the condition $\mathbf{D} \neq 0$ in the above equation follows that the determinant of L must vanish

$$\det \mathsf{L} = 0. \tag{6}$$

Under condition (6) the hypoelastic constitutive relation is not invertible so the yield condition is equivalent to the loss of invertibility. The flow rule resulting from (5) is determined by the eigenvector. For triaxial stress states, the response envelope of a hypoelastic constitutive equation is an ellipse with the initial stress in the centre. Furthermore, the response envelope at yield must go through the initial stress. It follows that this is only possible when the elliptical response envelope degenerates to a line or a point. In the former case the stress rate vanishes only for two specific directions of $\mathbf{D}$, while for the latter case the stress rate vanishes identically for all directions of $\mathbf{D}$. This is demonstrated with two specific hypoelastic constitutive equations in Figures 2. In passing, if the stress rate vanishes for all $\mathbf{D}$ the stress is locked.

3.2 Hyperplasticity

In hyperplasticity, a yield surface is specified a priori. Since more than one stress-strain relation is involved the response envelope is composed of several segments. This can be clearly observed from Figure 3. If the consistency condition is fulfilled the response envelopes will be continuous but not smooth. However, the response envelopes can be made smooth by choo-

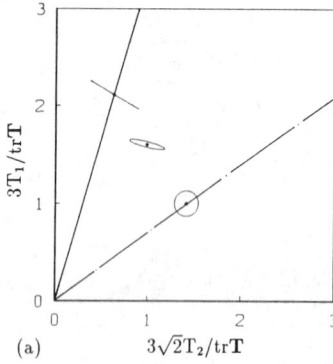 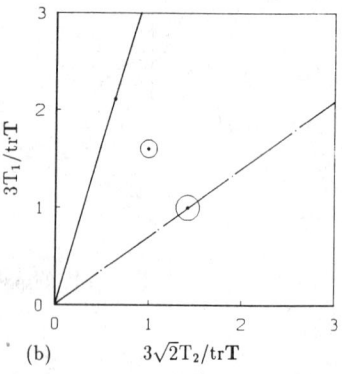

Figure 2: Typical response envelopes of hypoelastic constitutive equations: (a) the response envelope degenerates to a line at failure and (a) the response envelope degenerates to a point at failure

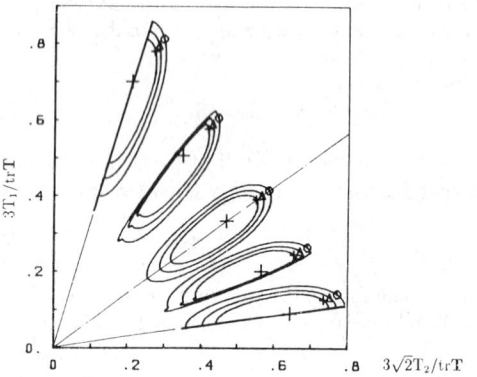

Figure 3: Response envelopes of the hyperplastic constitutive model proposed by Vermeer (1982) after Gudehus and Kolymbas (1985)

sing a particular flow rule (Niemunis 1993). When the stress is approaching the yield surface, one segment of the response envelope that corresponds to loading becomes flatter and coincides finally with the yield surface. The stress-strain relations, though different for loading and unloading, are incrementally linear. At yield the linear operator corresponding to loading becomes singular, which corresponds to the loss of invertibility of the hyperplastic equation.

The response envelopes are composed of segments of two or more ellipses. The compatibility condition guarantees the continuous transition from loading to unloading through the so-called neutral loading. The initial stress lies again in the centre of the ellipses. At yield the segment for loading degenerates to a line with the initial stress in the middle. For the so-called loading to the side, the behaviour predicted by the hyperplastic constitutive equations will be too stiff. Some modifications have been proposed to soften the response for the loading to the side, e.g. the stress vertex theory by Batdorf and Budiansky (1949). At yield the stress is allowed to move along the yield surface upon further loading but not across it (consistency condition). In other words, the yield surface bounds all accessible stress states. For the loading to the side, we have $\overset{\circ}{\mathbf{T}} \neq \mathbf{0}$. This does not contradict our definition of yield, since yield is not characterized by the stress alone but by the pair $(\mathbf{T}, \mathbf{D})$.

3.3 Hypoplasticity

Referring to constitutive equation (1), we derive explicit expressions of the yield surface. The derivations are similar to Chambon (1989) but more straightforward. For the vanishing stress rate we have

$$\overset{\circ}{\mathbf{T}} = \mathsf{L} : \mathbf{D} + \mathbf{N}\|\mathbf{D}\| = \mathbf{0}. \tag{7}$$

The above equation is an isotropic tensorial relation between the dynamical and kinematical quantities, $\mathbf{T}$ and $\mathbf{D}$. Due to rate independence, equation (7) is necessarily homogeneous of order zero in $\mathbf{D}$. Previous investigations by Sawczuk and Stutz (1968) indicate that the requirement of rate independence imposes a scalar function on the stress. This function is termed yield locus.

Let us first consider the flow rule at yield, namely the direction of stretching corresponding to the vanishing stress rate. Since the stretching in concern is other than zero, equation (7) can be divided by the norm of stretching to give

$$\frac{\mathbf{D}}{\|\mathbf{D}\|} = -\mathsf{L}^{-1} : \mathbf{N}. \qquad (8)$$

It is clear from (8) that only the direction of the stretching at yield is specified by equation (7). There is no one-to-one correspondence between $\mathring{\mathbf{T}}$ and $\mathbf{D}$. By making use of the definition of the norm $\|\mathbf{D}\|$ we have

$$\frac{\mathbf{D}:\mathbf{D}}{\|\mathbf{D}\|^2} = 1. \qquad (9)$$

Substitution of (8) into (9) gives the expression for the yield surface

$$f(\mathbf{T}) = \mathbf{N}^T : (\mathsf{L}^{-1})^T : \mathsf{L}^{-1} : \mathbf{N} - 1 = 0. \qquad (10)$$

The superscript T denotes a transposition. From the above derivations it may be seen that yield has two aspects, the kinematical one and the dynamical one. Consequenly, there are two equations which follow from the definition of yield. The first one specifies the direction of stretching and is called the flow rule and the second one concerns the stress and is termed the yield surface. It is worth noting that the yield surface and the flow rule in hypoplasticity emerge as *by-products* of the constitutive equation, whereas in hyperplasticity they must be prescribed a priori, e.g. the Mohr-Coulomb yield surface with the associated or nonassociated flow rule. A further remark is relevant to whether the flow rule (8), with reference to the yield surface (10), is associated. A perusal of (8) and (10) suggests that the flow rule is nonassociated, since in general $\partial f(\mathbf{T})/\partial \mathbf{T} \neq -\mathsf{L}^{-1}\mathbf{N}$.

Consider now invertibility of the hypoplastic constitutive equation (1) (cf. Niemunis (1993)). It is assumed that L is always positively definite in hypoplasticity. We denote $\mathbf{A} = \mathsf{L}^{-1} : \mathring{\mathbf{T}}, \mathbf{B} = \mathsf{L}^{-1} : \mathbf{N}, x = \|\mathbf{D}\|$ and solve (1) for $\mathbf{D}$. The unknown strain rate is

$$\mathbf{D} = \mathbf{A} - x\mathbf{B} \qquad (11)$$

and x can be found from the quadratic equation

$$\mathbf{D} : \mathbf{D} = x^2 = (\mathbf{A} - x\mathbf{B}) : (\mathbf{A} - x\mathbf{B}) \qquad (12)$$

The inversion of (1) is completed substituting

$$x_{1,2} = \frac{\mathbf{A}:\mathbf{B}}{\mathbf{B}:\mathbf{B}-1} \pm \sqrt{\left(\frac{\mathbf{A}:\mathbf{B}}{\mathbf{B}:\mathbf{B}-1}\right)^2 - \frac{\mathbf{A}:\mathbf{A}}{\mathbf{B}:\mathbf{B}-1}}, \qquad (13)$$

obtained from (12) into (11). Only positive root $x = \|\mathbf{D}\|$ can be admitted. Therefore, if $x_1 \cdot x_2 > 0$ the inverse solution does not exist or is not unique. This is the criterion of invertibility of (1). Using (13) we may write this criterion in a simpler form

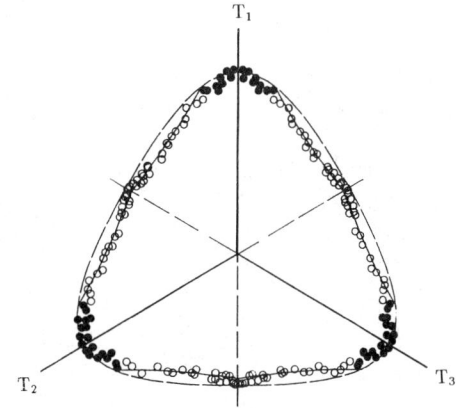

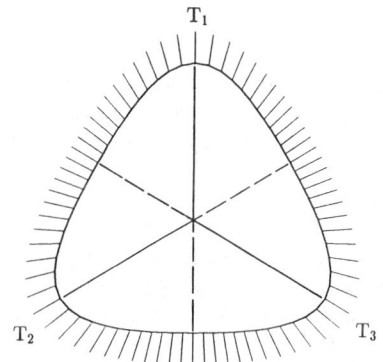

Figure 4: Yield surface and flow rule obtained with constitutive equation (2): (a) yield surface on a deviatoric plane, (b) flow rule on a deviatoric plane

$$\mathbf{B} : \mathbf{B} < 1 \qquad (14)$$

that is equivalent to the yield condition (10).

The yield surface calculated from constitutive equation (2) are shown on a deviatoric plane in Figure 4(a) together with the experimental data on Karlsruhe sand obtained with a true triaxial apparatus (Goldscheider 1976). Owing to the fact that constitutive equation (2) is homogeneous of first degree in $\mathbf{T}$, the tangential stiffness is proportional to $\mathrm{tr}\mathbf{T}$ and the friction angle is independent of $\mathrm{tr}\mathbf{T}$. In the principal stress space, the yield surface is a cone with the apex at the origin of the coordinate system (T_1, T_2, T_3). The calculated flow rule is shown on a deviatoric plane in Figure 4(b) together with the yield surface. A visual inspection of Figure 4(b) indicates that the derived flow rule is nonassociated. Since the nonlinear term in (1) depends only on the magnitude of $\mathbf{D}$ and not on its direction,

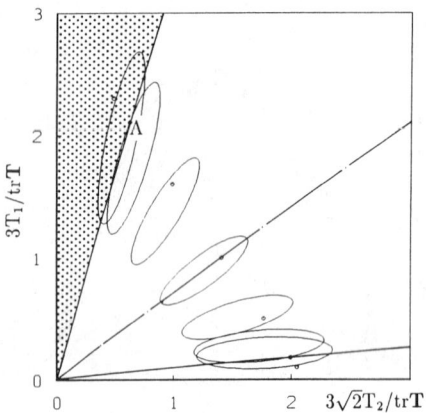

Figure 5: Response envelopes of hypoplastic constitutive equation (2)

the response envelope is shifted with the reference to the initial stress. The shape and orientation of the response envelope are fully determined by the linear term. At yield, the shift due to the nonlinear term is so large that the response envelope passes through the initial stress. It can be anticipated that the behaviour for loading to the side predicted by a hypoplastic model will be softer than that of a hyperplastic model. The response envelopes obtained with constitutive equation (2) are depicted in Figure 5 together with the yield surface. We observe from Figure 5 that the yield surface intersects the response envelope at yield. A small portion of the response envelope lies outside the yield surface. By choosing specific stress paths the yield surface can be surpassed. In other words, the yield surface does not coincide with the bound of the accessible stress states. This finding was quite surprising since it contradicts the classical concept of failure [1]. In the next section, we will investigate whether there exists a bound surface enclosing all accessible stress states for constitutive equation (2).

4 ACCESSIBLE STRESS STATES AND BOUND SURFACE

As it has been already shown in Section 3, some stress paths may surpass the yield surface. We proceed to investigate whether the accessible stress states can be

---

[1] During the second workshop in Gdańsk in 1989 we became aware that Prof. Chambon and Prof. Desrues came across the subject independently at almost the same time. However, they attempted to force the stress to fulfil the consistency condition (Desrues and Chambon 1993).

bounded by the so-called bound surface. We will demonstrate for constitutive equation (2) that all stress paths can be enclosed within the bound surface. In Figure 6 the bound surface obtained with constitutive equation (2) is shown on a deviatoric plane. This conical surface lies outside the yield surface. The bound surface is an intrinsic property of hypoplastic formulation (2) which means that we do not have to impose additional bounds as is the case in hyperplastic models on the stress state and to control the stress during the numerical calculation. Moreover, no procedure for projecting the stress back on this surface is needed. The stress states lying outside the bound surface, e.g. as a result of too large time increment, will be automatically reduced in the next time step. However, the existence of the bound surface for a different hypoplastic formulation within the framework of (1) is not a matter of course and should be verified. It should be pointed out that there are some versions of the hypoplastic model (1) for which the bound sur-

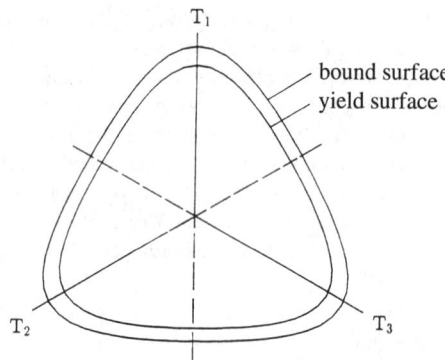

Figure 6: Cross-section of the yield surface and the bound surface with a deviatoric plane.

face does not exist. In what follows we formulate a useful criterion to find the bound surface numerically. Further, some experimental results supporting the theoretical studies are presented.

4.1 Theoretical Analysis

Suppose there exists a bound surface $b(\mathbf{T}) = 0$ with $b(\mathbf{T})$ being an isotropic function of stress. Consider the stress $\mathbf{T}^*$ lying on the bound surface so that $b(\mathbf{T}^*) = 0$. It is convenient to represent the stress $\mathbf{T}^*$ by its three principal components, i.e. in a diagonal form. It can be shown that the tensor normal and outer to the bound surface at $\mathbf{T}^*$

$$\mathbf{Z} = \left.\frac{\partial b(\mathbf{T})}{\partial \mathbf{T}}\right|_{\mathbf{T}=\mathbf{T}^*}, \tag{15}$$

must be diagonal as well. According to the definition of the bound surface all stress rates $\mathring{\mathbf{T}}$ calculated at $\mathbf{T}^*$ must be directed to the inside of the bound surface so that for any strain rate $\mathbf{D}$ the corresponding stress rate $\mathring{\mathbf{T}}$ must satisfy the inequality

$$\mathbf{Z} : \mathring{\mathbf{T}} \leq 0. \tag{16}$$

Note that in (16) it suffices to multiply only the diagonal components of $\mathring{\mathbf{T}}$ with $\mathbf{Z}$ since $\mathbf{Z}$ is diagonal. Thus, without loss of generality we may consider only the stress rates coaxial with the stress $\mathbf{T}^*$. From inequality (16) follows that if the stress state lies on the bound surface then the maximum of the scalar product vanishes, i.e.

$$\max \left(\mathbf{Z} : \mathring{\mathbf{T}}\right)_{\mathbf{T}=\mathbf{T}^*} = 0. \tag{17}$$

Substituting (1) into (17) and differentiating with respect to $\mathbf{D}$ leads to the following condition for maximum of $\mathbf{Z} : \mathring{\mathbf{T}}$

$$\frac{\partial \left[\mathbf{Z} : \mathsf{L} : \mathbf{D} + \mathbf{Z} : \mathbf{N}\|\mathbf{D}\|\right]}{\partial \mathbf{D}} = \mathbf{Z} : \mathsf{L} + \mathbf{Z} : \mathbf{N}\frac{\mathbf{D}}{\|\mathbf{D}\|} = 0. \tag{18}$$

Due to rate independence we may take $\|\mathbf{D}\| = 1$ and obtain the solution of (18) in the following form

$$\mathbf{D}_{\max} = -\frac{\mathbf{Z} : \mathsf{L}}{\mathbf{Z} : \mathbf{N}}. \tag{19}$$

Substituting (19) into (17) we arrive at the following criterion for the stress lying on the bound surface $b(\mathbf{T}) = 0$

$$(\mathbf{Z} : \mathsf{L}) : (\mathbf{Z} : \mathsf{L}) - (\mathbf{Z} : \mathbf{N})^2. \tag{20}$$

Clearly this criterion alone, contrary to the criterion of yield, does not lead to an algorithm to determine the bound surface since the normal direction $\mathbf{Z}$ still remains unknown.

This problem can be handled by taking into account that the constitutive relation (1) is positively homogeneous of the first degree in stress and that the bound surface is an isotropic function of stress. From the first property follows that $b(\mathbf{T}) = 0$ represents a conical surface with the vertex in the origin of the stress space. Indeed, by the definition of the positive homogeneity for any positive scalar α we have

$$\mathring{\mathbf{T}}(\mathbf{D}, \alpha \mathbf{T}) = \alpha \mathring{\mathbf{T}}(\mathbf{D}, \mathbf{T}). \tag{21}$$

As a consequence, if all stress rates $\mathring{\mathbf{T}}(\mathbf{D}, \mathbf{T}^*)$ satisfy the inequality (16) so do all $\mathring{\mathbf{T}}(\mathbf{D}, \alpha \mathbf{T}^*)$. Hence, if the stress $\mathbf{T}^*$ lies on the bound surface then also $\alpha \mathbf{T}^*$ does. This property can be formulated as the orthogonality condition between $\mathbf{T}^*$ and $\mathbf{Z}$

$$\mathbf{T}^* : \mathbf{Z} = 0. \tag{22}$$

From the assumption of the bound surface being an isotropic function of stress follows that for triaxial stress with $T_2^* = T_3^*$, the respective partial derivatives, here Z_2 and Z_3, are equal. Making use of this property under consideration of (22) we can find the points of the surface $b(\mathbf{T}) = 0$ for triaxial stress state by solving the following equation system

$$\begin{cases} \mathbf{T}^* - 1\dfrac{\mathbf{T}^* : \mathbf{T}^*}{\mathbf{T}^* : \mathbf{1}} = \mathbf{Z} \\ (\mathbf{Z} : \mathsf{L}) : (\mathbf{Z} : \mathsf{L}) = (\mathbf{Z} : \mathbf{N})^2 \end{cases} \tag{23}$$

We denote the triaxial stress $\mathbf{T}^*$ satisfying the above system of equations as $\mathbf{T}_0^*$. The next point $\mathbf{T}_1^*$ lying on the bound surface in the vicinity of the stress $\mathbf{T}_0^*$ may be found numerically. To this end, we replace the diagonal tensors $\mathbf{T}$ and $\mathbf{Z}$ with corresponding vectors $\mathbf{t}$ and $\mathbf{z}$ and calculate

$$\mathbf{t}^{*(i+1)} = \mathbf{t}^{*(i)} + \theta \ \mathbf{t}^{*(i)} \times \mathbf{z}, \tag{24}$$

where i is the number of step and θ the step length. The whole cross-section of the bound surface with a deviatoric plane can be found by repeating the calculation (24). During the numerical study with various versions within the framework of (1) we observed that only a suitable choice of the tensorial expressions for L and $\mathbf{N}$ leads to the existence of the bound surface. Unfortunately, at present we are not able to formulate criteria that would guarantee the existence of the bound surface. Therefore, any new version of the constitutive model and any new set of material parameters should be proved for the existence of the bound surface. One possible way would be to calculate and plot the bound surface according to the algorithm given above. Alternatively, one could test the constitutive model by trying out numerous randomly generated stress paths to see if the stress paths are bounded. A different algorithm that generates stress paths directed to the outside of the yield surface has been proved very helpful in testing the constitutive model. Such stress paths can be found by choosing such strain rate $\mathbf{D}$ that maximizes the product $\mathbf{M} : \mathring{\mathbf{T}}$, with

$$\mathbf{M} = \mathbf{T} - 1\frac{\mathbf{T} : \mathbf{T}}{\mathbf{T} : \mathbf{1}}.$$

Note that for triaxial stress states the direction $\mathbf{M}$ is identical with $\mathbf{Z}$. We have found out that $\mathbf{M}$ is a good

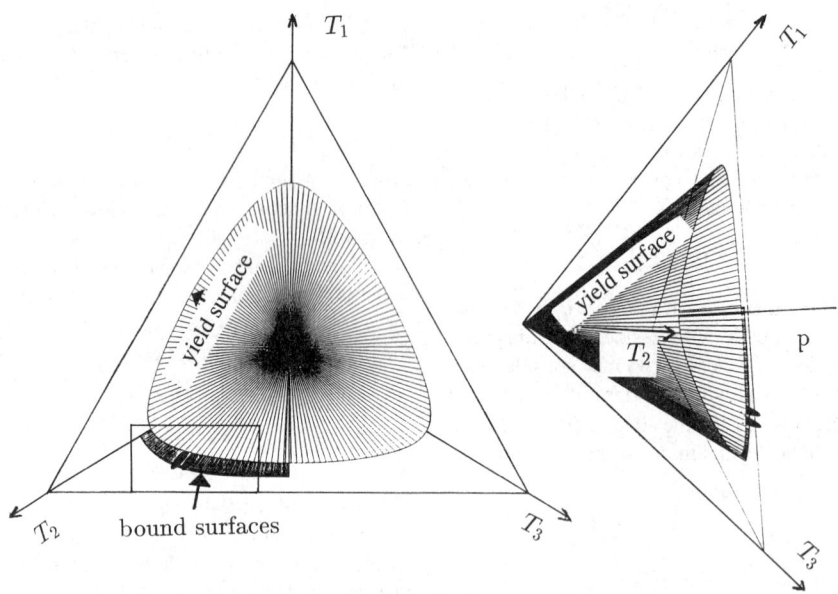

Figure 7: The yield surface and a fragment of the bound surface plotted in stress space according to equation (2).

approximation for Z also for other, non-symmetric stress states. An interesting aspect for the investigation of bound surface is the geometric analysis of the response envelopes. The response envelope at a stress point on $b(\mathbf{T}) = 0$ is an ellipsoid lying inside the bound surface and tangential to it. With the help of a graphical software it is possible to visualize the relative location of the response envelope with respect to the yield and the bound surface. In Figure 7 the yield surface, bound surface and two response envelopes are presented. The surface (the pseudo bound surface in Figure 8) between the yield surface and the bound surface was obtained with the method proposed by Kolymbas (1991). The stress $\mathbf{T}$ is said to lie on the pseudo bound surface if the equation $\alpha \cdot \mathbf{T} = \dot{\mathbf{T}}$ has only one solution $\alpha, \mathbf{D}$ for $\|\mathbf{D}\| = $ const. In other words, if the response envelope is tangential to (and not punctured by) the $\mathbf{0T}$ line. It can be seen from Figure 8(b) that the surface obtained according to Kolymbas (1991) can be surpassed, since the response envelope lies partially outside the surface.

4.2 Experimental Observation

The tests were conducted in the strain-controlled triaxial apparatus described by Wu and Kolymbas (1992). The so-called free-ends were used to reduce the influence of friction. The specimens with an initial void ratio of about 0.5 were prepared by air pluviation with Karlsruhe medium sand. The test process was controlled by a computer. A straight stress path could be followed by providing the inclination of the stress path as an input. If, however, the stress path is curved, it had to be approximated by several segments. A typical triaxial test is shown in Figure 9. The stress path together with the direction of strain increment along the stress path is depicted in Figure 9(a). In Figures 9(b) and 9(c) the stress-strain curves and the volumetric strain curves are given. The specimen was first brought to a hydrostatic stress of $\sigma_c = 1.0$ MPa and then loaded by increasing the axial stress while keeping the radial stress constant until yield point A was reached, which can be inferred from the plateau of the stress-strain curves in Figures 9(b) and 9(c). If the stress path was kept unchanged, the stress will remain stationary or showed moderate strain softening with further straining. For the test in Figure 9, the stress path was changed at point A following A⇒B. Different stress paths had been tried until the increase of the stress ratio T_1/T_3 reached its maximum. The following observation can be made from Figure 9. Along the stress path A⇒B the stress difference decreases monotonically whereas the stress ratio increases to its maximum and then decreases (Figures 9(b) and 9(c)). The yield criterion in this diagram is the straight line connecting the origin and the point A. Clearly the stress path A⇒B lies beyond the yield surface. It should be reminded that we do not intend to

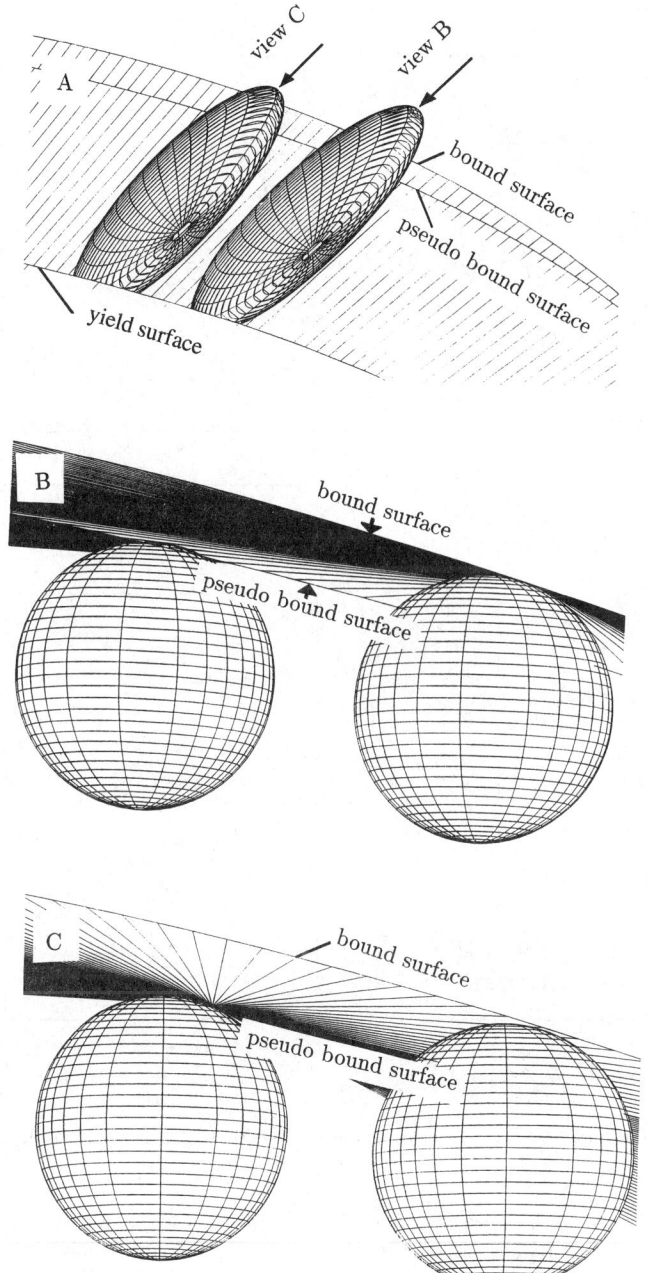

Figure 8: The enlarged portions of Figure 7: (a) the view point is located on the hydrostatic axis, (b) the view point is located along the direction of the initial stress of the left response envelope, (c) the view point is located along the direction of the initial stress of the right response envelope.

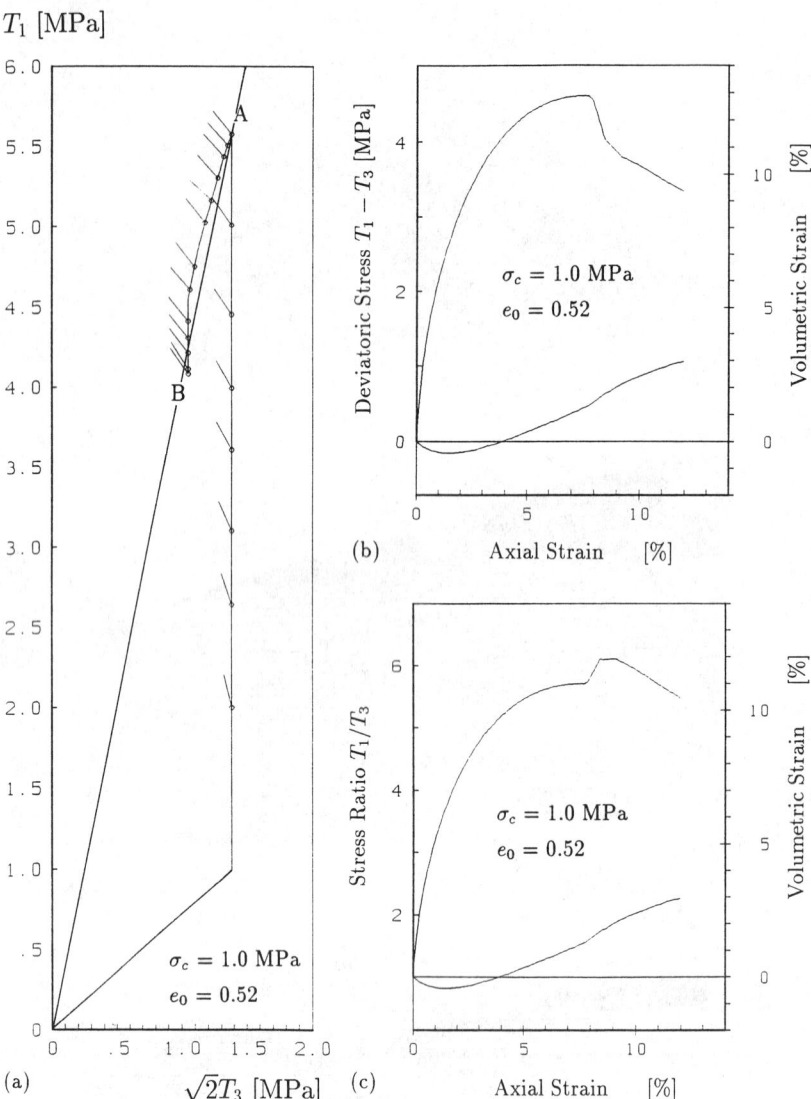

Figure 9: Triaxial compression test: (a) stress path, (b) stress difference over axial strain, (c) stress ratio over axial strain

make a quantitative comparison between the predicted and the experimental results. Rather, the experiment should be regarded as a qualitative verification of the prediction from hypoplastic constitutive equation (2). Another interesting phenomenon can be seen from Figure 9(a). Though there is a drastic change of the direction of stress increment, there is only minor variation of the direction of the strain increment. An increase of the volumetric strain can be also observed in Figure 9(b). The second order work density was calculated according to $d^2W = \Delta T_1 \Delta E_1 + 2\Delta T_2 \Delta E_2$ and is shown in Figure 10. Clearly, the magnitude of d^2W depends on the magnitude of the strain increment. For the results in Figure 10 an axial strain increment of $\Delta E_1 \approx 0.001$ was used. The second order work density is seen to decrease from an initially positive value to zero at yield and becomes even negative along the path A⇒B. An ensuing question will be whether it is possible to have $d^2W < 0$ before yield. This problem will be investigated together with the stability surface in the next section.

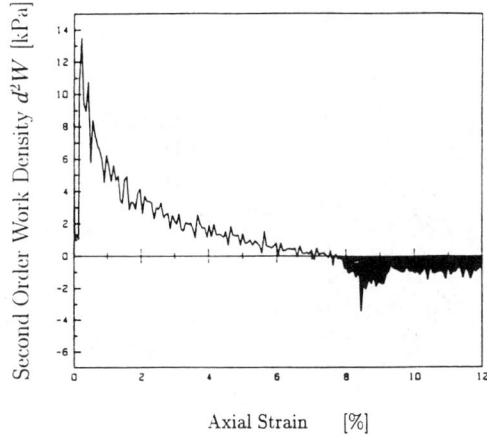

Figure 10: Second order work density over axial strain

5 STABILITY SURFACE AND STRAIN SOFTENING

The second order work density defined by $d^2W = \text{tr}(\mathring{\mathbf{T}}\mathbf{D})$ plays an important role in the discussion on stability and uniqueness (Hill 1978, Valanis 1985). There have been abundant investigations in the literature on the relevance of the second order density to stability and uniqueness. It seems to be difficult to draw definite conclusions on this issue. For the moment, we will leave the implication of the second work density for stability and uniqueness aside and investigate whether d^2W can be negative for constitutive equation (2). To this end, we search for the boundary between positive and negative second order work density by letting $d^2W = 0$. If this boundary builds a surface in the stress space, it will be called *stability surface*. Note that the stability surface here need not necessarily be equivalent to the definition by Hill (1978). For incrementally linear models, the requirement $d^2W > 0$ is equivalent to positive definitness of the tangential stiffness tensor. The proof of positive definiteness can be performed in a straightforward way. To show the difference between hypoplastic and incrementally linear models, constitutive equation (1) can be expressed in a more convenient form by virtue of Euler's theorem for homogeneous functions

$$\mathring{\mathbf{T}} = (\mathsf{L} + \mathbf{N} \otimes \vec{\mathbf{D}}) : \mathbf{D}, \qquad (25)$$

where $\vec{\mathbf{D}} = \mathbf{D}/\|\mathbf{D}\|$ stands for the direction of stretching; and the symbol $\otimes$ denotes a dyadic product between two tensors. Apparently the tangential stiffness tensor in the brackets in (25) depends not only on stress but also on the direction of stretching. The stability surface is given by the following equation

$$d^2W = \mathbf{D} : (\mathsf{L} + \mathbf{N} \otimes \vec{\mathbf{D}}) : \mathbf{D} = 0. \qquad (26)$$

Clearly, the above equation is satisfied by the pair (**T**,**D**) at yield. We search for such stress states, for which $\mathring{\mathbf{T}}$ and **D** can be orthogonal. The stress state satisfying equation (26) can be found as follows. Starting from a hydrostatic stress state, the second order work density is calculated for all possible directions of **D**. It is sufficient to restrict to stretching coaxial with the stress. To this end, the stretching **D** with $\|\mathbf{D}\| = 1$ can be expressed in a spherical coordinate system as follows

$$\mathbf{D} = \begin{pmatrix} \cos\zeta\sin\eta & 0 & 0 \\ 0 & \sin\zeta\sin\eta & 0 \\ 0 & 0 & \cos\eta \end{pmatrix}. \qquad (27)$$

d^2W is calculated by varying φ in the range ($0° \leq \zeta \leq 360°$) and η in the range ($-90° \leq \eta \leq 90°$). If d^2W is positive for all directions of **D**, the deviatoric stress is increased. The above procedure will be repeated until the stress satisfying equation (26) is found. In Figure 11(a) the second order work density calculated from constitutive equation (2) is depicted against ζ and η. The stress state corresponding to Figure 11(a) lies inside the yield surface. It is seen from Figure 11(a) that d^2W is negative only for specific directions of **D** (the dark region in Figure 11(a)). Figure 11(b) shows the stability surface according to constitutive equation (2) together with the yield and the bound surface on a deviatoric plane. It can be concluded that for constitutive equation (2) the second order work density may become negative before yield. Now, we try to investigate the physical meaning of the second order work density. A simple but non-trivial case in a triaxial apparatus is the undrained test, where the specimen undergoes isochoric deformation. Making use of the triaxial stress state with $T_2 = T_3$ the second order work density may be written as $d^2W = \dot{T}_1 D_1 + 2\dot{T}_2 D_2$. It follows from $d^2W = 0$ and incompressibility, $D_1 + 2D_2 = 0$, that $\dot{T}_1 - \dot{T}_2 = 0$, which means that the stress difference reaches its maximum. For loose sand, the maximum of the stress difference marks the beginning of strong reduction of the effective stress and hence the incipience of instability. A property that is closely connected with the second order work density is strain softening. Here, we will consider strain softening as a material property. Most of the experimental studies are limited to strain softening behaviour in either triaxial tests with constant confining pressure or in uniaxial tests. For these simple cases, strain softening

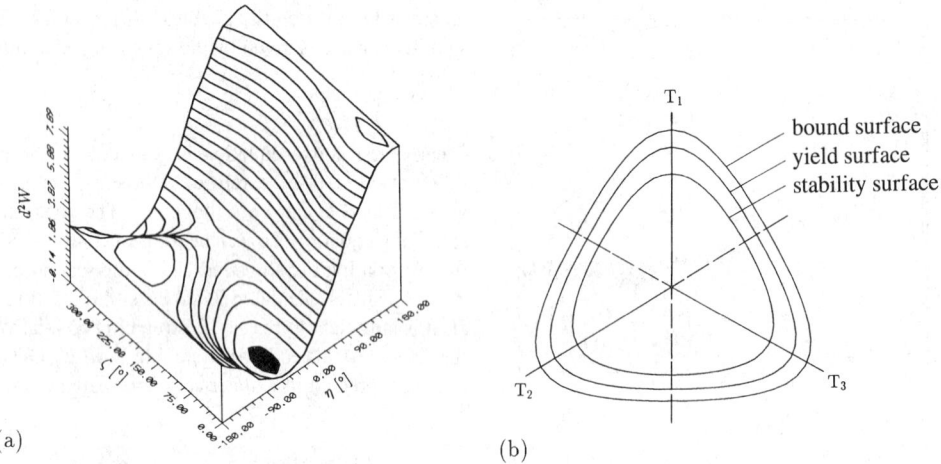

Figure 11: (a) second order work density against ζ and η. (b) cross-section of the stability surface together with the yield surface and the bound surface on the deviatoric plane.

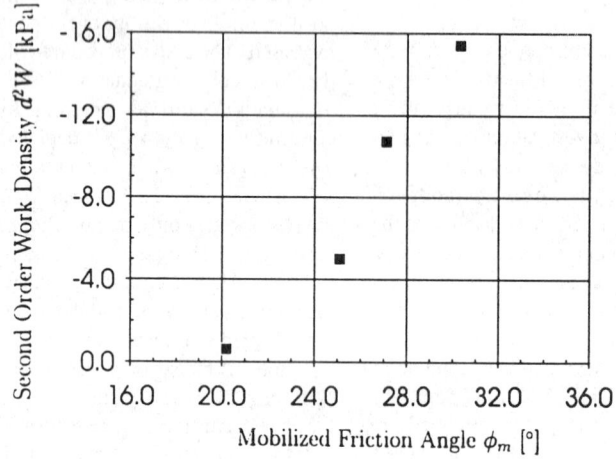

Figure 12: Second order work density over mobilized friction angle

is often referred to the so-called post-failure strain softening. After Valanis (1985) a material element under uniform stress **T** and subjected to homogeneous straining is said to exhibit strain softening if the second order work density is negative. According to this definition the stability surface is also the boundary for strain softening. Recent experiments on sand by Lade et al. (1987) and by Chu (1991) showed that there can be strain softening even prior to failure and called it pre-failure strain softening. This experimental finding agrees with the predictions of the hypoplastic model. Triaxial tests have been carried out on dense Karlsruhe medium Sand $e_0 \approx 0.5$ to capture the stability surface for the case of triaxial stress states with $T_2 = T_3$. The tests were similar to that described in the Section 4 with the difference that the change of the stress path was initiated at a mobilized friction angle $\phi_m = \arcsin[(T_1 - T_2)/(T_1 + T_2)]$ before failure. If it was possible to have $d^2W < 0$, another test was carried out with a smaller mobilized friction angle at which the stress path was changed. The results are shown in Figure 12. The mobilized friction angle at which the second order work density becomes zero is about $\phi_m \approx 20°$. This agrees with the prediction of the constitutive model.

Acknowledgement - The present work was financially supported by the German Research Community, (Project Ko 884/4-2 Hypoplasticity). W.W. is grateful to Lahmeyer International Ltd. for generous support to complete this work.

REFERENCES

Batdorf, S. B. and Budiansky, B. (1949), A mathematical theory of plasticity based on the concept of slip, *NACA TN 1871*

Chambon, R. (1989), Une classe de lois comportement incrementalement nonlinéaires pour sols non visqueux, resolution de quelques problemes de coherence, *C. R. Acad. Sci. Paris*, **t. 308**, Serie II, 1571-1576

Chu, J. (1991), Strain softening behaviour of granular soils under strain path testing, Dissertation, University of New South Wales, Australia

Davis, R. O. and Mullenger, G. (1978), A rate-type constitutive model for soils with a critical state, *Int. J. Num. Anal. Methods Geomech.*, **2**, 255-282

Desrues, J. and Chambon, R. (1993), A new rate type constitutive model for geomaterials: CloE, in: *Proc. Int. Workshop Modern Approaches in Plasticity*, Elsevier, Holland, 309-324

Goldscheider, M. (1976), Grenzbedingung und Flieregel von Sand, *Mech. Resear. Comm.*, **3**, 463-468

Gudehus, G. (1979), A comparison of some constitutive laws for soils under radially symmetric loading and unloading, in: *Proc. 3rd Int. Conf. Numer. Methods Geomech.*, Balkema, Holland, 1309-1323

Gudehus, G. and Kolymbas, D. (1985), Numerical testing of constitutive relations for soils, *5th Int. Conf. on Numer. Methods in Geomech.*, Balkema, Holland, 63-81

Hill, R. (1959), Some basic principles in the mechanics of solids without a natural time, *J. Mech. Phy. Solids*, **7**, 209-225

Hill, R. (1978), Aspects of invariance in solid mechanics, *Advances in Applied Mechanics*, **8**, Academic Press

Kolymbas, D. (1991), An outline of hypoplasticity, *Ing. Arch.*, **61**, 143-151

Lade, P. V., Nelson, R. B. and Y. M. Ito (1987), Instability of granular materials with nonassociated flow, *ASCE, J. Engr. Mech.*, **113**, 1302-1318

Niemunis, A. (1993), Hypoplasticity vs. elastoplasticity, selected topics, in: *Proc. Int. Workshop Modern Approaches in Plasticity*, Elsevier, Holland, 277-307

Romano, M. A. (1974), A continuum theory for granular media with a critical state, *Arch. Mech.*, **20**, 1011-1028

Royis, P. (1989), Interpolations and one–to–one properties of incremental constitutive laws, *Eur. J. Mech., A/Solids*, **8**, 385-411

Sawczuk A. and Stutz P. (1968), On formulation of stress-strain relations for soils at failure. *ZAMP*, **19**, 770-778

Stutz, A. (1973), Comportment elasto-plastique des milieux granularien, *Foundation of Plasticity*, Noordoff, Leyden, 33-49

Tokuoka, T. (1971), Yield conditions and flow rules derived from hypo-elasticity, *J. Rational Mech. Anal.*, **42**, 239-252

Truesdell, C. (1955), Hypo-elasticity, *J. Rational Mech. Anal.*, **4**, 83-133

Truesdell, C. (1956), Hypo-elastic shear, *J. Appl. Phy.*, **27**, 441-447

Valanis, K. C. (1985), On the uniqueness of solution of the initial value problem in softening materials, *ASME, J. Appl. Mech.*, **52**, 649-653

Vermeer, P. (1982), A five constant model unifying well established concepts, *Proc. Int. Workshop on Constitutive Relations for Soils*, Balkema, Holland, 175-197

Wu, W. and Kolymbas, D. (1990), Numerical testing of the stability criterion for hypoplastic constitutive equations, *Mech. Mater.*, **9**, 245-253

Wu, W. and Sikora, Z. (1991), Localized bifurcation in hypoplasticity, *Int. J. Eng. Sci.*, **29**, 195-201

Wu, W. and Kolymbas, D. (1991), On some issues in triaxial extension tests, *ASTM, Geotech. Testing J.*, **14**, 276-287

Wu, W. (1992), Hypoplasticity as a mathematical model for mechanical behaviour of granular materials, Dissertation, Karlsruhe University

Wu, W. and Bauer, E. (1993), A hypoplastic constitutive model for barotropy and pyknotropy of granular materials, in: *Proc. Int. Workshop Modern Approaches in Plasticity*, Elsevier, Holland, 225-258

Wu, W. and Niemunis, A. (1994), Failure criterion, flow rule and dissipation function derived from hypoplasticity, to appear in *Int. J. Numer. Anal. Methods Geomech.*

Wu W. and Bauer, E. (1994), A simple hypoplastic model for sand, submitted to *Int. J. Numer. Anal. Methods Geomech.*

Localised failure analysis using damage models
Étude de la rupture localisée avec des modèles d'endommagement

A. Dragon, F. Cormery, T. Désoyer & D. Halm
Laboratoire de Mécanique et de Physique des Matériaux, CNRS, ENSMA, Chasseneuil-du-Poitou, France

Abstract : The paper addresses the problem of strain and damage localization in the framework of microcrack-related damage inducing anisotropic response for rock-like solids. The relevant damage model introduces residual stress/strain effects governed by damage itself, without reference to plasticity-like mechanisms. It is complete in the sense that beyond evaluating effective elastic properties of a cracked solid, the damage evolution is settled through the normality law for the tensorial variable at stake. A three-dimensional computational procedure for detecting the onset of localization is then introduced and employed for studying localization geometry and mode(s) corresponding to the model in question. The results presented were obtained assuming (i) axisymmetric proportional strain-paths and particular axisymmetric stress-strain paths corresponding to the conventional 'triaxial' axisymmetric compression test, (ii) off-axes loading-paths for pre-strained and pre-damaged material, (iii) finite-element simulation for a boundary-value problem concerning a rock structure with a hole. Damage-induced anisotropy appears as a factor favouring a fixed localization surface for off-axes simulations compared to the multiplicity of solutions encountered otherwise. The study reveals the complexity of geometrical configurations of localization (characteristic) surfaces and completeness of related discontinuity modes when the three-dimensionality is admitted. The both indicate the shortage of common two-dimensional analyses of localization events. Some comparisons with localization mechanisms associated with damage isotropic modelling show insufficiency of the latter when oriented cracking becomes predominant.

On considère le problème de la localisation de la déformation et de l'endommagement dans le contexte de l'endommagement de type micro-fissures qui induit une réponse anisotrope des solides de type roche. Le modèle d'endommagement considéré introduit les effets résiduels induits par l'endommagement lui-même, sans référence à des mécanismes de type plastique. Il est complet en ce sens que l'évolution de l'endommagement, qui évalue les propriétés élastiques effectives d'un solide fissuré, est basée sur la loi de normalité pour la variable tensorielle. Une procédure numérique pour la détection de la naissance de la localisation dans le cas tridimensionnel est introduite et mise en oeuvre pour étudier la géométrie de la localisation pour ce modèle.

Les résultats présentés ont été obtenus en considérant (i) des chemins axisymétriques proportionnels, ainsi que le chemin triaxial axisymétrique conventionnel, (ii) des chemins hors-axes pour le matériau pré-déformé et pré-endommagé, et (iii) des simulations numériques aux éléments finis du problèmes aux limites représent une structure rocheuse avec un trou. L'anisotropie induite par l'endommagement apparaît comme étant un facteur qui favorise une surface de localisation donnée par rapport à la multiplicité de solutions rencontrées sinon. L'étude révèle la complexité de la géométrie des configurations des surfaces de localisation (caractéristiques) et la complétude des modes de discontinuité en relation lorsqu'on prend en compte le caractère tridimensionnel. Ces résultats montrent le caractère restrictif des analyses bidimensionnelles usuelles. La comparaison avec les mécanismes de localisation associés à un endommagement isotrope indiquent que ce dernier est insuffisant lorsque la fissuration orientée devient prédominante.

1. INTRODUCTION

Localization viewed in the context of rate-independent behaviour can be considered as a local bifurcation allowing two incremental solutions of the corresponding incremental equilibrium problem. One solution is exhibiting continuous deformation gradient increments (or rates) while the other solution (bifurcation branch) reveals a discontinuity. As such discontinuities in the deformation gradient increments are conceivable in some particular directions only – called characteristic directions –

the localization incipience corresponds to the loss of ellipticity of the incremental (rate) equations of a boundary value problem (Rice 1976, Vardoulakis 1976, Rice and Rudnicki 1980, Borré and Maier 1989, Benallal et al. 1989).

The nowadays classical form of the criterion of localization inside the solid reading :

$$\det (\underset{\sim}{N} \cdot \underset{\sim}{L} \cdot \underset{\sim}{N}) = 0 \tag{1}$$

involves the local direction of a characteristic surface S (its unit normal $\underset{\sim}{N}$) and the tangent stiffness matrix $\underset{\sim}{L}$ corresponding to an assumed constitutive law, so that $\underset{\sim}{\dot{\sigma}} = \underset{\sim}{L} : \underset{\sim}{\dot{\varepsilon}}$, $\underset{\sim}{\dot{\sigma}}$ and $\underset{\sim}{\dot{\varepsilon}}$ standing for the stress-rate and the strain-rate respectively. The associated localization-discontinuity mode is given by a vector $\underset{\sim}{G}$) colinear to a unit vector $\underset{\sim}{M}$. The pair $(\underset{\sim}{N}, \underset{\sim}{M})$ indicates the orientation and nature of discontinuity, i.e. 'opening' or 'shearing' mode, frequently a combination of the both. The influence of the tangential stiffness $\underset{\sim}{L}$ in the characteristic (acoustic) tensor $\underset{\sim}{Q} = \underset{\sim}{N} \cdot \underset{\sim}{L} \cdot \underset{\sim}{N}$ makes it possible to study the singularity of $\underset{\sim}{Q}$ corresponding to the loss of ellipticity through the spectral analysis for a given class of constitutive relations. One can then establish analytical expressions for the spectral properties of $\underset{\sim}{Q}$ allowing explicit relations concerning singularity and some features of the bifurcation at stake. This way has been exploited by Ottosen and Runesson (1991) who treat a class of non-associated plasticity laws regarding localization by spectral analysis approach. In fact the localization is taken up by these authors in a larger sense than in the classical approach recalled above : the discontinuity of displacement increment itself (or rate), i.e. strong discontinuity, is being incorporated in their bifurcation analysis. These authors argue that the classical bifurcation conditions linked to localization phenomena (under plastic/plastic and plastic/elastic response respectively on the characteristic surface and outside) are still valid under the restriction that the jump of displacement rate $[\dot{u}_i]$ remains constant along S. This step is one more argument allowing to link the study of localization to local failure, viz. macrocracking incipience.

The last authors show much favour to obtaining analytically critical localization directions and corresponding moduli and state that "it is desirable to avoid a numerical search..." in the context of finite element analysis. It seems however that the both should be developed and, concerning specific localization search algorithms associated as parallel processors with finite element analysis, their utility in engineering analysis of structures to detect actual zones and mechanisms of failure incipience is as evident as is the utility of analytical approach on the background level, see f.ex. Ortiz (1987).

This work is aiming at the strain and damage localization study in the context of a microcrack-related damage inducing anisotropic response. We discuss first (Section 2) a constitutive model accounting for oriented character of microcrack-set(s). The model is written in the framework of continuum damage theory but belongs otherwise to a class of more physically-oriented damage approaches through the signification of an internal variable adopted (Kachanov M., 1980). Suitable hypotheses lead to residual (irreversible) strain/stress effects without explicit reference to plasticity concepts. The model proposed is complete in the sense that beyond evaluating actual (effective) moduli of a cracked solid, the damage evolution concerning the splitting-like ('brittle') microcracking is being put forward in the proper thermodynamic framework.

The damage concept reflects explicitly deterioration phenomena in the framework of continuum modelling of nonlinear solid behaviour. Using this concept when studying localization, viz. transition from volume-diffused deterioration to surface-like localized one, takes on even more pertaining physical reference to failure incipience than localization study in the context of plasticity.

With this standpoint, involving furthermore, in the framework of the present model, some interaction between localization mode(s) and damage-induced anisotropy, we are proceeding here to an analysis of localization using a three-dimensional numerical localization processor. The latter has been conceived to detect – with high accuracy – the loss-of-ellipticity-threshold in the course of finite element analysis. The description of the processor is given in Section 3. The fitness of the computational procedure to predict configurations and modes of localization for a variety of loading paths is illustrated in Section 4. From these investigations for a rock-like solid appears clearly the complexity of geometrical configurations in three-dimensional situations. In the same time the related discontinuity modes are complete (full in the 3-D sense). The both indicate the shortage of common two-dimensional analyses of localization as failure-incipience phenomena. Detector potentialities concerning prediction of failure incipience through localization for more complex structural problems and, in particular, in the context of geomechanics are also discussed.

2. ANISOTROPIC DAMAGE MODEL

2.1 *Preliminaries. Damage variable*

A second-order tensor $\underset{\sim}{D}$ is chosen as damage internal variable to account for notable defect orientation. This is achieved by introducing the dyadic (tensor) product $\underset{\sim}{n} \otimes \underset{\sim}{n}$ with cartesian components $n_i n_j$, where $\underset{\sim}{n}$ is a unit normal to the crack (discontinuity) surface viewed as the plane one. The concept works as well for a set of parallel microcracks. The dyadic product appears instead of e.g. vectorial representation $\underset{\sim}{n}$ as the sense of $\underset{\sim}{n}$ is not significant and the elastic potential, whose existence is assumed in the foregoing, must be an even function of $\underset{\sim}{n}$.

To make distinction between a single microcrack and a set of microcracks and to account for the dissipative mechanism in question, namely, generation of microsurfaces by decohesion, some density concept should be put forward. Following Kachanov (1980, 1987) we consider a non-negative scalar non-dimensional function $d(s)$, corresponding to the extent of surfaces produced by microcracking. For example, accounting for micro-mechanical analyses concerning the influence of cracks on the elastic response, a reasonable choise is:

$$d(s) = \eta \frac{\sum_i s_i^{3/2}}{\overline{V}} \qquad (2)$$

with s_i standing for area of the i-th crack, $\overline{V}$ denoting the representative volume and the positive multiplier η representing a proportionality (norming) factor. When allowing a set of penny-shaped cracks with a_i standing for the radius of the i-th one, and taking $\eta = (\pi\sqrt{\pi})^{-1}$, $d(s) = \hat{d}(s)$ is found to coincide with the conventional crack density r^c:

$$\hat{d}(s) \equiv r^c = \frac{\sum_i a_i^3}{\overline{V}}, \quad 0 \leq \hat{d}(s) \leq 1 \qquad (2)'$$

In the spirit of continuum damage mechanics put forward in the present paper, the restriction to a particular crack shape can be relaxed and the interpretation (2) – more general than (2)' – accepted.

Thus we will use as an internal damage variable a symmetric tensorial density

$$\underset{\sim}{D} = d(s) \underset{\sim}{n} \otimes \underset{\sim}{n}, \quad (j = 1) \qquad (3)$$

where $j = 1$ indicates a single set of parallel microcracks. If there are several sets, the summation over the sets should be applied:

$$\underset{\sim}{D} = \sum_j d_j(s) (\underset{\sim}{n} \otimes \underset{\sim}{n})_j, \quad d(s) = \operatorname{tr} \underset{\sim}{D} = \sum_j d_j(s) \qquad (4)$$

with $j = (1, 2,..., k)$ for k systems. Note however that whatever the number k of crack sets being considered, $\underset{\sim}{D}$ is a sum of symmetric tensor products and thus a symmetric tensor itself. Consequently, it can be represented in its principal axes and corresponding principal directions $\underset{\sim}{v}_{(h)}$, $h = 1, 2, 3$ as follows:

$$\underset{\sim}{D} = D_{(1)}\underset{\sim}{v}_{(1)} \otimes \underset{\sim}{v}_{(1)} + D_{(2)}\underset{\sim}{v}_{(2)} \otimes \underset{\sim}{v}_{(2)} + D_{(3)}\underset{\sim}{v}_{(3)} \otimes \underset{\sim}{v}_{(3)}.$$

So, any system of crack-sets is equivalent to three orthogonal sets of parallel cracks.

The entity (4) can be otherwise viewed as a particular member of the family of damage tensors related to microcrack distribution examined recently and orderly described by Lubarda and Krajcinovic (1993).

2.2 Strain energy function and $\underset{\sim}{D}$-generated residual effects

We introduce the set $(\underset{\sim}{\varepsilon}, \underset{\sim}{D})$ of state variables and postulate the existence of thermo-dynamic potential (free energy per unit volume) $w(\underset{\sim}{\varepsilon}, \underset{\sim}{D})$. The material is assumed isotropic in the absence of damage, the only anisotropic properties arise from the presence of oriented microcracks, $\underset{\sim}{D} \neq \underset{\sim}{O}$. Since $w(\underset{\sim}{\varepsilon}, \underset{\sim}{D})$ is unchanged if both $\underset{\sim}{\varepsilon}$ and $\underset{\sim}{D}$ undergo a rotation, w is an isotropic invariant of $\underset{\sim}{\varepsilon}$ and $\underset{\sim}{D}$, i.e. can be regarded as a function of the independent invariants of $\underset{\sim}{\varepsilon}$ and $\underset{\sim}{D}$ including the simultaneous ones, Spencer (1984). For a given $\underset{\sim}{D}$, w represents the strain energy function defining elastic stress-strain relation

$$\sigma_{ij} = \frac{\partial w}{\partial \varepsilon_{ij}}(\underset{\sim}{\varepsilon}, \underset{\sim}{D}). \qquad (5)$$

Presuming the solid under consideration linearly elastic at constant $\underset{\sim}{D}$, $w(\underset{\sim}{\varepsilon}, \underset{\sim}{D})$ is at most quadratic in $\underset{\sim}{\varepsilon}$. Two additional hypotheses precise the influence of damage on stress-strain behaviour:

(i) w is a linear function of $\underset{\sim}{D}$. This statement may be regarded as equivalent to the hypothesis of a moderate density of non-interacting microcracks, Kachanov (1987).

(ii) $\underset{\sim}{D}$ is acting as a parameter modifying the strain energy function w. In particular, expressing w in the vicinity of $\underset{\sim}{\varepsilon} = \underset{\sim}{O}$ for previously irreversibly damaged material, one may expect non-zero residual stress components generated. Inversely, for $\underset{\sim}{\sigma} = \underset{\sim}{O}$ one may observe residual strains due to roughness and blocking on the microcrack lips.

In what follows the damage mechanism corresponding to the splitting-like ('brittle') microcracking will be considered. The splitting-like crack growth is a common mechanism in rocks, extensively observed experimentally, see f.ex. Peng and Johnson (1971), Lajtai (1974). When considering a conventional 'triaxial' axisymmetric compression test on a corresponding cyclindrical specimen undergoing quasi-homogeneous loading with $\sigma_3 < \sigma_2 = \sigma_1$, x_3 representing the cylinder axis and compression being assumed negative, the splitting mode of damage consists in microcracking oriented parallelly to x_3. The normal vectors n are thus all perpendicular to x_3, so we have correspondingly for (4): $D_{(1)} = D_{(2)} > 0$, $D_{(3)} = 0$, where $D_{(1)}$, $D_{(2)}$ and $D_{(3)}$ represent principal components of D. The directions x_1, x_2 and x_3 are principal directions for D as well as for the stress σ and the strain ε.

Anticipating the modelling of damage growth we may pursue by observing that in the foregoing confined 'triaxial' test as well as in the particular case of unconfined uniaxial compression the microcrack growth mode under consideration is correlated with lateral extensions (tensile lateral strains) orthogonally to crack-planes (parallelly to n). This leads to volume dilatancy and to an ultimate failure of the specimen. The statement that "tensile extensions in the principal directions contribute to microcrack growth" led Ju (1989) to introduce the fourth-order positive projection tensor P^+ that yields a new "tensile" strain tensor ε^+ from positive principal components of ε:

$$\varepsilon^+ = P^+ : \varepsilon, \quad \text{i.e.} \quad \varepsilon^+_{ij} = P^+_{ijkl} \varepsilon_{kl} \qquad (6)$$

The components of P^+ are defined as follows:

$$P^+_{ijkl} = \hat{Q}^+_{ia} \hat{Q}^+_{jb} \hat{Q}_{ka} \hat{Q}_{lb} \qquad (7)$$

with $\hat{Q} \equiv \sum_{i=1}^{3} q_{(i)} \otimes q_{(i)}$, $\hat{Q}^+ = \sum_{i=1}^{3} H(\varepsilon_i) q_{(i)} \otimes q_{(i)}$.

$H(.)$ is the Heaviside function, $q_{(i)}$ unit vector corresponding to the i-th principal direction of ε; $\varepsilon_{(i)}$ is the i-th principal strain. For the conventional 'triaxial' test $q_{(i)} = v_{(i)}$. $\hat{Q}$ and $\hat{Q}^+$ represent respectively the regular and positive spectral projections.

With the foregoing preliminaries and ex-emplification let us consider the foregoing free energy function (damage-dependent strain energy):

$$w(\varepsilon, D) = g \, \text{tr}(\varepsilon.D) + \frac{1}{2}\lambda(\text{tr}\,\varepsilon)^2 + \mu \, \text{tr}(\varepsilon.\varepsilon)$$
$$+ \alpha \, \text{tr}\,\varepsilon \, \text{tr}(\varepsilon.D) + 2\beta \, \text{tr}(\varepsilon.\varepsilon.D) \qquad (8)$$

where λ and μ represent classical Lame elastic constants defining the strain energy for non-damaged isotropic solid (the matrix), α and β are elastic coefficients related to damage-induced modification of the strain energy function and g is a constant relevant to damage-induced residual stresses.

The overall response dependent on elastic properties of the solid matrix and on averaged D-embodied intensity and orientation of microcrack field is elastically orthotropic, Kachanov (1980). The particular form (8) of the strain energy function, linearized with respect to D, yields effective elastic moduli expressions $E_{(i)}$, $v_{(ij)}$ and/or $G_{(ij)}$, functions of D and of four constants $\mu, \lambda, \alpha, \beta$. The form (8) gives different approximations of effective moduli than the complementary energy expression $u(\sigma,D)$ given by Kachanov, also linearized in D. An important difference resides in the presence of the term $g \, \varepsilon_{ij} D_{ij}$ relative to residual stress and the role played by this term in the context of microcrack-damage evolution treated below. The latter topic constitutes a fully original part of the present model.

Complete expressions of the elastic effective moduli are given and discussed in the more detailed paper, Dragon et al. (1994).

2.3 Elastic-damage response and damage growth

Making use of standard arguments relative to internal-variable framework, Kestin, Bataille (1977), we obtain from (8) the equations of state defining respectively the elastic behaviour and the thermodynamic force (damage energy release rate) relevant to D:

$$\sigma = \frac{\partial w}{\partial \varepsilon} = g \, D + \lambda (\text{tr}\,\varepsilon) \, 1 + 2\mu\varepsilon + \alpha \left[\text{tr}(\varepsilon.D)1 + (\text{tr}\,\varepsilon)D \right]$$
$$+ 2\beta (\varepsilon.D + D.\varepsilon), \qquad (9)$$

$$F^D = -\frac{\partial w}{\partial D} = -g\,\varepsilon - \alpha(\text{tr}\,\varepsilon)\,\varepsilon - 2\beta\,(\varepsilon.\varepsilon). \qquad (10)$$

We may distinguish in (10) the damage energy

release rate linked to the residual stress term of the strain energy w :

$$F^{D1} = -g\,\varepsilon = -g\,\varepsilon^+ - g\,\varepsilon^- = F^{D1+} + F^{D1-} \;;$$

$$\varepsilon^- = \varepsilon - \varepsilon^+, \qquad (11)$$

and the damage energy release rate relative to the remaining, reversible part :

$$F^{D2} = -\alpha\,(\operatorname{tr}\varepsilon)\,\varepsilon - 2\beta\,(\varepsilon.\varepsilon). \qquad (12)$$

The damage growth is supposed to be time-independent (non-viscous) and progressive one. The damage driving force F is not arbitrary but must belong to a convex reversibility domain C^D : $f\,(F^D, D) \le 0$, limited by the damage criterion $f\,(F^D, D) = 0$.

It is assumed that the evolution of damage D is governed in a determining manner by the driving force $F^{D1+} = -g\,\varepsilon^+$ corresponding to residual phenomena, i.e.

$$f\,(F^D, D) = f\,(F^D - F^{D2} - F^{D1-}, D). \qquad (13)$$

We assume furthermore that the evolution of D satisfies Hill's maximum dissipation principle equivalent to the normality rule

$$\dot{D} = \Lambda\,\frac{\partial f(F^D - F^{D2} - F^{D1-}, D)}{\partial F^D}, \quad \Lambda \ge 0 \qquad (14)$$

referred here to a regular (differentiable) damage loading function in f (equivalent of the yield function in plasticity).

We propose in particular the following form for f in the F^D-space :

$$f\,(F^D - F^{D2} - F^{D1-}, D) =$$

$$\sqrt{\tfrac{1}{2}\operatorname{tr}\!\left[(F^D - F^{D2} - F^{D1-}).(F^D - F^{D2} - F^{D1-})\right]}$$

$$-(C_0 + C_1 \operatorname{tr} D) \le 0. \qquad (15)$$

It can be seen from (13) that D appears explicitly in the term defining actual damage threshold $k\,(D) = C_0 + C_1 \operatorname{tr}\,(D)$ with C_0 and C_1 denoting positive definite material constants. In particular C_0 represents an initial damage limit and C_1 indicates material ductility in relation to damage progressiveness.

Accounting for (15) in (14), the damage process is characterized by the following evolution equation :

$$\dot{D} = \begin{cases} 0, & \text{if } f<0 \text{ or } f=0,\,\dot{f}<0 \\[4pt] \Lambda\left[\dfrac{\tfrac{1}{2}g(P^+:\varepsilon)}{\sqrt{\tfrac{1}{2}g^2\operatorname{tr}(\varepsilon^+:\varepsilon^+)}}\right], & \text{if } f=0 \text{ and } \dot{f}=0 \end{cases} \qquad (16)$$

The 'directional' term in (16) is $g\,P^+_{ijkl}\,\varepsilon_{kl}$, thus emphasizing the contribution of tensile extensions (when viewed in the principal directions) to microcrack growth. The components of $\dot{D}_{ij}$ are coaxial with those of ε^+_{ij}.

Remark : One can write (15) as a function of strain :

$$f\left[F^{D1+}(\varepsilon), D\right] = \hat{f}\,(\varepsilon, D). \qquad (15)'$$

Furthermore, the reduced representation of $\hat{f} = 0$ in the strain space corresponding to particular loading can be obtained as f.ex. the one for axisymmetric compression/tension with $\varepsilon_1 = \varepsilon_2$. It can be likewise transferred to the stress-space, through (9). The initial damage threshold, i.e. for $D = 0$ and subsequent ones relative to axisymmetric loading are shown in Fig. 1. One can see, in particular, different limits in uniaxial tension and compression respectively.

The typical stress-strain behaviour corresponding to the present model is illustrated in Fig. 2 for uniaxial tension. In this case the non-zero residual stress $\sigma^r = g\,D$ for $\varepsilon = 0$ after unloading appears in the axial direction and is a compressive one ($g < 0$) as the only non-zero component of D is the axial one according to (16). Under uniaxial compression loading the transversal (lateral) residual stress components appear under the splitting-like damage growth (16). The qualitative agreement with experimental data for a large class of rock-like solids is thus obtained.

2.4 Modelling options. Comments on evaluation of material parameters

The model introduced is an application-oriented one and constitutes a basis for more involved (coupled) modelling incorporating alternatively a form of anisotropic poroelasticity for a fluid-saturated rock (Charlez, 1993a) or plasticity-like effects relevant to frictional sliding in a microcracked medium. It is obvious that damage-plasticity coupled models are

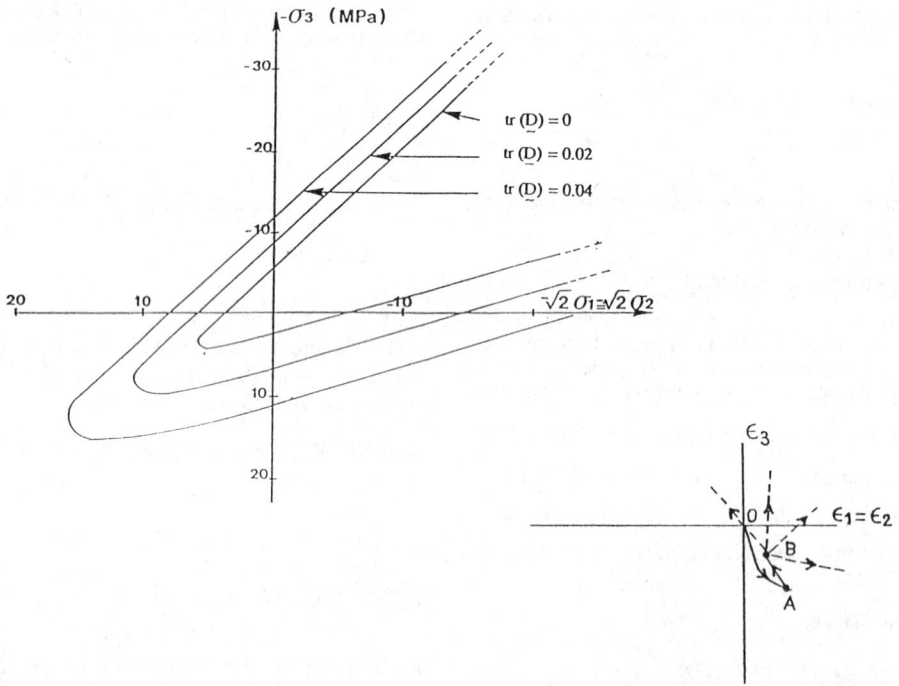

Fig. 1 Initial and subsequent simulated damage loci in stress-space corresponding to axisymmetric loading. The curves corresponding to tr D = 0.02 and tr D = 0.04 were obtained after relevant initial damage produced under uniaxial (unconfined) compression (the strain path O–A). After unloading A–B, radial strain paths starting from B were simulated until respective damage criterion was reached.

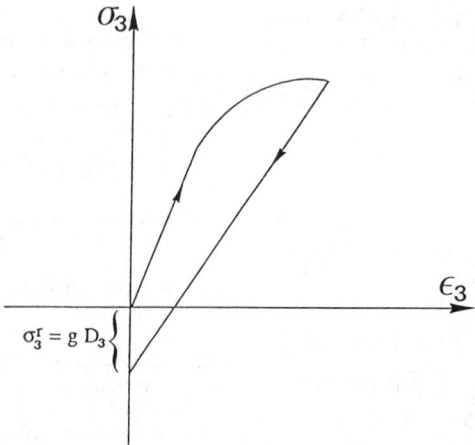

Fig. 2 Typical stress-strain behaviour according to the model ; uniaxial tension.

more complex (see f.ex. Andrieux et al., 1986, Gambarotta and Lagomarsino, 1993, in the similar context) and not easy to handle. So, we will proceed here with localization analyses using the basic version introduced above. Its another advantage is that one has to identify seven material constants : λ, μ, α, β, g, C_0, C_1 and seven is not a high number of constants when considering nonlinear and anisotropic material behaviour.

Pursuing furthermore in the application-oriented spirit we tend to evaluate the constants mentioned from a widely available engineering test. From such a standpoint, the axisymmetric triaxial apparatus, constituting a standard equipment of any geomechanical laboratory, offers sufficient feasibility provided both loading and unloading data are explored. In the corresponding conventional test with unloading, with splitting microcracking predominant, the principal axes of damage tensor defining the axes of induced orthotropy are fixed and the same ones as those of stress and strain tensors. Moreover, orthotropy reduces to the transversal isotropy with respect to the x_3-axis (the vertical axis of the sample) so that any two orthogonal axes in the horizontal plane may be regarded as principal ones since $\sigma_2 = \sigma_1$, $\varepsilon_2 = \varepsilon_1$, $D_{(2)} = D_{(1)}$.

Considering the 'virgin' rock as approximately undamaged and isotropic, one can determine the conventional elastic constants λ and μ from E_0 and ν_0 (initial Young modulus and Poisson's ratio) relevant to the stress-strain relationships $(\sigma_3 - \sigma_2)$ vs. ε_3 and $(\sigma_3 - \sigma_2)$ vs. ε_2 respectively, in the elastic range without damage growth $(f < 0)$.

To determine the values α, β and g, the non-linear, i.e. damage affected portions of the experimental curves mentioned and the subsequent unloading portions are exploited. The point at which unloading is performed should correspond to damage $D_{(2)} = D_{(1)}$ advanced enough. On the other hand it should be chosen reasonably far from the anticipated stress-deviator peak-value to avoid any interference with eventual bifurcation phenomena. The unloading portion corresponding to $\underset{\sim}{D}$-modified degraded moduli $E_{(3)}$, $E_{(2)}$, $\nu_{(32)}$ are linked to α, β, $D_{(2)}$. The detailed procedure concerning the latter constants as well as the one concerning the damage locus constants C_0 and C_1 is described elsewhere (Pham, 1994), see also Dragon et al. (1993).

In the last paper a preliminary version of the present model was given. This preliminary version allowed to reproduce correctly the complex stress-strain behaviour for $\underset{\sim}{P}^+$-proportional loading paths, i.e. the paths corresponding to a fixed configuration of principal strain directions and components (including their positiveness and negativeness). Difficulties appeared for strain paths implying variations in $\underset{\sim}{P}^+$, i.e. for $\underset{\sim}{P}^+$-non-proportional strain-paths. In this case stress discontinuities appeared during local strain and damage evolution. As shown by Cormery (1994), the origin of such discontinuities was the presence of $\underset{\sim}{P}^+$ in the first term of the strain energy $w(\underset{\sim}{\varepsilon}, \underset{\sim}{D})$ reading : $g\,(\underset{\sim}{P}^+ : \underset{\sim}{\varepsilon})$: $\underset{\sim}{D}$. The explicit occurrence of $\underset{\sim}{P}^+$ in this term was redundant as the actual cumulated damage $\underset{\sim}{D}$ contains already (see Eqn (16)) – throughout its step-by-step evolution – the $\underset{\sim}{P}^+$ effect governing the splitting-like cracking mechanism. The redundant presence of $\underset{\sim}{P}^+$ in the strain energy expression has been suppressed in the present model, see Eqns (8)-(10). Consequently, it allows to follow the $\underset{\sim}{P}^+$-non-proportional and cyclic strain-paths as well.

Regarding the evaluation of material parameters from loading-unloading 'triaxial' test as discussed above, it should be stressed that the complete set of constitutive parameters may be formally determined through a single conventional test (i.e. a couple of curves $(\sigma_3-\sigma_2)$ vs. ε_3 and ε_2 respectively, under a given confining pressure). In fact it is preferable to cover a reasonable range of confining pressure values using several tests and settle ultimate values of the constants through averaging process.

Furthermore, the hypothesis of an initially undamaged and isotropic solid can be circumvented and any initial microcrack-damage pattern (if known) may be accounted for. An example of experimental vs. model stress-strain curves for Vosges sandstone under confining pressure of – 50

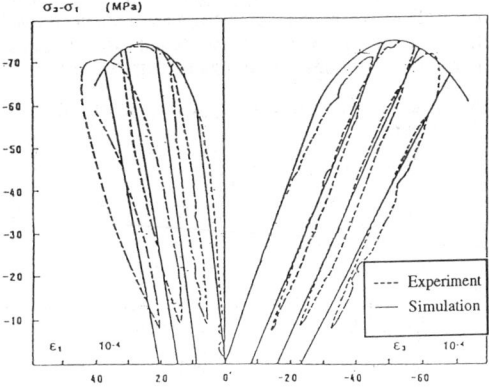

Fig. 3 Comparison of the experimental data by Charlez (1993b) and anisotropic model simulation for Vosges sandstone (axisymmetric 'triaxial' test ; $p_c = -50$ MPa).

MPa is shown in Fig. 3. The experimental data by Charlez (1993b) have been used for identification. The set of material parameters adopted further in the localization analysis is as follows:

λ = 14 141 MPa
μ = 7 954 MPa
α = 19 000 MPa
β = – 14 280 MPa
g = – 179 MPa
C_0 = 0.015 MPa
C_1 = 0.96 MPa

3. TANGENT STIFFNESS TENSOR AND 3-D LOCALIZATION DETECTOR

3.1 Tangent stiffness. Localization problem.

The tangent stiffness corresponding to the anisotropic damage model introduced in the foregoing (Eqns (8)-(16)) has the following form:

$$\underset{\sim}{L} = \underset{\sim}{U} + \bar{\eta}\, \underset{\sim}{V} \qquad (17)$$

where $\bar{\eta}$ equals to 1 or 0 according to whether the damage loading or unloading is considered respectively, as indicated in (16). The damage multiplier $\Lambda \geq 0$ is being correspondingly 'activated' or 'desactivated', as in the similar inviscid plasticity formalism. The tensor $\underset{\sim}{L}$ is endowed with the minor symmetries: $L_{ijkl} = L_{jikl} = L_{ijlk}$ but not with the major one: $L_{ijkl} \neq L_{klij}$. This is here a specific consequence of $\underset{\sim}{P}^+$–governed splitting– like damage growth, but illustrates, in the same time, a more general affinity of a class of damage models with non-associated plasticity models. In other terms, normality characterizing eventually damage growth (as in the present model) does not necessarily lead to normality-like configurations between the stress rate $\underset{\sim}{\dot{\sigma}}$ and damage-related non-elastic strain rate $\underset{\sim}{\dot{\varepsilon}}^D$. Referring to geometrical features in the stress/strain space it can be stated that there is no reason to have acute-angle-configurations between $\underset{\sim}{\dot{\sigma}}$ and $\underset{\sim}{\dot{\varepsilon}}^D$ (analogous to those between $\underset{\sim}{\dot{\sigma}}$ and $\underset{\sim}{\dot{\varepsilon}}^P$ in associated plasticity) in the typical framework of damage modelling. This is obvious if one realizes that no evolution law is being postulated for $\underset{\sim}{\dot{\varepsilon}}^D$ itself.

The lack of the major symmetry above is a particular expression of structural resemblance of damage constitutive response (even if D-normality exists) and non-associated plasticity response, see also Désoyer and Cormery (1994).

It should be added however that some authors incorporate a supplementary evolution equation for damage-related strain rate $\underset{\sim}{\dot{\varepsilon}}^D$ in damage modelling, see f.ex. Chow and Yang (1991), Laborde and Michrafy (1991). If, specifically, such an evolution law satisfies normality, the analogy with associated plasticity exists, contrarily to the foregoing, more common case.

The components of $\underset{\sim}{U}$ and $\underset{\sim}{V}$ are given as follows:

$$U_{ijpq} = \lambda\, \delta_{ij}\, \delta_{pq} + \mu\, (\delta_{ip}\, \delta_{jq} + \delta_{iq}\, \sigma_{jp})$$
$$+ \alpha\, (\delta_{ij}\, D_{pq} + D_{ij}\, \delta_{pq}) + \beta\, (\delta_{ip}\, D_{iq}$$
$$+ D_{ip}\, \delta_{jq} + \delta_{iq}\, D_{jp} + D_{iq}\, \delta_{jp})\, ; \qquad (18)$$

$$V_{ijpq} = M\left[\alpha\, (\varepsilon_{mm}\, \varepsilon^+_{nm})\, \delta_{ij} + \alpha \varepsilon_{mm}\, \varepsilon^+_{ij}\right.$$
$$\left. + 2\beta\, (\varepsilon_{im}\, \varepsilon^+_{mj} + \varepsilon^+_{im}\, \varepsilon_{mj}) + g\, \varepsilon^+_{ij}\right] \varepsilon^+_{pq}$$

with $\quad M = \dfrac{|g|}{C_1\, \mathrm{tr}\, \underset{\sim}{\varepsilon}^+\, \sqrt{2\, \mathrm{tr}\, (\varepsilon^+.\varepsilon^+)}}$.

When considering the localized bifurcation in an infinitesimal neighbourhood of a point in a damaging solid (or in a homogeneous system, i.e. a domain with homogeneous stress/strain field and past damage history) it is useful to insist on simultaneity of strain and damage localization (Billardon and Doghri, 1989). In other terms the strain localization (classically: onset of a strain-rate discontinuity) brings about damage localization via damage-rate vs. strain-rate connection. The latter can be viewed explicitly for the present model when expressing the damage multiplier Λ (from consistency equation $\dot{f}(\varepsilon,D) = 0$):

$$\Lambda = \dfrac{|g|\, \mathrm{tr}\, (\underset{\sim}{\varepsilon}^+ : \underset{\sim}{\dot{\varepsilon}}^+)}{C_1\, \mathrm{tr}\, \underset{\sim}{\varepsilon}^+} \qquad (19)$$

Following Borré and Maier (1989) we can approach the localization problem considering the actual solid with the nonlinear constitutive response:

$$\underset{\sim}{\dot{\sigma}} = \underset{\sim}{L}(\varepsilon, D) : \underset{\sim}{\dot{\varepsilon}} \quad \text{with}\quad \underset{\sim}{L} = \underset{\sim}{U} + \bar{\eta}\, \underset{\sim}{V},\ \bar{\eta} = 0\ \text{or}\ 1 \qquad (20)$$

i.e. with the non-linear damage evolution law (16), with both loading-loading and loading-unloading bifurcations allowed. The criterion:

$$\det Q \leq 0, \qquad Q \equiv N.L.N \qquad (21)$$

represents a sufficient and necessary condition for bifurcation. In the same time the localization threshold itself is correctly determined from (1) for a linear comparison material and loading-loading bifurcation, see Proposition 6, Borré and Maier (1989), p. 40.

The advantage of the analysis by Borré and Maier resides evidently in embracing more general context of incremental non-linearity (see also Chambon and Desrues (1984)). In the context of numerical search algorithms aiming at minimizing $\det \underset{\sim}{Q} (\underset{\sim}{N})$, the condition (21) validates (i) normals $\underset{\sim}{N}$ for which $\det \underset{\sim}{Q}$ becomes negative and (ii) eventual non-continuous variations of $\det \underset{\sim}{Q} (\underset{\sim}{N})$ in the course of numerical search. In some way it legitimates numerical procedures like the one by Ortiz (1987). The analysis below modifies and generalizes the technique of Ortiz.

3.2 Computational procedure for localization detection

We consider $F(\underset{\sim}{N}) = \det \underset{\sim}{Q} (\underset{\sim}{N})$. By expressing $\underset{\sim}{N}$ in terms of spherical angles Θ and Φ (see Fig. 4), the problem of minimizing $F(\underset{\sim}{N})$, subject to $(\underset{\sim}{N}) = 1$ consists in minimizing the expression:

$$F(\underset{\sim}{N}) = \hat{F}(\Theta, \Phi)$$

$$= \sum_{p,q,r} A_{pqr} (\sin \Phi)^{p+q} (\cos\Theta)^p (\sin\Theta)^q (\cos\Phi)^r \quad (22)$$

with $\Phi \in \left[0, \frac{\pi}{2}\right]$, $\Theta \in [0, 2\pi]$, $p,q,r = 1,2,...,6$ and $p + q + r = 6$. The coefficients A_{pqr} are determined by actual tangent stiffness (Eqns (17), (18)).

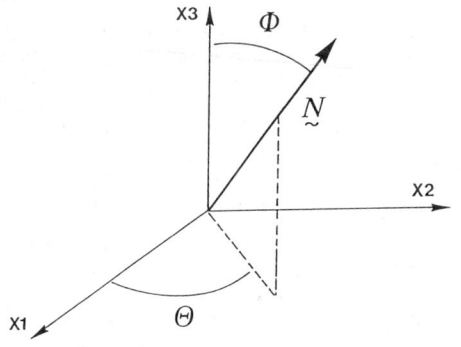

Fig. 4 Configuration of the spherical angles Θ, Φ with respect to cartesian axes x_1, x_2, x_3.

The first approach, following Ortiz (1987), consists in discretizing both Φ and Θ and checking $\hat{F}(\Theta_i, \Phi_i) = 0$ by sweeping the corresponding range of variations of Θ and Φ. This is to be done at each step during the integration process at each quadrature point for current tangent stiffness $\underset{\sim}{L}$ at stake. Practically, sweeping the space gives a first approximation of the minima of $\hat{F}$, say (Θ_m^o, Φ_m^o), that is subsequently refined using e.g. one of iterative gradient schemes. The method of conjugated gradient was found to give satisfactory results for refinement.

The approach was tested for a family of elastic-damage and elastic-plastic-damage models. For two simple isotropic damage models considered by Désoyer and Cormery (1994) the spectral analysis was parallelly performed. It showed that some solutions could not be detected by the algorithm above. A second algorithm was then proposed with a single argument, e.g. Θ, being discretized. By sweeping likewise the respective range of Θ at, say 5-degree increments, the equation $\hat{F}(\Theta_i, \Phi) = 0$ was solved at each Θ_i with respect to Φ by an iterative scheme. This kind of location method, though evidently dependent on the choice of a set $\{\Theta_i\}$ was found to be well adapted to the search of localization bands. It could be perfected by repeating computations in the alternative way i.e. by discretizing Φ and then solve the equation $\hat{F}(\Theta, \Phi_j)$ with respect to Θ at each Φ_j.

The second approach was finally retained as it proved fully satisfactory when compared to analytical solutions (via spectral properties of $\underset{\sim}{Q}$ for isotropic damage models). In what follows, some selected results are presented for localization detection combined with anisotropic damage.

After determining the orientations $\underset{\sim}{N}$ of the local discontinuity planes, the corresponding failure modes $\underset{\sim}{M}$ (solving $\underset{\sim}{Q}.\underset{\sim}{M} = \lambda \underset{\sim}{M}$ for $\lambda = 0$) are computed.

4. FAILURE INCIPIENCE THROUGH STRAIN AND DAMAGE LOCALIZATION

4.1 Results for axisymmetric proportional strain paths and conventional 'triaxial' loading

The localization detection procedure discussed in Section 3 combined with the anisotropic damage model introduced in Section 2 was applied to study the localized failure modes for Vosges sandstone under various loading conditions. The set of material constants given at the end of Section 2 has been employed in this analysis.

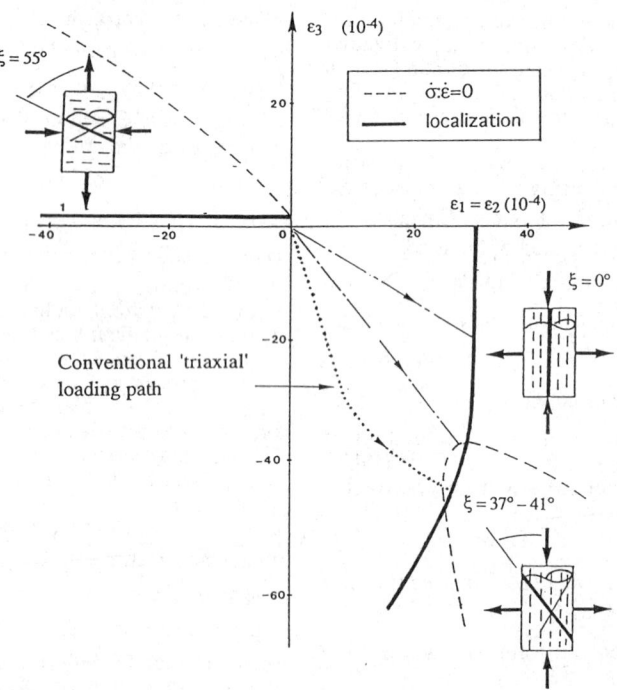

Fig. 5 Strain space representation of results corresponding to the localization onset (solid lines) for strain-controlled radial paths and for a series of conventional 'triaxial' loading paths. Localization vs. loss of positiveness of the second-order work.

The results obtained for a large number of proportional (radial) strain paths in the space ($\varepsilon_1=\varepsilon_2,\varepsilon_3$) corresponding to axisymmetric straining are shown in Fig. 5. The solid lines indicate the localization onset while the dashed lines give strains corresponding to peak-stress values, i.e. to the softening incipience relative to zero value of second-order work $\overset{\cdot}{\underset{\sim}{\sigma}}:\overset{\cdot}{\underset{\sim}{\varepsilon}}$. Strong regularity of strain level coinciding with the localization onset can be observed in the quarter ($\varepsilon_1 < 0, \varepsilon_3 > 0$). Localization occurs systematically prior to the peak in the stress-strain curve. The cone-shaped localization locus corresponding to 'infinity' of solutions is obtained. This means that localization onset appears here for $\Phi = 35°$ and for any Θ. Note that in Figure 5 the angle ξ between the plane of failure and x_3 is indicated instead of Φ ; $\xi = \pi/2 - \Phi$. A mixed discontinuity mode is being observed with the separation ('opening') mechanism prevailing over the 'shearing' one. The radial paths considered are strain-controlled here ; they can be seen as qualitatively corresponding to axisymmetric 'triaxial' extension testing programmes.

A closer relationship with experimental data has been established in the quarter ($\varepsilon_1 > 0, \varepsilon_3 < 0$) where both radial strain controlled paths as well as genuine conventional 'triaxial' loading paths (axial compression increasing after some initial uniform pressure) were considered. Two localization failure patterns are obtained for radial strain paths. Axial splitting is observed following distributed damage pattern for smaller absolute values of ε_3 ; once again multiple solutions (any Θ) are obtained. For increasing axial compression the infinity of localization loci forming a cone-shaped surface are obtained. The angle ξ varies in a small interval ($37° \leq \xi \leq 41°$). Again a mixed discontinuity mode is found but the 'shearing' mechanism is more pronounced that separation. For majority of radial paths localization occurs prior to the peak except for a narrow segment where the inverse effect is observed.

A loading path relevant to the conventional 'triaxial' compression test corresponds to a curved trajectory as depicted by the dotted line in Fig. 5. A set of loading paths assuming six different levels of confining pressure (from 0 to 50 MPa) was examined, see Fig. 6. Despite of some difference between radial and curved trajectories we obtain the analogous localized failure pattern. Moreover, the simulations performed fall into the 'inversion segment' mentioned above giving localization onset

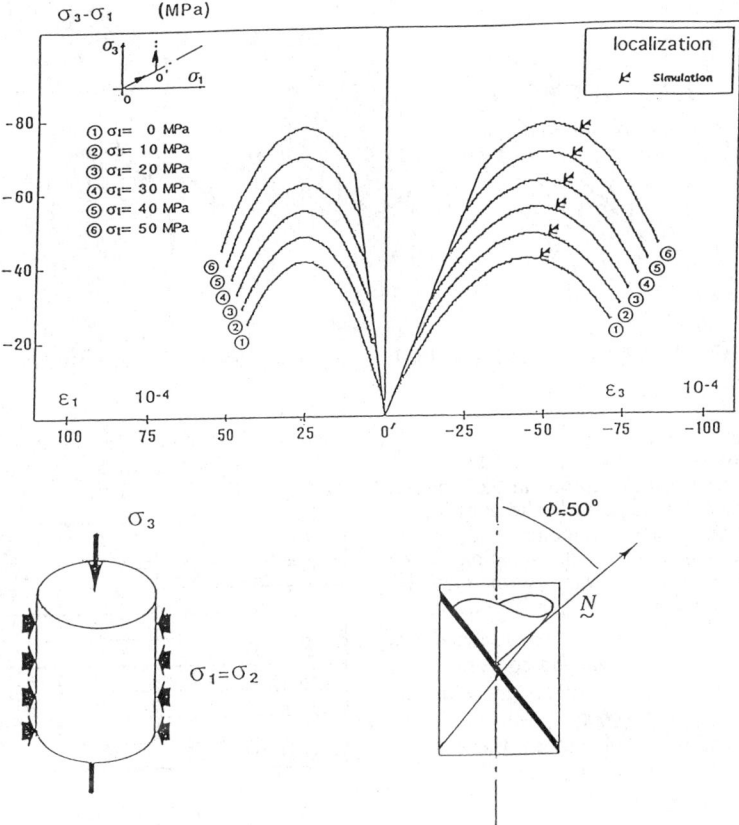

Fig. 6 Stress-strain curves and localization onset for a set of conventional 'triaxial' loading paths. Localized failure pattern.

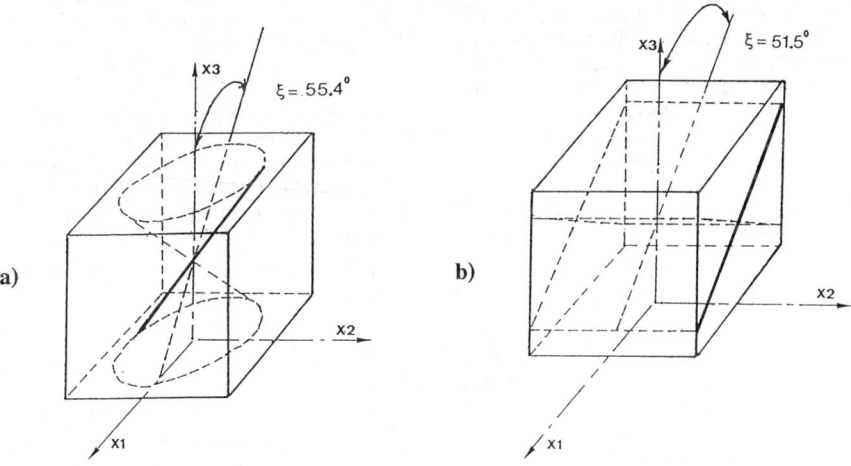

Fig. 7 Localized failure patterns for a class of plane-strain radial paths.

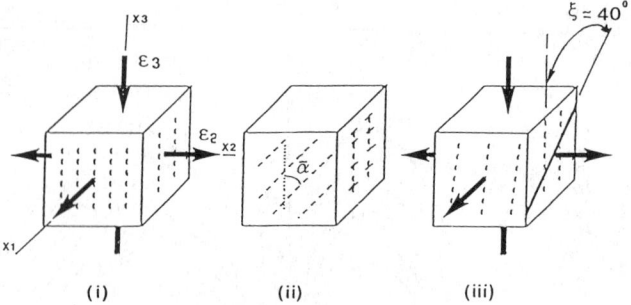

Fig. 8   Off-axes strain paths : loading programme and localization pattern.

after the stress-strain peak.

It can be reminded that localization features are known to be sensitive to variations in the material constants and to subtle aspects of the constitutive modelling. A premiliary parametric study in the present context indicates that the position of the 'inversion segment' may vary with material constants ; it seems to be particularly sensitive to variations of α. On the other hand experimental data available for Vosges sandstone and concerning the localization in the 'triaxial' test do not coincide. Some of them (Santarelli (1990), Berthaud (1992)) indicate localization onset before the loss of positiveness of second-order work.

Table 1   Computed values of Θ and ξ defining the orientation of localization planes under off-axes simulations.

tr D	γ	$\bar{\alpha}$	ξ	Θ
0.2	−2	30°	41.7°	176.6°
0.2	−2	60°	39.7°	187.7°
0.2	−0.7	30°	40.8°	178.2°
0.2	−0.7	60°	38.4°	187.6°
0.1	−2	30°	41.1°	174.7°
0.1	−2	60°	41.1°	178.3°
0.1	−0.7	30°	40.6°	177.5°
0.1	−0.7	60°	40.2	179.3°

4.2 Effects of anisotropy : off-axes straining and structural analysis

Advantages of a genuine three-dimensional search algorithm as the one applied here could be seen in the foregoing as multiple localization directions in space forming cone-shaped loci have been dectected. It can be reminded that the applications given by Ortiz (1987) are limited to two-dimensional configurations, though the search technique in three dimensions was postulated by this author. More examples showing utility of systematic 3D-search can be cited. One case of interest is the following set of plane strain radial paths :

$$\varepsilon_1 = 0, \quad \varepsilon_2 = a\,\varepsilon_3, \quad -1 \leq a \leq 1$$

For $-1 \leq a \leq 0$ infinity of bands is obtained (localization onset at any Θ) forming a cone-shaped locus characterized by a semi-angle ξ (with respect to x_3) equal 55.4° (see Fig. 7a). For $0 < a < 1$ the localization pattern is different : two locally planar bands appear ; their configuration is depicted in Fig. 7b. A particular feature is that both are found away from the plane (x_2, x_3) of straining. It has been shown parallelly that by imposing $\Theta = \pi/2$, i.e. bifurcation in the plane (x_2,x_3), there was no solution involving the anisotropic model under consideration. This result has been corroborated analytically using MATHEMATICA software. It seems to reflect effects of damage-induced anisotropy since for the basic isotropic damage material considered by Désoyer and Cormery (1994) solutions exist in the plane of straining.

For the simulations quoted up to now, the loading-axes (principal strain/stress axes) were considered to coincide with principal axes of $\underset{\sim}{D}$, i.e. with anisotropy axes. Such coaxiality is well known to 'mask' certain salient anisotropy effects. The latter are largely pronounced when off-axes test and/or simulations are effected. These considerations have motivated the loading programme involving (i) some initial damage simulated e.g. by compression along x_3 : $\varepsilon_2 = \varepsilon_1 = \gamma\,\varepsilon_3, \gamma < 0$, followed by (ii) rotation of the principal damage axes in the (x_2, x_3)-plane (the angle $\bar{\alpha}$ with respect to x_3), and (iii) reloading preserving the same ratio γ along x_1, x_2 and x_3.

There is some re-orientation of damage pattern during the reloading stage (according to the splitting-like mechanism, Eqn (16)). As concerns the localization pattern, a single plane of localization is detected, see Fig. 8, instead of fully symmetric

a)

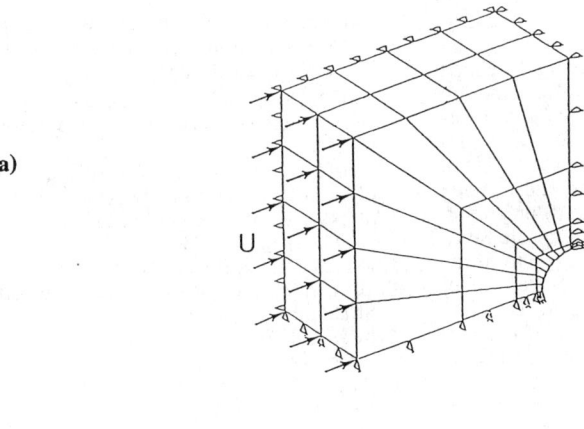

b)

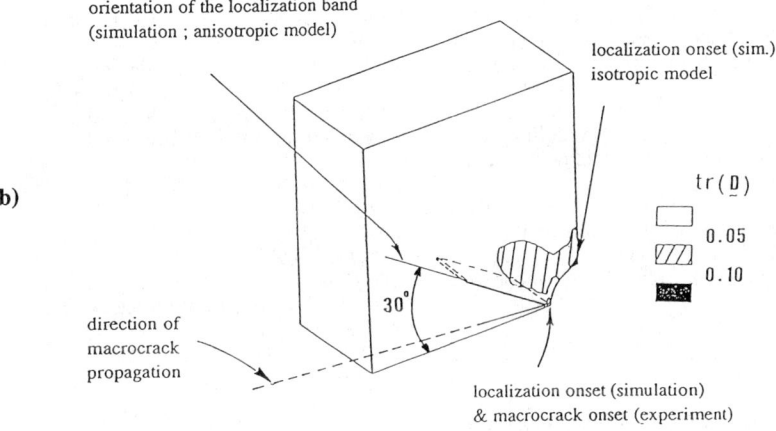

Fig. 9 Finite-element analysis of a rock-structure with a hole : discretization and boundary conditions for a quarter of the block. Damage localization vs. macrocrack onset (Onaisi, IFP (1989)).

pattern characterizing 'in-axes' monotonic simulations. This seems to be relevant to damage-induced anisotropy, well-pronounced in the course of off-axes reloading stage. Table 1 gives the computed values of Θ and ξ defining roughly constant orientation of localization in spite of varying initial damage, the strain-path parameter γ and the rotation angle $\bar{\alpha}$.

After presenting some predictions of localization analysis for loading paths at the elementary level (homogeneous states in the vicinity of a material point), the last point to be approached concerns an application on the level of structural analysis. A rock block with a hole (a quarter is shown in Fig. 9a) was considered. Careful experimental analysis of cracking onset about the hole has been performed by Onaisi (1989) at the Institut Français du Pétrole for different ratios of block's dimensions. The influence of slenderness on failure mechanisms was examined in particular. For a particular block considered (with dimensional proportions 2:2:1) we have performed finite element analysis involving (i) the basic elastic-damage isotropic model mentioned above and (ii) the anisotropic damage model under consideration. In both cases the numerical 3-D localization detector was employed for each step as a post-processor. The analysis involving the damage isotropic model indicated the localization onset in the area near the pole of the hole (see Fig. 9b). By contrast, the computations applying the anisotropic model indicate the localized failure onset at the prolongation of the 'equator' of the hole in accordance with experimental observations by Onaisi. The experimentally observed propagation was nearly horizontal (at about 10°) while the slope of the bifurcation plane was about 30° ; it is also slightly oblique with respect to block's front wall.

5. CONCLUSION

Attention has been focused on microcrack-related anisotropic damage in rock-like solids and on transition of damage evolution from a volume-distributed one to surface-localization resulting from the loss of ellipticity of the associated rate-problem. It makes sense to consider the strain-and-damage localization as a failure (macrocracking) precursor. To this purpose a numerical (finite-element-implemented) detector of localization in three dimensions has been applied using the model presented here. The model attempts to associate application-oriented spirit, accessibility of material parameters through common experimental testing and consistency of constitutive framework.

The numerical results concerning localized failure initiation for a wide range of loading paths can be considered to be in accordance with essential failure modes for rocks. The basic interest has been to study localization vs. anisotropy effects including a boundary-value problem for rock structure. Comparisons with localization mechanisms associated with damage isotropic modelling show insufficiency of the latter when oriented cracking becomes predominant.

The fair predictions obtained yield the anisotropic model and the technique of detection of localized bifurcation suited for engineering applications concerning f.ex. borehole-related rock mechanics problems.

REFERENCES

ANDRIEUX S., BAMBERGER Y., MARIGO J.-J., 1986. Journal de Mécanique théorique et appliquée, 5, 471-513.
BENALLAL A., BILLARDON R., GEYMONAT G. 1989. C.R. Acad. Sci. Paris, 308, Sér. II, 893-898.
BERTHAUD Y. 1992. Research Rep., LMT-Cachan.
BILLARDON R., DOGHRI I. 1989. C.R. Acad. Sci. Paris, 308, Sér. II, 347-352.
BORRE G., MAIER G. 1989. Meccanica, 24, 36-41.
CHAMBON R., DESRUES J. 1984. Mech. Res. Comm., 11, 145-153.
CHARLEZ P. 1993a. Private communication.
CHARLEZ P. 1993b. Research Rep. TOTAL.
CHOW C.L., YANG F. 1991. Engineering Fracture Mechanics, 40, 335-343.
CORMERY F. 1994. Doctoral Thesis, University of Poitiers-ENSMA.
DESOYER T., CORMERY F. 1994. Int. J. Solids Structures, 31, 733-744.
DRAGON A., CHARLEZ P., PHAM D., SHAO J.F. 1993. [in] Assessment and Prevention of Failure Phenomena in Rock Engng., Pasamehmetoglu et al. eds, Balkema, 71-78.
DRAGON A., CORMERY F., CHARLEZ Ph., DESOYER T. 1994. [to be published].
GAMBAROTTA L., LAGOMARSINO S. 1993. Int. J. Solids Structures, 30, 177-198.
JU J.W. 1989. Int. J. Solids Structures, 25, 803-833.
KACHANOV M. 1980. J. Engng Mech. Div., Proc. ASCE, 106, 1039-1051.
KACHANOV M. 1987. [in] Damage Mechanics in Composites-Proc. ASME Ann. Meeting, ASME, 99-105.
KESTIN J., BATAILLE J. 1977. [in] Continuum Models of Discrete Systems, Univ. Waterloo Press., 39-67.
LABORDE P., MICHRAFY A. 1991. Eur. J. Mech., A/Solids, 10, 213-536.
LAJTAI E.Z. 1974. Int. J. Fracture, 10, 525-536.
LUBARDA V.A., KRAJCINOVIC D. 1993. Int. J. Solids Structures, 30, 2859-2877.
ONAISI A. 1989. Doctoral Thesis, Ecole Centrale de Paris.
ORTIZ M. 1987. Mechanics of Materials, 6, 159-174.
OTTOSEN N.S., RUNESSON K. 1991. Int. J. Solids Structures, 27, 401-421.
PENG S., JOHNSON A.M. 1971. Int. J. Rock Mech. Min. Sci., 9, 37-86.
PHAM D. 1994. Doctoral Thesis, University of Poitiers-ENSMA.
RICE J. 1976. [in] Theoretical and Applied Mechanics, W.T. Koiter ed., North-Holland, 207-220.
RICE J., RUDNICKI J.W. 1980. Int. J. Solids Structures, 16, 597-605.
SANTARELLI F.J. 1990. Rev. Française Géotechnique, 50, 61-70.
SPENCER A.J.M. 1984. [in] Continuum Theory of the mechanics of fibre-reinforced composites, Springer-CISM n°282, 1-32.
VARDOULAKIS J. 1976. Mech. Res. Comm., 3, 209-214.

Essential features of a Cosserat continuum in interfacial localisation
Localisation d'interface dans les milieux de Cosserat

P. Unterreiner & J. Sulem
CERMES, Ecole Nationale des Ponts et Chaussées, Paris, France

I. Vardoulakis
National Technical University of Athens, Greece

M. Boulon
Laboratoire Sols, Solides, Structures (3S), UJF – INPG – CNRS, Grenoble, France

ABSTRACT : Interfacing of a granular material with a structural member involves the localisation of the deformations within a few grain diameter thick layer, which is called interface layer. This layer has been shown experimentally to depend mostly on the roughness characteristics and on the grain diameter but not on the geometrical dimensions of the sample or structural member. Among all the interface shear tests available in the literature for studying the formation of interfaces in a granular medium, we have selected the plane simple shear test and the ring simple shear test since they allow the sample to deform freely in its volume before interface localisation starts.

When modelling the plane simple shear test with a classical continuum, one cannot explain the formation of an interface layer unless one introduces a strong heterogeneity with a softer material near the interface. The analysis of the ring simple shear test with a classical continuum and rigid plastic constitutive equations yields a boundary layer of plastic material near the interface which has either a zero thickness for an associated material or a thickness directly proportional to the radius of the interface cylinder for a non-associated material.

Within a granular material, individual grains can translate and rotate. A classical continuum takes into account only the transitional degrees of freedom, while a Cosserat continuum considers the additional rotational degrees of freedom. Thus limitations of a classical continuum can be overcome within the framework of a Cosserat continuum.

The plane and ring simple shear tests are analysed using linear elastic and rigid plastic constitutive equations respectively. Both types of constitutive equations introduce a material length which is shown to control the thickness of the interface layer independently of the geometric dimensions of the sample. Moreover, it is shown that localisation of deformation is directly related to the existence of antisymmetric stresses in the localised zones.

L'interaction d'un milieu granulaire avec un élément de structure met en jeu une localisation de la déformation au sein d'une couche de quelques grains, la couche d'interface. On a pu monter expérimentalement que cette couche dépend essentiellement des caractéristiques de rugosité et du diamètre des grains, mais pas des dimensions de l'échantillon ni de la structure. Parmi les dispositifs d'essai possibles, nous avons choisi l'essai de cisaillement simple plan et l'essai de cisaillement simple annulaire parce qu'ils laissent libres la déformation en volume avant la naissance de la localisation d'interface.

Dans une modélisation de l'essai de cisaillement plan avec un continu classique, on ne peut pas expliquer la formation d'une couche d'interface à moins d'introduire une hétérogénéité forte sous la forme d'un matériau moins résistant au voisinage de l'interface. Dans le cas de l'essai de cisaillement annulaire, la modélisation avec un milieu continu classique rigide plastique donne une couche plastifiée d'épaisseur nulle pour un matériau standard, et une épaisseur fonction du rayon du cylindre pour un matériau non-standard.

Dans un milieu granulaire, les grains peuvent se déplacer individuellement en translation et en rotation. Le milieu continu classique ne prend en compte que la liberté en translation, tandis que le milieu de Cosserat considère des degrés de liberté en rotation. Ainsi les limitations des milieux continus classiques peuvent être dépassées.

On analyse les essais de cisaillement plan et annulaire avec respectivement un modèle élastique linéaire et un modèle rigide plastique. Ces deux modèles introduisent une longueur matérielle et on montre que cette longueur contrôle l'épaisseur de l'interface indépendamment de la dimension de l'échantillon. On montre en outre que la localisation est directement liée à l'existence de contraintes antisymétriques dans les zones localisées.

1. INTRODUCTION

Most civil engineering constructions involve the interfacing of soil and structural members. For various materials such as granular soils (Vesic, 1977; Clichy et al., 1989; Boulon, 1989) or model materials like Schneebeli rods (Bogdaneva-Bontcheva and Lippmann, 1975; Löffelmann, 1989), it has been observed that very soon after shearing starts across an interface, the shear deformations concentrate into a thin layer, a few grains diameter thick, at the contact with the structural member. This thin layer can be called the interface layer and is characterised by high displacement gradients, strong local dilatancy and significantly high grain rotations (Boulon, 1989; Hoteit, 1990).

The interface thickness depends on the grain size and the roughness of the contact surface. At the scale where such phenomena occur, i.e., at the scale of a few grains, the interface surface of the structural member is viewed by the grains as a plane. The only remaining space variable of interest is thus the distance to the interface surface. Among all interface shear tests, we may select the plane simple shear test (Kishida and Uesugi, 1987) and the ring simple shear test (Unterreiner et al., 1994) (Figure 1). First, they induce strain and stress fields in the sample, which depend only on the distance to the interface. Secondly, they do not impose the location of the localisation like it is the case in direct shear tests.

The plane simple shear test simulates the shearing by a plane structural member of a long layer of soil, of finite thickness (Figures 1a and 1b). On the other hand, the ring simple shear test simulates the shearing of a thick wall cylinder in plane strain by an internal cylindrical inclusion submitted to torsion (Figures 1c and 1d).

Analysis of the two simple shear tests, using a "classical" continuum model does not yield any of the localisation phenomena actually observed. A "classical" continuum is defined as a continuum where only degrees of freedom in translation are taken into account for defining the kinematics. For such a continuum, the displacement vector at a given point M is determined experimentally by averaging the displacements of the grains over a sufficiently large surface or volume as compared to the grain size. The individual grain rotations are ignored for defining the kinematics.

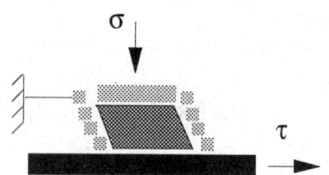

Figure 1a : Plane simple shear apparatus

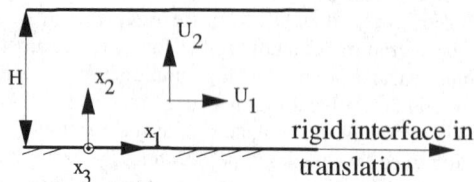

Figure 1b : Modelling of plane simple shear

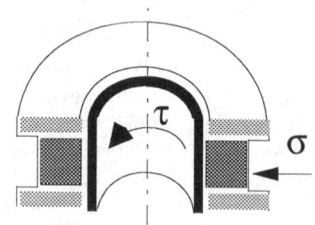

Figure 1c : Ring simple shear apparatus

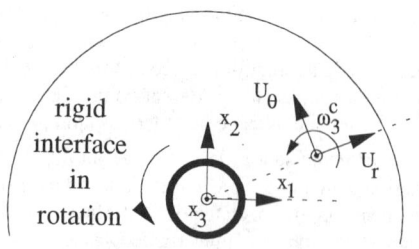

Figure 1d : Modelling of ring simple shear

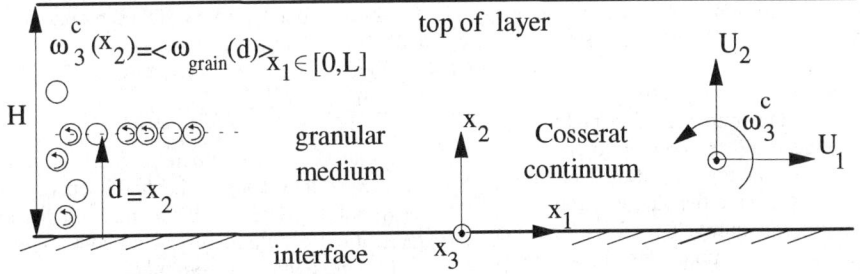

Figure 2 : Modelling of a granular medium with a Cosserat continuum

When modelling plane simple shear of an homogeneous sample using a classical continuum, it is impossible to explain the formation of an interface layer even when one assumes different boundary conditions (at the interface and at the top of the sample) and whatever the constitutive equations are. The only solution to this problem is to assume that the material is heterogeneous with a strong gradient of properties near the interface surface (Vardoulakis and Unterreiner, 1994).

An analysis for the ring simple shear test, based on a classical continuum and using a Mohr-Coulomb rigid plastic non-associated behaviour, yields a solution with a zone of plastified soil in contact with the interface. The thickness of this layer in contact with the interface is directly proportional to the internal diameter of the sample (Bogdaneva-Bontcheva and Lippmann, 1975; Unterreiner, 1994). However, experiments show evidence of an interface layer, where tangential displacements concentrate and whose thickness is independent of the diameter of the sample. The use of more complex constitutive equations does not alleviate these limitations of the classical continuum approach for modelling such phenomena and thus other approaches must be developed.

Laboratory experiments, analytic computations and discrete numerical modelling of granular materials undergoing localisation tend to show consistently that strain localisation in shear bands (Mühlhaus and Vardoulakis, 1987; Ord et al., 1991; Bardet and Proubet, 1991) or at interfaces (Teichman, 1989; Kishida and Uesugi, 1987; Boulon and Hassan, 1993) is systematically accompanied by significant individual grain rotations. What should be understood by significant individual grain rotations is that the grains, on average, tend to rotate more than would be expected if they were embedded in a classical continuum medium and would follow its rotation Ω_{ij} defined by the antisymmetric part $U_{i,j}^a$ of the displacement gradient $U_{i,j}$.

Therefore, it seems natural when modelling granular media where, for the considered range of stresses, the grains behave almost as rigid bodies, to use a generalised continuum which takes into account not only the translational degrees of freedom of the individual material points but also their rotational degrees of freedom (Mindlin, 1964; Eringen, 1966; Germain, 1973). In the case of simple interface shear tests, where the only space variable of interest is the distance to the interface, the experimental values of the degrees of rotation in a given point, at a distance $d = x_2$ (plane shear) or $d = r-r_{int}$ (ring simple shear) of the interface surface, are determined by averaging the individual grain rotations over a sufficiently large length parallel to the interface (Figure 2).

Similar averaging procedures have already been performed in discrete numerical simulations of an assembly of grains (Bardet and Proubet, 1991). Experimental measurements of individual grain rotation have been reported only recently for polycrystalline metals (Diepolder and Lippmann, 1991) and for sand grains within an interface layer (Boulon and Hassan, 1993). Within the experimental range of error, all these measures confirm that grains may rotate differently than the rotation of the classical continuum medium in which they are assumed to be embedded.

A Cosserat continuum owns two main features which are of major interest in modelling interfacial localisation. First, the constitutive equations introduce a material length, which can be related to the grain size based on micro-mechanical considerations (Mühlhaus and Vardoulakis, 1987; Kanatani, 1979) and which controls the interface thickness independently of any other geometrical length of the system. Secondly, corresponding to the extra degrees of freedom, additional boundary conditions stem naturally out of the principle of virtual work. Those boundary conditions have to be defined with respect to the surface of the interface and in particular its roughness as previously proposed within the framework of second gradient plasticity theories (Vardoulakis et al., 1992).

In this article, we will analyse the plane simple shear test and the ring simple shear test of a granular medium modelled as a Cosserat continuum. The emphasis will be on the features of the Cosserat continuum as compared to the classical continuum, rather than on the constitutive equations. Therefore, two limiting cases of isotropic constitutive equations will be studied, i. e., linear elastic and rigid plastic behaviour. It will be shown how a Cosserat continuum can be successful in reproducing the salient properties of an interface layer.

2 KINEMATICS AND STATICS OF A COSSERAT CONTINUUM

Derivation of the equilibrium equations for a Cosserat continuum has been done by many authors (Mindlin, 1964; Eringen, 1966, 1968, 1970). In a 1972 paper, Germain advocates the systematic use of the method of virtual work in continuum mechanics to derive the fundamental equations of a given continuum and he applies it to the theory of second gradient (Germain, 1973a) as well as to a continuum with microstructure (Germain, 1973b). A continuum with a rigid microstructure which is allowed only to rotate and not to deform, is defined as a Cosserat continuum. The kinematics of such a continuum is completely defined when one knows the macro-displacement field U_i and the micro-displacements gradient which is a micro-rotation tensor ω_{ij}^c in the present case. The corresponding micro-rotation vector is defined with:

$$\omega_{ij}^c = -e_{ijk}\omega_k^c \tag{1}$$

where e_{ijk} is the alternating tensor.

Concerning deformations, one has to generalise the classical symmetric strain tensor $\varepsilon_{ij} = U_{i,j}^s$ as follows:

$$\varepsilon_{ij} = U_{i,j}^s + \left(U_{i,j}^a - \omega_{ij}^c\right) \tag{2}$$

where the symmetric and antisymmetric parts of the displacement gradient are defined respectively as follows:

$$U_{i,j}^s = \frac{1}{2}\left(U_{i,j} + U_{j,i}\right) \tag{3}$$

$$U_{i,j}^a = \frac{1}{2}\left(U_{i,j} - U_{j,i}\right) \tag{4}$$

In addition, one has to consider the rotation gradient, called curvature tensor:

$$\kappa_{ij} = \omega_{i,j}^c \tag{5}$$

Application of the principle of virtual work yields the statical quantities conjugated in energy with the above kinematical variables. The stress tensor is a general second order tensor with both a symmetric and an antisymmetric component:

$$\sigma_{ij} = \sigma_{ij}^s + \sigma_{ij}^a \tag{6}$$

The symmetric or Cauchy stress tensor σ_{ij}^s is conjugated in energy with the symmetric or classical deformation tensor ε_{ij}^s. The antisymmetric stress tensor σ_{ij}^a is conjugated with the antisymmetric deformation tensor ε_{ij}^a which measures the relative rotation of the grains with the continuum. In addition to the general stress tensor, there is a couple stress tensor μ_{ij} which is conjugated with the curvature tensor κ_{ij}. Accordingly, the expression of the work of internal forces becomes:

$$w^{(i)} = \sigma_{ij}^s\,\varepsilon_{ij}^s + \sigma_{ij}^a\,\varepsilon_{ij}^a + \mu_{ij}\,\kappa_{ij} \tag{7}$$

The limiting case of a classical continuum is obtained by assuming a zero relative rotation ε_{12}^a or a zero antisymmetric stress tensor σ_{12}^a. The antisymmetric tensors σ_{ij}^a and ε_{ij}^a measure thus how strong the Cosserat effects are.

In the general case of a Cosserat continuum, one has to introduce volume couples c_i in addition to volume forces f_i and surface couples m_i, in addition to surface tractions t_i. For the present applications, we will assume that the volume forces f_i and volume couples c_i are neglectable.

However, both tractions t_i and surface couples m_i will be considered hereafter as they play a major role in interfacial localisation. Accordingly, the expression for the work of external forces becomes:

$$W^{(e)} = \int_V \left\{f_i U_i + c_i \omega_i^c\right\} \rho\, dV + \int_S \left\{t_i U_i + m_i\,\omega_i^c\right\} dS \tag{8}$$

When compared to a classical continuum, the refinement of the kinematics in a Cosserat continuum introduces new kinematical variables ω_i^c, ε_{ij}^a and κ_{ij}, new statical variables σ_{ij}^a and μ_{ij}, as well as extra boundary conditions which will be called Cosserat boundary conditions. They can be statical and formulated in terms of σ_{ij}^a or μ_{ij}. They can also be kinematical and formulated in terms of ω_i^c, ε_{ij}^a or κ_{ij}. Those extra boundary conditions can be called micro-boundary conditions since they involve micro-variables. They have to be defined in correspondence with the microscopic features of the interface surface and in particular its roughness measured at the scale of the grains.

3 PLANE SIMPLE SHEAR OF A LINEAR ELASTIC MATERIAL

3.1 Classical continuum

When modelling the plane simple shear test with a classical continuum, one cannot explain the formation of an interface layer unless one introduces a strong heterogeneity with a softer material near the interface (Vardoulakis and Unterreiner, 1994).

3.2. Cosserat continuum

3.2.1. Constitutive equations

The constitutive equations of a linear elastic isotropic Cosserat continuum can be written under the following form (Mindlin, 1964; Koiter, 1963; Schaeffer, 1967):

$$\{\sigma\} = [L]\{\varepsilon\} \qquad (9)$$

$$L_{ij} = \begin{vmatrix} K+G & K-G & & & \\ K-G & K+G & & & \\ & & 2G & & \\ & & & 2G_c & \\ & & & & 2N \end{vmatrix} \qquad (10)$$

where the vectors $\{\sigma\}$ and $\{\varepsilon\}$ are the normalised generalised stress and strain vectors respectively, which in the case of the plane simple shear test are equal to:

$$\{\sigma\}^t = \left\{\sigma_{11}, \sigma_{22}, \sigma_{12}^s, \sigma_{12}^a, \frac{\mu_{32}}{R}\right\} \qquad (11)$$

$$\{\varepsilon\}^t = \left\{\varepsilon_{11}, \varepsilon_{22}, \varepsilon_{12}^s, \varepsilon_{12}^a, R\,\kappa_{32}\right\} \qquad (12)$$

The constitutive equations introduce an elastic material length, noted by R, through the ratio of the bending modulus namely $\frac{2N}{R^2}$ relating μ_{32} with κ_{32}

and the classical shear modulus G, where N and G both have dimensions of a stress. The necessary condition of stability of such an elastic material, in the sense of Hadamard, requires that : $G > 0$, $3K - G > 0$, $G_c > 0$ and $N > 0$.

3.2.2. Boundary conditions

In a Cosserat continuum, the two classical boundary conditions for the tangential displacement U_1 at both boundaries :

$$U_1(0) = f_1\,W_I \qquad (13)$$

$$U_1(H) = 0 \qquad (14)$$

have to be complemented by two extra boundary conditions, corresponding to the new degree of freedom in rotation (Teichman, 1989). For the present analysis, we will choose the two following kinematical Cosserat boundary conditions :

$$\omega_3^c(0) = f_2\,\frac{W_I}{R} \qquad (15)$$

$$\varepsilon_{12}^a(H) = 0 \qquad (16)$$

where W_I is the tangential displacement of the structural member whose interface is being tested, $f_1 \in [0,1]$ is the fraction of W_I which is transmitted to the soil in translation (partial stick), and $f_2 > 0$ is the fraction which is transmitted in rotation. The boundary condition (16) dictates that the soil behaves "classically", at least near the upper boundary. This corresponds to the experimental observations, when a sample is sheared, of the formation of a "plug zone" where rotations are not predominant and the assumption of classical behaviour is reasonable.

Table 1 : Boundary conditions used for the analysis of the plane simple shear.

Interface $x_2 = 0$	Top of layer H
$U_2 = 0$	$\sigma_{22} < 0$
rigid interface	normal stress
$U_1 = f_1\,W_I > 0$	$U_1(H) = 0$
tangential displacement	zero tangential displacement
$R\,\omega_3^c = f_2\,W_I > 0$	$\varepsilon_{12}^a = \Omega_3 - \omega_3^c = 0$
Cosserat rotation	zero relative rotation "classical solution"

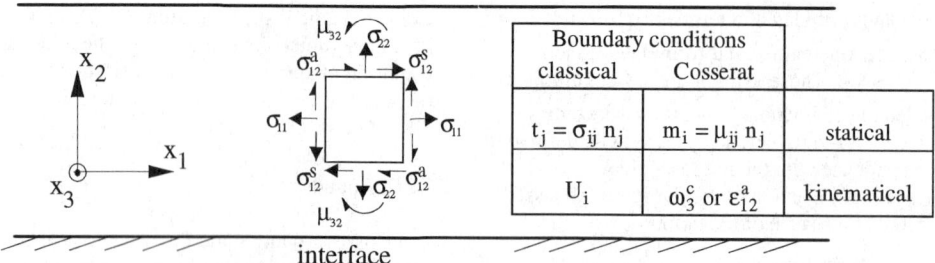

Figure 3 : Plane simple shear of an homogeneous Cosserat linear elastic material

3.2.3. Solution in displacements and deformations

When solving the problem of the simple plane shear, it is interesting to use, as the unknown variable, the relative rotation ε_{12}^a which measures the Cosserat effects. After combining the equilibrium and constitutive equations together, one obtains the following differential equation in ε_{12}^a :

$$\frac{d^2}{dx_2^2}(\varepsilon_{12}^a) - \frac{\alpha^2}{R^2}\varepsilon_{12}^a = 0 \qquad (17)$$

where the coefficient α is a dimensionless constitutive coefficient :

$$\alpha = \sqrt{\frac{2GG_c}{N(G+G_c)}} \qquad (18)$$

The necessary condition of stability for the elastic material imposes α^2 to be always positive and the solution of the differential equation is of exponential type. The Cosserat solution is equal to the classical one plus two additional terms proportional to $\exp(\pm\alpha\, x_2/R)$, which will be called Cosserat terms, and decrease or increase very rapidly over a few material lengths R (Figure 4). The exact solutions in displacement U_1 and Cosserat rotation ω_3^c are :

$$U_1(x_2) = U_1(0) + \frac{\sigma_{12}}{G}x_2$$
$$-\frac{2\beta}{X}R\Delta\omega\left(\text{ch}\left(\alpha\frac{x_2-H}{R}\right) - \text{ch}\left(-\alpha\frac{H}{R}\right)\right) \quad (19)$$

$$\omega_3^c(x_2) = \omega_3^c(0)$$
$$+\frac{1}{X}\Delta\omega\left(\text{sh}\left(\alpha\frac{H}{R}\right) + \text{sh}\left(\alpha\frac{x_2-H}{R}\right)\right) \quad (20)$$

$$\beta = \sqrt{\frac{NG_c}{2G(G+G_c)}} \qquad (21)$$

$$X = -\text{sh}\left(\alpha\frac{H}{R}\right) + \frac{R}{H}\beta\left(\text{ch}\left(\alpha\frac{H}{R}\right) - 1\right) \qquad (22)$$

$$\Delta\omega = \omega_3^c(0) - \frac{U_1(0)}{2H} \qquad (23)$$

The solution in antisymmetric shear deformations is:

$$\varepsilon_{12}^a(x_2) = \frac{G}{G+G_c}\frac{1}{X}\Delta\omega\,\text{sh}\left(\alpha\frac{x_2-H}{R}\right) \qquad (24)$$

It should be noticed that all the Cosserat terms are multiplied by the factor $\Delta\omega$ which measures the difference between the Cosserat rotation $\omega_3^c(0)$ at the interface and its value for the classical case : $\Omega_{12}(0) = U_1(0)/2H$. It is thus clear that if the extra boundary conditions do not prompt the degree of freedom in rotation, more than the classical ones do it, the Cosserat solution will degenerate to the classical one. For example if one changes the Cosserat boundary condition chosen at the interface into either $\Delta\omega = 0$ or $\varepsilon_{12}^a(0) = 0$, the classical solution will be obtained.

The material length R is of the order of the grain radius and thus it is typically much smaller than the height of the layer H. The above exact formulas thus reduce to :

$$U_1(x_2) = U_1(0) + \frac{\sigma_{12}}{G}x_2$$
$$+ 2\beta R\Delta\omega\left(\exp\left(-\alpha\frac{x_2}{R}\right) - 1\right) \quad (25)$$

$$\omega_3^c(x_2) = \omega_3^c(0) + \Delta\omega\left(\exp\left(-\alpha\frac{x_2}{R}\right) - 1\right)$$
$$\qquad (26)$$

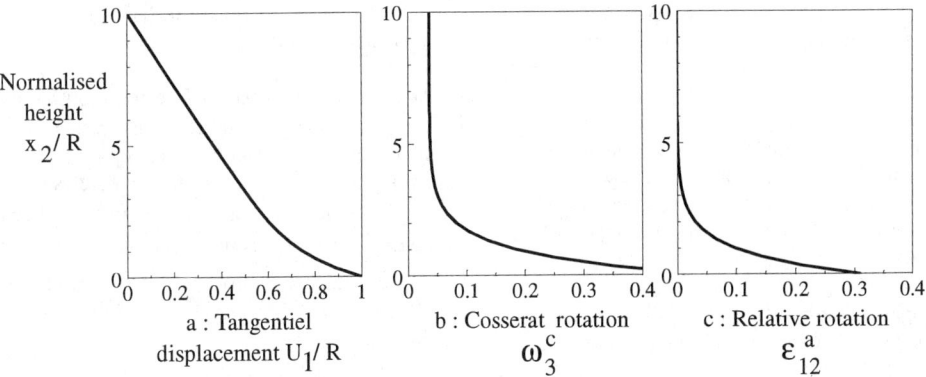

Figure 4 : Solution of the plane simple shear of a Cosserat linear elastic material

$$\varepsilon_{12}^a(x_2) = \frac{G}{G+G_c} \Delta\omega \exp\left(-\alpha \frac{x_2}{R}\right) \quad (27)$$

Using these formulae, it is possible to calculate the distance d_I to the interface over which the kinematical variables U_1, ω_3^c and ε_{12}^a decrease and reach only a small fraction p of their values at the interface. Those distances are given by :

$$\frac{d_I(U_1)}{R} = (1-p)\frac{G+G_c}{G_c}\frac{U_1(0)}{R\,\omega_3^c(0)} \quad (28)$$

$$\frac{d_I(\omega_3^c)}{R} = \frac{d_I(\varepsilon_{12}^a)}{R} = -\sqrt{\frac{N(G+G_c)}{2\,GG_c}} \ln p \quad (29)$$

All these distances are independent of the height H of the sample. They are directly proportional to the material length R which is introduced by the ratio of the bending modulus $2\,N/R^2$ and the classical shear modulus G . Only $d_I(U_1)$ depends on the value of the boundary conditions and, precisely, the ratio between the displacement $U_1(0)$ and the Cosserat rotation R $\omega_3^c(0)$ which are imposed at the interface. Numerical applications with p=1% and a ratio of $U_1(0)$ by R $\omega_3^c(0)$ equal to 2, yield, for the statical model, $d_I(U_1) = 4\,R$ and $d_I(\omega_3^c) = d_I(\varepsilon_{12}^a) = 3\,R$, while for the kinematical model they are equal to 5.6 and 1.5 R, respectively (Mühlhaus and Vardoulakis, 1987).

It should be noticed that the thickness of the interface layer is directly controlled by the Cosserat shear modulus G_c. If G_c is set equal to zero, the antisymmetric stress tensor disappears, the coefficients α and β are equal to zero and the Cosserat solution reduces to the classical one with no interface layer. Localisation of deformations is therefore directly related to the existence of antisymmetric stresses in the localised zones.

4. RING SIMPLE SHEAR OF A RIGID PLASTIC MATERIAL

4.1. Classical continuum

The solution of the ring simple shear for a classical rigid plastic material with a Mohr-Coulomb yield criterion is composed of two zones : a plastic zone in contact with the interface and a rigid zone extending to the external radius. In the case of associated plasticity, the plastic zone has a zero thickness (Bogdaneva-Bontcheva and Lippmann, 1975). In the case of non-associated plasticity, the plastic zone has a finite thickness which is directly proportional to the interface cylinder radius (Unterreiner, 1994). These results, which can be developed analytically for a rigid perfectly plastic material, are still valid for more general elasto-plastic constitutive equations. This dependency of the "interface layer" on the geometric dimensions of the sample comes from the classical continuum description which does not introduce any material length to scale down the problem. In this case, the layer thickness is controlled by the geometric dimensions.

4.2. Cosserat continuum

4.2.1. Constitutive equations

While there has been a lot of research on elastic Cosserat continuum (Koiter, 1963; Schaeffer, 1967), relatively little has been done on developing yield criteria. Concerning plasticity, some previous work has been done for granular materials modelled as isotropic Cosserat continuum (Lippmann, 1969; Bogdaneva-Bontcheva and Lippmann, 1975; Mühlhaus and Vardoulakis, 1987; Diepolder et al., 1991), while in Rock Mechanics, some work has been done to develop anisotropic Cosserat type models for layered and blocky rocks (Besdo, 1974; Mühlhaus, 1993).

Two approaches have been investigated for isotropic plastic Cosserat continuum. On one hand, one can start from the von Mises or Drucker-Praguer criterion and generalise it for a Cosserat continuum based on micro-mechanical considerations Mühlhaus and Vardoulakis, 1987; Kanatani, 1979). This leads to a plasticity theory with one yield criterion based on the definition of a generalised shear stress invariant τ_c :

$$\tau_c^2 = \eta_1 \, s_{ij} \, s_{ij} + \eta_2 \, s_{ij} \, s_{ji} + \frac{\eta_3}{R^2} \, \mu_{ij} \, \mu_{ij} \qquad (30)$$

where $\eta_1 + \eta_2 = 1/2$. On the other hand, one can start from the Mohr-Coulomb criterion and generalise it to a Cosserat continuum by including in it the antisymmetric stresses σ_{ij}^a. However, this criterion has to be completed by a second one which is function at least of the couple stresses μ_{ij}. Such an approach has already been tried and for the considered yield criteria yields an interface layer with a zero thickness despite the introduction of a material length through the second criterion (Bogdaneva-Bontcheva and Lippmann, 1975). Hereafter, a different set of yield criteria will be chosen. The classical Mohr-Coulomb criterion is kept as it is :

$$F_1 = q + p \sin\phi - c \cos\phi = 0 \qquad (31)$$

It is completed with a second criterion F_2 which is given by a function of only the antisymmetric and couple stresses. This is the simplest possible choice. Different types of such a criterion have been proposed (Lippmann, 1969).

Similarly to the Mohr-Coulomb criterion which is a linear combination of the 1^{rst} and 2nd invariants p and q of the symmetric stress tensor σ_{ij}^s, one can think of F_2 as a linear function of the two new invariants $\sqrt{J_2(\sigma_{ij}^a)}$ and $\sqrt{J_2(\mu_{ij})}$, which, in the case of the ring simple shear, reduce to $|\sigma_{\theta r}^a|$ and $|\mu_{zr}|/\sqrt{2}$ respectively. The analysis of the ring simple shear with a linear criterion can be found in (Unterreiner, 1994).

Another type of criterion F_2 can be considered starting from the generalised shear stress invariant τ_c which is physically founded on micro-mechanical considerations. After separation of the classical terms from the Cosserat ones, τ_c can be written as a function of q, which is the second invariant of σ_{ij}^s, and of q_c which is an invariant of σ_{ij}^a and μ_{ij} :

$$\tau_c^2 = q^2 + q_c^2 \qquad (32)$$

$$q_c^2 = 2 \, (\eta_1 - \eta_2) \left(\frac{\sigma_{ij}^a \, \sigma_{ij}^a}{2} \right) + \frac{2 \, \eta_3}{R^2} \left(\frac{\mu_{ij} \, \mu_{ij}}{2} \right) \qquad (33)$$

Since q is already incorporated in F_1, it will not be introduced into F_2 which will be chosen simply as a function of q_c :

$$F_2 = q_c - N = 0 \qquad (34)$$

where N is a sort of "rotational cohesion". Such a choice allows to uncouple the macroscopic terms σ_{ij}^s from the relative and microscopic terms σ_{ij}^a and μ_{ij}. For the ring simple shear, F_2 can be written as follows :

$$F_2 = \sqrt{\left(\frac{\sigma_{\theta r}^a}{\sigma_{max}^a} \right)^2 + \left(\frac{\mu_{zr}}{\mu_{zr}^{max}} \right)^2} - 1 = 0 \qquad (35)$$

with $\sigma_{max}^a = N/\sqrt{\eta_1 - \eta_2}$ and $\mu_{zr}^{max} = N R/\sqrt{\eta_3}$.
The two constants σ_{max}^a and μ_{zr}^{max} can be interpreted physically as the maximum (yield) antisymmetric and couple stresses allowable in the material with respect to the microscopic criterion. Their ratio, which has the dimension of a square length, introduces a plastic material length which will be denoted by the symbol R but it is different from the elastic material length R introduced earlier.

With the present choice of criteria, the Mohr-Coulomb criterion in the (q, p) plane is not modified. Since graphical representation of the criteria in the 4 dimension space (q, p, $\sigma_{\theta r}^a$, μ_{zr}) is not easy to visualise, we will restrict ourselves to the 3 dimension space (q, $\sigma_{\theta r}^a$, μ_{zr}) for a given value of the pressure p. In this space, the Mühlhaus and Vardoulakis criterion is represented by an ellipsoid (figure 5a) while the quadratic micro-criterion is represented by a cylinder of axis parallel to the q axis and of section an ellipse (Figure 5b). Therefore,

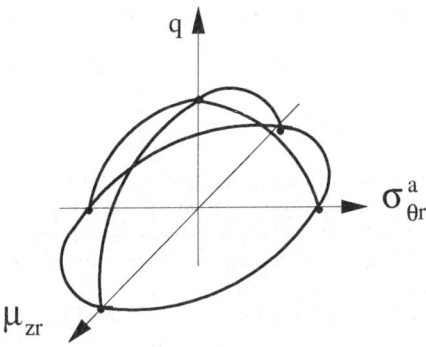

Figure 5a : Mühlhaus and Vardoulakis criterion

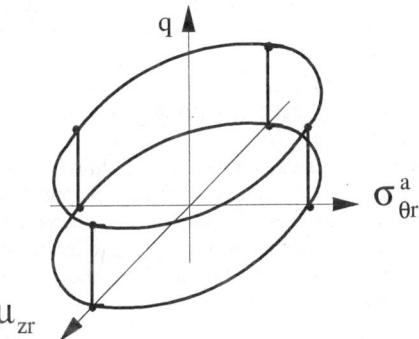

Figure 5b : Quadratic criterion

the quadratic criterion F_2 can be considered as a simple approximation of the Mühlhaus and Vardoulakis criterion with uncoupling of the macroscopic terms from the microscopic ones.

In the following lines we will analyse the ring simple shear of a Cosserat material with rigid plastic constitutive equations and two yield criteria : the Mohr Coulomb criterion F_1, which is termed macroscopic criterion, and the quadratic criterion F_2, which is termed microscopic criterion. The first flow rule G_1 will be non-associated ($\beta < \phi$) while the second flow rule will be associated $G_2 = F_2$. The calculations will be performed within the framework of the J_2 theory of rigid plasticity. With such choices the problem becomes reasonably simple to be solved analytically (Unterreiner, 1994).

4.2.2. Boundary conditions

The solution will be developed only at a given point of the loading for the set of boundary conditions summarised in Table 2. When compared to the modelling with a classical continuum, only three new parameters are introduced : two for the microscopic yield criterion and one for the Cosserat boundary condition in couple stress. The couple stress imposed at r_{int} by the interface surface measures its roughness with respect to the maximum allowable couple stress in the material μ_{zr}^{max}. For very smooth interface surfaces, the roughness defined by the ratio $|\mu_{zr}(r_{int})|/\mu_{zr}^{max}$ will be equal to 0 while for very rough interface surfaces, it will be equal to 1.

Table 2 : Boundary conditions for the statically determined ring simple shear problem

Interface radius r_{int}	External radius r_{ext}
$U_r = 0$	$\sigma_{rr} < 0$
rigid interface	confining pressure
$\sigma_{\theta r}^a < 0$	$U_\theta = 0$
interface shear stress	zero tangential displacement
$\mu_{zr} > 0$	$\mu_{zr} = 0$
measure of interface roughness	"classical solution"

Different types of solution can be developed depending on the intensity of the loading. For the sake of simplicity, we will assume that for the chosen set of boundary conditions, a zone plastified for both criteria forms in contact with the interface. This zone will be called the interface layer since deformations concentrate inside. It can then be shown mechanically that the next zone is rigid for F_2 and plastic for F_1 while the outer zone is rigid for both criteria.

4.2.3. Interface layer thickness

Within the interface layer, the second criterion F_2 is reached. Elimination of the antisymmetric stress $\sigma_{\theta r}^a$ in F_2 using the equilibrium equations yields a non linear differential equation of first order in μ_{zr}.

At the interface contact, the couple stress $\mu_{zr}(r_{int})$ is known. We are looking for a transition between a Cosserat zone to a classical one where the couple stress μ_{zr} is identically zero. Therefore, μ_{zr} is taken equal to zero in $r_{int} + e$ and is continuous. The boundary values in r_{int} and $r_{int} + e$ are known and one can thus solve for the thickness e. In the case of the ring simple shear, this non linear equation cannot be solved analytically. The plane simple shear can be obtained as a limiting case of the ring simple shear when r tends toward infinity and the difference r_{ext}-

r_{int} stays constant and is equal to the sample thickness H. In that case and for the quadratic criterion, the differential equation in μ_{zr} can be integrated as:

$$\frac{\mu_{zr}}{\mu_{zr}^{max}} = \sin\left(\operatorname{asin}\left[\frac{\mu_{zr}(r_{int})}{\mu_{zr}^{max}}\right] + 2\frac{\sigma_{max}^a}{\mu_{zr}^{max}}(r - r_{int})\right) \quad (36)$$

The thickness e of the interface layer is given by:

$$e = \frac{1}{2}\frac{\mu_{zr}^{max}}{\sigma_{max}^a}\operatorname{asin}\left(\frac{\mu_{zr}(r_{int})}{\mu_{zr}^{max}}\right) \quad (37)$$

and its maximum value is equal to:

$$e_{max} = \frac{\pi}{4}\frac{\mu_{zr}^{max}}{\sigma_{max}^a} = \frac{\pi}{4}\sqrt{\frac{\eta_1 - \eta_2}{\eta_3}}\,R \quad (38)$$

Figure 6 presents the variations of the interface layer thickness for the plane simple shear test as a function of the interface surface roughness measured by the normalised boundary couple stress $|\mu_{zr}(r_{int})|/\mu_{zr}^{max}$. The shape of the figure is very similar to the experimental measures (Kishida and Uesugi, 1987).

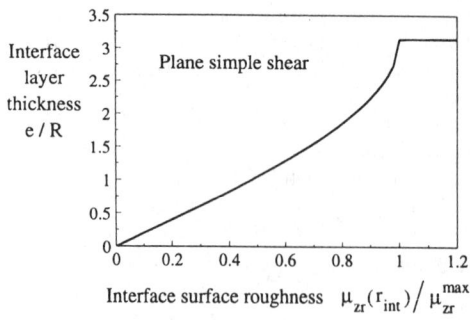

Figure 6 : Variation of the interface layer thickness in function of the interface surface thickness

Knowing the variations of the couple stress μ_{zr} in the interface layer, one can construct the complete statical solution within this zone and in the outer zones. Once the statical solution is developed, construction of a kinematical solution can be done. This solution is characterised, in the interface layer, by very strong relative rotations and antisymmetric stresses. The tangential displacement decreases very rapidly within this layer and its gradient is discontinuous at the limit with the classical plastic zone.

5 CONCLUSION

A classical continuum, whatever its degree of heterogeneity and constitutive equations is not appropriate for modelling interface layers since it does not introduce a material length which can scale down the interface layer thickness. This phenomenon exists also for volume localisation phenomena in shear bands. A Cosserat continuum which describes both at a microscopic and macroscopic level the kinematics of a granular material with rigid grains, is able to overcome those limitations. The linear elastic constitutive equations for an isotropic Cosserat continuum introduce an elastic material length R through the ratio of the bending modulus N by the shear modulus G. During shearing, the shear deformations concentrate in an interface layer of finite thickness which is a function of R and the Cosserat boundary condition. Although the decrease of the antisymmetric stress and couple stresses is very fast over a few material length distance, they never reach exactly zero in the elastic case. In a Cosserat rigid plastic isotropic continuum, the constitutive equations introduce a material length through the ratio of $\partial F_2/\partial \mu_{ij}$ and $\partial F_2/\partial \sigma_{ij}^a$ which has the dimension of a square length. This plastic material length controls directly the thickness of the interface layer where both the macroscopic and microscopic criteria are reached. It has been shown that, whatever the constitutive equations, elastic or rigid plastic, localisation of deformation is directly related to the existence of antisymmetric stresses in the localised zones.

THANKS

The authors want to acknowledge the Commission of the European Communities who sponsored this research through the SCIENCE Project No. SC1*-CT91-0659.

Part of the present research was developed within the ALERT network (CHM Project No. CHRX-CT93-0217).

REFERENCES

Bardet, J.P., and Proubet, J. 1991. The structure of persistent shear bands in idealized granular media. *Computer Methods and Advances in Geomechanics*, Balkema.

Besdo, D. 1974. Ein Beitrag zur nichtlinearen Theorie des Cosserat-Kontinuums. *Acta Mechanica* Vol. 20.

Bogdaneva-Bontcheva, N., and Lippmann, H. 1975. Rotationssymmetrisches ebenes Fliessen eines granularen Modellmaterials. *Acta Mechanica*, Vol. 21.

Boulon, M. 1989. Basic features of soil-structure interface behaviour. *Computer and Geomechanics*, Vol. 7.

Boulon, M., and Hassan, H. 1993. Development of a visualisation of the movement of the grains within a soil-structure interface. *CEC SCIENCE Program No. 659, Internal Report*.

Chen, Z., and Schreyer, H. L. 1987. Simulation of Soil-Concrete Interfaces with Nonlocal Constitutive Models. *ASCEE Journal Engineering Mechanics*, Vol. 113, No. 11.

Clichy, W., Boulon, M., and Desrues, J. 1987. Etude expérimentale stéréophotogrammétrique des interfaces. *Proc. 4th French-Polish Conf.*, Grenoble, France.

Diepolder, W., Mannl, V., and Lippmann, H. 1991. The cosserat continuum, a model for grain rotations in metals. *International Journal of Plasticity*, Vol. 7.

Eringen, A. C. 1966. Linear theory of micropolar elasticity. *J. Math. Mech.*, Vol. 15.

Eringen, A. C. 1966. Linear theory of micropolar elasticity. *J. Math. Mech.*, Vol. 15.

Eringen, A. C. 1968. Mechanics of micromorphic continua. *Mechanics of Generallized Continua. Proc. IUTAM Symposium*, E. Kröner ed., Springer-Verlag.

Eringen, A. C. 1970. Balance laws of micromorphic mechanics. *Int. J. Eng. Sc.*, Vol. 8.

Germain, P. 1972. Sur l'application de la méthode des puissances virtuelles en mécanique des milieux continus. *CRAS*, Paris, Série A, No. 274.

Germain, P. 1973a. La méthode des puissances virtuelles en mécanique des milieux continus-1ère partie, Théorie du second gradient. *Journal de Mécanique*, Vol. 12.

Germain, P. 1973b. The method of virtual power in continuum mechanics. Part 2 : Microstructure. *SIAM J. Appl. Math.*, Vol. 25, No. 3.

Hoteit, N. 1990. Contribution à l'étude du comportement d'interface sable-inclusion et application au frottement apparent. *Ph. D. Thesis*, I. N. P. G., Grenoble.

Kanatani, K. I. 1979. A micropolar continuum theory for the flow of granular materials. *Int. Journal Engineering Science*, Vol. 17.

Kishida, H., and Uesugi, M. 1987. Test of the interface between sand and steel in the simple shear apparatus. *Géotechnique*, Vol. 37, No. 1.

Kishida, H., and Uesugi, M. 1987. test of the interface between sand and steel in the simple shear apparatus. *Géotechnique* 37, No. 1.

Koiter, W. T. 1963. Couple-stresses in theory of elasticity. *Proceedings Koninklijke Nederlands Akademie Van Wetenshcaffen*, Series B, Vol. 67.

Lippmann, H. 1969. Eine Cosserat Theorie des plastischen Fliessens. *Acta Mechanica* Vol. 8.

Löffelmannn, G. 1989. Theorische und experimetelle Untersuchungen zur Schüttgut-Wand - Wechslwirkung von Granulaten. *Dr.-Ing. Dissertation*, Universität Karlsruhe.

Mindlin, R. D. 1964. Microstructure in linear elasticity. *Arch. Rat. Mech. Anal.*, Vol. 16.

Mühlhaus, H. B. 1993. Continuum models for layered and blocky rock. in *Comprehensive Rock Engineering*, Vol. 2, Hudson Editor, Pergamon.

Mühlhaus, H. B. and Vardoulakis, I. 1987. The thickness of shear bands in granular materials. *Géotechnique*, Vol. 37, N°. 3.

Neuber, H. 1966. Über Probleme der Spannungkonzentration im Cosserat-Körper. *Acta Mecanica*, Vol. 2.

Ord, A., Vardoulakis, I., and Kajewski, R. 1991. Shear band formation in Gosford sandstone. *Int. J. Rock Mech. Min. Science & Geomechanics Abstracts*, Vol. 28, N°. 5.

Schaeffer, H. 1967. Das Cosserat-Kontinuum. *Zeitschrift für Angewandte Mathematik und Mechanik*, Band 47, Heft 8.

Teichman, J. 1989. Scherzonenbildung und Verspannungseffekte in Granulaten unter Berücksichtigunug von Kerndrehungen. *Dr.-Ing. Dissertation*, Universität Karlsruhe.

Unterreiner, P. 1994. Modélisation des interfaces en mécanique des sols: applications aux calculs en déformation des murs en sol cloué. *Thèse de Doctorat*, ENPC, Paris.

Unterreiner, P., and Vardoulakis, I. 1994. Interfacial localisation in granular media. Proceedings 8th Int. Conf. of the Association for Computer Methods and Advances in Geomechanics, West Virginia University, USA.

Unterreiner, P., Lerat, P., Vardoulakis, I., Schlosser, F., De Laure, E., and Belmont, G. 1994. Brevet sur l'Appareil de Cisaillement Simple Annulaire (ACSA) (patent).

Vardoulakis, I., and Unterreiner, P. 1994. Interfacial localisation in simple shear tests on granular medium modelled as a Cosserat continuum in

Mechanics of Geomaterials Interfaces, Editors Selvadurai and Boulon, Balkema.

Vardoulakis, I., Shah, K. R., and Papanastasiou, P. 1992. Modelling of Tool-Rock Shear Interfaces using Gradient dependent Flow Theory of Plasticity. *Int. J. Rock Mech. Min. Science & Geomechanical Abstracts*. Vol. 29, No. 6.

Vesic, A. 1977. Design of pile foundations; *NCHRP report N°. 42*.

4 Experiments
  Expérimentation

Some observations of zones of localisation in model tests on dry sand
Observations de zones de localisation dans un sable dense

D. Muir Wood
Department of Civil Engineering, University of Glasgow, UK

K.J.L. Stone
Department of Civil Engineering, University of Western Australia, W.A., Australia

ABSTRACT: Results of model tests are reported in which boundary slope discontinuities are imposed on dense sand beds without discontinuities of boundary displacement. Zones of localisation are detected by visual and radiographic techniques. A contrast is drawn between weak zones of continuum localisation and strong zones of shear discontinuity. The ability of numerical analysis to estimate the probable region of development of shear discontinuities is demonstrated in preliminary studies.

L'article présente les résultats d'essais sur des modèles de sable déposé en couches, dans lesquels on impose une discontinuité d'inclinaison à la frontière sans discontinuité de déplacement. Les zones de localisation sont détectées par observation visuelle ou en utilisant la technique radiographique. On met en évidence de manière contrastée l'existence de zones faibles de localisation continue et de zones fortes de cisaillement discontinu. La capacité d'une approche numérique à prédire la zone probable de développement de cisaillement discontinu est montrée au travers d'études préliminaires.

1 INTRODUCTION

Laboratory observations of zones of localised deformation in soils have been made using element tests in which the onset of inhomogeneous deformations occurs from a quasi-uniform state of stress, possibly triggered by the deliberate presence of a small weak region (Desrues, Lanier and Stutz, 1985; Vardoulakis, 1988; Arthur, Dunstan, Al-Ani and Assadi, 1977). These tests have attempted to identify or confirm the constitutive requirements for the development of nonuniform response so that appropriate bifurcation conditions can be incorporated in numerical studies of boundary value problems. Zones of localised deformation have also been observed in model tests in which boundary displacement discontinuities are deliberately introduced in order to force the initiation of a rupture surface from a specified location (James and Bransby, 1970; Scarpelli and Wood, 1982; Stone and Muir Wood, 1992). Such observations tend to be supported by purely kinematic evidence, but can provide data against which numerical analyses can be compared. However, it is not necessary for there to be discontinuities of boundary displacement in order to generate localisations (Airey, Budhu and Wood, 1985; Stone and Wood, 1988). In this paper some observations are presented of patterns of localisations induced in model tests in which a discontinuiuty of boundary slope but not of boundary displacement is introduced.

2 MODEL TESTS

2.1 *Apparatus*

The mode of boundary deformation to be applied to the base of beds of dry sand is shown in Fig 1. The tests were conducted in a centrifuge strong box (Fig 2) as part of a wider programme of model tests studying the effects of subterranean movements on ground behaviour (Stone, 1988). Some initial results of centrifuge model tests conducted at 100 gravities will be reported but most of the results come from tests conducted in the same apparatus but at one gravity, on the laboratory floor. Beds of dense sand were prepared in the strong box to a depth of about 200mm by pluviation. The base of the strong box incorporated a piston (p) supporting two flaps (ff).

Fig 1 : Basement movement with discontinuity of slope but not of displacement.

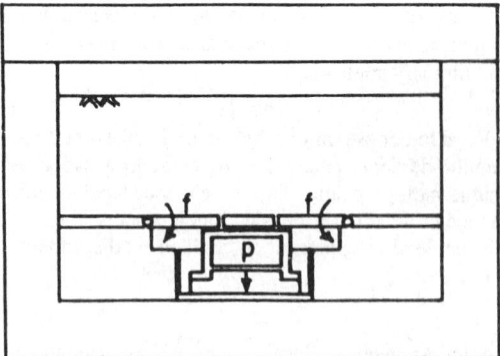

Fig 2 : Displacing piston (p) and rotating flaps (ff) in centrifuge strong box.

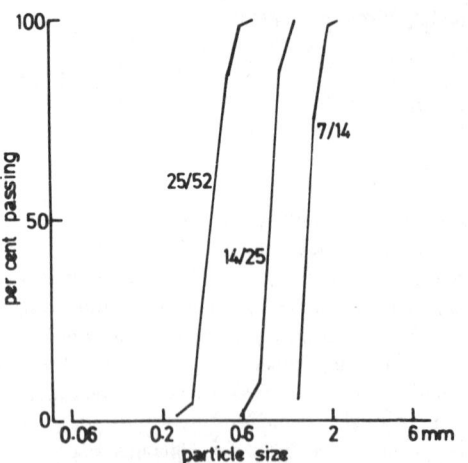

Fig 3 : Grading curves for sands used in model tests.

By lowering the piston the flaps were able to rotate. A thin sheet of rubber was placed over the flaps so that there was a continuous boundary profile at all times corresponding to the desired boundary displacement mechanism shown in Fig 1.

Model tests were performed using three different but nearly single sized gradings of Leighton Buzzard sand, which will be referred to by the sieve sizes which passed and retained them : 7/14, 14/25, 25/52 (Fig 3) with corresponding average particle sizes d_{50} about 1.5mm, 0.85mm, and 0.4mm respectively. Most of the data presented have been obtained from tests using the finest sand, 25/52. For all three sands the maximum void ratio is about 0.8 and the minimum void ratio about 0.49. Sand beds were prepared with a uniform void ratio of about 0.53 in all tests, corresponding to a relative density of about $I_D = 0.87$.

2.2 *Observation of localisations*

Zones of localisation have traditionally been observed using radiographic techniques to detect the regions of reduced density that result from the volumetric expansion associated with shearing of dense granular materials (Roscoe, 1970). Such an observation indicates that there has been a change in volumetric packing, and development of volumetric strain, but gives no further clues concerning the deformation mechanism which the soil has experienced. If, in addition to observing density changes, radiography is used to monitor the locations of lead markers within the sand, then an indication of the strain field within the sand can be obtained and movements around the zones of localisation can be deduced (Stone and Wood, 1988).

Insofar as zones of localisation are also zones of concentrated shearing, then they can also be spotted by looking for breaks in marker bands inserted within the soil. In beds of clay it is possible to inject thin threads of material which is opaque to X-rays (Airey, Budhu and Wood, 1985; Stone and Wood, 1988); in sand beds it is easier to place marker beds of a different coloured sand as the bed is being prepared. Visual or photographic observation of marker bands can reveal the presence and location of ruptures (provided that the relative orientations of rupture and marker band are appropriate) but cannot easily reveal density changes within the soil.

Other techniques, such as those based on photogrammetry (Butterfield, Harkness and Andrawes, 1970; Desrues et al, 1985) can be used to identify fields of deformation and localisations in granular materials. These have not been used in the present work.

2.3 *Patterns of localisations : centrifuge models*

Radiographic observation of models on the geotechnical centrifuge is not feasible. Photographs of models showing the marker bands were taken with the models in flight and also after the tests. The pattern of deduced ruptures shown in Fig 4 is typical for the fine sand. The main rupture appeared around point A at a basement rotation of about 12° and then extended upwards and downwards into the sand at an angle of about 25° to the vertical with further basement movement. As basement rotation continued, development of this rupture ceased and other ruptures began to appear, one on each side, around B and C. It does not appear that the alignment and orientation of any of the ruptures are directly linked with the position of the point of rotation of the basement. Ruptures A and B, in particular, develop within the zone of sand that is

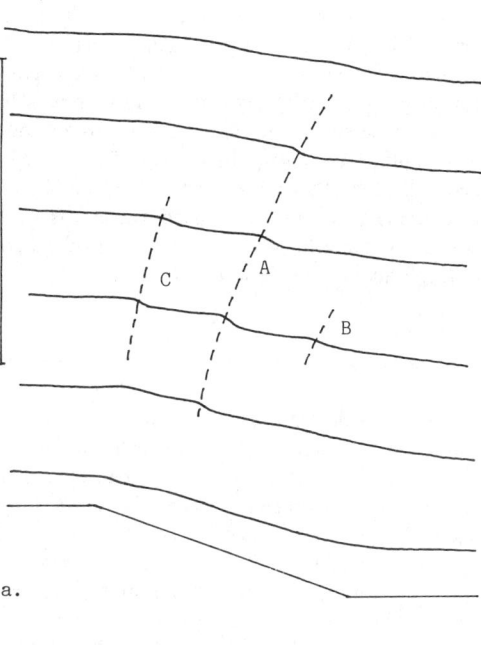

a.

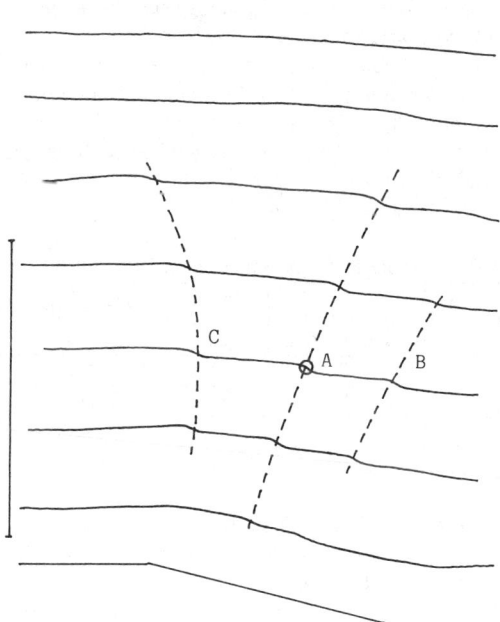

Fig 4 : Marker bands and displacement discontinuities in 25/52 sand tested at 100g (vertical bar = 100mm).

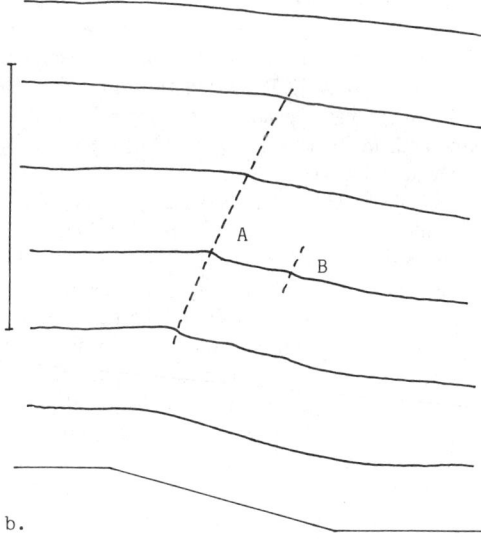

b.

Fig 5 : Marker bands and displacement discontinuities in (a) 14/25 and (b) 7/14 sand tested at 100g (vertical bar = 100mm).

being subjected to a sort of simple shear state of deformation over the descending boundary.

Centrifuge models performed using the two coarser sands show essentially similar response, but, by the time that the maximum displacement of the boundary has been reached, the full mechanism is less extensively developed : the main rupture A has not extended so far, and the side ruptures are much less significant or even absent (Figs 5a, b). It has been noted by Stone and Muir Wood (1992) that displacement across shear localisations has to be scaled according to particle size if similar patterns of localisations are to be obtained.

2.4 *Patterns of localisations : single gravity models*

The radiographic studies of single gravity models show that the mode of response deduced from the observations of marker bands in the centrifuge model tests gives an incomplete picture. By a rotation of about 2.8° (corresponding to about 0.05^c) a thin region of dilated sand (D) has appeared extending up about two thirds of the thickness of the sand bed from the break of slope of the boundary, at an angle of about 15° to the vertical (Fig 6a - actually sketched for a rotation of about 8.3°). A series of faint dilated zones are also just visible above the sloping part of the boundary. With further rotation, nothing much happens to the initial localisation, but one of the other ones develops as a shear localisation (E) and breaks through to the sand surface at a rotation of about 13.7° (0.24^c) (Fig 6b). It is this localisation that appears to correspond to that (A) seen both in flight and at the end of the centrifuge model test (Fig 4).

It should be noted that the shading in Figs 6a and 6b, and in Figs 7 and 8, is intended to indicate the location of the zones of dilated sand. The width of these zones is harder to determine convincingly from the radiographs. The density of the sand seems to vary rather continuously (though periodically) and it is not easy to indicate boundaries of dilated regions.

It might be suggested that some of the difference between the patterns of localisations observed in the centrifuge tests (Figs 4, 5) and in the single gravity tests (Figs 6, 7, 8) was in some way connected with the very different stress levels in these two sets of models. It will be argued subsequently that in fact two quite different sets of observations are being made, and that the technique used to monitor localisation in the centrifuge models was not capable of detecting the periodic zones of dilatancy that were found from radiographic observation of the single gravity models. On the other hand, differences in inclination between localisations A and B (Fig 4) and localisations D and E (Fig 6b) might be attributed to the difference in stress level. The higher stress level on the centrifuge would tend to reduce angles of friction and angles of dilatancy by a few degrees.

With the coarser sands a series of dilated zones are rather clearly seen extending up into the sand over the inclined boundary (Figs 7, 8) with the spacing of these zones being roughly proportional to the particle size of the sand being tested. For the 14/25 sand, with $d_{50} = 0.85$mm, the distance between the centrelines of the zones over the rotating boundary is about 13mm, giving a ratio to particle size of about 15. For the 7/14 sand, with $d_{50} = 1.5$mm, the distance between the centrelines of the zones is about 22.5mm, giving a ratio to particle size again of about 15. (It is possible to convince oneself that the faint dilated zones for the finest (25/52) sand have a centre-to-centre spacing of about 6.7mm which would imply a ratio to particle size of about 17.) Eventually one of these zones (E as marked in Figs 7, 8), emerging from a point about 20-30% of the distance from the top of the slope becomes dominant and extends towards the free surface.

As noted above, the shading in the figures should not be taken to provide an accurate indication of the width of the dilating zones - this is difficult to estimate with any degree of confidence from the available radiographs.

Table 1 : Shear box tests on 14/25 sand

σ_V kPa	ϕ deg	ν deg	$90-\phi$ deg	$90-\frac{1}{2}(\phi+\nu)$ deg
35	48.5	14.6	41.5	58.5
178	45.0	13.7	45.0	60.6
325	43.8	12.5	46.2	61.8

For the 14/25 sand there is a hint at an early stage of a set of complementary zones of dilation making an angle of about 50-60° with the first set (Fig 7). Results from shear box tests on this sand are summarised in Table 1. Tests have been performed on the dry sand with three different values of normal stress σ_V, and the values of angle of friction ϕ and angle of dilation ν are quoted. The Coulomb orientations of ruptures are at $45\pm\frac{1}{2}\phi$ to the direction of major principal stress, so that the angle between

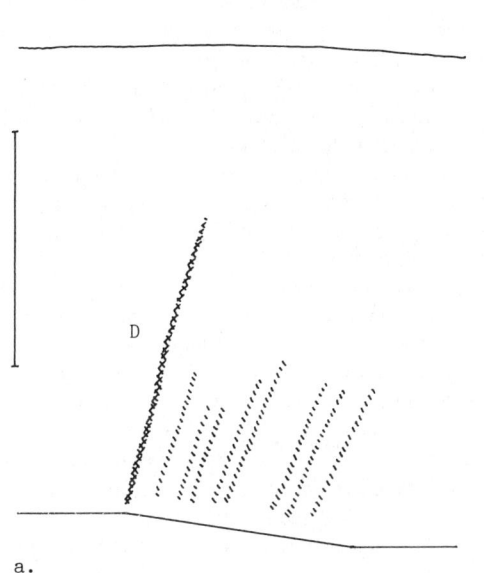

a.

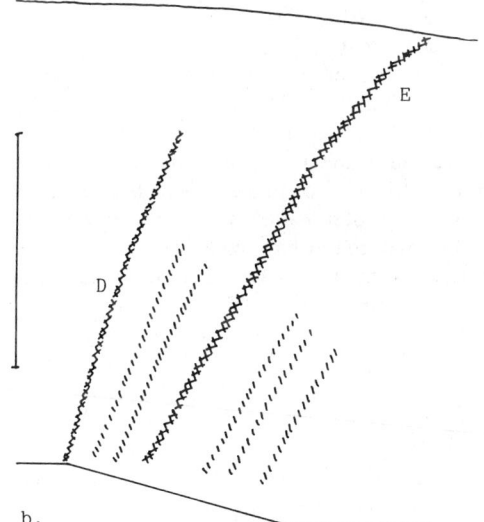

b.

Fig 6 : Localisations for 1g test on 25/52 sand, (a) rotation 8.3° (0.14ᶜ), (b) rotation 13.7°, 0.24ᶜ. Sketches from radiographs (vertical bar = 100mm).

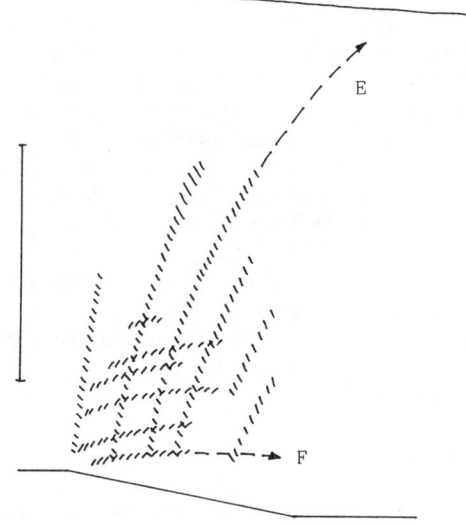

Fig 7 : Localisations for 1g test on 14/25 sand, rotation 10.9° (0.19ᶜ). Sketch from radiograph (vertical bar = 100mm).

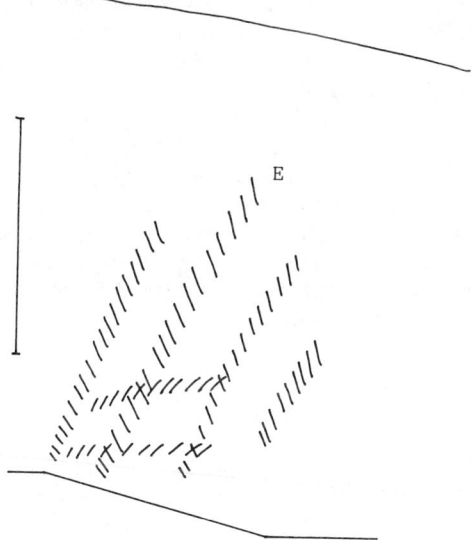

Fig 8 : Localisations for 1g test on 7/14 sand, rotation 15.1° (0.26ᶜ). Sketch from radiograph (vertical bar = 100mm).

ruptures is 90-φ. The Vardoulakis orientations of ruptures are at 45±¼(φ+ν) to the direction of major principal stress, so that the angle between ruptures is 90-½(φ+ν). There is some evidence that these latter are the preferred orientations that are adopted when there is no particular overall kinematic constraint on the deformation of the sand. The stress levels in the shear box are probably much higher than those obtaining in the single gravity model tests but it is clear that the angle between the observed ruptures matches rather well with the Vardoulakis angle. At large rotations a member of the complementary set of zones of dilation develops as a rupture extending from the crest of the slope (F in Fig 7).

3 NUMERICAL ANALYSES

A limited number of finite element analyses of the problem illustrated in Fig 1 were performed using the finite element program ICFEP (used, for example, by Potts, Dounias and Vaughan, 1990) with a simple elastic-plastic soil model incorporating strain softening and non-associated flow. The model, appropriate for monotonic loading, assumes initial elastic response with a frictional limiting stress condition with mobilised friction decreasing with plastic deviatoric strain from a specified peak to a critical state value (Fig 9). The values of parameters used in the analysis are :

    peak angle of friction $\phi_p = 45°$
    critical state angle of friction $\phi_{cs} = 30°$
    deviator strain to peak $\varepsilon_{peak} = 2\%$
    deviator strain to critical state $\varepsilon_{cs} = 22\%$

The current angle of dilation during plastic deformation is assumed to be equal to the difference between the current mobilised angle of friction and the critical state value.

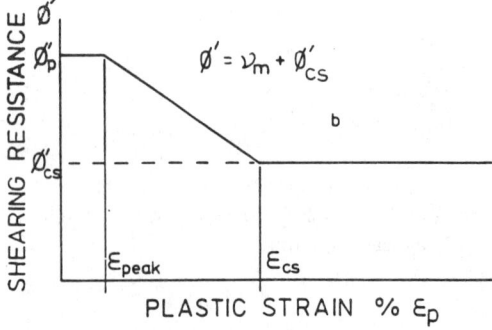

Fig 9 : Strain-softening soil model

Eight noded isoparametric elements were used (Fig 10) with reduced integration, and an accelerated Newton-Raphson technique was used to solve the non-linear finite element equations. Results are shown for purely illustrative purposes, and no attempt has been made to perform an exhaustive numerical study of development of localisation for this boundary value problem, and, in particular, mesh sensitivity has not been explored.

In the initial stages of deformation (Fig 11a), the shear strain varies rather smoothly over the rotating boundary. At later stages (Fig 11b), concentration into three roughly vertical zones has developed. As noted, the link between the pattern of strain concentration and mesh size has not been explored. The model used is a continuum model, in which dilation is introduced as a link between first and second invariants - volumetric strain and deviatoric strain. Kinematic compatibility does not then require any link between shear displacement and normal displacement on a shear surface but merely some overall volume change as a result of the shearing strain that has developed. The necessary volumetric expansion is provided for by the rotation and lowering of the boundary, and by the stage illustrated in Fig 11b the sand has been brought to a constant volume critical state condition in the zones of concentrated shearing. The books in the bookstack (De Josselin de Jong, 1972) are sliding and simultaneously extending, but without separating (Fig 12).

It is, however, worth noting that, leaving aside the obvious concentrations of strain that occur close to the discontinuities of imposed boundary slope, the highest levels of deviatoric strain occur in the centre of the sand bed, which could be seen as being consistent with the initiation of shear discontinuities in real particulate soil in this general region of the model, as was actually observed.

4 DISCUSSION

It is clear that the experimental observations have shown the presence of two quite different types of localisations. There is a pattern of what might be described as weak, continuum localisations which are zones of dilation but which lead to no major discontinuities of shear displacement and result from the development within the soil of a critical quasi-uniform state of stress and strain. Then there is a much more skeletal pattern of strong shear discontinuities which may be triggered from

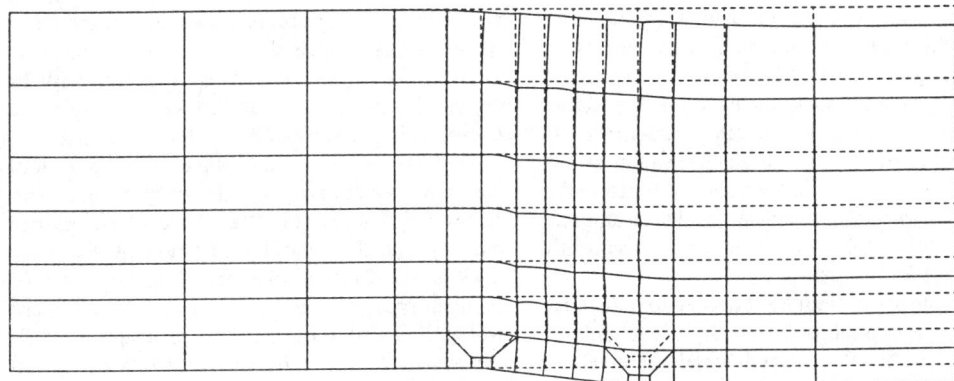

Fig 10 : Finite element mesh

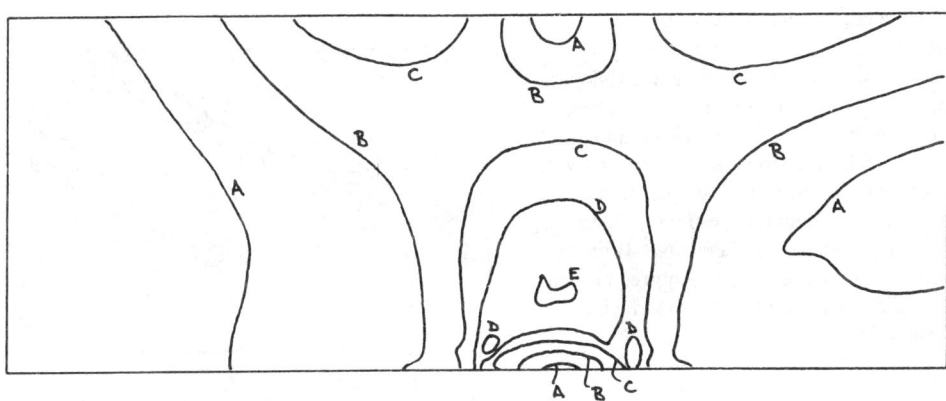

a.

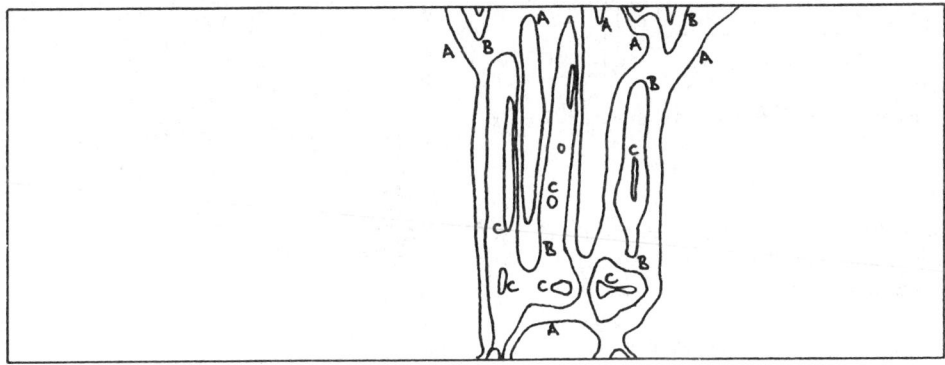

b.

Fig 11 : Contours of deviatoric strain from finite element analysis
(a) for base rotation 0 to 0.64° (contour values : A 0.001, B 0.002, C 0.003, D 0.004, E 0.005);
(b) for base rotation 9.95 to 17.1° (contour values : A 0.04, B 0.12, C 0.20).

particular members of the set of weak localisations. Radiographic studies of sand models are most useful for detecting the zones of volumetric expansion; information about displacements inevitably comes from discrete markers which can be interpreted to give a field of strains for a continuous soil mass, but may not provide useful data for shear discontinuities unless the markers are providentially placed around a region where the discontinuity subsequently develops.

The average shear strain level needed to initiate the continuum localisations is small - and this is consistent with other experimental observations (James and Bransby, 1970; Budhu and Wood, 1978) where engineering shear strains of the order of 5-10% are sufficient to produce zones of dilation. The passive wall experiments described by James and Bransby provide another example of the subsequent formation of a shear discontinuity in the interior of the sand mass, and again this discontinuity is not aligned with the discontinuity of boundary slope - which in their tets would be the toe of the rotating wall (Fig 13). Practically this may be an important result : damage to buried structures or services will be greater where there are significant discontinuities of displacement; continuum response, even if periodic in nature, may be less damaging. Boundary displacement discontinuities are not required in order to generate mode II shear ruptures within the interior of the soil mass.

some degree of uniformity), with the occurrence of narrow zones of high shear strain in the numerical model. It is likely that such zones will be encouraged to develop if the constitutive model for the soil incorporates significant strain softening.

There has been discussion about the orientation of zones of localisation in laboratory model tests. James and Bransby (1970) and others have asserted strongly that they are aligned with zero extension lines (strain characteristics: the Roscoe orientation) in their passive wall tests, whereas Bransby and Milligan (1975) show results which appear to lie between strain characteristics and stress characteristics (the Coulomb orientation) in active wall tests (although information concerning the direction of major principal stresses is usually lacking and the assumption of coaxiality of principal

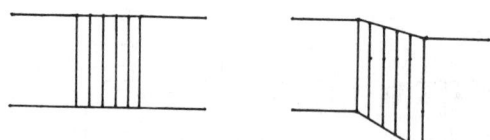

Fig 12 : Sketch of compatible mode of sliding and extension.

Numerically, the prediction of the development and orientation of continuum localisations requires the testing of some criterion of bifurcation of mechanical response to be superimposed on the calculated continuous strain fields. Even without such a bifurcation analysis, it may be possible to associate the probable development of shear discontinuities in a physical model of a boundary value problem, in which the soil is subjected to a non-uniform strain field (in contrast with a laboratory single element test in which there is at least an attempt to maintain

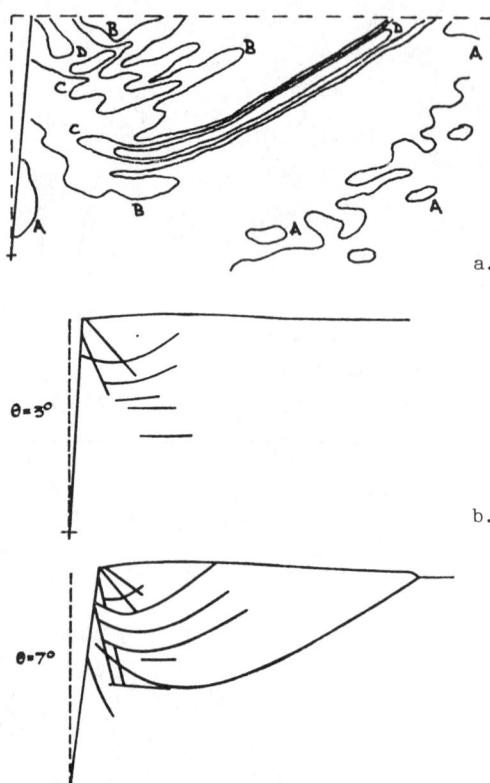

Fig 13 : Shear discontinuity in passive wall test on dense 14/25 Leighton Buzzard sand, wall height 307mm, rotation 5° : (a) contours of shear strain (contour values : A 0.025, B 0.1, C 0.2, D 0.3); (b) zones of localisation for wall rotations of 3° and 5° (after James and Bransby, 1970)

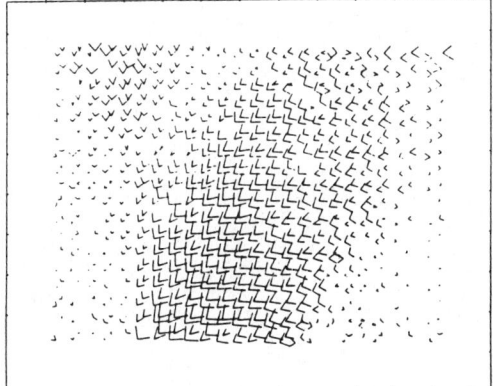

Fig 14 : Zero extension directions for 1g test on 14/25 sand, rotation 5.4 to 10.9°.

stresses and principal strains is not one for which there is great support in complex stress states). Theoretical studies (Vardoulakis, 1980; Vermeer, 1982) have tended to favour an orientation which roughly bisects the Roscoe and Coulomb orientations - again assuming coaxiality. In the present tests, the data are as usual scattered, but, as has been noted, it appears that the angle between the continuum localisations is smaller than would be plausibly calculated using the Roscoe orientation, and the orientations of these localisations do not appear to be aligned with either set of zero extension directions (Fig 14).

5 CONCLUSIONS

Two distinct sets of localisations have been observed in model tests on dense sand in which discontinuities of boundary slope but not of boundary displacement have been imposed. Different techniques may be more appropriate for observation of the two sets. Radiography provides a way to detect the zones of continuum localisation which are associated with volumetric expansion and fall of density (and hence of absorption of X-rays). Direct observation of breaks in marker bands may be a better method of detection of major shear discontinuities.

Continuum numerical analyses of boundary value problems may be able to reveal the probable occurrence of shear discontinuities even if they make no attempt to predict the bifurcation of response that is associated with the continuum localisations.

Shear discontinuities will be more damaging than continuum localisations in their effect on buried structures and services.

ACKNOWLEDGEMENTS

The experimental work was performed by the second author while a research student in the Cambridge Soil Mechanics Group. He is grateful to the Science and Engineering Research Council and to Cambridge University Engineering Department for financial support which enabled him to undertake and complete this research.

REFERENCES

Airey, D.W., Budhu, M. and Wood, D.M. (1985) Some aspects of the behaviour of soils in simple shear. Chapter 6 in P.K. Banerjee and R. Butterfield (eds), *Developments in soil mechanics and foundation engineering 2 : Stress-strain modelling of soils:* 185-213. London : Elsevier Applied Science Publishers

Arthur, J.R.F., Dunstan, T., Al-Ani, Q.A.J.L. and Assadi, A. (1977) Plastic deformation and failure in granular media. *Géotechnique* 27:1:53-74.

Bransby, P.L. and Milligan, G.W.E. (1975) Soil deformations near cantilever sheet pile walls. *Géotechnique* 25:2:175-195.

Budhu, M. and Wood, D.M. (1978) *Report on simple shear tests on fine sand performed for Delft Soil Mechanics Laboratory.* Cambridge University Engineering Department (unpublished).

Butterfield, R., Harkness, R.M. and Andrawes, K.Z. (1970) A stereo-photogrammetric method for measuring displacement fields. *Géotechnique* 20:3:308-314.

De Josselin de Jong, G. (1972) Discussion : The meaning and measurement of basic soil parameters. In R.H.G. Parry (ed), *Stress-strain behaviour of soils (Proc. Roscoe Memorial Symposium):* 258-261. Henley-on-Thames : G.T. Foulis & Co.

Desrues, J., Lanier, J. and Stutz, P. (1985) Localisation of the deformation in tests on sand sample. *Engineering Fracture Mechanics* 21:4:909-921.

James, R.G. and Bransby, P.L. (1970) Experimental and theoretical investigations of a passive earth pressure problem. *Géotechnique* 20:1:17-37.

Potts, D.M., Dounias, G.T. and Vaughan, P.R. (1990) Finite element analysis of progressive failure of Carsington embankment. *Géotechnique* 40:1:79-101.

Roscoe, K.H. (1970) The influence of strains in soil mechanics. 10th Rankine Lecture. *Géotechnique* 20:2:129-170.

Scarpelli, G. and Wood, D.M. (1982) Experimental observations of shear band patterns in direct shear tests. In P.A. Vermeer and H.J. Luger (eds), *Deformation and failure of granular materials (Proc. IUTAM Symposium, Delft):* 473-484. Rotterdam : A.A. Balkema.

Stone, K.J.L. (1988) *Shear band formation in granular materials*. PhD thesis, Cambridge University.

Stone, K.J.L. and Muir Wood, D. (1992) Effects of dilatancy and particle size observed in model tests on sand. *Soils and Foundations* 32:4:43-57.

Stone, K.J.L. and Wood, D.M. (1988) Model studies of soil deformations over a moving basement. In F.G. Bell, M.G. Culshaw, J.C. Cripps and M.A. Lovell (eds), *Engineering geology of underground movements :* 159-165. London : Geological Society Engineering Geology Special Publication 5.

Vardoulakis, I. (1980) Shear band inclination and shear modulus of sand in biaxial tests. *International Journal for Numerical and Analytical Methods in Geomechanics* 4:103-119.

Vardoulakis, I. (1988) Theoretical and experimental bounds for shear-band bifurcation in biaxial tests on dry sand. *Res Mechanica* 23:239-259.

Vermeer, P.A. (1982) A simple shear-band analysis using compliances. In P.A. Vermeer and H.J. Luger (eds) *Deformation and failure of granular materials (Proc. IUTAM Symposium, Delft)* : 493-499. Rotterdam : A.A. Balkema.

Shear banding in sands observed in plane strain compression

La déformation des bandes de cisaillement dans les sables soumis à une compression en déformation plane

T. Yoshida, F. Tatsuoka, M.S.A. Siddiquee & Y. Kamegai
University of Tokyo, Japan

C.-S. Park
Chang-Wan National University, Korea (Formerly: University of Tokyo, Japan)

ABSTRACT: Plane strain compression tests of dense specimens of quartz- or silica-rich poorly graded sands having different grain sizes and different degrees of particle angularity were performed to observe the deformation of shear band by a photogrametric method. A rather unique relationship between the normalized shear stress level and the shear displacement across a shear band divided by each mean particle size was obtained. The possible effects of the shape and rigidity of particle were noticed. The dilatancy rate of shear band is largest near the peak state, while it decreases to nearly zero at the residual state.

Pour explorer la dépendance des caractéristiques de la déformation en bandes de cisaillement dans le sable vis à vis du niveau de contrainte de cisaillement, une série d'essais de déformation plane bien instrumentés a été réalisée. Les échantillons testés étaient pour la plupart des échantillons de sable riche en quartz ou silice, à granulométrie étroite, avec des grains de taille et d'angularité diverses, provenant de nombreux pays, ainsi que des billes de verre. Les déformations locales de la face de déformation plane étaient observées par stéréophotogrammétrie pour obtenir la déformation de la bande de cisaillement. Une relation assez unique est mise en évidence entre le niveau de contrainte de cisaillement et le déplacement de cisaillement à travers la bande rapporté à la taille moyenne des grains. Ces résultats suggèrent que la déformation de cisaillement cumulée nécessaire pour atteindre un état résiduel augmente avec le degré d'angularité, la résistance et la rigidité des particules. Le taux de dilatance de la bande est maximum vers le pic, mais il décroît dans la partie radoucissante de l'essai, pour presque s'annuler à l'état résiduel.

INTRODUCTION

It has been found that when proto-type sand is used in a centrifuge model, even if the model test properly simulates the pressure level in the given proto-type ground, the bearing capacity of a larger proto-type strip footing is over-estimated (Tatsuoka et al., 1991, 1992, 1994a). Fig. 1 shows a typical comparison of the behaviour of strip footing between corresponding tests in 1g and in a centrifuge performed under plane strain conditions on air-dried Toyoura sand with a void ratio of 0.69. The equivalent footing width B was equal to $B_0 \cdot n = 50$ cm, while the physical model footing width B_0 was 50 cm in the 1g tests, and 2 cm and 3 cm in the centrifuge tests. n is the acceleration level and n=1 means the gravitational acceleration. Fig. 2 shows the values of N_γ plotted against B. N_γ is the peak values of $2q/\gamma \cdot B$ (q is the average footing pressure, see Fig. 1). Solid data points represent the results from a series of 1g tests changing B_0 for a range from 0.5 cm to 50 cm, which are the proto-type tests in this case, and the corresponding centrifuge tests with a constant $B_0 = 3$ cm. The so-called scale effect can be seen for the results from the 1g tests changing B_0. Importantly, for a given B, a centrifuge test with a smaller B_0 exhibits a larger N_γ when compared to its corresponding 1g test. This difference is due to the effect of sand particle size relative to B_0 (the particle size effect). The pressure-level effect is typically observed in this series of centrifuge tests changing n with the same B_0. Accordingly, the scale effect consists of the particle size effect and the pressure level effect. Note that the particle size effect could be observed also in tests using a same footing but using different types of sand having different particle sizes but having the same pre-peak deformation properties and peak strength.

FEM SIMULATIONS CONSIDERING SHEAR BANDING

The authors simulated those results of model bearing

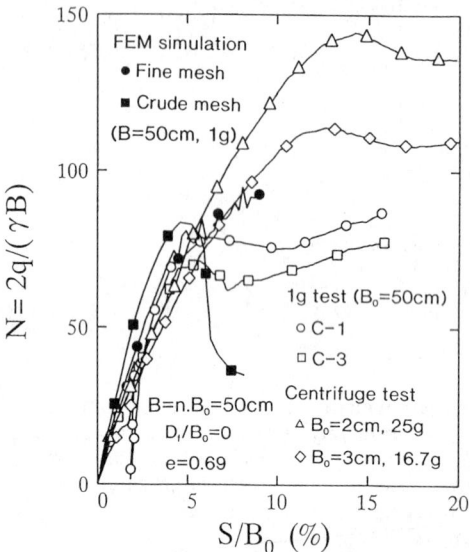

Fig. 1 Comparison of load-settlement relation for an equivalent footing width B= n·B$_0$= 50 cm between 1 g and centrifuge model tests on sand under plane strain conditions and their FEM simulations (Siddiquee, 1991, Tatsuoka et al., 1994a).

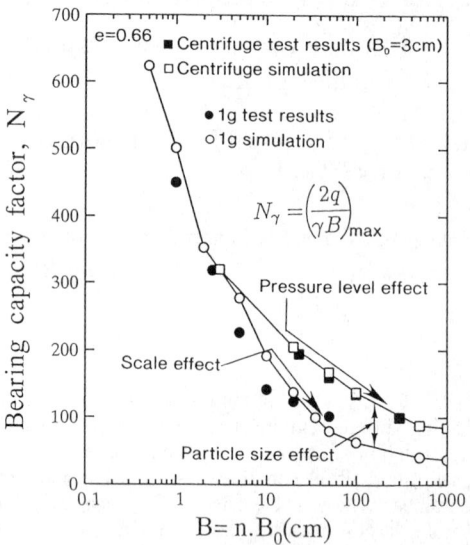

Fig. 2 Summary of plane strain model bearing capacity tests and their FEM simulations (Siddiquee, 1991, Tatsuoka et al., 1994a).

capacity tests by the plane strain FEM developed by Tanaka and Kawamoto (1989) (Siddiquee, 1991, Siddiquee et al., 1993, Tatsuoka et al., 1991, 1992, 1994a). In these analyses, the following simplified assumptions were used:
a) In the pre-peak regime, deformation of a given sand element under a uniform boundary stress condition is homogeneous.
b) Strain localisation starts suddenly at the peak stress state (n.b. in the present study, the assumptions (a) and (b) were not used to obtain the deformation of shear band in plane strain compression tests on sands).
c) The angle of internal friction ϕ_{max}=arcsin{$(\sigma_1-\sigma_3)/(\sigma_1+\sigma_3)$}$_{max}$ is a fixed material property for a given mass of sand at a given stress state; ϕ_{max} is independent of the boundary condition. Yet, ϕ_{max} is functions of many other factors including b= $(\sigma_2-\sigma_3)/(\sigma_1-\sigma_3)$ and inherently anisotropic structure.
d) In the post-peak regime, the material outside a shear band(s) exhibits only elastic rebound. In actual cases, plastic deformation also starts to occur in the course of unloading.
e) The stress-strain relationship inside a shear band is independent of the size, shape and boundary conditions of a given element containing the shear band. This relation and shear band thickness are unique for a given mass of sand, and independent of density and pressure level.
f) For both pre- and post-peak regimes, a given FEM element has the average stress-average strain relationship which is the same with that of the corresponding PSC test specimen having the same area with this FEM element. In the local scale inside each FEM element, the strain rate is continuous even after the onset of shear banding, while ignoring the interaction between its neighbouring elements due to shear banding.
g) The dilatancy characteristics in both pre- and post-peak regimes follow the same flow rule, which is Rowe's stress-dilatancy relation under plane strain conditions: $\sigma_1/\sigma_3= -K(d\varepsilon_3^P/d\varepsilon_1^P)$.

According to the assumptions (e) and (f), when a single shear band is involved in a given element, with the increase in the element size, the rate of strain-softening in the average stress-average strain relationship becomes larger (i.e., a larger decrease in σ_1/σ_3 for a given increment of average shear strain). Fig. 3 illustrates two footings **A** and **B** having the same shape but different sizes, both placed on the same sand ground. Elements **a** and **b** are located inside the ground at the same proportional points with respect to each footing, while the size of element is also in proportional to the footing size. Then, even when the pre-peak average stress-average strain relationships for the two elements are the same, the rate of strain-softening in the post-peak relationship for a given average strain is faster for the larger element **b**. This is the key feature to explain the particle size effect. In the FEM analyses, each of elements **a** and **b** may consist of several sub-elements.

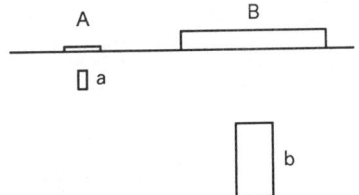

Fig. 3 Schematic figure of two similar footings on the same ground.

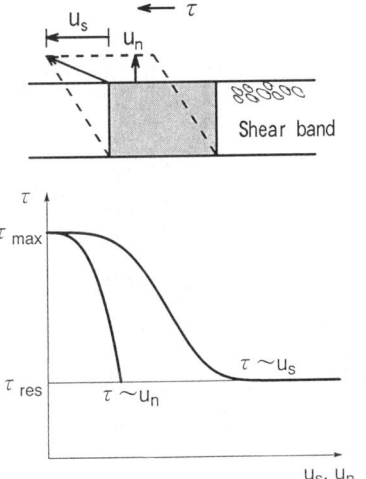

Fig. 4 Illustration of post-peak deformation of shear band.

simulate successfully the whole of the experimental results of both 1g and centrifuge tests (at least up to the peak footing load) and all of the scale effect, the particle size effect and the pressure-level effect.

The assumption (e) may not be valid for a wide range of pressure including high pressure levels at which the effect of crushing of sand particles on shear band behaviour cannot be ignored. Even in case the deformation characteristics of shear band are pressure level-dependent, particle size effects should be observed between a pair of corresponding 1g and centrifuge tests with an identical B (i.e., at an identical pressure level in the ground). When needed, this pressure-level dependency should be taken into account in relevant FEM analyses.

For this type of FEM analysis, the deformation characteristics of shear band as illustrated in Fig. 4 are one of the basic information. It is considered that when the shear band deformation characteristics specific for a given type of sand can be estimated, the effect of particle size in a given boundary value problem can be reasonably evaluated by a relevant numerical analysis (such as the method used in the present study). As the first step to this end, it was attempted to obtain the relationships among the shear stress level, the shear displacement across a shear band u_s and the change in the shear band thickness u_n by observing shear banding in a series of plane strain compression tests (PSC tests) performed using several different types of sands. Similar measurements have been performed by Drescher et al. (1990), Han and Vardoulakis (1991) and Han and Drescher (1993) for Ottawa Sand (D_{50}= 0.72 mm) and St Peter Sandstone Sand (D_{50}= 0.165 mm).

For the result of FEM analysis to be objective, the result should be independent of the total number of elements. It has been confirmed that this is the case when the total number of elements is larger than a certain limit (Siddiquee, 1991). The effects of other factors including inherent material anisotropy, the effects of pressure level on strength and deformation characteristics, the pre-peak non-linearity and the non-associated flow characteristics were also taken into account.

The results by the FEM simulations are shown in Fig. 1. They were obtained by using either relatively crude meshing without using an anti-hour-glass mode technique or relatively fine meshing with using an anti-hour-glass mode technique (Siddiquee et al. 1993). In the former case, the post-peak part includes an effect of numerical instability, thus not reliable. However, the values of N_γ by these two methods are nearly the same. The results from these FEM analyses are summarized in Fig. 2. It may be seen from Figs.1 and 2 that the FEM analyses

TESTING METHOD

<u>Plane strain compression apparatus and testing method:</u> Specimens were 20 cm high, 16 cm long and 8 cm wide (in the direction of the minor principal stress σ_3) (Fig. 5). The lateral surface of specimen was covered with a 0.3 mm-thick latex rubber membrane. The top and bottom surfaces of σ_1 (the major principal stress) were in contact with well polished stainless steel boundaries of the cap and pedestal (Nos. **8** and **9** in Fig. 5). They were well lubricated by means of a 0.3mm-thick latex rubber sheet smeared with an initial thickness (= 0.05 mm) of a selected type of silicon grease (Tatsuoka et al. 1984, Goto et al. 1993). The surfaces of the two confining platens (No. **12**) were lubricated by using a similar silicon grease layer. One of the two confining platens was a stainless steel platen having a polished surface with a load cell on its back face to measure σ_2. The other was a transparent Acryle platen to observe the deformation of specimen. The

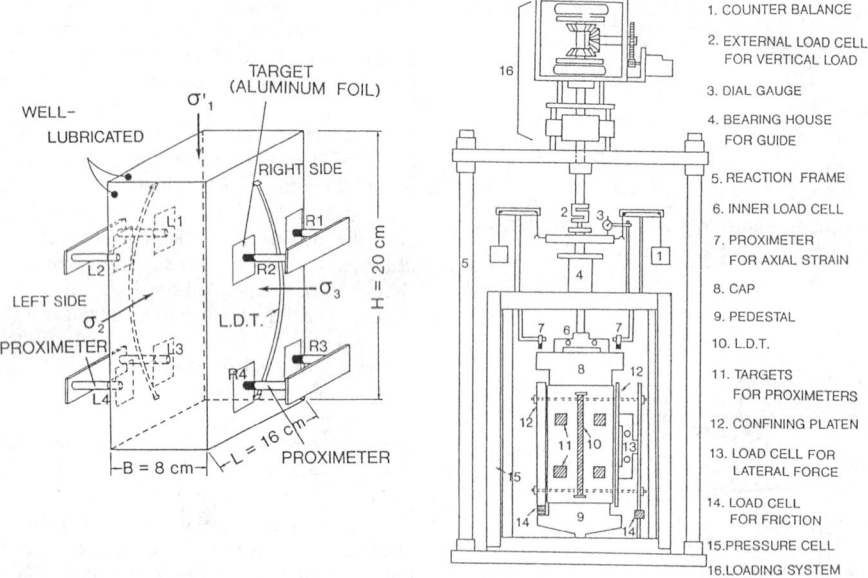

Fig. 5 Setup of plane strain compression tests (Park and Tatsuoka, 1994, Tatsuoka et al., 1994b).

coefficient of friction between this type of lubrication layer and Toyoura sand and Silver Leighton Buzzard sand (SLB sand) is as small as 0.003~0.004 at a normal stress of about 300 kPa (Goto et al., 1993). This value is equivalent to that of a high-quality linear motion steel bearing. In fact, very small restraint to the homogeneous deformation of specimen and the rigid displacement of specimen at the specimen ends was confirmed by detailed photogrametric analyses of such pictures as shown in Fig. 7.

Axial compression was applied at a constant axial strain rate of 0.125 %/min. In each test, the minor principal stress σ_3 was constant, equal to 0.8 kgf/cm² (78.4 kPa), 2.0 kgf/cm² (196.1 kPa) or 4.0 kgf/cm² (392.2 kPa). For the tests at σ_3=0.8 kgf/cm², partial vacuuming was applied without using a pressure cell, while for the tests at σ_3=2.0 and 4.0 kgf/cm², in addition to partial vacuum of 0.8 kgf/cm² applied to the inside of specimen, air confining pressure was applied.

Axial strain ε_1, averaged for the whole height of specimen, was obtained by measuring the axial displacement of the specimen cap with a pair of proximeters (No. 7 in Fig. 5) (i.e. external axial strain). Axial strain was measured also locally along both σ_3 planes by means of a pair of LDTs (Local Deformation Transducers; No. 10, Goto et al., 1991). The axial strains presented herein are those measured externally, which thus include some effects of

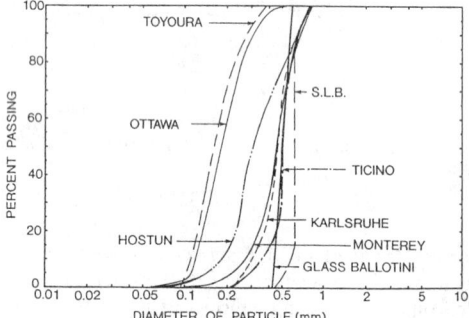

Fig. 6 Grain size distribution curves of the tested materials.

bedding error; locally measured axial strain are presented in Tatsuoka et al. (1993, 1994b) and Park and Tatsuoka (1994). This local strain measuring method was used to obtain the deformation of shear band in triaxial compression tests on sedimentary soft rock (mudstone) (Tatsuoka and Kim, 1994; this volume). Until the onset of obvious shear banding, lateral strain ε_3 was obtained from the change in the specimen width measured by means of four proximity transducers set on each σ_3 plane (thus eight in total). The targets of aluminum foil piece were attached on the σ_3 surfaces with silicon grease (No. 11 in Fig. 5).

The deviator stress σ_1-σ_3 was obtained by dividing

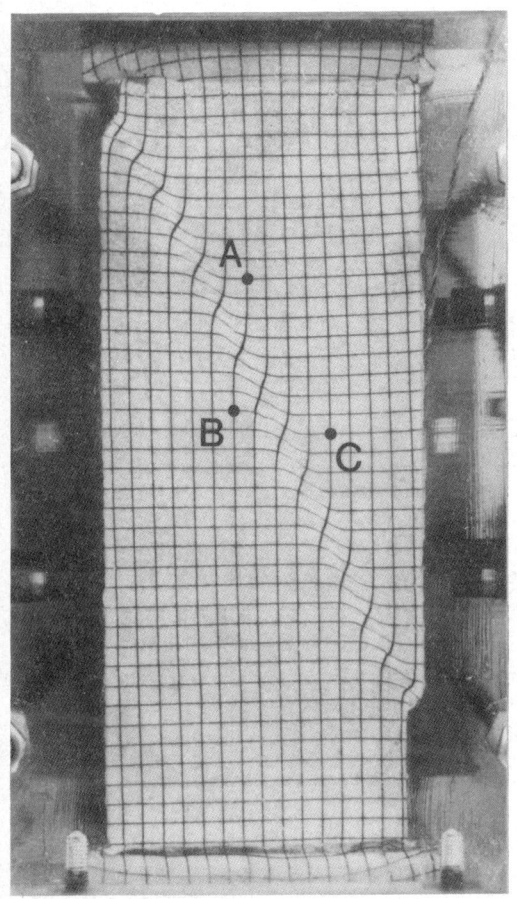

the axial load measured with a load cell placed immediately above the specimen cap (No. 6) by the current average cross-sectional area of specimen. This stress value was corrected for the friction on the σ_2 surfaces measured with a pair of load cells set at the bottoms of the confining platens (No. 14). This correction was in fact very small. After the onset of obvious shear banding, the average cross-sectional area was assumed unchanged. The detail of the apparatus and the testing method are given in Tatsuoka et al. (1986, 1994b), Goto et al. (1993) and Park and Tatsuoka (1994).

Testing materials and specimen preparation: Some physical properties of the sands tested are listed in Table 1. Hostun, Toyoura, Monterey, Silver Leighton Buzzard sands are quartz-rich sands, Ottawa sand is silica-rich sand, and Ticino sand is a mixture of quartz (28 %), feldspar (30 %), mica (5 %) and others. These sands have sub-angular to sub-round particle shapes, and all poorly-graded (Fig. 6). The effects of the grading, the type of parent rock, the density of sand are beyond the scope of the present study. These sands have been used extensively in research works in many countries. As a reference, glass ballotini having a sphere shape was also used.

These materials are listed in Table 1 in the sequence of particle angularity (more to less angular from the top to the bottom). The degree of particle angularity was evaluated by automatic analyses of particle shape by using a newly developed digital processing method, which will be published in the near future.

Hostun sand (Sable d'Hostun) was provided by Prof. Darve of Institut de Mecanique de Grenoble, France, which is a quarry sand sold by SIKA Ltd. As it was obtained by washing and sieving this residual soil to remove fines, this sand has most angular particles. Toyoura sand is a sieved dune sand from Toyoura Town in Yamaguchi Prefecture near the west end of the main island of Japan, sold by Toyoura Koseki Co. Toyoura sand has been used as the standard sand in the community of concrete engineering in Japan. Ticino sand (batch TS-4) was provided by Prof. Jamiolkowski of Politechnico di Torino, Italy. This sand was from a quarry about 20 km south of Milan existing along the bank of Ticino River (one of the affluence of Po River). The sand after having removed fines was provided. Monterey No. 0 sand is a sieved dredged beach sand, sold by Monterey Sand Co., California. Ottawa F-75 Banding Sand is a mined sand from St. Peter glacial formation (a sandstone) in Ottawa, Illinois, provided by U.S. Silica Co. in Ottawa. This sand has been blasted, washed, dried and graded. These two types of sands were provided by Prof. Wu, Jonathan T.H. of University of Colorado at Denver. Silver Leighton

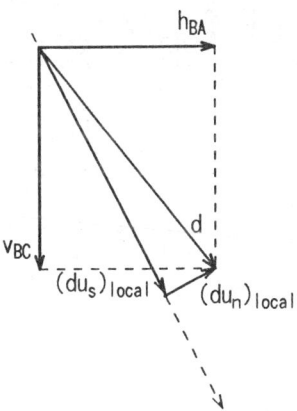

Direction of shear band

Fig. 7 Method to obtain the relative displacement increment vector **d** across a shear band (the picture was taken at an average shear strain γ=14.1 % and u_s=8.58 mm for Ticino Sand, σ_3=0.8 kgf/cm^2).

Table 1 List of test materials.

Material (Origin)	D_{50} (mm)	U_c	G_s	e_{max}	e_{min}	Grain shape
Hostun Sand (France)	0.31	1.94	2.65	0.95	0.55	Subangular
Toyoura Sand (Japan)	0.162	1.46	2.636	0.973	0.612	Subangular
Ticino Sand (Italy)	0.502	1.33	2.68	0.96	0.59	Subangular
Monterey No.0 Sand (USA)	0.44	1.74	2.64	0.86	0.55	Intermediate
Silver Leighton Buzzard Sand (UK)	0.62	1.11	2.66	0.79	0.49	Subround
Karlsruhe Sand (Germany)	0.45	1.65	2.65	0.87	0.54	Subround
Ottawa Sand (USA)	0.182	1.70	2.665	0.864	0.515	Subround
Glass Ballotini (Japan)	0.505	1.21	2.49	0.713	0.563	Sphere

Listed in the order of grain shape angularity from the top.

Table 2 List of PSC tests (1.0 kgf/cm² = 98 kN/m²).

Material[1]	σ_3' (kgf/cm²)	e_{test}[2]	D_r (%)	ϕ_{peak}[3] (deg)	ϕ_{res}[4] (deg)	θ[5] (deg)	t_0[6] (mm)	t_0/D_{50}	γ_{peak}[7] (%)	$(u_s^*)_{res}$[8] (mm)	$(u_s^*)_{res}/t_0$[9] (%)	$(\gamma^P)_{res}$[10] (%)
Hostun Sand	0.8	0.616	84	47.6	35.9	63	6.2	20	6.0	3.36	54.1	60.1
	4.0	0.648	76	44.8	34.2	58	2.8	9.3	9.5	2.02	72.3	81.8
Toyoura Sand	0.8	0.694	77	45.7(45.3)	35.5(36.9)	66	3.5	22	4.1(4.8)	5.82(2.24)	166.4(63.9)	170.5(68.7)
	2.0	0.660	85	45.9	35.0	65	3.1	19	5.5	2.65	85.4	90.9
	4.0	0.661	85	45.0	33.7	66	2.5	15	7.4	2.51	100.4	107.7
Ticino Sand	0.8	0.657	82	48.1	34.8	61	5.1	10	7.2	4.62	90.6	97.8
	4.0	0.679	76	45.7	34.5	60	3.6	7.2	11.1	3.06	85.1	96.2
Monterey Sand	0.8	0.604	83	47.8(47.7)	34.4	66	5.3	12	3.8(4.6)	4.16(3.69)	78.5(69.6)	82.3(74.2)
	4.0	0.643	70	45.5	34.7	59	3.6	8.2	7.7	2.56	71.2	78.9
S.L.B. Sand	0.8	0.549	80	44.7	32.7	59	6.1	9.8	7.4	4.50	73.8	81.2
	2.0	0.548	81	43.4	31.3	62	5.8	9.4	6.9	5.90	101.8	108.7
	4.0	0.547	81	42.5	30.8	61	5.5	8.9	7.6	4.57	83.1	90.8
Karlsruhe Sand	0.8	0.621	75	43.8	33.0	59	4.5	10	7.1	3.64	80.8	87.9
	4.0	0.636	71	42.8	31.0	58	4.2	9.3	8.9	4.04	96.1	105.0
Ottawa Sand	0.8	0.598	76	43.4	34.2	70	3.7	20	4.1	1.97	53.2	57.3
	4.0	0.608	73	44.3	32.2	65	3.6	20	6.4	2.04	56.8	63.2
Glass Ballotini	0.8	0.573	93	35.7	26.5	54	9.8	19	4.0	5.03	51.3	55.3
	4.0	0.621	61	32.3	26.2	53	9.0	18	5.8	8.29	92.1	97.9

1) Listed in the order of grain shape angularity from the top.   2) Void ratio at the start of loading.
3) The figures in ( ) mean the second peak appeared after the first peak.
4) ϕ_{res} is defined as the smallest stress ratio during the residual state. The figure in ( ) is the first minimum value (not the smallest one) used in the analysis.
5) The angle of shear band relative to the σ_3 direction at the residual condition.   6) The width of shear band at the residual state.
7) Homogeneous shear strain at the peak.   8) Increment of u_s between the peak state and the start of residual state, $(u_s^*)_{res}=(u_s)_{res}-(u_s)_{peak}$.
9) Plastic shear strain increment between the peak state and the start of residual state.   10) $\gamma_{peak}+(u_s^*)/t_0$.

Buzzard sand was from an aeolian deposit in Leighton Buzzard, U.K. The batch used was provided by Prof. J.R.F. Arthur of University College London (n.b. the properties of this sand may differ noticeably from a batch to another). This sand is coarse, closely graded between BS sieve sizes 18-25 (650-800 μm) and 98 % component is sub-round quartz particles. Karlsruhe (medium) sand was provided by Prof. G. Gudehus and Dr. W. Wu, University of Karlsruhe, Germany. This sand consists of sub-round mostly quartz grains (82 % quartz, 15 % feldspar and 3 % calcite; Wu and Kolymbas, 1991).

All the specimens were prepared by pluviating air-dried particles through air. Therefore, the direction of σ_1 during the PSC tests coincided with the direction of pouring. The specimens were relatively dense, except for one glass ballotini specimen tested at σ_3= 4.0 kgf/cm^2 (392 kPa) (Table 2). The specimens were kept air-dried throughout the tests. The density imperfection technique (Vardoulakis and Graf, 1985) to enhance shear banding from a specified point within a specimen was not used.

Method of shear band observation: Grids made of dyed latex rubber had been printed in advance on the outer surface of the latex rubber membrane on the σ_2 plane (Fig. 7). Many pictures of the σ_2 surfaces as shown in Fig. 7 were taken during each test through a transparent confining platen. In the tests at σ_3= 2 and 4 kgf/cm^2 (196 and 392 kPa), pictures were taken from outside the pressure cell (No. **15** in Fig. 5). The distortion of picture due to the annular shape of the pressure cell was properly corrected. As cell water was not used, the degree of distortion was only about 1 %.

The coordinates of the nodes of grid were read automatically to an accuracy of 0.03 mm or less by means of a newly developed photogrametric system. Strain fields on the σ_2 surface were constructed from these readings. Increments between two successive loading stages of the shear displacement across a shear band u_s and the change in the shear band thickness u_n were obtained from the changes in the coordinates of nodes located along two lines in parallel, located immediately outside the boundaries of the finally appeared single shear band (Fig. 7). They were obtained as follows. First, the horizontal component of the relative displacement h_{BA} between the vertically neighbouring points **A** and **B**, for example, and the vertical component v_{BC} between the horizontally neighbouring points **B** and **C** were obtained. If the specimen deforms perfectly homogeneously and the axial and lateral strains are exactly principal strains, both components h_{BA} and v_{BC} are always zero. Then, the local relative displacement increment vector **d** across the shear band was obtained by combining the two components h_{BA} and v_{BC}. The components of the local vector **d** in the directions in parallel to and normal to the shear band direction are $(du_s)_{local}$ and $(du_n)_{local}$. Then, the values of du_s and du_n for the whole of each shear band were obtained by averaging these local values for a length of the central part of the shear band (about 50 to 70% of the full width of specimen). The accumulated values of u_s and u_n at a given loading stage were obtained by integrating the increments du_s and du_n from the start of loading (i.e., from σ_1/σ_3= 1.0).

In the present analysis, it was assumed that the shear band pattern is homogeneous in the σ_1 direction. This assumption becomes more acceptable as the effects of membrane forces and end friction become smaller, namely as the pressure level increases.

Drescher et al. (1990) and Han and Vardoulakis (1991) have performed similar observations of the deformation of shear band in plane strain compression tests on sand. In their tests, the cap was fixed for lateral and rotational displacements and the pedestal was supported with a linear steel bearing. They measured the relative axial and lateral displacements between the cap and pedestal. They assumed that only a single shear band appears upon strain localisation, and upon the onset of strain localisation, the material outside a shear band becomes rigid. The authors did not use their method considering that the relative displacement between the specimen cap and pedestal may not be very representative of the deformation of shear band, particularly at the beginning stage of shear banding. The reasons are as follows:

a) At early stage of strain localisation, which is usually immediately before the peak stress state, multiple shear bands which have not well developed yet may appear.
b) Even by using high-quality lubrication layers, the deformation in the zones outside a shear band may be distorted to some extent due to the restraint at the σ_1 and σ_2 surfaces.
c) On the other hand, it is not certain whether overall slipping may occur at the top and bottom ends of specimen in either lateral direction due to end lubrication made for enhancing homogeneous deformation of specimen.
d) The effects of bedding error at the top and bottom ends may not be always negligible; its accurate value is usually very difficult to evaluate due to its very complicated mechanism.
e) Even after the onset of strain localisation, the material outside a shear band does not become perfectly rigid, and their strains may not be negligible. Namely, in the zones outside a shear band, at early stages of strain localisation in the pre-peak regime, essentially plastic shear strain increments, which are

positive, and associated dilative volumetric strain increments occur. On the other hand, in the post-peak regime, negative and essentially elastic shear strain increments occur followed by the occurrence of negative elasto-plastic shear strain increments and associated positive volumetric strain increments.

TEST RESULTS

Average stress-strain relations: Fig. 8 shows the relationships between the boundary stress ratio σ_1/σ_3 and the average shear strain $\gamma=\varepsilon_1-\varepsilon_3$ at three different σ_3s (see Table 2). In some tests, two peaks having similar ϕ were observed. In all the tests, before γ becomes 15 %, the specimens have reached the residual state. A rather constant stress ratio was attained at the residual state in each test, which indicates that a single shear band developed rather freely. In fact, in most of the tests, the shear band did not intersect with the top and bottom rigid boundaries. Yet, the residual principal stress ratio $(\sigma_1/\sigma_3)_{res}$ in each test was defined as the smallest stress ratio during the recorded residual state.

Start of shear banding: Figs. 9(a) and (b) show the contours of accumulated local shear strain by a 10 %

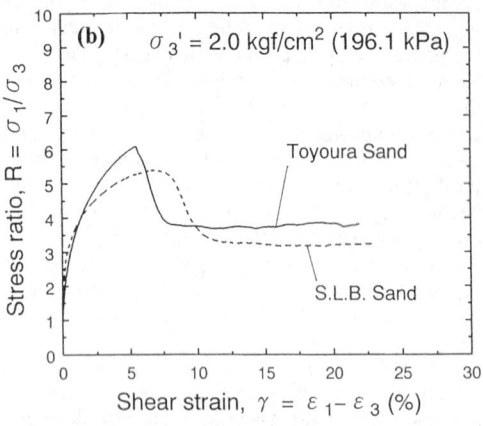

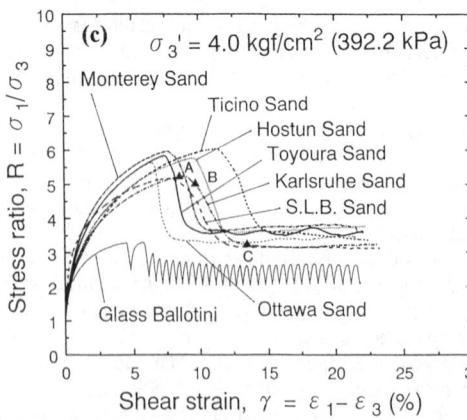

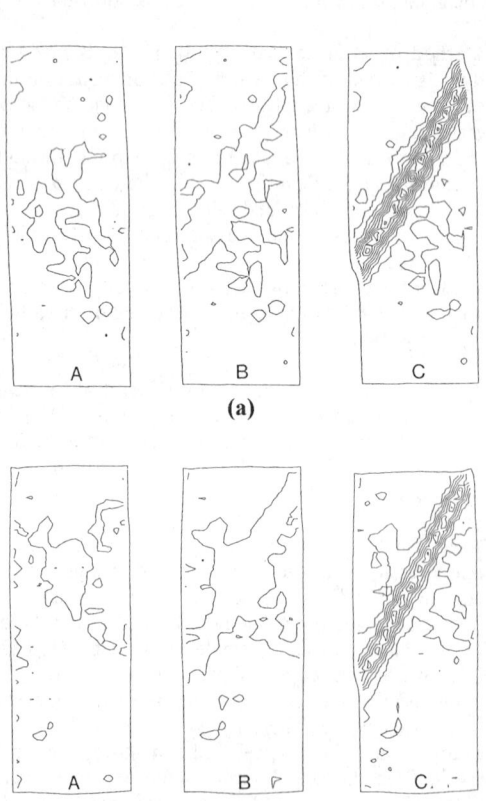

Fig. 8 Relationships between σ_1/σ_3 and average shear strain $\gamma=\varepsilon_1-\varepsilon_3$; (a)$\sigma_3=$ 0.8 kgf/cm^2, (b)$\sigma_3=$ 2.0 kgf/cm and (c)$\sigma_3=$ 4.0 kgf/cm^2.

Fig. 9 Contours of local shear strain by a 10 % increment: (a) SLB sand, $\sigma_3=$0.8 kgf/cm^2 (Fig. 8a) and (b) Karlsruhe sand, $\sigma_3=$ 4.0 kgf/cm^2 (Fig. 8c).

increment which occurred from the start of loading. The loading stages **A, B** and **C** in the test on SLB sand at $\sigma_3'=0.8$ kgf/cm^2 and in the test on Karlsruhe sand at $\sigma_3'=4.0$ kgf/cm^2 are indicated in Figs. 8a and 8c, respectively. The boundary lines in each figure in Fig. 9 represent the outmost lines of grid located inside the specimen periphery (see Fig. 7). It may be seen by inspecting the changes of their shapes that the effects of the boundary restraint to uniform deformation of specimen was very small. These local strain distributions were obtained from strains defined in each 0.5 cm × 0.5 cm element by assuming uniform strains in each element. By this reason, as seen from figures at stage **C**, some fictitious strain fields were obtained in very intensely strained regions inside the shear band.

In both tests, already at stage **A**, which is immediately before the peak state, a sign of multiple shear bands has appeared, which is, in Fig. 9(a), two shear bands in the direction from the left top to the right bottom and the other one in the opposite direction, and in Fig. 9(b), two bands in the opposite directions. Then, by stage **B**, which is immediately after the peak state, only one of them has further developed in each test, while the deformation of the other band(s) has stopped. Subsequently in the post-peak regime, this tendency was substantially accelerated. Stage **C** is the beginning of the residual state.

The strain fields shown above and the other ones indicate that the start of strain localisation is not very

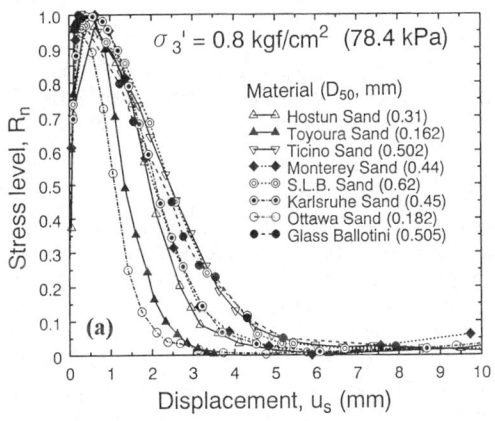 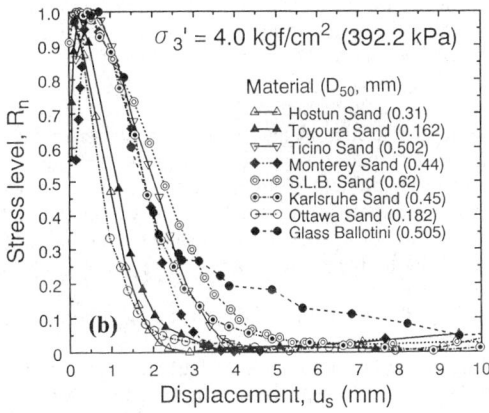

Fig. 10 Relationships between stress state parameter R_n and accumulated shear displacement across the shear band u_s; (a) $\sigma_3'= 0.8$ kgf/cm^2 and (b) $\sigma_3'= 4.0$ kgf/cm^2 (1.0 kgf/cm^2= 98.0 kPa).

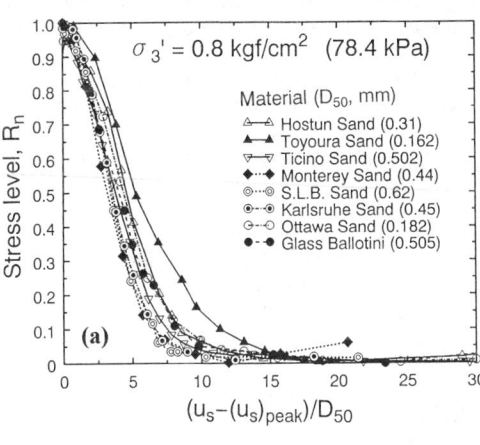 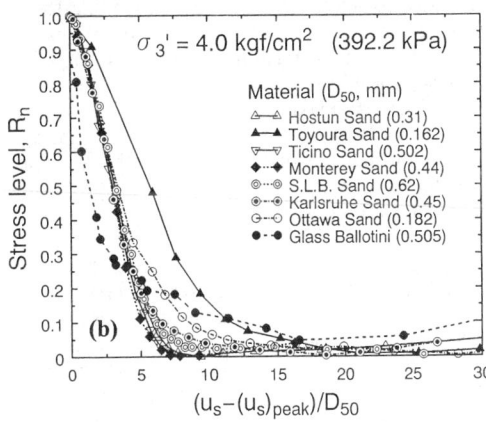

Fig. 11 Relationship between R_n and $\{u_s-(u_s)_{peak}\}/D_{50}$; (a) $\sigma_3'= 0.8$ kgf/cm^2 and (b) $\sigma_3'= 4.0$ kgf/cm^2 (1.0 kgf/cm^2= 98.0 kPa).

spontaneous, and therefore it is not simple to identify its exact moment. This fact indicates that at earlier stages of strain localisation, the relative displacement between the top and bottom ends of specimen may not be very representative of the deformation of single shear band.

Shear deformation of shear band: Fig. 10 shows the relationships between shear stress level R_n and the accumulated shear displacement across the shear band u_s at $\sigma_3 = 0.8$ kgf/cm² and 4.0 kgf/cm². The parameter R_n is defined as;

$$R_n = \frac{(\sigma_1/\sigma_3) - (\sigma_1/\sigma_3)_{res}}{(\sigma_1/\sigma_3)_{peak} - (\sigma_1/\sigma_3)_{res}}$$

where ($\sigma_1/\sigma_3)_{peak}$ and ($\sigma_1/\sigma_3)_{res}$ are the peak and residual stress ratios listed in Table 2. $R_n=1.0$ and 0.0 mean the peak and residual states. For glass ballotini, the stress ratio σ_1/σ_3 was defined for the top envelope of the measured stress-strain curve exhibiting stick-slip behaviour (see Fig. 8). The origin for u_s is defined at the start of each PSC test. It may be seen from Fig. 10 that the pattern of the curve for glass ballotini at $\sigma_3= 4$ kgf/cm² is somewhat different from the others. In the other tests, the pattern is very similar. It may be noted that the amount of u_s to reach the residual state tends to increase with the particle size.

In Fig. 10, the amount of u_s which occurs before the peak stress state is not consistent among these different tests. This could be attributed to that the deformation of shear band is not stable before the peak stress state, which is probably affected by uncontrollable initial local variations in each specimen. Considering the above, the increment from the peak state $u_s^* = u_s - (u_s)_{peak}$ was obtained (Table 2). In so doing, the value of u_s at the peak state was obtained between the two loading stages where u_s was measured on the both sides of peak by assuming a linear relation between u_s and the boundary shear strain. For the tests in which two nearly identical maximum stresses were observed, $(u_s)_{peak}$ was defined at the second peak. In addition, considering that the first primary factor controlling u_s be the particle size, $u_s - (u_s)_{peak}$ was divided by the mean grain size D_{50} (Figs. 11a and b).

It may be seen from Fig. 11 that despite a wide range of particle angularity among the test materials from a spherical one to a sub-angular one, the general pattern is very similar for all the tests (except for glass ballotini at $\sigma_3= 4$ kgf/cm²). Yet, some scatter exists in the relations, in particular between Toyoura sand and the other materials. This point will be discussed again later.

The shear band widths t_0 measured at the start of residual state are listed in Table 2. Each t_0 value was defined as the distance between two points on the opposite sides of a shear band where the rate of shear distortion along grid lines crossing the shear band was the maximum. The ratio t_0/D_{50} ranges from about 10 to about 30. A tendency that t_0 decreases as σ_3 increases and as D_{50} increases may be seen, but this point is not conclusive yet.

The plastic shear strain $(\gamma^p)_{res}$ within a shear band to reach the residual state was obtained by summing "the average shear strain γ_{peak} developed until the peak state" and "u_s^*/t_0 developed until the start of the residual state" (Table. 2). The start of the residual state was defined as the moment when $R_n = 0.05$ in the course of stain-softening (n.b. the value of 0.05 was chosen arbitrarily, but this state is very close to the residual state). This method ignores elastic strains. These values range between about 40 % and 110 %, which are similar to or larger than about 50 % for Ottawa Sand reported by Drescher et al. (1990). Of course, these values are much larger than the corresponding boundary shear strains (see Fig. 8).

It is seen that the relations shown in Fig. 11 are not very unique and $(\gamma^p)_{res}$ listed in Table 2 are not very constant among the different tests. It was found that the particle shape alone is not sufficient to explain the difference. Then, "the ratio of the particle strength and rigidity to the pressure level", or simply the effect of pressure level was additionally considered. To this end, $R_n \sim \{u_s-(u_s)_{peak}\}/D_{50}$ relations at different σ_3s were summarized for each type of sand (Fig. 12). It is seen that for Toyoura and SLB sands, the effects of σ_3 on the relation are very small. A faster drop in R_n at a larger σ_3 was observed in the relations for the other types of sands, which was considered due to the effects of particle breakage at higher pressure levels in a shear band.

Now, the scatter in the $R_n \sim \{u_s-(u_s)_{peak}\}/D_{50}$ relations of these different materials (Fig. 11) may be explained in such a way as that the rate of strain softening in the relation becomes slower as the particles become more angular and as the particles become stronger and more rigid; namely:
1) Toyoura sand has the lowest rate of strain softening. This would be because the particles are relatively angular (the second most angular) and the particles are relatively strong and rigid. This feature is suggested by a negligible change in the relation with σ_3 (Fig. 12b).
2) A larger rate of strain softening for the other materials may be due to either a less degree of particle angularity or lower strength and rigidity of particles or both. For example, Hostun sand has the

most angular particles, but the particles may be relatively crushable, as suggested by the fact that this sand was a residual sand and as supported by a relatively large effect of σ_3 on the relation (Fig. 12a). On the other hand, SLB sand has relatively round particles, but it is likely that the particles are relatively less crushable as suggested by a negligible effect of σ_3 on the relation (Fig. 12e).

3) It may be seen from Fig. 11 that at each σ_3, the relations for the materials other than Toyoura sand are very similar. This trend would be fortuitous, caused by balance between the effect of particle

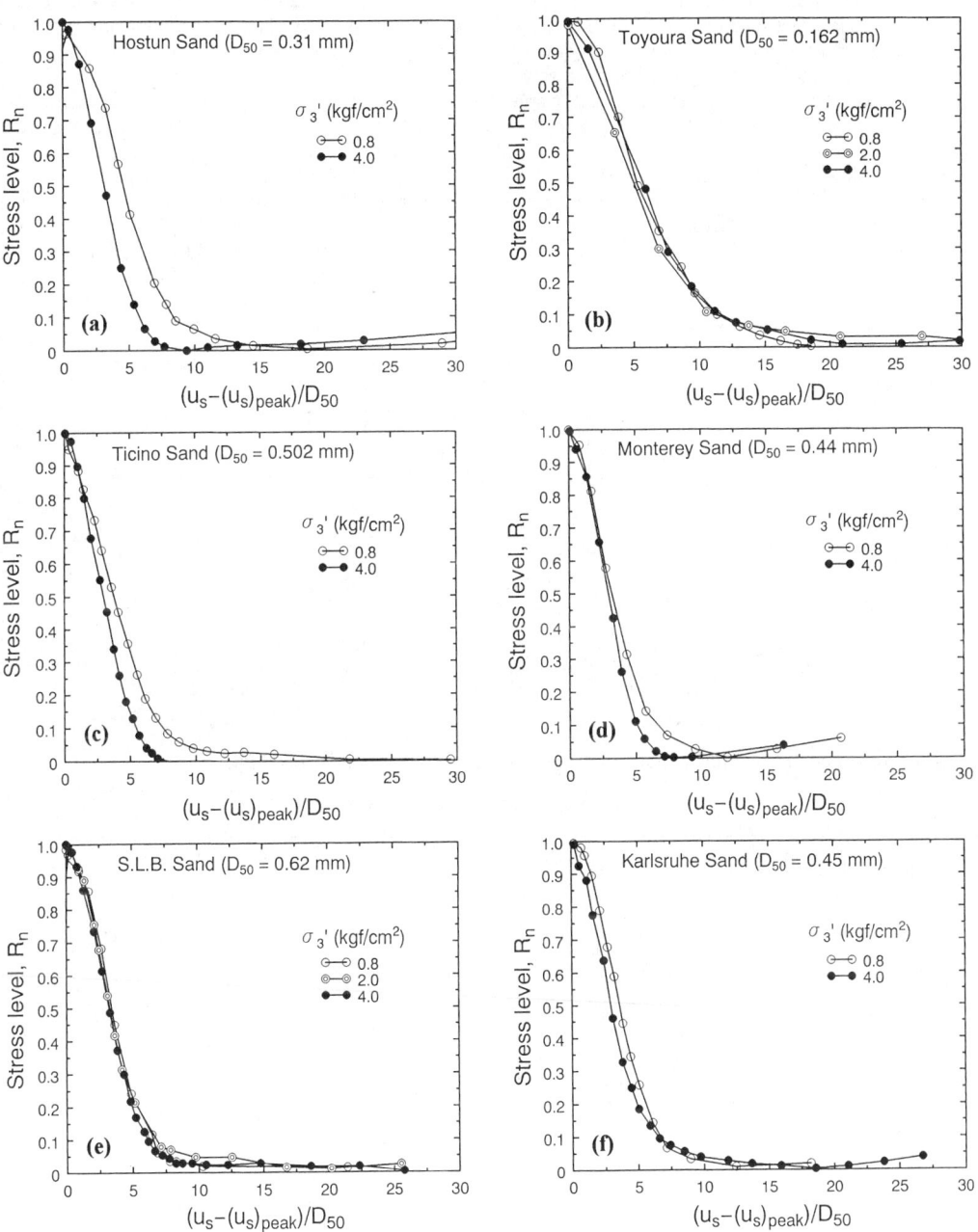

Fig. 12 $R_n \sim \{u_s - (u_s)_{peak}\}/D_{50}$ relations for different σ_3s; **(a)** Hostun sand, **(b)** Toyoura sand, **(c)** Ticino sand, **(d)** Monterey No.0 sand, **(e)** SLB sand and **(f)** Karlsruhe sand (to continue).

shape and that of the strength and rigidity of particle.
4) The behaviour of Glass ballotini at σ_3= 4.0 kgf/cm^2 is erratic. It seems that this is due to that the accuracy of R_n is not enough because of a small post-peak drop of σ_1/σ_3 due to the relatively low density of the specimen.

Judging from very small effects of σ_3 on the relations for both relatively fine sand (Toyoura sand; Fig. 12b) and relatively coarse sand (SLB sand; Fig. 12e), it is very likely that the effects of the restraint at the specimen boundary on the shear band deformation were very small.

The stress state in a shear band may not be very uniform along its lengthwise direction due to the lack of complimentary shear stress on the σ_3 planes (Vermeer, 1990). In addition, strain localisation starts always from near the center of specimen (see Fig. 9). It follows from the above-mentioned two factors that the stress state in the shear band be non-uniform in its lengthwise direction. Therefore, the true local relations in a shear band free from the end effects and non-homogeneous strain localisation, if they exist, may be slightly different from those presented in the above. Yet, the present authors consider that these relations presented above provide very useful information as to shear banding.

Dilatancy of shear band: Fig. 13 shows the relationships between R_n and the accumulated increase in the shear band thickness u_n at σ_3= 0.8 kgf/cm^2 (78.4 kPa) and 4.0 kgf/cm^2 (392.2 kPa). Comparing Fig. 10 with Fig. 13, it is seen that in each test, the largest rate of dilatation is attained near the peak stress state, and the rate of dilatation

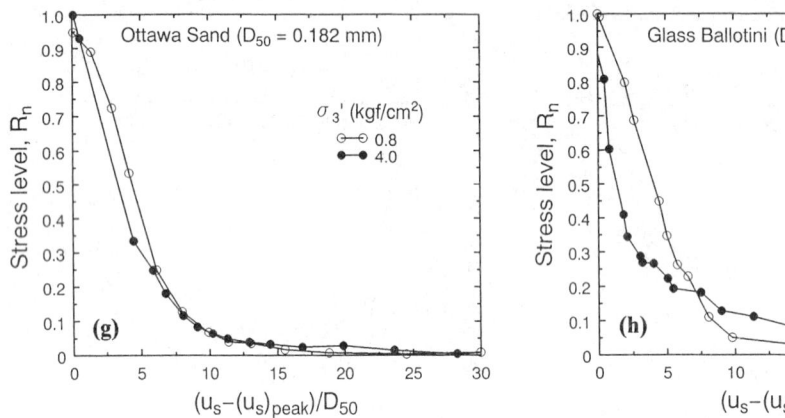

Fig. 12 (continued) (g) Ottawa sand and (h) glass ballotini (1.0 kgf/cm^2= 98.0 kPa).

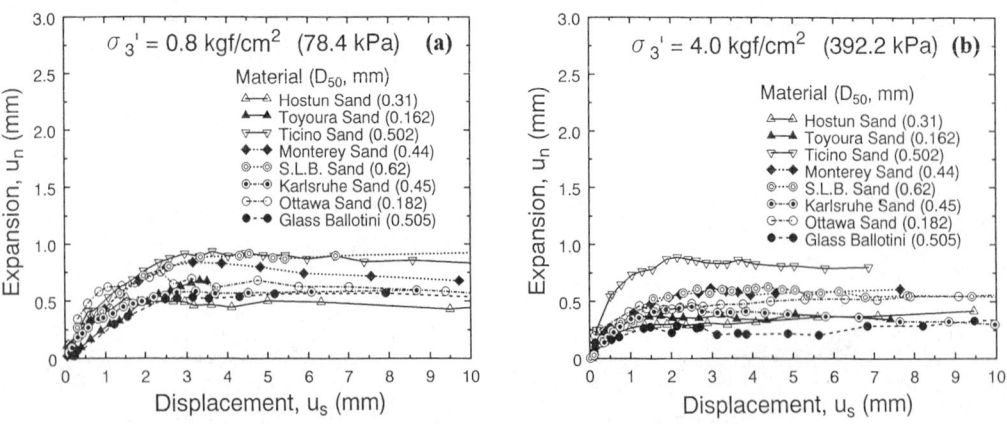

Fig. 13 Relationships between accumulated change in shear band thickness u_n and u_s; (a)σ_3= 0.8 kgf/cm^2 and (b)σ_3= 4.0 kgf/cm^2 (1.0 kgf/cm^2= 98.0 kPa).

decreases monotonously towards zero at the residual state. Namely, the shear band thickness increases continuously after the onset of shear banding. The above-mentioned tendency can be seen also in the relationships between R_n and $\{u_n - (u_n)_{peak}\}/D_{50}$ (Fig. 14).

This dilatancy behaviour of shear band described above is different from that reported by Drescher et al. (1990), Han and Vardoulakis (1991) and Han and Drescher (1993). They reported that in their tests, a large rate of increase in the shear band thickness occurred immediately after the onset of shear banding, which was near the peak stress state, and subsequently the shear band thickness decreased at a large rate, exhibiting large contractive behaviour. The present authors wonder whether some of the factors pointed out in the end of the section TESTING METHOD are responsible for this unusual contractancy of shear band in the post-peak regime.

The final amount of u_n at the residual state is controlled not only by the particle size (or the shear band thickness), but also by the initial density of specimen. Namely, for a given type of sand, if the void ratio in a shear band at the residual state is a unique function of the pressure level, the amount of u_n at the residual state is approximately equal to $\{\Delta e/(1 + e_i)\} \cdot t_0$. Here, $\Delta e = e_{critical} - e_i$ ($e_{critical}$ is the void ratio at the critical or residual state and e_i is the void ratio at the onset of shear banding) and t_0 is the shear band thickness at the residual state. This is the largest factor for the large variations in the relations shown in Figs. 13 and 14.

Shear band direction: The angles θ relative to the σ_3 direction of the average direction of shear band measured at the start of residual state are listed in Table 2. It may be noted that the angle θ decreases as the particle size increases, as has been suggested by Arthur et al. (1982), Desrues et al. (1989) and Tatsuoka et al. (1990). This effect was explained theoretically by Vermeer (1990). Namely, at the both lateral surfaces of specimen, the stress condition is uniform with the confining pressure $\sigma_c = \sigma_3$ without shear stress. Accordingly, this stress condition is common for the inside and outside of shear band (i.e., end effect). Then, when the stress condition is uniform in the shear band, the stress component on the vertical planes in the shear band is equal to σ_c. Further, between inside and outside the shear band, the stress components on the planes in parallel to the shear band direction should be the same to maintain the stress equilibrium. Accordingly, there exists no stress discontinuity between inside and outside the shear band. At the same time, during the post-peak regime, the zones outside the shear band are in the unloading regime while the shear band is in the loading (strain-softening) regime. When elastic shear strain increments can be ignored relative to large plastic strains occurring in the shear band, the zones outside the shear band become essentially rigid. Then, the kinematic constraint forces the direction of shear band, which is the direction of the discontinuity of strain rate, to be equal to the direction of zero-extension in the shear band, the angle of which is equal to 45°+ (the dilatancy angle ν in the shear band)/2 relative to the direction of the plastic minor principal strain rate $d\varepsilon_3$. When the directions of $d\varepsilon_3$ and $\sigma_c = \sigma_3$ are the same with each other, the angle θ of shear band relative to the σ_3 direction is equal to 45°+ ν/2. This value is theoretically the lowest possible angle of shear band direction in that when stress discontinuity exists between the inside and outside a shear band, angles of θ larger than 45°+ ν/2 are theoretically possible. It is conceivable that the above mentioned end effect

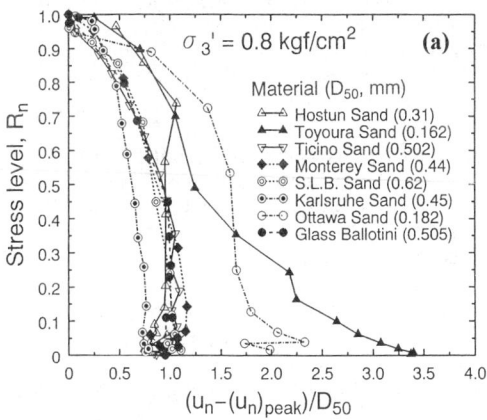
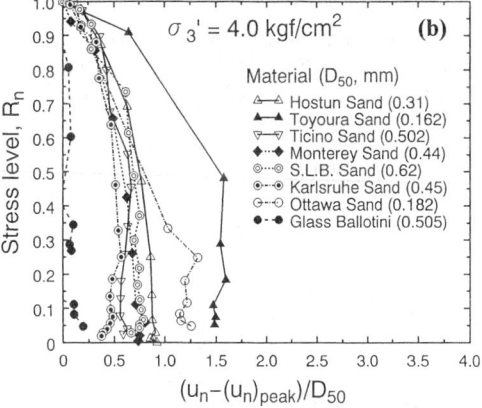

Fig. 14 $R_n \sim \{u_n - (u_n)_{peak}\}/D_{50}$ relations; (a)$\sigma_3 = 0.8$ kgf/cm² and (b)$\sigma_3 = 4.0$ kgf/cm² (1.0 kgf/cm²= 98.0 kPa).

becomes more dominant as the shear band thickness relative to the specimen size increases; namely, as the particle size increases. Accordingly, the angle θ decreases as the particle size increases (as observed in the test results).

It may also be seen from Table 2 that the angle θ decreases as σ_3 increases, as has been suggested by Tatsuoka et al. (1990), Desrues and Hammad (1989) and Han and Drescher (1993). More detailed observation and analysis of shear band direction will be required.

CONCLUSIONS

In the plane strain compression tests performed on many different types of sands and glass ballotini, the following observations were obtained:

1) Strain localisation started immediately before the peak state in all of the tests. In most of the tests, a sign of multiple shear bands was observed. The development of a single well-defined single shear band was attained only after the peak state.

2) Among different types of sands having different particle sizes, a rather unique relationship between the shear stress level R_n and "u_s/D_{50}= the shear displacement across a shear band divided by the mean particle size" was obtained. Namely, the most important material property controlling the deformation of shear band is the particle size.

3) Yet, the above-mentioned R_n–u_s/D_{50} relations scattered to some extent. Probably, it is due to the effects of other secondarily important factors including the angularity and the strength and rigidity of particles.

4) In all of the tests, the rate of the increase in the shear band thickness was largest near the peak stress state, and it decreased monotonously with shear deformation towards nearly zero at the residual state.

ACKNOWLEDGEMENTS: The authors gratefully acknowledge the help of many researchers of many countries for obtaining sands used for this study. They also thank the Japanese Ministry of Education, Science and Culture and Makita Scholarship Foundation for financial support. The cooperation by Mr. S. Nakamura, Mr. T. Sato and Dr. Y. Kohata are also appreciated.

REFERENCES

Arthur,J.R.F. and Dunstan,T. (1982). Rupture layers in granular media, Proc. IUTAM Conf. Deformation and Failure of Granular Materials, Delft (eds P.A.Vermeer & H.J.Luger), Balkema, pp.453-459.

Desrues,J. and Hammad,W. (1989). Etude experimentale de la localisation de la deformation sur sable: influence de la contrainte moyenne, Proc. 12th International Conference on Soil Mechanics and Foundation Engineering, Rio de Janeiro, Vol.I, pp.31-32.

Desrues,J., Mokni,M., and Mazaerolle,F. (1991). Tomodensitometrie et la localisation sur les sables, Proc. 10th European Regional Conf. on SMFE, Florence, Vol.I, pp.61-64.

Drescher,A., Vardoulakis,I. and Han,C. (1990). A biaxial apparatus for testing soils, Geotechnical Engineering Journal, ASTM, Vol.13, No.3, pp.226-234.

Goto,S., Tatsuoka,F., Shibuya,S., Kim,Y.-S. and Sato,T. (1991). A simple gauge for local small strain measurements in the laboratory, Soils and Foundations, Vol.31, No.1, pp.169-180.

Goto,S., Park,C.-S., Tatsuoka,F., and Molenkamp,F. (1993). Quality of the lubrication layer used in element tests on granular materials, Soils and Foundations, Vol.33, No.2, pp.47-59.

Han,C. and Vardoulakis,I. (1991). Plane strain compression experiments on water-saturated fine-grained sand, Geotechnique, Vol.41, No.1, pp.49-78.

Han,C. and Drescher, A. (1993). Shear bands in biaxial tests on dry coarse sand, Soils and Foundations, Vol.33, No.1, pp.118-132.

Park,C.-S. and Tatsuoka,F. (1994). Anisotropic strength and deformation characteristics of sands in plane strain compression, Proc. of 13th International Conference on Soil Mechanics and Foundation Engineering, New Delhi, Vol.1, pp.1-4.

Siddiquee,M.S.A. (1991). FEM analysis of settlement and bearing capacity of footing on sand, Master of Engnrg thesis, Univ. of Tokyo.

Siddiquee,M.S.A., Tanaka,T. and Tatsuoka,F. (1993). Objectivity of FEM solutions for bearing capacity of footing on sand, Proc. 28th JSSMFE, Kobe, Japan, pp.1565-1568.

Tanaka,T. and Kawamoto,O. (1989). Plastic collapse analysis of strain-softening materials, Proc. 3rd Int Conf on Numer. Models in Geomechanics, Niagara Falls, Canada, pp.667-674.

Tatsuoka,F., Molenkamp,F., Torii,T. and Hino,T. (1984). Behaviour of lubrication layers of platens in element tests, Soils and Foundations, Vol.24, No.1, pp.113-128.

Tatsuoka,F., Sakamoto,M., Kawamura,T. and Fukushima,S. (1986). Strength and deformation characteristics of sand in plane strain compression at extremely low pressures, Soils and Foundations, Vol.26, No.1, pp.65-84.

Tatsuoka,F., Nakamura,S., Huang,C.-C. and Tani,K. (1990). Strength anisotropy and shear band direction in plane strain tests of sand, Soils and Foundations, Vol.30, No.1, pp.35-54.

Tatsuoka,F., Okahara,M., Tanaka,T., Tani,K., Morimoto,T. and Siddiquee,M.S.A. (1991). Progressive failure and particle size effect in bearing capacity of a footing on sand, Proc. ASCE Geotechnical Engineering Congress 1991, Geotechnical Special Publication No.27, Vol.II, pp.788-802.

Tatsuoka,F., Siddiquee,M.S.A., Tanaka,T. and Okahara,M. (1992). A new aspects of a very old issue: Bearing capacity of footing on sand, Discussion for Session 3, Proc. 9th Asian Regional Conf. on Soil Mechanics and Foundation Engineering, Bangkok, 1991, Vol.2, pp.358.

Tatsuoka,F., Siddiquee,M.S.A., Park,C.-S., Sakamoto,M. and Abe,F. (1993). Modelling stress-strain relations of sand, Soils and Foundations, Vol.33, No.2, pp.60-81.

Tatsuoka,F., Siddiquee,M.S.A. and Tanaka,T. (1994a). Link among design, model tests, theories and sand properties in bearing capacity of footing on sand, Panel Discussion for Plenary Session B. Foundations, Proc. 13th International Conference on Soil Mechanics and Foundation Engineering, New Delhi, Vol.5, pp.87-88.

Tatsuoka,F., Sato,T., Park,C.-S., Kim,Y.-S., Mukabi,J.N. and Kohata,Y. (1994b). Measuring of elastic properties of geomaterials in laboratory compression tests, Geotechnical Testing Journal, ASTM, Vol.17, No.1, pp.80-94.

Tatsuoka,F. and Kim,Y.-S. (1994). Deformation of shear zone in sedimentary soft rock observed in triaxial compression, Localisation and Bifurcation Theory for Soils and Rocks (eds Chambon, Desrues & Vardoulakis), Balkema (this volume).

Tatsuoka,F., Siddiquee,S.A. and Tanaka,T. (1994c). Link among design, model tests, theories and sand properties in bearing capacity of footing on sand, Proc. XIII ICSMFE, New Delhi, Vol.5, pp.87-88.

Vardoulakis,I. and Graf,B. (1985). Calibration of constitutive models for granular materials using data from biaxial experiments, Geotechnique, Vol.35, No.3, pp.299-317.

Vermeer,P.A. (1990). The orientation of shear bands in biaxial tests, Geotechnique, Vol.40, No.2, pp.223-236.

Wu,W. and Kolymbas,D. (1991). On some issues in triaxial extension tests, Geotechnical Testing Journal, Vol.14, No.3, pp.276-287.

Deformation of shear zone in sedimentary soft rock observed in triaxial compression

Déformation des zones de cisaillement dans les roches sédimentaires tendres observées en compression triaxiale

F.Tatsuoka
Institute of Industrial Science, University of Tokyo, Japan

Y.-S. Kim
Ministry of Supply, Seoul, Korea (Formerly: University of Tokyo, Japan)

ABSTRACT: Shear rupturing in two consolidated drained triaxial compression tests of sedimentary soft rock (mudstone) is described. As the global axial compression of specimen in the post-peak regime is a result of the elastic rebound of specimen and the localised plastic deformation in the shear zone, strain averaged for the whole specimen is a structural response of the specimen. The relationships between the shear stress and the displacement across a shear zone (shear band) evaluated from axial strains locally measured at diagonally opposite ends of the specimen diameter are presented.

On décrit la rupture de deux échantillons de roche tendre sédimentaire (mudstone) soumis à un essai triaxial consolidé drainé en compression. Dans la mesure où le raccourcissement axial de l'échantillon en régime post-pic résulte de la combinaison de la déformation localisée dans la zone de cisaillement et du retour élastique de l'échantillon, la déformation moyenne définie pour l'échantillon complet est une réponse de structure. On présente les relations entre la contrainte de cisaillement et le déplacement relatif de part et d'autre de la bande évalué à partir de mesures locales de déformation.

INTRODUCTION

For granular materials, strain is localized into a shear zone(s), or shear band(s), upon failure. The shear zone has a characteristic thickness, which is of the order of 10~30 times the mean diameter (e.g. Yoshida et al. 1994, this volume). In stiff clay and sedimentary soft rock, shear rupture occurs upon failure associating a very thin shear zone or zones. Since it is not simple to identify its very small thickness, the shear zone is often called the shear plane.

Several attempts to measure shear rupturing in a geomaterial are reported in the literature. Burland (1990) reported the result of one unconsolidated undrained triaxial compression test on Todi clay performed by Rampello and Georgiannou. Viggiani et al. (1993) reported similar test results for three types of Italian clays. In those tests, as a single shear zone developed outside the gauge lengths of two local gauges (Fig. 1), the local gauges sensed only the deformation of the unfailed zones outside the shear zone. They showed that in the post-peak regime, the deformation outside a shear zone rebounded due to unloading in the axial stress. Similar post-peak rebound has been observed in

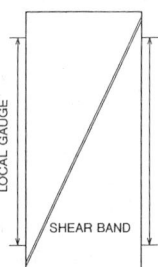

Fig. 1 Relative positions of a pair of local gauges used in the study of Burland (1992) and Viggiani et al. (1993)

consolidated undrained triaxial compression tests on a sedimentary soft rock (tuff) (Akai et al. 1981), in which local strains were measured with electrical resistance strain gauges attached on the lateral surface of specimen. Also for a cohesionless soil (Toyoura sand), similar post-peak rebound behaviour was observed on the σ_2 surface (plane strain surface) by means of the laser speckling method in plane strain compression tests (Nakamura et al., 1988) and by means of a photogrametric

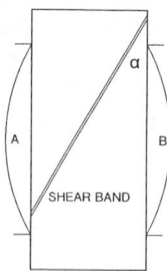

Fig. 2 Relative positions of a pair of local gauges used in this investigation.

method in torsional simple shear tests using a hollow cylindrical specimen (Pradhan et al. 1988).

In plane strain compression tests of many different types of sands, the shear deformation and dilatation of shear zone was measured by means of a photogrametric method (Yoshida et al. 1994). They showed that in the post-peak regime, the relationship between the ratio of "shear displacement across a shear zone u_s" to "the mean diameter D_{50}" and "the normalized shear stress level $R_n = [\sigma_1/\sigma_3 - (\sigma_1/\sigma_3)_{res}]/[(\sigma_1/\sigma_3)_{peak} - (\sigma_1/\sigma_3)_{res}]$" is rather unique for a range of particle sizes examined. Here, σ_1/σ_3 is the boundary stress ratio, and $(\sigma_1/\sigma_3)_{peak}$ and $(\sigma_1/\sigma_3)_{res}$ are the peak and residual values.

Labuz (1991) and Labuz and Biolzi (1991) reported some relationships between the shear stress and the net slip across a shear zone (i.e., slip-weakening constitutive relations) of Indiana limestone (intact hard rock core) in triaxial compression tests. These relations were obtained from local measurements of both the change in diameter and the axial compression of specimen.

The deformation of shear zone may be obtained also from a shear box test by measuring the relative displacement between the upper and lower boxes. However, the possible progressive nature of failure in the shear box tests may obscure the relationship.

The detailed deformation characteristics of shear zone for sedimentary soft rocks have not been reported in the literature. The authors performed many consolidated undrained and drained triaxial compression tests on sedimentary soft rock (mudstone) (Kim et al. 1994). Fortunately, in two consolidated drained triaxial compression tests performed at a constant effective lateral stress, a pair of local gauges (gauges **A** and **B** in Fig. 2) measured the local deformations either including or excluding the deformation of shear zone. From these measurements, both elastic rebound of the specimen and net shear deformation across a shear zone could be evaluated. Although similar measurements were obtained in some undrained tests, only the two drained tests were analyzed, since in drained tests, the deformation of shear zone can be related in a more straightforward way to the shear stress level in terms of effective stresses.

TESTING METHOD

Many undisturbed samples were taken by rotary core tube sampling from a thick deposit of sedimentary soft rock (mudstone) of about two million years old of the Late Neogene epoch to the Early Quaternary epoch at a test site in Sagamihara City near Tokyo (Tatsuoka et al. 1993, Ochi et al. 1993, Kim et al. 1994). The ranges of physical properties of the many samples were a total density of 1.972~2.01 g/cm³, a specific gravity of 2.69~2.72, and a mean diameter D_{50} of 0.0057~0.023 mm (obtained by the method usually used for uncemented fines after having been oven-dried and crushed thoroughly).

The two specimens, 5.5 cm in diameter and 15 cm in height, were isotropically consolidated to $\sigma_c' =$ either 627 kPa (6.4 kgf/cm²) (specimen A86) or 980 kPa (10.0 kgf/cm²) (specimen A81), which are the in-situ effective overburden pressures. The top and bottom ends of specimen were not lubricated, but they were capped with gypsum. A filter paper sheet was used for side drainage. As the loading piston was fixed to the cap, the cap was unable to rotate vertically. Therefore, the effect of the inclination of the cap during a test on the local axial measurements was very small. Local axial strains will be denoted as ε_1 (local), while those obtained from the axial displacement of the loading piston will be denoted as ε_1(external). The rate of external axial strain was 0.01 %/min.

The local gauge used was a pair of Local Deformation Transducers (LDTs, Goto et al. 1991) having a gauge length of 10 cm. The LDT detects the axial compression between its top and bottom ends attached on the lateral surface of specimen from bending strains occurring in it. With its very high resolution, local axial strains can be measured from about 0.0001 % (10^{-6}) to several percent, which enables us to measure elastic properties at strains less than about 0.001 % (10^{-5}).

In the measurements illustrated in Fig. 1, the axial compression resulting solely from the deformation of shear zone deformation is obtained from locally measured axial compression and displacement of loading piston (or the specimen cap). However, the latter quantity may include the effects of bedding error at the ends of specimen and system compliance. Therefore, it would be difficult to obtain accurate

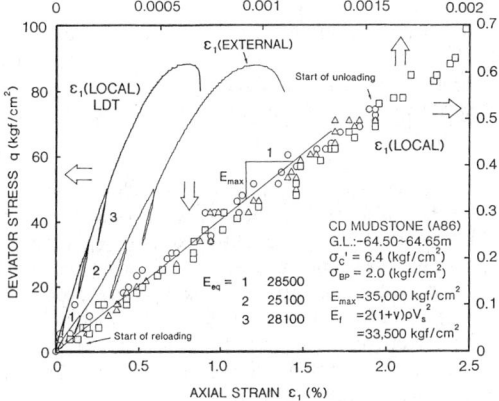

Fig. 3 Stress-strain relations until immediately after the peak state (specimen A86) (1.0 kgf/cm² = 98 kPa).

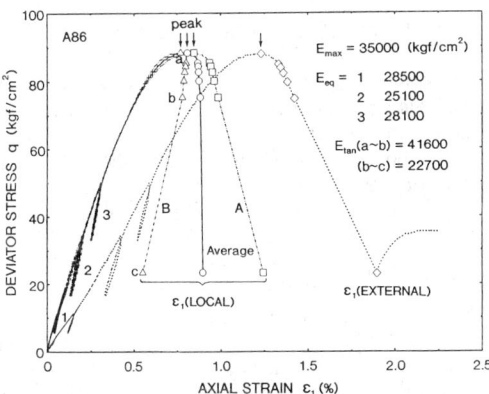

Fig. 4 Full-range stress-strain relations (specimen A86) (1.0 kg/cm² = 98.0 kPa).

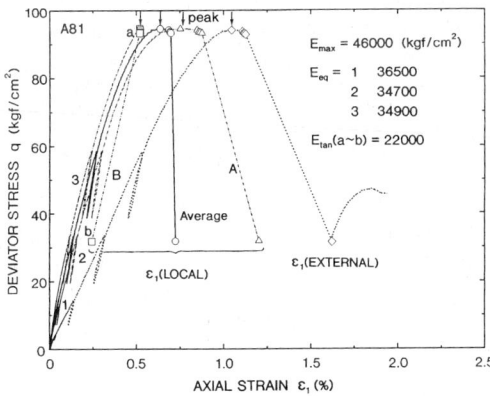

Fig. 5 Full-range stress-strain relations (specimen A81) (1.0 kg/cm² = 98.0 kPa).

deformation of shear zone, unless externally measured axial compression is properly corrected for those errors.

TEST RESULTS

Fig. 3 shows the test result for specimen A 86. The two solid curves, which are actually continuous data points, are the relationships between the deviator stress $q=\sigma_a-\sigma_r$ and the local and external axial strains. A large difference between them is due to the effects of system compliance and bedding error (mostly due to the latter). This result indicates that the external axial strain measurements are totally unreliable for this type of soft rock (Tatsuoka and Shibuya, 1992, Tatsuoka et al. 1994, Kim et al. 1994). Discrete data points show the relation at very small strains less than 0.002% obtained by LDT measurements, plotted in a scale different from that for the two solid curves. A very small cycle of unload and reload was applied at this very small strain level. It is seen that the relations for primary loading, unloading and reloading almost overlap, which means that the deformation is virtually recoverable. Further, it has been known that the dependency on strain rate of the stiffness at these small strains is very low for this type of soft rock (Tatsuoka and Shibuya, 1992, Tatsuoka et al. 1993, Kim et al. 1994). Therefore, the deformation at these small strains is virtually elastic. The maximum and elastic Young's modulus E_{max} at the initial state for specimen A86 is 35,000 kgf/cm² (3.6 MPa). Although the values of E_{max} of the many samples scattered to some extent, the average was very close to that obtained from the field shear wave velocity, 33,500 kgf/cm² (3.4 MPa) (Kim et al. 1994).

In the same way, the Young's moduli E_{eq} were obtained for several small unload/reload cycles applied during triaxial compression (Fig. 3). Fig. 4 shows the full range relations for which the axial strains are those obtained from each readings of the two LDTs **A** and **B** and their average and external measurements. Fig. 5 shows a similar result for specimen A81.

For the both tests, the number of data points in the strain-softening regime was limited compared to those before the peak state. This is because although the external axial strain rate was set constant, the actual rate of average axial strain in the strain-softening regime was accelerated by the release for a short duration of elastic energy stored in the loading frame. A temporary minimum stress condition was recorded immediately before reaching the residual condition (n.b. the relation after this point is continuous data points). This momentary minimum stress value is likely due to that the effects of the

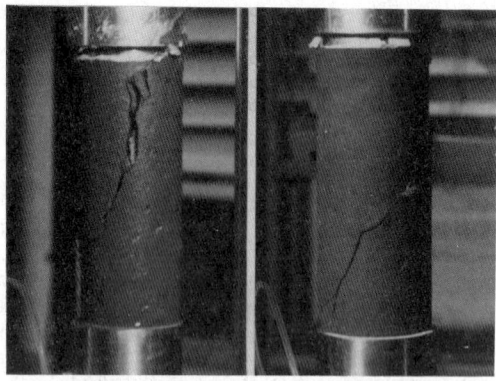

(a) A86 **(b)** A81
Plate 1 Shear rupturing in mudstone specimens.

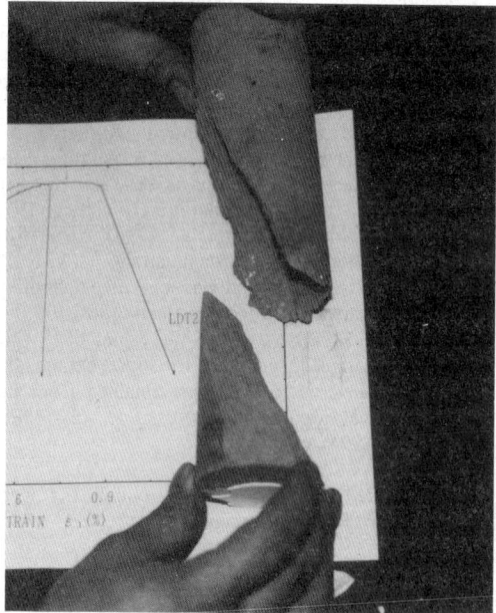

Plate 2 Splitted specimen A86.

inertia of the loading piston and specimen cap and their attachments during the very fast post-peak straining were not totally registered with the load cell located immediately above the specimen. Despite the above, the post-peak relation could be reasonably defined in the both tests.

For specimen A86 (Fig. 4), the two readings of LDTs **A** and **B** are almost the same until immediately before the peak stress state. This behaviour indicates very homogeneous pre-peak deformation. On the other hand, for specimen A81 (Fig. 5), the pre-peak deformation was less homogeneous, as seen from noticeably different readings of a pair of LDTs. Commonly for the both tests, the two readings had become obviously different before the peak stress state was attained, in particular for specimen A81 (Fig. 5).

In the post-peak regime, a single shear zone was developed in each specimen (Plates 1a and b). The planes of shear zone observed after splitting the specimen were not planar, but curved in the lateral direction forming part of cylinder (Plate 2). By this reason, the intersection of the shear zone with the lateral surface of specimen is not straight (Plates 1a and b). In addition, the planes of shear zone were not very smooth with an undulation of the order of 2 mm. However, in their lengthwise direction, the planes of shear zone were rather straight as a whole (see Plate 2). The average angle α (in rad.) between the overall direction of shear zone and the direction of σ_1 is about $\arccos(1/1.17)$ for the both specimens.

Due to their different positions relative to the shear zone, the readings of LDTs **A** and **B** became drastically different after the start of shear rupturing; the axial strain measured with LDT **A** increased at a large rate, whereas that measured with LDT **B** decreased showing rebound deformation. By the following reason, this rebound can be considered to be virtually elastic, which represents an objective and essential material property. The tangent Young's moduli E_{tan} observed in the post-peak unloading regime gauged with LDT **B** are listed in Figs. 4 and 5, together with the initial Young's modulus E_{max} and the E_{eq} values measured for several small unload/reload cycles applied prior to the peak state. Note that E_{eq} decreased slightly with straining. For specimen A86 (Fig. 4), in the post-peak regime, the E_{tan} value defined for a range between points **a** and **b** immediately after the peak state was very large. This would be due to the effects of creep shear deformation which was still continuing after the peak state. However, the subsequent relation between points **b** and **c** is approximately in parallel to the relations for the pre-peak unload/reload cycles, and the E_{tan} value defined for the range between points **b** and **c** is similar to the pre-peak values of E_{eq}. This similarity can be seen also for specimen A81 (see Fig. 5).

On the other hand, it is seen that this post-peak modulus E_{tan} is noticeably smaller than the pre-peak moduli E_{max} and E_{eq}. It is perhaps due to a continuous damaging by shear deformation process for the micro-structure of specimen.

The axial deformation measured with LDT **A** is a result of both the plastic deformation in the shear zone and the axial elongation due to the elastic

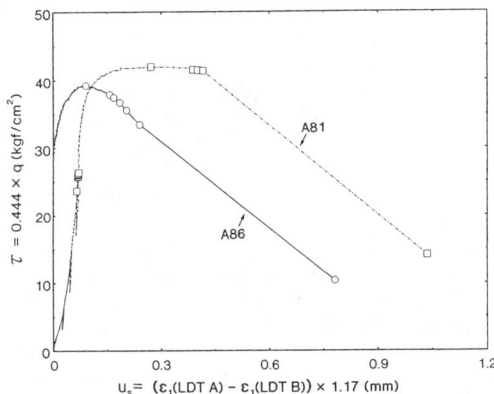

Fig. 6 Two relationships between shear stress τ in the shear zone direction and relative displacement u_s across a shear zone.

rebound of the whole specimen, while that measured with LDT **B** is a result of only the latter factor. Accordingly, "the axial strain increment in the post-peak regime obtained from the average of two readings of LDTs **A** and **B**" is equal to a half of "the axial strain increment due to the plastic deformation of shear zone" plus "the axial strain increment (negative value) due to the elastic rebound of the whole specimen." Therefore, in the post-peak regime, "the average strain increment from the readings of LDTs **A** and **B**" could happen to be nearly zero. However, this relation is not any objective material property. Note also that the post-peak behaviour of the external axial strain and that by LDT **A** are similar to each other, since both measurements include the deformation of shear zone and the elastic rebound of specimen in a similar way. Obviously, these two relations are not any essential and objective property of the test material either, but a structure response of the specimen.

The relationships between the shear stress τ in the shear zone direction and the relative displacement u_s across a shear zone were obtained from the measurements shown above (Fig. 6). The shear rupturing was assumed to occur in a simultaneous manner everywhere along its extending across the whole specimen. The effects of the end restraint and the possible dilatancy in the shear band were ignored. Then, defining ΔH as the increase in the specimen length, the relative displacement u_s across a shear zone was obtained as "[{axial compression measured with LDT **A** ($= -\Delta H_A$)} - {axial compression measured with LDT **B** ($= -\Delta H_B$)}]/cosα". As cosα= 1/1.17 and $-\Delta H_A = (\varepsilon_1$ measured with LDT **A**) × (a gauge of 100 mm for the LDTs) (similarly for $-\Delta H_B$), u_s is equal to "{(ε_1 measured with LDT **A**) - (ε_1 measured with LDT **B**)}(%) × 1.17 (mm)". The increment $d(-\Delta H_A)$ is always positive, while the increment $d(-\Delta H_B)$ is positive and negative in the pre-peak and post-peak regimes, respectively. It was considered that errors in these simplified calculations are not particularly large when compared with those associated with the level of accuracy in the measurement of the shear zone direction. To obtain both dilatancy and shear deformation of shear zone, in addition to the local axial strain measurements, it is necessary to measure locally the change in the specimen diameter at several elevations. However, the latter measurement was not performed in the present study. The average shear stress τ along the shear zone was obtained as $(q/2)\cdot\sin(2\alpha)$.

For specimen A81, a noticeable displacement u_s occurred until τ became about 30 kgf/cm² (3 MPa), which was well prior to the peak stress state (see Fig. 6). This is certainly due to the heterogeneous deformation, not by shear rupturing. The post-peak relation is very similar between the two tests. The increment of the relative displacement u_s for a range from the peak state to the residual state is very small, about 0.7 mm, which is very close to "the order of 0.7 mm" reported for Indiana limestone (Fig. 8b of Labuz and Biolzi, 1991). Indeed, in many boundary value problems in both field and laboratory, the ratio of this characteristics length of shear zone relative to the structure size (e.g., footing width) should have some important effects on the failure mechanism. For granular materials, this effect has been known as the particle size effect (Tatsuoka et al. 1991, 1994a). In any reasonable modeling of the post-peak behaviour of sedimentary soft rocks (and other similar materials), such relationships as shown in Fig. 6 should be incorporated. On the other hand, any post-peak modeling based on strain(s) averaged for the whole specimen having a certain size is subjective.

In the two tests described above, upon shear rupturing, the mode of deformation changed from an axially symmetry one in triaxial compression to a quasi-two dimensional one in plane strain compression. Therefore, the shear zone properties presented in Fig. 6 are not consistently purely two-dimensional ones. It was considered, however, that the post-peak deformation of shear zone is reasonably two-dimensional. Similar measurements in plane strain compression tests on sedimentary soft rocks as those performed on sands by Yoshida et al. (1994) will be needed in the future.

Based on the above observation, it can be conceivable that the total length of specimen can increase during the post-peak strain softening regime if the axial elongation of specimen due to the post-peak elastic rebound is larger than the axial compression resulting from the plastic deformation of shear zone. This situation can be realized if the shear displacement across a shear zone to reach the residual state is very small due to a small thickness of shear zone and/or if the total elastic rebound deformation of specimen is large due to a large post-peak drop in the axial stress and/or a large specimen height. The post-peak increase in the total specimen height in a compression test on a hard rock specimen has been called "snap back." (Labuz, 1991). A meaningful analysis of snap-back can be performed based on such local deformation measurements as described in this paper.

CONCLUSIONS

The observation of shear rupturing in two consolidated drained triaxial compression tests on sedimentary soft rock (mudstone) showed the following:
1) The localisation of plastic strains into a shear zone(s), or shear band(s), started immediately before the peak stress state.
2) Post-peak deformation in the zones outside a shear zone was nearly elastic rebound.
3) The relationships between the shear stress and the relative shear displacement across a shear zone were obtained from the overall direction of a shear zone and the readings of a pair of local gauges which measured local axial compressional deformation either including or excluding the deformation of the shear zone. The shear displacement across the shear zone to reach the residual state from the peak state was very small, less than 1 mm.

ACKNOWLEDGEMENTS: The authors appreciate the critical review of the manuscript by Mr. H. Nakase of Tokyo Electric Power Service Co. Ltd. They also thank the Japanese Ministry of Education, Science and Culture and Makita Scholar Foundation Report for financial support. The cooperation by Mr. T. Sato, Dr. Y. Kohata and Mr. T. Yoshida is also appreciated.

REFERENCES

Akai,K., Ohnishi,Y. and Yashima,A. (1981). Strain-softening behavior of soft sedimentary rock, Proc. of Int. Sympo. on Weak Rock, Tokyo, Balkema, Vol.1, pp.81-86.
Burland,J.B. (1990): On the compressibility and shear strength of natural clays, Geotechnique, Vol.41, No.3, pp.329-378.
Goto,S., Tatsuoka,F., Shibuya,S., Kim,Y.-S. and Sato,T. (1991). A simple gauge for local small strain measurements in the laboratory, Soils and Foundations, Vol.31, No.1, pp.169-180.
Kim,Y.-S., Tatsuoka,F. and Ochi,K. (1994). Deformation characteristics at small strains of sedimentary soft rocks by triaxial compression tests, Geotechnique, No.2 (in press).
Labuz,J.F. (1991). The problem of machine stiffness revisited, Geophysical Research Letter, Vol.18, No.3, March, pp.439-442.
Labuz,J.F. and Biolzi,L. (1991). Class I vs class II stability: a demonstration of size effect, Int. J. Rock Mech. Min. Sci. & Geomechani. Abstr. Vol.28, No.2/3, pp. 199-205.
Nakamura,S., Tatsuoka,F. and Shinno,F. (1988): Strain distribution in sand specimens subjected to plane strain compression, Proc. 22th Japan Conf. on SMFE, JSSMFE, Niigata, pp.349-352 (in Japanese).
Ochi,K., Tatsuoka,F. and Tsubouchi,T. (1993). Stiffness of sedimentary soft rock from in situ and laboratory tests and field behaviour, Geotechnical Engineering of Hard Soils-Soft Rocks (eds. Anagnostopoulos et al.), Balkema, Vol.1, pp.707-714.
Pradhan,T.B.S., Tatsuoka,F. and Horii,N. (1988): Simple shear testing on sand in a torsional shear apparatus, Soils and Foundations, Vol.28, No.2, pp.95-112.
Tatsuoka,F., Nakamura,S., Huang,C.-C., and Tani,K. (1990): Strength anisotropy and shear band direction in plane strain tests of sands, Soils and Foundations, Vol.30, No.1, pp.35-54.
Tatsuoka,F., Okahara,M., Tanaka,T., Tani,K., Morimoto,T. and Siddiquee,M.S.A. (1991). Progressive failure and particle size effect in bearing capacity of a footing on sand, Proc. ASCE Geotechnical Engineering Congress 1991, Geotechnical Special Publication No.27, Vol.II, pp.788-802.
Tatsuoka, F. and Shibuya,S. (1992). Deformation characteristics of soils and rocks from field and laboratory tests, Keynote Lecture (Session No.1), Proc. 9th Asian Regional Conf. on SMFE, Bangkok, 1991, Vol.2, 101-170.
Tatsuoka,F., Kohata,Y., Mizumoto,K., Kim,Y.-S., Ochi,K. and Shi,D. (1993). Measuring small strain stiffness of soft rocks, Geotechnical Engineering of Hard Soils-Soft Rocks (eds. Anagnostopoulos et al.), Balkema, Vol.1, pp.809-816.
Tatsuoka,F., Sato,T., Park,C.-S., Kim,Y.-S., Mukabi,J.N. and Kohata,Y. (1994): Measurements of elastic properties of geomaterials in laboratory compression tests, Geotechnical Testing Journal, Vol.17, No.1, pp.80-94.
Tatsuoka,F., Siddiquee,S.A. and Tanaka,T. (1994a): Link among design, model tests, theories and sand properties in bearing capacity of footing on

sand, Proc. XIII ICSMFE, New Delhi, Vol.5, pp.87-88.

Viggiani,G., Rampello,S. and Georgiannou,V.N. (1991). Experimental analysis of localisation phenomena in triaxial tests on stiff clays, Geotechnical Engineering of Hard Soils/Soft Rocks, Balkema, Vol.1, pp.849-856.

Yoshida,T., Tatsuoka,F., Kamegai,Y., Siddiquee,M.S.A. and Park,C.-S. (1994). Shear banding in sands observed in plane strain compression, Localisation and Bifurcation for Soils and Rocks (eds Chambon, Desrues & Vardoulakis), Balkema (this volume).

Experimental observations of strain localisation in plane strain compression of a stiff clay

Observations expérimentales de la localisation de la déformation dans des essais de compression en déformation plane sur une argile raide

G. Viggiani
University of Rome 'La Sapienza', Italy

R.J. Finno & W.W. Harris
Northwestern University, Evanston, Ill., USA

ABSTRACT: Plane strain compression tests performed with local strain devices and pore pressure probes allowed for detailed observations of the onset and progressive development of a shear surface within undisturbed samples of an Italian stiff overconsolidated clay. The tests were performed from isotropic stress conditions; the shearing stresses were applied under both undrained and drained strain-controlled compression. Internally-measured displacements and local pore water pressure measurements indicated the points in the loading process where the localisation began. Globally-measured pore pressures and displacements did not yield the same information. In all the tests the onset of localisation occurred before the peak load was attained, the global strain softening being a result of the inhomogeneous deformations which arose as the slip surface formed within the specimen. It was observed that this onset occurred quite close to the peak load for the undrained test results. Recognising the importance of the two-phases nature of the clay, these results must be analysed as a boundary value problem wherein the effective stress response is coupled with fluid flow equations.

Une série d'essais de déformation plane en compression instrumentés avec des capteurs de déformation et de pression interstitielle locales a permis des observations détaillées de la naissance et du développement progressif de surfaces de cisaillement dans des échantillons intacts d'une argile raide Italienne. Les essais étaient réalisés en partant d'un état de contrainte isotrope; le chargement déviatoire était effectué à déformation contrôlée, en conditions drainées dans certains essais, non-drainées dans d'autres. Les mesures internes de déplacement et de pression interstitielle ont permis d'indiquer le moment de déclenchement de la localisation. Les mesures globales des mêmes quantités ne fournissaient pas cette information. Dans tous les essais, le début de la localisation s'est produit avant que la charge de pic soit atteinte, l'adoucissement global étant une conséquence des déformations hétérogènes qui surviennent lorsque la surface de glissement se forme dans l'échantillon. Dans les essais non-drainés, le déclenchement est survenu très près du pic. Compte tenu de l'importance du caractère biphasique de l'argile, ces résultats doivent être analysés comme un problème aux limites dans lequel la réponse en contrainte effective est couplée avec les équations d'écoulement du fluide interstitiel.

1 INTRODUCTION

Instability phenomena of natural slopes and deep cuts in stiff overconsolidated clays are often associated with the development of narrow shear zones. Most of the deformation is concentrated in these zones, in which relative movement of nearly rigid regions takes place.

Laboratory evidence of localisation phenomena in soils has existed for a long time; the classical paper by Hvorslev (1960) discusses these effects in overconsolidated clays. However, published laboratory data can be ambiguous and at times contradictory. Perhaps a major reason for these contradictions is that test results were presented in terms of nominal stress and strain; possible effects of non-uniformities or even discontinuities that may arise as the peak stress is approached were not extensively studied in the past.

In conjunction with major developments in both theoretical and numerical aspects of modelling, an interest has arisen in detailed experimental observations of strain localisation within a soil specimen subjected to shear. Experimental results have been recently presented for sandstones (Ord et al. 1991), sands (Mokni 1992, Harris 1994), soft clays (Finno and Rhee 1992, 1993), marls (Tillard-Ngan et al. 1993) and stiff clays (Viggiani et al. 1993, Viggiani 1994).

The analysis of the onset and propagation of shear bands in porous media such as saturated clays poses a formidable problem, because of coupled effects relating to fluid flow in the soil mass and in the localised zone which have to be taken into account. In such conditions, significant progress can only be achieved by considering accurate experimental observations together with the theoretical treatment of the phenomena.

A collaborative effort between the University of Rome and Northwestern University has been recently initiated to investigate the conditions leading to strain localisation in saturated stiff clays and to examine the behaviour in the post-localisation regime. A preliminary testing programme was carried out at Northwestern University in a plane strain compression device specifically instrumented to study the patterns of localisation. This specific device had been formerly used for evaluating the effects of shear banding on the behaviour of soft Chicago glacial clays (Rhee 1991, Finno and Rhee 1992, 1993) and loose Mason sand (Harris 1994).

One drained and two undrained shear tests have been performed on a natural stiff clay coming from Vallericca, Italy. Local measurements of both strain and pore water pressure have been made, which greatly enhance the ability to capture the onset of strain localisation during a test. This paper describes the experimental apparatus, presents the results of the preliminary testing programme and briefly discusses the implications of these results.

2 SOIL TESTED

The clay deposit located at Vallericca, Italy, has been selected for this work because of its homogeneity and absence of major macro-structure. Moreover, Vallericca is only a few kilometres north of Rome and the existence of a brick pit in the deposit offers an opportunity to extract, as needed, high quality block samples of the clay from vertical faces of deep cuts.

Vallericca clay was deposited in a shallow marine environment during the Plio-Pleistocene age. The soil is a CH clay according to the Unified Soil Classification System with a liquid limit of 59 and a plasticity index of 32. The soil consists of about half clay size and half silt size particles and has a calcium carbonate content as high as 30 %.

The natural clay is heavily overconsolidated (overconsolidation ratio (OCR) varies between 7 and 10) and quite stiff, with a liquidity index slightly less than zero. The vertical yield stress from oedometer tests is about 1900 kPa and the mean effective stress $p' = (\sigma'_1 + \sigma'_2 + \sigma'_3)/3$ in specimens after they are taken from the ground is 430 kPa. This latter value is measured in the triaxial apparatus by applying a cell pressure under undrained conditions and monitoring the pore water pressure until an equilibrium value is reached.

In the last five years, the mechanical properties of this clay have been extensively studied at the University of Rome, and comparisons have been made between the behaviour of natural soil and that of reconstituted specimens of the same clay (Georgiannou et al. 1991, Viggiani 1991, Rampello and Silvestri 1993, Rampello et al. 1993, Viggiani et al. 1993).

3 LABORATORY EQUIPMENT AND TESTING PROGRAMME

The shear tests on Vallericca clay were performed in a servo-controlled plane strain compression device at the Northwestern University Geotechnical Laboratory. The nucleus of the device is based on the system developed by Vardoulakis and Drescher (1988). The major advantage of this device over a conventional plane strain apparatus is its ability to allow free shear band formation and to measure the load-displacement characteristics of the failure zone. Kinematically unconstrained formation of a planar shear band is possible because of the presence of a free-sliding bottom plate (sled, hereafter). Displacement measurements of the sliding bottom plate and internal lateral displacements allow accurate determination of the onset of shear banding. Stress-strain response in the pre-localisation regime is also observed.

A clay specimen 135 mm high, 40 mm deep and 81 mm wide is enclosed in a rubber membrane and mounted between two rigid walls inducing plane strain conditions (Figure 1). Axial load is kinematically applied by an enlarged upper plate guided by an adjustable bushing. The two exposed vertical sides of the specimen are subjected to confining pressure. All the surfaces in contact with the specimen are glass-lined and lubricated to minimise friction. The assemblage is placed into a large confining cell for consolidation and set into a load frame. The computer programmed electronic signal for frequency and magnitude of loading is applied to an electro-pneumatic transducer which controls pneumatic amplifiers for the application of cell pressure and, if desired, vertical load.

A total of 21 sensors are currently used in the system. Four internally located load cells allow for accurate measurements of axial force, its eccentricity and the friction along the side walls. Four load cells are embedded in an aluminium side plate to measure out-of-plane forces, allowing a full evaluation of the boundary stress during consolidation and shearing. Seven displacement transducers monitor the axial and lateral displacements of the specimen and the horizontal movement of the

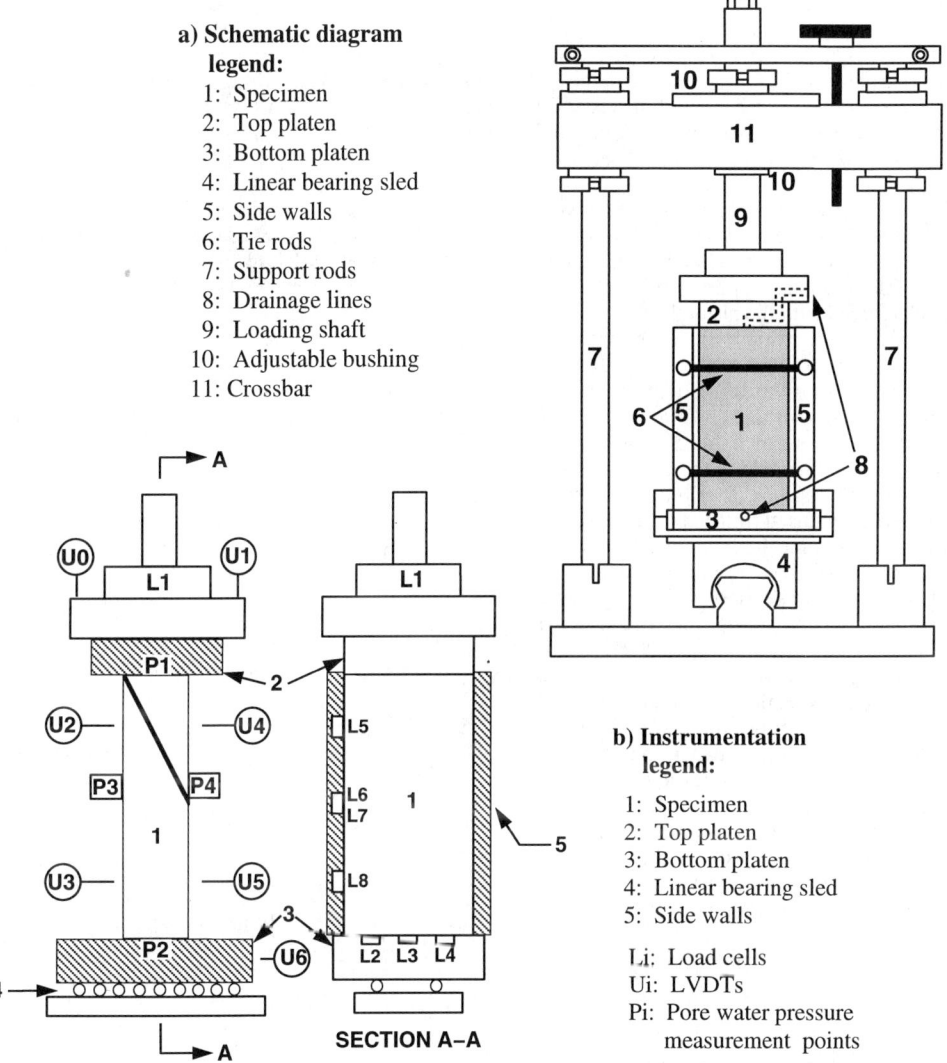

Figure 1. Biaxial apparatus: schematic diagram (a) and arrangement of measuring devices (b)

base plate. Two pressure transducers monitor pore water pressure at the top and at the bottom of the specimen. Two miniature pore pressure probes (Hight 1982) are also mounted on the sides of the specimen subjected to confining pressure, at mid-height. Global volume change is monitored with a burette system which incorporates a sensitive differential pressure transducer at its base to monitor the height of the column of fluid in the burette. Cell pressure is continuously monitored with a pressure transducer. The output signals of these transducers are conditioned and received by a process interface unit. This unit is linked to a microcomputer which controls all the operations.

Software has been developed to apply any stress path in plane strain conditions, including isotropic and K_0 consolidation, and drained and undrained strain-controlled compression.

Furthermore, stereophotogrammetry can be used to monitor internal displacements during consolidation and shear. One of the biaxial device side plates is in fact a plexiglas window which permits a view of the specimen side during the test so that the uniformity of deformations and evolution of a shear band can be monitored. The principle of stereophotogrammetry is used by maintaining a fixed camera location while the specimen is deforming in a displacement-controlled mechanism. Although some preliminary analyses of the photographs taken in the tests described herein have been performed (in cooperation with J. Desrues at the Laboratoire 3S in Grenoble, France), further work needs to be done to improve the accuracy of the technique, and these results will not be presented in the paper.

One drained and two undrained shear tests were carried out on Vallericca clay. Prior to shearing, the specimens were placed in the chamber and a cell pressure of 600 kPa was applied under globally undrained conditions with the piston locked in place to prevent axial deformations. From the measured pore water pressure response to these conditions, a degree of saturation close to 100 % (full saturation) was inferred for all the soil specimens. A back pressure of 200 kPa was then applied while opening the drainage lines and unlocking the piston, thereby allowing axial deformations. Swelling occurred as the pore pressure equalised in the specimen. As the imposed value for the mean effective stress is very close to the pre-existing value in the sample, only minor volume change was observed. Each specimen was in an approximately isotropic state of effective stress at the end of the swelling stage, which took as much as nine days. It is worth noting that it was difficult to completely deair the drainage lines and pore pressure measuring system. As a result, the volume change (in drained shear) and the pore pressure (in undrained shear) measurements at the ends of the specimens were not reliable. This was not the case for the pore pressure measurement by means of the piezometer probes; these are not affected by the presence of air in the drainage lines because the sensing element is adjacent to the specimen inside the cell.

4 RESULTS FROM UNDRAINED COMPRESSION

Both undrained tests were performed under displacement control with a rate of applied displacement of 0.15 mm/hour, which corresponds to a

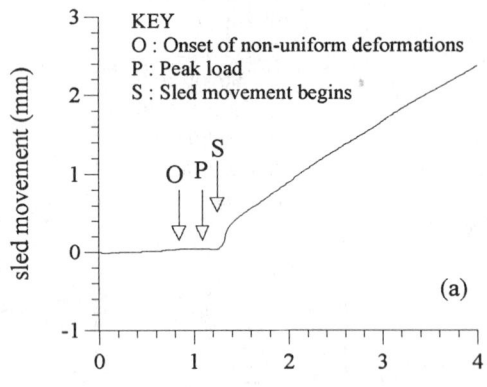

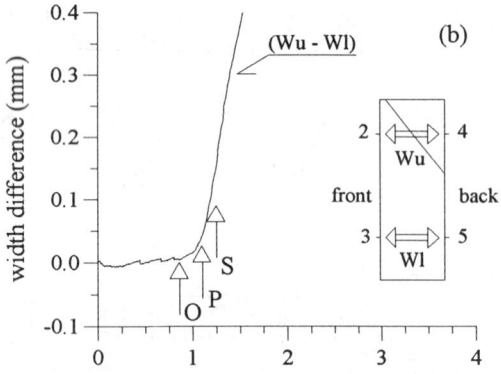

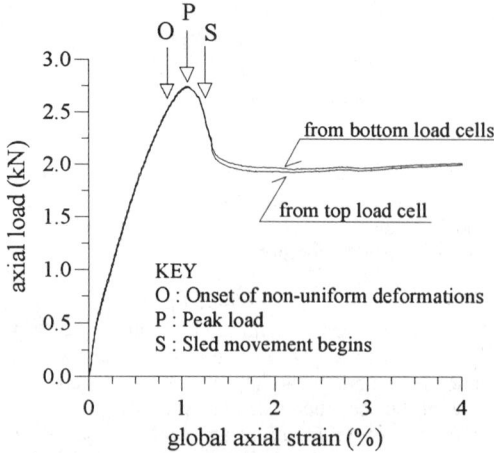

Figure 2. Nominally undrained test on Vallericca clay: vertical load vs. global axial strain

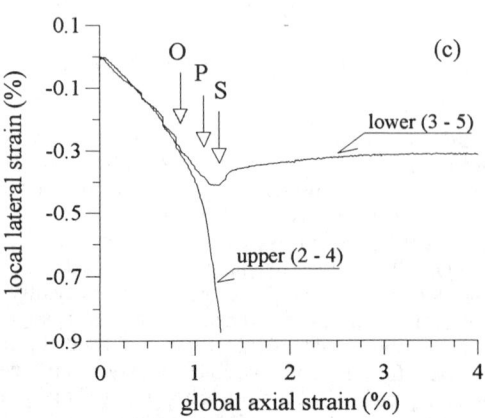

Figure 3. Nominally undrained test on Vallericca clay: uniformity of deformations

nominal axial strain rate of 0.11 %/hour for a 135 mm specimen height. Because similar responses were observed in both tests, only one result will be presented in detail.

Figure 2 shows the load versus global axial strain for one of the tests. The load monotonically increases until 1.1 % strain whereupon it decreases sharply. At about 1.5 % strain, the load essentially levels off. A striking feature of this experiment is the very low friction developed along the side plates, as noted by the similarity of the loads from the top and the three bottom load cells.

Three different measures of the uniformity of deformation during the test are shown in Figure 3. Figure 3(a) shows the bottom sled movement as a function of global axial strain. The bottom sled did not move until 1.25 % strain (point S), or after the peak axial load had been attained at 1.1 % strain (point P on Figure 2). Figure 3(b) shows the difference of widths as measured by the pairs of LVDTs mounted on the sides of the specimen. For perfectly uniform deformations, the difference should be zero as the specimen deforms rectilinearly. This uniform mode of deformation is maintained until about 0.85 % strain (point O) whereupon the width difference rapidly increases until an approximately constant value of rate of change is attained. These same LVDT data can be examined as local lateral strain rates at the two elevations of the LVDTs. Figure 3c shows a plot of the local lateral strain versus the global axial strain. The rates of local lateral strain are noted by the slope of the curves because the axial deformation was applied at a constant rate throughout the test. The rates of local lateral strain are essentially equal until 0.85 % (point O) whereupon they begin to diverge. The local lateral strain rate at the lower LVDT essentially becomes zero soon after the sled begins to move (point S), whereas that at the upper LVDT location across which the band has traversed increases rapidly after point O and eventually attains a constant value. These observations are consistent with the development of a sliding surface on which almost all of deformations are concentrated.

Additional measures of a non-uniform loading process are shown in Figures 4 and 5. The horizontal loads measured by the four load cells embedded in the side plate are plotted versus global axial strain in Figure 4. The load remains essentially the same until point O is attained. Figure 5(a) shows the difference in excess pore pressure measured by

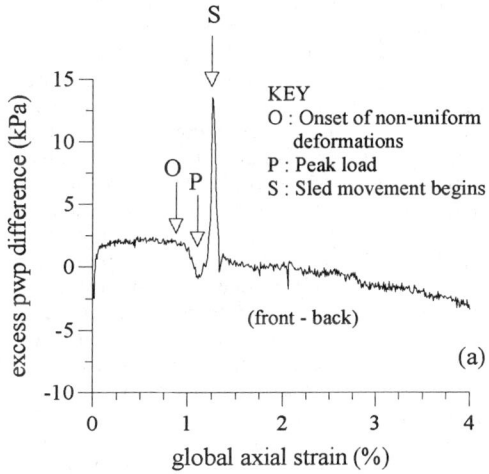

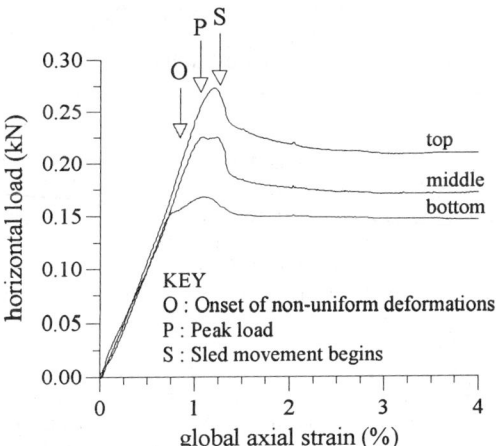

Figure 4. Nominally undrained test on Vallericca clay: horizontal loads vs. global axial strain

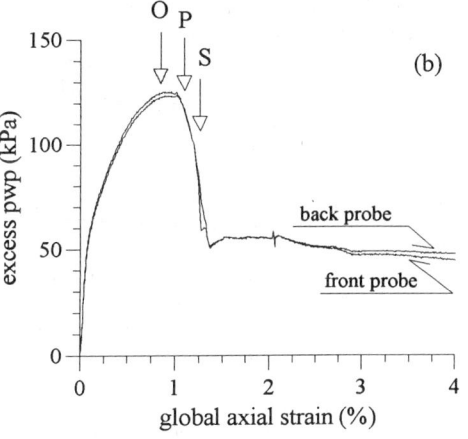

Figure 5. Nominally undrained test on Vallericca clay: pore water pressures difference (a) and individual measurements (b)

the two internal pore pressure probes; both the individual measurements are plotted in Figure 5(b). By coincidence, the surface of discontinuity came out from the specimen just below the back pore pressure probe. The probes measured essentially the same positive excess pore pressure until point O, whereupon the pore pressure began to increase at the back probe (i.e., the "future" location of the discontinuity) relative to that measured by the front probe. Immediately after the peak load had been attained (point P) the difference in pore pressures rose rapidly until the sled began to move (point S); thereafter the pore pressure equalised rapidly. Possible explanations for this rapid equalisation include details of the band as it comes out from the specimen or an increase of permeability along the band, when it is completely formed. The former refers to the fact that the shear band came out from the specimen just below the probe location. As the top "block" slid over the bottom "block", the membrane could not follow the sharp break between the blocks and a small reservoir of water accumulated near the miniature pore pressure transducer. The presence of this reservoir so close to the diaphragm of the miniature pressure transducer would tend to reduce the excess pore pressure measured by the transducer. Further work needs to be done to clarify this rapid equalisation response. In any case, the pore pressure initially increases (at point O) suggesting a concentration of shearing strains near the back probe, then after the peak has been attained, the pore pressure decreases to a value lower than the prevailing conditions away from the band implying that the soil is locally tending to dilate as a band propagates through the specimen. The pore pressure response is clearly an indication of band development, and is further evidence that the internal displacement measurements can accurately define the onset of localised deformation.

Figure 6 shows a plot of q/p' versus global axial strain, the deviator stress, q, being defined as:

$$q = \sqrt{\tfrac{1}{2}\left[(\sigma_1-\sigma_2)^2 + (\sigma_2-\sigma_3)^2 + (\sigma_3-\sigma_1)^2\right]}.$$

The plot is based on effective stresses computed from measurements of pore water pressures by both transducers at the mid-height of the specimen. As seen, the peak load, P, is attained at the top of this effective stress-strain curve. Note that point O is quite close to point P. If point O is interpreted to be the point where localisation is initiated within the specimen, then this point on the effective stress response is quite close both to the point where q/p' is maximum and to the peak of the load-deflection response, point P. While questions remain related to the values of the pore water pressures at the location of the initiation point within the clay specimen at

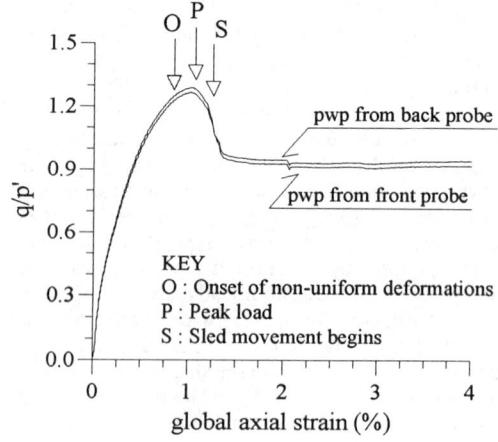

Figure 6. Nominally undrained test on Vallericca clay: effective stress ratio vs. global axial strain

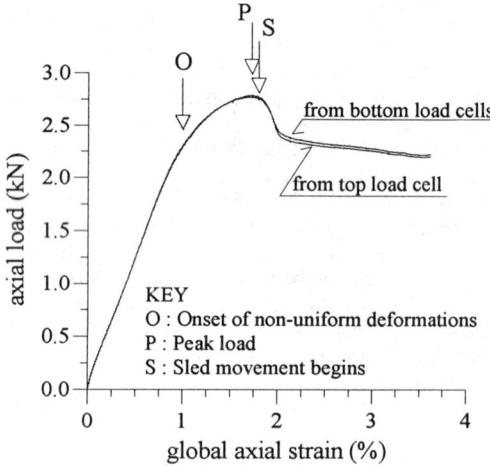

Figure 7. Nominally drained test on Vallericca clay: vertical load vs. global axial strain

the time of initiation, and therefore the exact values of effective stress at the onset of localisation, the localisation initiates quite close to the limit point in this test. Note that pore pressures increase monotonically up to point O (Figure 5b) indicating that the material behaviour is contractant. A drop in the pore pressures at the band after point P (Figure 5a) indicates a very localised dilatant response associated with the formation of the band. The peak load is apparently the result of the formation of a shear band within the specimen which occurs nearly coincident with the peak of the effective stress ratio q/p'.

Observations of internal deformations, horizontal

loads and pore water pressures indicate that the uniformity of the loading process was maintained until point O, whereupon the strains began to localise. This occurred before the peak load was attained and before the discontinuity was completely formed. The point where the sled began to move seems to represent the point during the test when the shear band intersected the entire specimen.

5 RESULTS FROM DRAINED COMPRESSION

After allowing a specimen of natural Vallericca clay to swell under 600 kPa cell pressure and 200 kPa back pressure, a shear test was performed with both drainage lines open under displacement control with a rate of applied displacement of 0.026 mm/hour, which corresponds to a nominal axial strain rate of 0.02 %/hour for a 135 mm specimen height.

Figure 7 shows the load versus global axial strain. The load monotonically increases until 1.73 % strain whereupon it decreases. At about 2.1 % strain, the load essentially levels off. As in the undrained tests, there is very low friction developed along the side plates.

Three different measures of the uniformity of deformation during the test are shown in Figure 8. Figure 8(a) shows the bottom sled movement as a function of global axial strain. After an initial adjustment, the bottom sled did not move until 1.8 % strain (point S), or just beyond the point where the peak load had been attained (point P at 1.73 % strain). Figure 8(b) shows the difference of widths as measured by the pairs of LVDTs mounted on the sides of the specimen. A uniform mode of deformation is maintained until about 1.0 % strain (point O) whereupon the width difference decreases until 1.4 % strain whereupon it rapidly increases until an approximately constant value of rate of change is attained. Figure 8(c) shows a plot of the local lateral strain versus the global axial strain. The rates of local lateral strain, noted by the slope of the curves, are essentially equal until 1.0 % (point O) whereupon they begin to diverge. The local lateral strain rate at the lower LVDT pair essentially becomes zero soon after the sled begins to move (point S), whereas the strain at the upper LVDT location across which the band has traversed increases rapidly after 1.2 %, soon after non-uniformities were observed in the width differences at point O. These observations are again consistent with the development of a sliding surface on which most of deformations are concentrated.

The test was performed with free draining boundaries, but with pore pressure measurements at the mid-height of the specimen. The global volumetric strain (positive values indicating compression) computed from the volume of water expelled

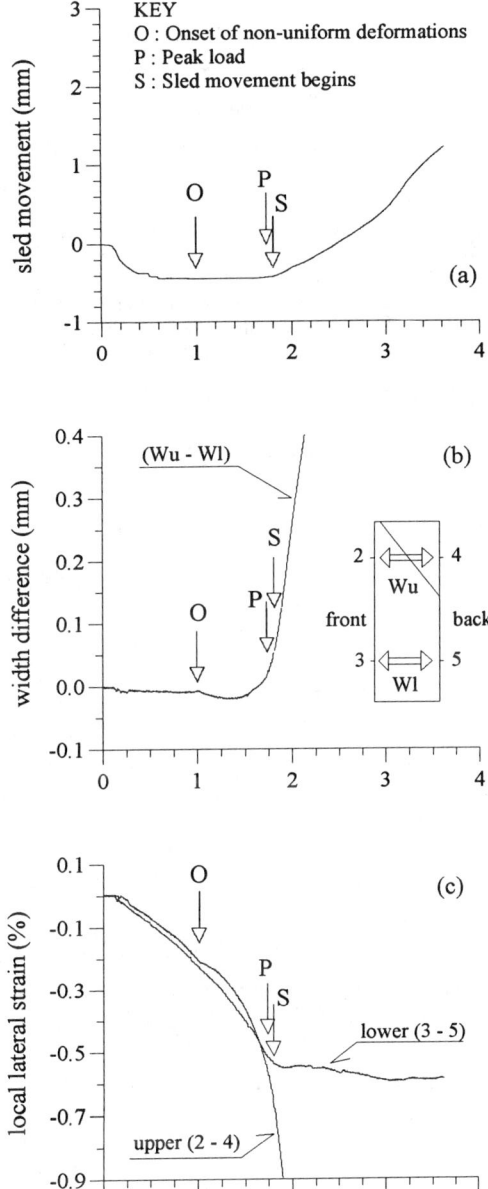

Figure 8. Nominally drained test on Vallericca clay: uniformity of deformations

from the specimen is plotted versus global axial strain on Figure 9(a). The sample compressed until the peak load was attained (point P); this volume change behaviour is consistent with the positive excess pore pressure developed during the undrained test discussed before. Figure 9(b) shows a

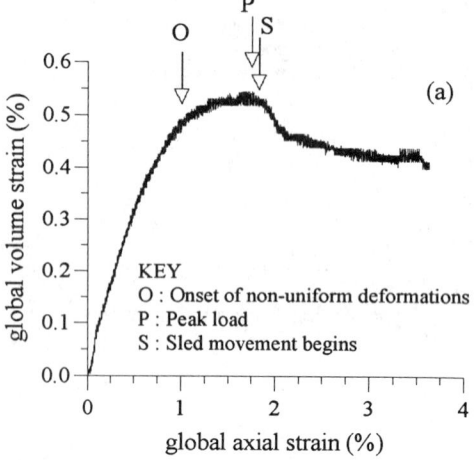

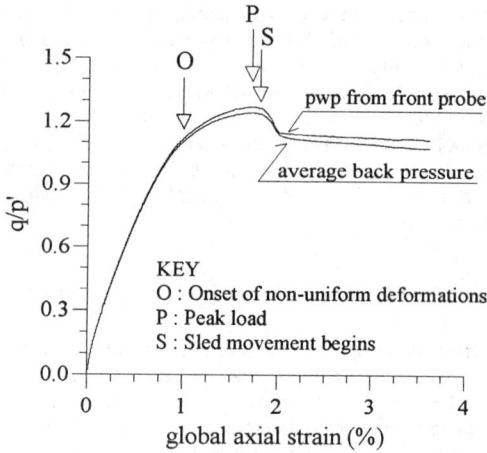

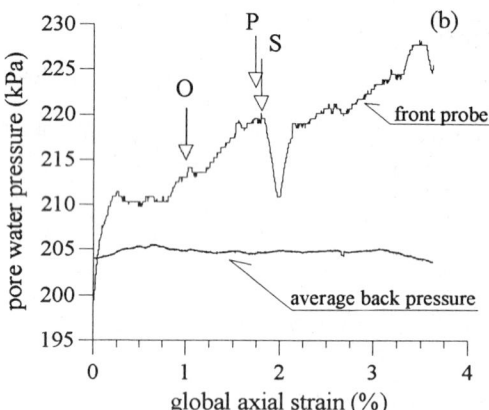

Figure 9. Nominally drained test on Vallericca clay: global volumetric strain (a) and pore pressure (b)

Figure 10. Nominally drained test on Vallericca clay: effective stress ratio vs. global axial strain

plot of the average back pressure measured at the top and bottom of the specimen and the pore pressure measured at the probe furthest from the location of the band. Unfortunately, the probe closest to the band location did not yield consistent results as a result of testing difficulties. After an initial adjustment, the pore pressure remained constant until about 0.9 % global axial strain, just before the point where nonuniformities were first observed in the displacements (point O). The pore pressures near the mid-height of the specimen gradually rose until 1.8 %, whereupon the sled began to move. Thereafter the pore pressures sharply decreased (until 2 % strain or the point where the load began to become constant) then increased. Even though this test was carried out at a nominal strain rate of 0.02 %/hour, slower than the strain rate typically associated with a drained test, excess pore pressures developed which seem to be related to the development of the shear band in the specimen. Because the probe was no closer than 30 mm from the eventual location of the band, the excess pore pressure measured at the probe location lagged behind the development of a shear surface; if excess pore pressures are generated where the shear strain is concentrating, dissipation of these excess pore pressures will cause a time lag until they are sensed at some distance from the band. This hypothesis is consistent with the results of the undrained tests described previously wherein one probe was located at the band surface and the changes in pore pressures were more pronounced and were observed as the band formed.

Figure 10 shows a plot of q/p' versus global axial strain based on effective stresses computed from measurements of pore water pressures by a transducer at the mid-height of the specimen. As was the case with the undrained test results, the peak load is attained very near the top of the effective stress-strain curve, or at the limit point in the constitutive behaviour. However the onset of localisation has occurred much earlier. Figure 9a indicates that the material is globally compressing right up to the point of the peak load. Note that the local pore pressures increase beyond point O and thus the specimen exhibits a contractant response. This response is contrary to what was observed in the undrained tests. A local dilatant response was not observed until after the sled had moved (point S in Figure 9b). Because the location of the pore pressure transducer was far from the band in the drained test, the effect of dilation on the pore pressure response was delayed by the low permeability of the clay. Thus it is believed that the trend of the pore pressure response near the band is similar in both the drained and undrained test. However the same uncertainties of the exact values of the effective stress response exist as discussed for the undrained test, although probably to a lesser extent due to the slower rate of loading which would allow for a greater degree of pore pressure equalisation in the drained test.

Observations of internal deformations, global volume changes and internal pore water pressures indicate that the uniformity of the loading process was maintained until point O, whereupon the strains began to localise. This occurred before the peak load was attained and before the discontinuity was completely formed.

6 DISCUSSION OF THE RESULTS

There are similarities between the results of the undrained and drained tests of Vallericca clay. In both cases, nonuniformities in the deformation process were observed before the peak load, maximum shear stress and maximum effective stress ratio were attained, although the non-uniformities developed earlier in the drained test than the undrained tests. Therefore, the peak shear stress is apparently a consequence of the formation of a discontinuity, and constitutive behaviour can only be extracted from these results based on globally-derived behaviour in the pre-localisation regime. In both cases, pore water pressures were different near the band and away from the band once the non-uniform deformation began. The concentration of straining at the band location gives rise to excess pore pressure concentrations at these locations (even for the low rate of deformation imposed) which makes it difficult to evaluate the effective stress conditions in the band, and impossible to do so based on standard measurements of pore pressures.

There are differences between the observed pore pressures related, to some extent, to the distance of the probe from the shear band. The different pore pressure response close to the band establishes a hydraulic gradient within the specimen resulting in non-uniform pore water pressures during the test. The "drained" test is in reality a partially drained test where the effects of the generation of excess pore pressures at the band are masked to a greater extent than in a (globally) undrained test, by the greater amount of pore pressure equalisation which occurs in this latter test due to the smaller rate of imposed deformation and the drained boundaries.

7 CONCLUDING REMARKS

On the basis of the experimental results described herein, the following preliminary conclusions may be drawn regarding the behaviour of stiff Vallericca clay in plane strain compression:

1. The data imply that the global strain softening of the samples was at least partly a result of the inhomogeneous deformations which arise as a thin shear band forms within the specimen. Therefore, after the onset of non-uniform deformation within the clay specimen, it must be thought of as a structure (or a system) and not as a single element.
2. The sled begins to move well after the onset of localisation. The first movement of the sled seems to correspond to the moment when the band is completely formed and comes out from the boundaries of the specimen. Thereafter, the deformation consists in a near-rigid body sliding on the shear zone.
3. There is a significant difference between a globally drained and a locally drained response. The same concept applies to globally undrained response, where local drainage and volume changes can occur close to the localised zone. Theoretical and numerical work concerning localisation of soils has for the most part dealt with a one phase material model. The experimental observations presented in this paper emphasise the importance of considering pore water pressures: a coupling between the effective stress and pore water responses must be incorporated in any model to understand strain localisation phenomena in saturated clays.

REFERENCES

Finno R.J., Rhee Y. (1992) - Kinematically unconstrained compression of soft clay. *Proceedings of a Specialty Conference on Stability and Performance of Slopes and Embankments*, ASCE, Berkeley, CA, Vol. II, pp. 142-157.

Finno, R.J., Rhee Y. (1993) - Consolidation, pre- and post peak shearing responses from internally instrumented biaxial compression device. *Geotechnical Testing Journal*, ASTM, Vol.16, No. 4, pp. 496-509.

Georgiannou V.N., Rampello S., Silvestri F. (1991) - Static and dynamic measurements of undrained stiffness on natural overconsolidated clays. *Proceedings 10th ECSMFE*, Florence, Italy, 1991, Vol. 1, pp. 91-95.

Harris W.W. (1994) - *Localization of loose granular soils and its effect on undrained steady state strength*. Ph.D. Dissertation, Northwestern University, Evanston, IL.

Hight D.W. (1982) - A simple piezometer probe for the routine measurement of pore pressure in triaxial tests on saturated soils. *Géotechnique*, **32**, pp. 396-401.

Hvorslev M.J. (1960) - Physical components of the shear strength of saturated clays. *Proceedings of the Research Conference on Shear Strength of Cohesive Soils*, ASCE, Boulder, CO, pp. 169-273.

Mokni M. (1992) - *Relations entre déformations en masse et déformations localisées dans les matériaux granulaires*. Thèse de doctorat, Université J. Fourier - Institut National Polytechnique de Grenoble.

Ord A., Vardoulakis I., Kajewski R. (1991) - Shear band formation in Gosford Sandstone. *International Journal of Rock Mechanics and Mining Sciences & Geomechanics Abstracts*, Vol. 28, No. 5, pp. 397-409.

Rampello S., Georgiannou V.N., Viggiani G. (1993) - Strength and dilatancy of natural and reconstituted Vallericca clay. *Proceedings of the International Symposium on Hard Soils and Soft Rocks*, Athens, Vol. 1, pp. 761-768.

Rampello S., Silvestri F. (1993) - The stress-strain behaviour of natural and reconstituted samples of two overconsolidated clays. *Proceedings of the International Symposium on Hard Soils and Soft Rocks*, Athens, Vol. 1, pp. 769-778.

Rhee Y. (1991) - *Experimental evaluation of strain-softening behavior of normally consolidated Chicago clay in plane strain compression*. Ph.D. Dissertation, Northwestern University, Evanston, IL.

Tillard-Ngan D., Desrues J., Raynaud S., Mazerolle F. (1993) - Strain localisation in Beaucaire marl. *Proceedings of the International Symposium on Hard Soils and Soft Rocks*, Athens, Vol. 2, pp. 1679-1686.

Vardoulakis I., Drescher A. (1988) - Development of a biaxial apparatus for testing frictional and cohesive granular media. *Report to the National Science Foundation*, NSF Grant No. CEE 84-06500, University of Minnesota.

Viggiani G. (1991) - Instability phenomena and propagation of discontinuities in stiff clays. *Proceedings of the Symposium on Experimental Characterization and Modelling of Soils and Soft Rocks*, Napoli, 1991, pp. 225-278.

Viggiani G. (1994) - *Localizzazione delle deformazioni e fenomeni di rottura nelle argille consistenti sovraconsolidate*. Tesi di dottorato in Ingegneria Geotecnica, Università di Roma " La Sapienza", Rome, Italy.

Viggiani G., Rampello S., Georgiannou V.N. (1993) - Experimental analysis of localisation phenomena in triaxial tests on stiff clays. *Proceedings of the International Symposium on Hard Soils and Soft Rocks*, Athens, Vol. 1, pp. 849-856.

5 Micromechanics of granular media
Micromécanique des milieux granulaires

A gradient elasticity model for granular materials
Un modèle élastique à gradient pour les milieux granulaires

H.-B. Mühlhaus
CSIRO, Division of Exploration and Mining, Nedlands, W.A., Australia

F. Oka
Department of Civil Engineering, Gifu University, Japan

ABSTRACT: A generalised continuum model for granular media is derived by direct homogenisation of the discrete equations of motion. In contrast to previous works on this topic, continuum concepts such as stress and moment stress are introduced after homogenisation. The resulting continuum theory is a combination of a Cosserat Continuum and a higher order deformation gradient continuum. The salient features of the theory are illustrated by means of the dispersion relations for planar wave propagation.

Un modèle de milieu continu généralisé est déduit par homogénéisation directe des équations discrètes du mouvement. Contrairement aux travaux antérieurs sur ce sujet, les concepts tels que la contrainte et les moments de contrainte sont introduits après homogénéisation. La théorie de milieux continus qui en résulte est un combinaison de continu de Cosserat et de continu à gradient d'ordre supérieur. Les points marquants de la théorie sont illustrés par une étude des relations de dispersion pour la propagation d'une onde plane.

1 INTRODUCTION

In most previous works on continuum models for random granular assemblies (Jenkins, 1991; Walton, 1987; Digby, 1981; Mühlhaus and Vardoulakis, 1987; Mühlhaus et al. 1991) certain prior assumptions are made with respect to relationships between the statical and kinematical quantities of the continuum model envisaged and the original discrete system, the granulate. These assumptions are not critical in the case of homogeneous or almost homogeneous deformations, where higher order deformation gradients or higher order rotation gradients do not play a role. For strongly inhomogenous deformations however (e.g. upon shear banding, see Mühlhaus and Vardoulakis, 1987) the situation is different. In this case higher order deformation gradients have to be considered leading to a generalised continuum theory of some kind. In an attempt to extend the validity of his standard continuum model to strongly inhomogenous deformation, Jenkins adopted a nonlocal interpretation of the Cauchy-Love relation (Love, 1927). Vardoulakis and Aifantis (1990) have included gradients of the plastic strain into the dilatancy constraint equation in order to account for the strong spatial nonuniformity of the deformation upon shear banding. Mühlhaus et al. (1991) defined average stress and moment-stress tensors by equating the virtual work of a Cosserat Continuum to the corresponding expression of the discrete system. There are many possibilities. The question now is which model comes closest to the behaviour of the original particulate material. In this paper continuum relations are derived by directly homogenising the discrete equations of motion. In this way any bias towards a particular continuum theory is avoided. Not too surprisingly the result turns out to be a combination of a Cosserat theory and a strain gradient theory. The relative importance of the nonstandard terms, namely the deformation gradient terms in the constitutive relations, is elucidated by means of the dispersion relation for planar wave propagation.

Throughout the paper we assume that the deformation is infinitesimal and for simplicity we consider an idealised material consisting of identical spherical grains.

2 THREE DIMENSIONAL CONTINUUM MODEL

2.1 Formulation

We consider a three dimensional assembly of identical, spherical grains. In view of the envisaged continuum formulation, summations over grain contacts occurring in the discrete equations of motion are replaced by a corresponding integral, that is we replace

$$\sum_{\text{contacts}} (\cdot) \to \int dn A(\mathbf{r},\mathbf{n})(\cdot) \qquad (1)$$

where

$$\mathbf{n} = (\sin\theta\cos\phi, \sin\theta\sin\phi, \cos\theta) \qquad (2)$$

is the unit vector from the centre of a sphere to a contact on its surface, θ and ϕ are coordinate angles of a spherical coordinate system with the origin at the sphere (grain) centre, $\mathbf{r}$ is the position vector with respect to a spatially fixed frame of reference and

$$\int d\mathbf{n} = \int_0^{2\pi}\int_0^{\pi} \sin\theta\, d\theta\, d\phi \ . \qquad (3)$$

The function $A(\mathbf{n})$ accounts for the orientational distribution of contacts so that $A(\mathbf{r},\mathbf{n})d\mathbf{n}$ is the probable number of contacts in the element $d\mathbf{n}$ centred at $\mathbf{n}(\theta,\phi)$. When the distribution of contacts is isotropic and independent of the position then $A(\mathbf{r},\mathbf{n}) = \kappa/4\pi$, where κ is the coordination number, the average number of contacts per grain.

Using the relation (21) the equations of motion can be written as

$$\rho\ddot{\mathbf{u}} = \frac{6v_s}{\pi D^3}\int_\alpha^{\mathbf{n}} A\mathbf{F}\, d\mathbf{n}$$

$$\text{and}\quad \frac{\rho}{10}D^2\ddot{\omega} = \frac{3v_s}{\pi D^2}\int_\alpha^{\mathbf{n}} A\mathbf{n}\times\mathbf{F}\, d\mathbf{n} \qquad (4)$$

where $\mathbf{u}(\mathbf{r},t)$ and $\omega(\mathbf{r},t)$ are the translations and rotations of a grain centred at the position $\mathbf{r}$; $\mathbf{F}^\mathbf{n}$ is the force exerted by the sphere at the contact at $\mathbf{r}+\frac{D}{2}\mathbf{n}$ and as before, v_s designates the solid volume fraction and D is the sphere diameter.

The components normal and parallel to the tangential plane of a contact depend upon the relative displacements and rotations between adjacent grains. The relative displacement $\Delta\mathbf{u}^\mathbf{n}$ between the contact points of two spheres centred at $\mathbf{r}$ and $\mathbf{r}+D\mathbf{n}$ respectively reads

$$\Delta\mathbf{u}^\mathbf{n} = \mathbf{u}^\mathbf{n} - \mathbf{u} + D\mathbf{n}\times\omega + \frac{D}{2}\mathbf{n}\times(\omega^\mathbf{n} - \omega) \qquad (5)$$

where $\mathbf{u}^\mathbf{n} = \mathbf{u}(\mathbf{r}+D\mathbf{n})$. Assuming linear elasticity the $\mathbf{F}$ - $\Delta\mathbf{u}$ relation reads

$$\mathbf{F}^\mathbf{n} = \mathbf{K}\Delta\mathbf{u}^\mathbf{n}, \quad K_{ij} = (k_n - k_s)n_i n_j + k_s \delta_{ij} \ , \qquad (6)$$

where k_n and k_s are the normal and tangential contact stiffnesses and the indices refer to a spatially fixed cartesian coordinate system. To simplify the algebra of the derivation it is assumed from now on that A is independent of position.

Then, because a contact is common to two spheres, $A(-\mathbf{n}) = A(\mathbf{n})$. Inserting (6) into (4) the equations of motion are obtained as:

$$\rho\ddot{\mathbf{u}} = \frac{6v_s}{\pi D^3}\int_{\alpha/2} A\mathbf{K}(\mathbf{u}^\mathbf{n} - 2\mathbf{u} + \mathbf{u}^{-\mathbf{n}})d\mathbf{n}$$

$$+\frac{3v_s}{\pi D^2}\int_{\alpha/2} A k_s \mathbf{n}\times(\omega^\mathbf{n} - \omega^{-\mathbf{n}})d\mathbf{n} \qquad (7)$$

$$\frac{\rho}{10}D^2\ddot{\omega} = \frac{3v_s k_s}{\pi D^2}\int_{\alpha/2} A\mathbf{n}\times(\mathbf{u}^\mathbf{n} - \mathbf{u}^{-\mathbf{n}})d\mathbf{n}$$

$$+\frac{6v_s k_s}{\pi D}\int_{\alpha/2} A\mathbf{n}\times\mathbf{n}\times\omega\, d\mathbf{n}$$

$$+\frac{3v_s k_s}{\pi D}\int_{\alpha/2} A\mathbf{n}\times\mathbf{n}\times(\omega^\mathbf{n} - 2\omega + \omega^{-\mathbf{n}})\, d\mathbf{n} \qquad (8)$$

We have used the symmetry property $A(-\mathbf{n}) = A(\mathbf{n})$ so the integration extends only over half of the solid angle, i.e.

$$\int_{\alpha/2} d\mathbf{n} = \int_0^{2\pi}\int_0^{\pi/2}\sin\theta\, d\theta\, d\phi \ . \qquad (9)$$

Next $\mathbf{u}^{\pm\mathbf{n}} = \mathbf{u}(\mathbf{r} \pm D\mathbf{n})$ and $\omega^{\pm\mathbf{n}}$ are replaced by the Taylor expansions

$$u_i^{\pm\mathbf{n}} = u_i \pm \frac{D}{1!}u_{i,j}n_j + \frac{D^2}{2!}u_{i,jk}n_j n_k \pm \ldots\ , \qquad (10)$$

and

$$\omega_i^{\pm\mathbf{n}} = \omega_i \pm \frac{D}{1!}\omega_{i,j}n_j + \frac{D^2}{2!}\omega_{i,jk}n_j n_k \pm \ , \qquad (11)$$

where $(\cdot)_{,i} = \frac{\partial}{\partial x_i}(\cdot)$. Neglecting terms of higher than fourth order in D yields

$$u_i^\mathbf{n} - 2u_i + u_i^{-\mathbf{n}} \simeq$$

$$D^2\left(u_{i,jk}n_j n_k + \frac{D^2}{12}u_{i,jklm}n_j n_k n_l n_m\right) \qquad (12)$$

and

202

$$u_i^n - u_i^{-n} \simeq 2D\left(u_{i,j} n_j + \frac{D^2}{6} u_{i,jk} n_j n_k\right). \quad (13)$$

Analogous expressions are obtained for the ω-differences. What remains is to insert the Taylor expansions into the eqs (7) and (8). The result reads:

$$\rho \ddot{u}_i = \sigma_{ij,j}, \quad (14)$$

$$\sigma_{ij} = \frac{6v_s}{\pi D}\Big((k_n - k_s) A_{ijlm} \varepsilon_{lm}$$

$$+ k_s A_{lj} \varepsilon_{il} + k_s A_{lj}(W_{il} - W_{il}^c)$$

$$+ \frac{D^2}{12}\big((k_n - k_s) A_{ijklmn} \varepsilon_{kl,mn} + k_s A_{ljmn} \varepsilon_{il,mn}$$

$$+ k_s\big(A_{ljmn} W_{il,mn} - 2 A_{ljmn} W_{il,mn}^c\big)\Big), \quad (15)$$

where

$$A_{i_1 i_2 \cdots i_{2n}} = \int_{\alpha/2} A n_{i_1} n_{i_2} \cdots n_{i_{2n}} \, dn, \quad (16)$$

ε_{ij} and W_{ij} are the symmetric and skew symmetric part of the displacement gradient respectively,

$$W_{ij}^c = -e_{ijk} \omega_k, \quad (17)$$

e_{ijk} with $e_{123} = 1$ is the permutation symbol and the superscribed c (for Cosserat) is used to distinguish the particle spin from the spin W_{ij} of an infinitesimal element dV.

In similar fashion one obtains for eq (18):

$$\frac{\rho D^2}{10} \ddot{\omega}_i = \frac{6v_s k_s}{\pi D}\Bigg[e_{ijk} A_{lj}\Big(\varepsilon_{kl} + W_{kl} - W_{kl}^c\Big)$$

$$+ e_{ijk} A_{\ell jmn} \frac{D^2}{6} u_{k,lmn}$$

$$+ e_{ijk} e_{klm} A_{jlrs} \frac{D^2}{4} \omega_{m,rs}\Bigg] \quad (18)$$

or

$$\frac{\rho D^2}{10} \ddot{\omega}_i = \mu_{is,s} + e_{ijk} \sigma_{kj}, \quad (19)$$

where the moment stress tensor μ introduced in this way is obtained as

$$\mu_{is} = \frac{6v_s k_s}{\pi D} \frac{D^2}{12} e_{ijk}\Big[A_{ljms}\big(u_{k,lm} - W_{kl,m}^c\big)\Big] \quad (20)$$

Using the standard formula

$$e_{ijk} e_{lmk} = \delta_{il} \delta_{jm} - \delta_{im} \delta_{jl}, \quad (21)$$

eq (20) can be written somewhat more explicitly as

$$\mu_{is} = \frac{6v_s k_s}{\pi D} \frac{D^2}{12}\Big[e_{ijk} A_{ljms} \varepsilon_{kl,m} -$$

$$- A_{inms}(\Omega_{n,m} - \omega_{n,m}) + A_{ms}(\Omega_{i,m} - \omega_{ij,m})\Big] \quad (22)$$

where

$$\Omega_{n,m} = -\frac{1}{2} e_{nkl} W_{kl,m}. \quad (23)$$

In the following we discuss a number of special cases for which the coefficients are evaluated in detail. First the standard continuum is considered where higher order deformation gradients are neglected.

2.2 Standard continuum

In order to obtain explicit expressions for the elasticities we have to make an assumption concerning the form of the contact distribution function $A(\mathbf{n})$. For orthotropic depositional anisotropy $A(\mathbf{n})$ can be expressed in terms of a symmetric, traceless tensor $\mathbf{A}$ [Kanatani, 1984; Cowin, 1985]:

$$A(\mathbf{n}) = \frac{\kappa}{4\pi}\left(1 + A_{ij} n_i n_j\right) \quad (24)$$

For transversely isotropic materials, $\mathbf{A}$ may be expressed in terms of the unit vector $\mathbf{h}$ in the direction of the axis of anisotropy and the strength of the anisotropy:

$$A_{ij} = -a\left(\delta_{ij} - 3 h_i h_j\right), \quad (25)$$

where $0 \leq a \leq 1$. Then

$$A(\mathbf{n}) = \frac{\kappa}{4\pi}\Big[(1-a) + 3a(h_i n_i)^2\Big] \quad (26)$$

Next the components of the tensors $A_{i_1 i_2 \cdots i_{2n}}$ are to be evaluated. Note that according to the definition (16) all integrations over the angular domain of (θ, ϕ) extend over half of the solid angle only. Integrals that facilitate this calculation are (Kanatani, 1984; Jenkins, 1991):

$$I_{i_1 i_2 \cdots i_{2n}} = \int_{\alpha/2} n_{i_1} n_{i_2} \cdots n_{i_{2n}} \, dn = \frac{1}{2} \frac{4\pi}{2n+1},$$ (27)

for $i_1 = i_2 = \ldots i_{2n}$,

and

$$2I_{ij} = \frac{4\pi}{3}\delta_{ij}, \; 2I_{ijkl}$$

$$= \frac{4\pi}{15}\left(\delta_{ij}\delta_{kl} + \delta_{ik}\delta_{jl} + \delta_{il}\delta_{jk}\right), I_{ijklmn}$$

$$= \frac{1}{7}\left(\delta_{in} I_{jklm} + \delta_{jn} I_{klmi} + \delta_{kn} I_{lmij}\right.$$

$$\left. + \delta_{ln} I_{mijk} + \delta_{mn} I_{ijkl}\right)$$ (28)

The result for the stress-deformation relation (15) reads:

$$\sigma_{ij} = \frac{v_s \kappa}{5\pi D} \left\{ \frac{1}{7}(k_n - k_s)\left[(7-4a)(2\varepsilon_{ij} + \varepsilon_{kk}\delta_{ij})\right. \right.$$

$$+ 18a\left(\varepsilon_{kk} h_k h_l \delta_{ij} + \varepsilon_{kk} h_i h_j + \right.$$

$$\left. + 2h_i \varepsilon_{jk} h_k + 2h_j \varepsilon_{ik} h_k\right)\right]$$

$$+ k_s\left[(5-2a)\left(u_{i,j} - W^c_{ij}\right)\right.$$

$$\left.\left. + 6a h_l h_j\left(u_{i,l} - W^c_{il}\right)\right]\right\}$$ (29)

When the deformation is homogeneous and body couples are absent, the stress must be symmetric. Symmetry of σ_{ij} requires that

$$3a h_l\left(\varepsilon_{il} h_j - \varepsilon_{jl} h_i\right) + 3a h_l\left[\left(W_{il} - W^c_{il}\right)h_j \right.$$

$$\left. - \left(W_{jl} - W^c_{jl}\right)h_i\right] + (5-2a)\left(W_{ij} - W^c_{ij}\right) = 0$$ (30)

Solving for W^c_{ij} yields

$$W^c_{ij} = W_{ij} + \frac{3a}{5+a}\left(\varepsilon_{il} h_j - \varepsilon_{jl} h_i\right)h_l,$$ (31)

where the identity

$$h_l\left[\Omega_{il} h_j - \Omega_{jl} h_i\right] = \Omega_{ij},$$

$$\Omega_{ij} = -\Omega_{ji}, \quad h_i h_i = 1$$ (32)

has been used.
For $a = 0$ the material is isotropic. The Lamé coefficients μ and λ are related to the contact stiffnesses as

$$\mu = \frac{v_s \kappa}{5\pi D}\left(k_n + \frac{3}{2}k_s\right)$$

$$\text{and } \lambda = \frac{v_s \kappa}{5\pi D}\left(k_n - k_s\right)$$ (33)

Combination of the relations (33) yields $k_s = 2\pi D$ $(\mu - \lambda) / (v_s \kappa)$, and, since $k_s \geq 0$ it follows that the Poisson ratio of a random packing of spherical grains has to be less than or equal to 0.25 (Walton, 1987). Jenkins (1988) has derived relations between k_n, k_s and the moduli of the grain material for an infinitesimal deviation from an homogeneously and isotropically prestrained ground state. The result reads

$$k_n = \frac{9}{2D} M\Delta^{\frac{1}{2}}, \; k_s = 2\frac{(1-v_g)}{(2-v_g)}k_n,$$

$$\text{with } M = \frac{2}{9\sqrt{3}} \frac{\mu_g D^2}{(1-v_g)},$$ (34)

where v_g and μ_g are the Poisson ratio and the shear modulus respectively of the grain material and Δ is the magnitude of the volume strain of the ground state.

2.3 Isotropic fabric, general case

Inserting the relations (25) through (28) into the constitutive relationships (15) and (20) yields, for a = 0 (isotropic contact distribution):

$$\sigma_{ij} = 2\mu\varepsilon_{ij} + \lambda\varepsilon_{kk}\delta_{ij} + 2(\mu-\lambda)\left(W_{ij} - W^c_{ij}\right)$$

$$+ \frac{D^2}{12}\left\{\frac{6}{7}\lambda\left(\varepsilon_{ij,kk} + \varepsilon_{kk,ij} + \frac{1}{2}\varepsilon_{kk,mm}\delta_{ij}\right)\right.$$

$$+ \frac{6}{5}(\mu-\lambda)\varepsilon_{ij,kk} + \frac{6}{5}(\mu-\lambda)\left(W_{ij,kk} - W^c_{ij,kk}\right)$$

$$\left. - \frac{8}{5}(\mu-\lambda) W^c_{im,mj} + \frac{2}{5}(\mu-\lambda)W^c_{ij,kk}\right\}$$ (35)

and

$$\mu_{is} = \frac{2}{5}(\mu-\lambda)\frac{D^2}{12} e_{ijk}$$

$$\left(\gamma_{kj,s} + \gamma_{km,m}\delta_{js} + \gamma_{ks,j}\right),$$ (36)

where γ_{ij} designates the relative deformation

$$\gamma_{ij} = u_{i,j} - W^c_{ij} \quad . \tag{37}$$

We note that, as a consequence of the number of terms included in the Taylors expansions eqs (30) and (31), the moment stresses μ_{is} depend on the gradients of the relative deformation tensor γ_{ij} rather than on the gradients of W^c_{ij} alone as in the classical Cosserat theory (see Günther, 1958; Schäfer; 1967 for reviews on this subject). As in the Cosserat models for granular media (e.g. Mühlhaus et al 1991) it is the grain diameter which provides an intrinsic length scale which becomes crucial upon strain localisation, in boundary layers or in high frequency wave-propagation and related phenomena.

2.4 Dispersion relations

First the equations of motion are specialised to one dimensional deformations, deformations where the fields depend on one coordinate only. Let $x = x_1$ be this coordinate, and for deformations in the (x_1,x_2) plane we have

$$\mathbf{u} = u_1(x)\mathbf{e}_1 + u_2(x)\mathbf{e}_2 \quad , \quad \omega = \omega_3 \mathbf{e}_3 \tag{38}$$

Inserting (64) into the equations of motion, gives for isotropic fabric:

$$c^2_{0L}\left(\frac{\partial^2 u_1}{\partial x^2} + \frac{d^2}{12}\frac{\partial^4 u_1}{\partial x^4}\right) = \ddot{u}_1 \quad , \tag{39}$$

$$d^2 = \frac{3}{35}\frac{14\mu + 11\lambda}{2\mu + \lambda}D^2 \quad , \quad c^2_{0L} = \frac{1}{\rho}(2\mu + \lambda) \tag{40}$$

for the longitudinal wave along the x-axis, and

$$2(\mu - \lambda)\left(\frac{\partial u_2}{\partial x} + \frac{D^2}{12}\frac{6}{5}\frac{\partial^3 u_2}{\partial x^3}\right)$$

$$-4(\mu - \lambda)\left(\omega_3 + \frac{D^2}{12}\frac{6}{5}\frac{\partial^2 \omega_3}{\partial x^2}\right) = \frac{\rho}{10}D^2 \ddot{\omega}_3 \quad , \tag{41}$$

$$-2(\mu-\lambda)\left(\frac{\partial \omega_3}{\partial x}+\frac{D^2}{12}\frac{6}{5}\frac{\partial^3 \omega_3}{\partial x^3}\right)+(2\mu-\lambda)\frac{\partial^2 u_2}{\partial x^2}$$

$$+\frac{D^2}{12}\left(\frac{6}{5}\mu-\frac{27}{35}\lambda\right)\frac{\partial^4 u_2}{\partial x^4} = \rho\ddot{u}_2 \quad , \tag{42}$$

for the shear and micro-rotation waves respectively. In eq (39) the influence of the gradient term is strongest if $\mu = \lambda$, that is, if $k_s = 0$. In this case it is $d = \frac{5}{7}D$. For $D = 0$ one obtains the longitudinal and shear wave equations of the classical continuum, as it must be. Also for $D = 0$ it follows that $\frac{\partial u_2}{\partial x} = 2\omega_3$ (cp. eq 42); c_{0L} designates the longitudinal wave speed of the classical continuum which is obtained here in the limit for infinitely long wave length.

For the derivation of the dispersion functions for the rotational and shear waves we consider travelling wave solutions of the form

$$\begin{bmatrix} u_2 \\ \omega_3 \end{bmatrix} = \exp i(qx - \omega t)\begin{bmatrix} U_2 \\ i\Omega_3 \end{bmatrix} \quad , \tag{43}$$

Where $i = \sqrt{-1}$ and q is the wave-number. Inserting (43) into (41) and (42) yields the homogeneous system of equations

$$\begin{bmatrix} A_{uu} & A_{u\omega} \\ A_{\omega u} & A_{\omega\omega} \end{bmatrix}\begin{bmatrix} U_2 \\ D\Omega_3 \end{bmatrix} = 0 \quad , \tag{44}$$

where

$$A_{uu} = \left(2-\frac{\lambda}{\mu}\right)\gamma^2 - \frac{1}{12}\left(\frac{6}{5}-\frac{27}{35}\frac{\lambda}{\mu}\right)\gamma^4 - \tilde{\omega}^2 \quad , \tag{45}$$

$$A_{\omega\omega} = 4\left(1-\frac{\lambda}{\mu}\right)\left(1-\frac{1}{10}\gamma^2\right) - \frac{1}{10}\tilde{\omega}^2 \quad , \tag{46}$$

$$A_{\omega u} = A_{u\omega} = 2\left(1-\frac{\lambda}{\mu}\right)\left(1-\frac{1}{10}\gamma^2\right)\gamma \quad , \tag{47}$$

where

$$c_{0S} = \sqrt{\frac{\mu}{\rho}} \quad , \quad \gamma = qD \text{ and } \tilde{\omega} = \frac{D}{c_{0S}}\omega \quad , \tag{48}$$

The equations (44) are nontrivially soluble if

$$\tilde{\omega}^2_{1/2} = -\frac{b}{2a} +/- \sqrt{\left(\frac{b}{2a}\right)^2 - \frac{c}{a}} \quad , \tag{49}$$

where

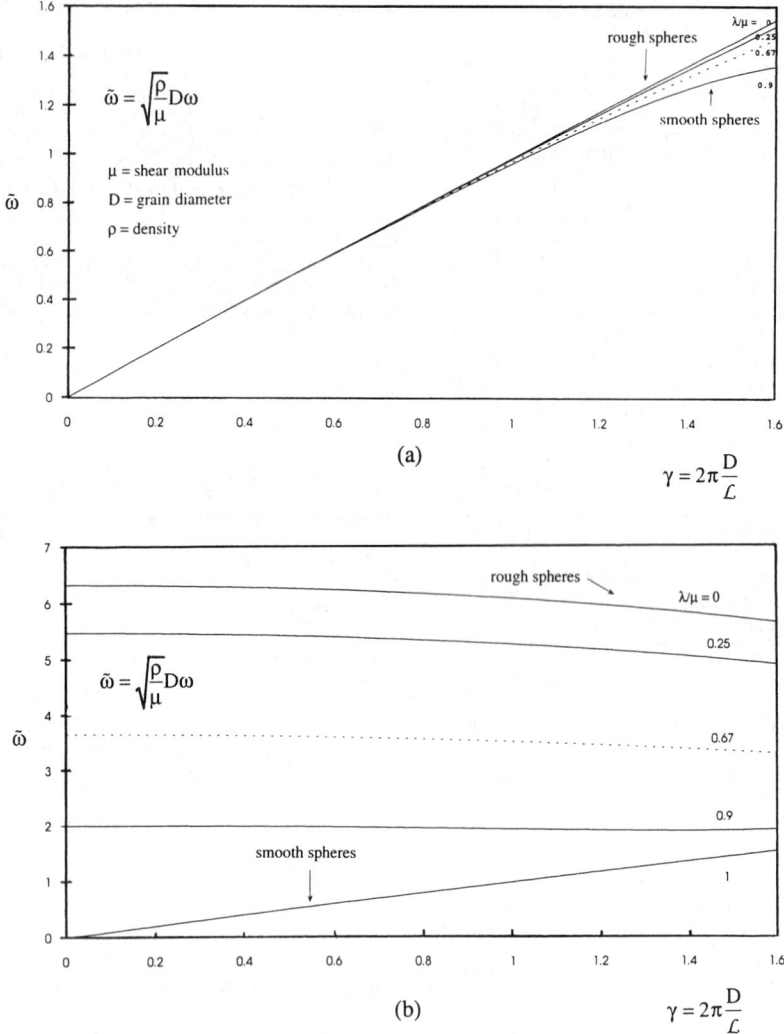

Fig. 1. Angular frequency as function of the dimensionless wave number $\gamma = 2\pi\frac{D}{L}$.
(a) Shear wave (b) Micro-rotation wave

$a = \frac{1}{10}$, $b = -4\left(1 - \frac{\lambda}{\mu}\right)$

$+ \frac{1}{10}\left(2 - 3\frac{\lambda}{\mu}\right)\gamma^2 + \frac{1}{120}\left(\frac{6}{5} - \frac{27}{35}\frac{\lambda}{\mu}\right)\gamma^4$,

$c = 4\left(1 - \frac{\lambda}{\mu}\right)\gamma^2 - \frac{4}{12}\left(1 - \frac{\lambda}{\mu}\right)\left(\frac{6}{5} + \frac{3}{7}\frac{\lambda}{\mu}\right)\gamma^4$. (50)

By expansion of (49) in powers of γ, the leading terms are obtained as

$$\tilde{\omega}_1 = \frac{4\left(1 - \frac{\lambda}{\mu}\right)}{\frac{1}{10}}\left(1 - \frac{3}{40}\gamma^2\right) \text{, and } \tilde{\omega}_2 = \gamma^2 \quad (51)$$

By inspection of the amplitude ratio one finds that the plus sign in (49) corresponds to a micro-rotation wave (that is $|D\Omega_3/(\gamma U_2)| \gg 1$ within

206

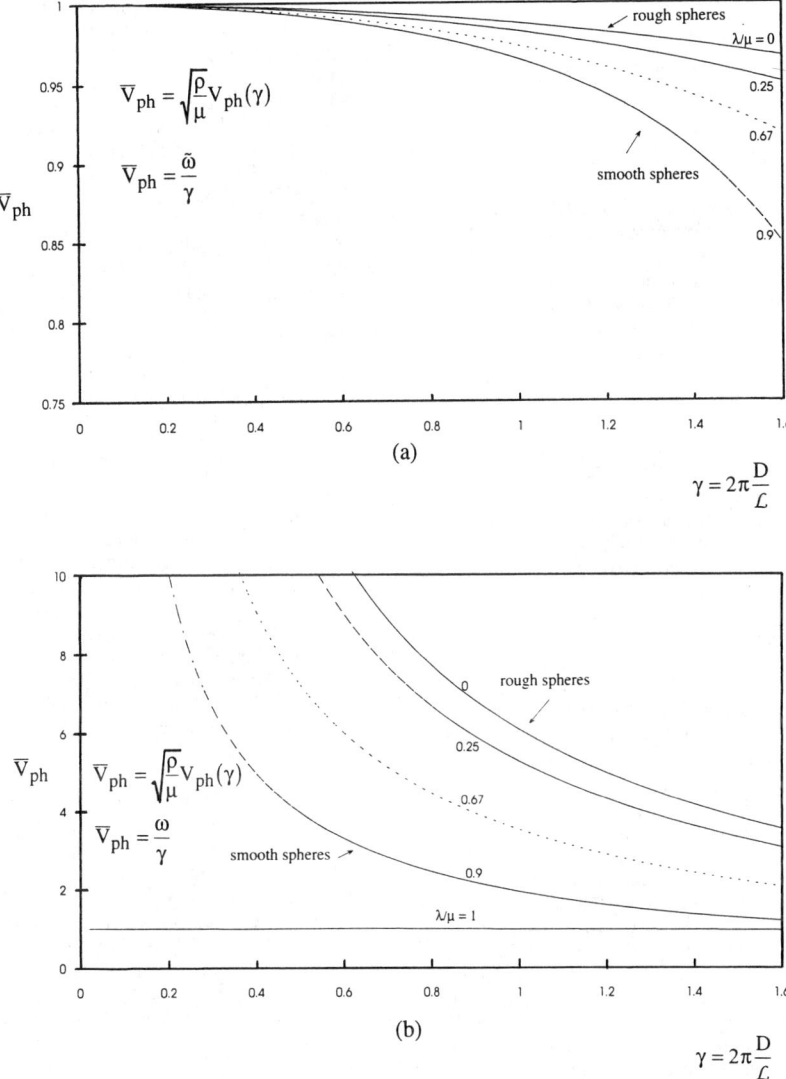

Fig. 2. Phase velocities (a) Shear wave (b) Micro-rotation wave

range of wave numbers considered here), a wave type which does not exist in a standard continuum. At least in the range $\gamma \leq 1$ the minus sign clearly corresponds to a conventional shear wave. It should be mentioned that the existence of micro-rotation waves is typical for continua with extra degrees of freedom such as the Cosserat Continuum (Suhubi and Eringen, 1964; Sluys, 1992) and Mindlin's (1964) generalisation of it. The frequency and phase velocity functions as functions of the dimensionless wave number γ are represented in the Figures 1 and 2. The present theory is an approximation of the actual behaviour of a granular material. One cannot expect the dispersion relations (49) to yield reasonable values over the whole range of γ. The maximum value of γ (=1.6) shown in the figures corresponds to a grain diameter, wavelength ratio of about $1/4$. A special case occurs if the spheres are ideally smooth so that $k_s = 0$ and accordingly $\lambda = \mu$. One then expects an instability which, as a matter of fact, is reproduced by the model. From (49) and (50) it follows that for $\lambda = \mu$:

$$\tilde{\omega}_2 = 0 \text{ with } D\Omega_3 / U_2 = \infty \text{ and} \qquad (52)$$

$$\tilde{\omega}_1 = \gamma^2\left(1 - \frac{1}{28}\gamma^2\right) \text{ with } D\Omega_3 / U_2 = 0 \qquad (53)$$

3 CONCLUSION

A three dimensional continuum model for random packings of elastic spheres has been derived. The starting point for the derivation of the continuous model are Newton's equations of motion. Similar to the situation in a Cosserat Continuum the equations of motion contain an additional independent, rotational degree of freedom, the grain rotation. However, different from the Cosserat Continuum, a consequent approximation scaled by powers of the grain diameter requires the inclusion of higher order displacement gradients as well. If higher order displacement and rotation gradients are neglected the additional degree of freedom, the grain spin, becomes an internal variable. The evolution equation for the grain spin is obtained from the moment equilibrium condition. For isotropic granulates it follows that the grain spin has to be equal to the nonsymmetric part of the displacement gradient. In general however, for anisotropic fabrics, the relation is less trivial. Finally we have evaluated the dispersion relations for the propagation planar waves. For the shear waves, two dispersion relations are obtained. In the one case the main carrier of the energy is the displacement and in the other case the main carrier is the grain rotation. The latter wave type, which does not exist in standard continua, is typical for micropolar theories of all kinds. In the present model all parameters of the dispersion relations are determined in the sense that they can be expressed in terms of solid volume fraction, normal contact stiffness, coordination numbers and average grain diameter. It should therefore be possible, at least in principle, to verify, falsify or determine a range of validity of the model by careful measurements of the various wave speeds either in physical experiments or, for a start, by means of discrete element simulations. We wish to emphasise that here the situation is different from previous developments in this area where it was suggested to calibrate certain nonstandard moduli by means of wave speed measurements. In the latter case the verification of the model would require an additional experiment.

4 REFERENCES

Cowin, S.C. (1985). The relationship between the elasticity tensor and the fabric tensor. *Mechanics of Materials*, 4, pp. 137-147.

Cundall, P.A., Jenkins, J.T. and Ishibashi, I. (1988). Evolution of elastic moduli in a deforming granular assembly., in *Powders and Grains* (Biarez, J. and Gourvès, R., Eds) pp. 319-322, Balkema: Rotterdam.

Digby, P.J. (1981). The effective elastic moduli of porous granular rocks. *Journal of Applied Mechanics*, 16, pp. 803-808.

Fermi, E. (1965). Collected papers of Enrico Fermi. Univ. of Chicago Press, Chicago. Vol II, p. 978.

Günther, W. (1958). Zur Statik und Kinematik des Cosseratschen Konrinuums. Abh. Braunschweig. *Wiss. Ges.*, 10, 195-213.

Jenkins, J.T. (1988). Volume change in small strain axisymmetric deformations of granular material in, *Micromechanics of Granular Materials* (Satake, M. and Jenkins, J.T., Eds.) pp. 245-252, Elsevier: Amsterdam.

Jenkins, J.T. (1991). Anisotropic elasticity for random arrays of identical spheres in, *Modern Theory of Anisotropic Elasticity and Applications* (J.Wu, Ed.), SIAM, Philadelphia.

Jenkins, J.T., Cundall, P.A. and Ishibashi, Il (1988). Microchemical modelling of granular materials with the assistance of experiments and numerical simulations in, *Powder and Grains* (Biarez, J. and Gourvès, R., Eds.) pp. 257-264, Balkema: Rotterdam.

Kanatani, K. (1984). Distribution of directional data and fabric tensors. *International Journal of Engineering Science*, 22, pp. 149-164.

Love, A.E.H. (1927). A Treatise on the Mathematical Theory of Elasticity. Cambridge University Press: Cambridge.

Mindlin, R.D. (1964). Microstructure in linear elasticity. *Arch. Rat. Mech. Anal.* 16, 51-78.

Mindlin, R.D. (1942). Compliance of elastic bodies in contact. *Journal of Applied Mechanics*, 16, pp. 259-268.

Mühlhaus, H.-B. and Vardoulakis, I. (1987). The thickness of shear bands in granular materials. *Géotechnique*, 37, pp. 271-283.

Mühlhaus, H.-B., de Borst, R. and Aifantis, E.C. (1991). Constitutive models and numerical analyses for inelastic materials with microstructure, *Comp. Meth. and Adv. in Geomech.* Beer, Booker and Carter (eds.), pp. 377-384, Balkema: Rotterdam.

Schäfer H. (1967). Das Cosserat Kontinuum. *ZAMM*, 47, pp. 485-498.

Schwartz, L.M., Johnson, D.L. and Feng, S. (1984). Vibrational modes in granular materials. *Physical Review Letters*, 52, pp. 831-804.

Sluys, L.J. (1992). Wave propagation, localisation and dispersion in softening solids. Ph.D Thesis, Civil Engng. Dept., TU Delft.

Suhubi, E.S. and Eringen, A.C. (1964). Nonlinear Theory of Micro-Elastic solids - II. *Int. J. Engng. Sci.*, 2, 389-404.

Toupin, R.A. (1964). Theories of elasticity with couple stress. *Archive for Rational Mechanics and Analysis*, 11, pp. 385-414.

Vardoulakis, I. and Aifantis, E.C. (1989). Gradient dependent dilatancy and its implications in shear banding and liquefaction. Ingenieur Archiv 59, pp. 197-208.

Walton, K. (1987). The effective elastic modulus of a random packing of spheres. *Journal of Mechanics and Physics of Solids,* 35, pp. 213-216.

Walton, K. (1988). Wave propagation within random packings of spheres. *Geophysical Journal,* 92, pp. 89-97.

Analytical solutions of deformation in gradient dependent model
Solutions analytiques de déformation avec des modèles dépendant du gradient

F. Oka
Department of Civil Engineering, Gifu University, Japan

H.-B. Mühlhaus
CSIRO, Division of Exploration and Mining, Nedlands, W.A., Australia

ABSTRACT: Recently, gradient dependent constitutive models have been proposed and studied for solving the post strain localization problem. However, the physical meaning of the higher strain gradient term is still an open problem. In the present paper, a strain gradient dependent constitutive model has been derived from the micro-structural consideration of particulate materials. Then, the analytical solutions of gradient dependent constitutive models have been found using the Hirota's direct method to solve a partial differential equations. Localization solutions and periodic solutions are obtained.

Récemment, des lois de comportement faisant intervenir le gradient de la déformation ont été proposées pour résoudre le problème de la post-localisation. Cependant, la signification physique du terme de gradient d'ordre supérieur reste un problème ouvert. Dans le présent article, on fait découler un modèle à gradient de déformation de considérations micro-structurales propres aux matériaux granulaires. Les solutions analytiques de modèles de comportement à gradient ont pu être obtenus en utilisant la méthodes d'Hirota, une méthode directe de résolution des équations aux dérivées partielles. On obtient des solutions avec localisation, et des solutions périodiques.

1 INTRODUCTION

Recently, the gradient dependent constitutive model has been proposed and studied for solving the post strain localization problem (Aifantis[1],[2], Vardoulakis and Aifantis[3], Mühlhaus and Aifantis[4], Zbib and Aifantis[5]). The problem of strain localization has been studied as a bifurcation problem based on the works by Thomas[6] and Hill[7] et al. Within the framework of a bifurcation theory, the condition for the initiation of shear bands are derived from the view point of the loss of ellipticity for quasi-static problems. However, these approaches are not appropriate for predicting the thickness of a shear band or simulating the post localization behavior. To overcome these shortcomings, Coleman[8] and Aifantis[1] advocated a new approach for dealing with the localization problem which uses the gradient dependent constitutive model. The feature of this method is known that a characteristic length scale appears explicitly in the model like the particle size of the soil. Although the effectiveness of the gradient dependent model has been studied for the prediction of thickness of shear band and the continuation of the numerical simulation in the post localization regime, the physical meaning the higher order strain gradient term is still a open problem.

In the first part of the present paper, we focus on the derivation of of gradient dependent constitutive model from a micro-structural consideration of particulate materials. In the second part of the present paper, the analytical solutions of gradient dependent constitutive models are derived to illustrate a characteristics of the material model with gradient term. The results also allow the evaluation of the numerical solution for the boundary value problems with gradient dependent constitutive models.

2 MICRO-STRUCTURAL NATURE OF GRADIENT DEPENDENT CONSTITUTIVE MODELS

Although gradient dependent constitutive models have been applied to granular materials and clay (e.g. Oka et al.[9]) recently, the physical nature has not yet sufficiently clarified. Therefore, the nature of gradient term should be more discussed within the micro-structural consideration. In the present chapter, we focus on the physical nature of gradient term for soil. Let us derive a gradient dependent constitutive model from a particulate discrete model with a nonlinear elastic spring.

As for the nonlinear lattice, Fermi-Pasta-Ulam model[10] is famous by the discovery of soliton in

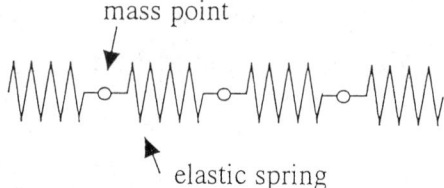

Figure 1. Discrete model with nonlinear spring.

the numerical simulation of vibration of nonlinear lattice (Figure 1). They used a following nonlinear spring as:

$$F = \kappa(r_i + \alpha r_i^2) \tag{1}$$

where F is the force, y_i is the displacement of the ith mass point (or particle), $r_i = y_i - y_{i-1}$ is the relative displacement between two particles and κ and α are material constants. We only deal with the quasi-static case in which the acceleration term is neglected.

$$F = \kappa[y_{n+1} - y_n + \alpha(y_{n+1} - y_n)^2]$$
$$-\kappa[y_n - y_{n-1} + \alpha(y_n - y_{n-1})^2] \tag{2}$$

where y_n is the displacement.

Assuming that $x = nh$ (x:coordinate, n:number of particle, h is the distance of two near-by particles), the continuum model can be derived through the use of the following Taylor series expansion to Eq.(2) as:

$$y_{n\pm 1} = y \pm h y_x + \frac{h^2}{2!} y_{xx} \pm \frac{h^3}{3!} y_{xxx} + \frac{h^4}{4!} y_{xxxx} + \ldots \tag{3}$$

By substituting Eq.(3) into Eq.(2) and truncating the higher order term,

$$\frac{\partial \sigma}{\partial x} = F \tag{4}$$

$$\sigma = \kappa h^2 \varepsilon + \frac{1}{12}\kappa h^4 \varepsilon_{xx} + \alpha h^3 \varepsilon^2 \tag{5}$$

where $\varepsilon = y_x$ is the strain and σ is the stress.

From Eq.(5) we obtain the following gradient dependent elastic model,

$$\sigma = a\varepsilon + b\varepsilon_{xx} + c\varepsilon^2 \tag{6}$$

where $a = \kappa h^2, b = \frac{1}{12}\kappa h^4, c = \alpha h^3$ are positive constants.

It is interesting that in the derivation of Eq.(6), the gradient term comes from the discrete nature of the physical model. In addition, it becomes evident that the material parameters depends on the distance of particles h, in other words, material parameters have a dimension of characteristics length scale. We call this model the first model in the present paper. In is worth noting that the gradient dependent term in Eq.(6) works as a destabilizer because the sign of this term is negative at the localized zone of the strain. On the other hand, the last term of Eq.(6) plays a hardening role.

Next, we will derive the other type of gradient dependent model. We consider the chain with two neighbor interactions (Figure 2) which is similar interactions of the elastic model with microstructure by Kunin[11]. In this model, we assume that the spring for the next-nearest neighbor interactions has a opposite sign to the one for the only-nearest neighbor interactions.

$$F_1 = \kappa r_i + \kappa \alpha r_i^2 \tag{7}$$

$$F_2 = -\kappa' r_i' \tag{8}$$

where $F_i, i = 1, 2$ is the external force and κ' is a constant of the interaction between the mass point and the next nearest neighbor mass point as $r_i' = y_i - y_{i-2}$. Considering two interactions, the equilibrium equation becomes

$$\kappa(y_{n+1} - 2y_n + y_{n-1})[1 + \alpha(y_{n+1} - y_{n-1})]$$
$$-\kappa'(y_{n+2} - 2y_n + y_{n-2}) = F \tag{9}$$

where F is the external force. From the Taylor series expansion,

$$y_{n\pm 1} = y \pm h y_x + \frac{h^2}{2!} y_{xx} \pm \frac{h^3}{3!} y_{xxx} + \frac{h^4}{4!} y_{xxxx} + \ldots \tag{10}$$

$$y_{n\pm m} = y \pm H y_x + \frac{H^2}{2!} y_{xx} \pm \frac{H^3}{3!} y_{xxx} + \frac{H^4}{4!} y_{xxxx} + \ldots \tag{11}$$

where $mh = H$, $m > 1$. The case of Figure 2 corresponds to $m = 2$.

By substituting Eqs.(10) and (11) into Eq.(9) and truncating the higher order terms, we have

$$\frac{\partial \sigma}{\partial x} = F \tag{12}$$

$$\sigma = (\kappa - \kappa' m^2) h^2 \varepsilon + \frac{1}{12}(\kappa - \kappa' m^4) h^4 \varepsilon_{xx} + \alpha h^3 \varepsilon^2 . \tag{13}$$

Eq.(13) is rewritten by

$$\sigma = h^4 B[\frac{(\kappa - \kappa' m^2)}{Bh^2}\varepsilon - \varepsilon_{xx} - \frac{\alpha}{Bh}\varepsilon^2] \tag{14}$$

The following conditions are required:

$$(\kappa - \kappa' m^2) > 0 \tag{15}$$

$$B = -\frac{\kappa - \kappa' m^4}{12} > 0 \tag{16}$$

$$\alpha > 0 . \tag{17}$$

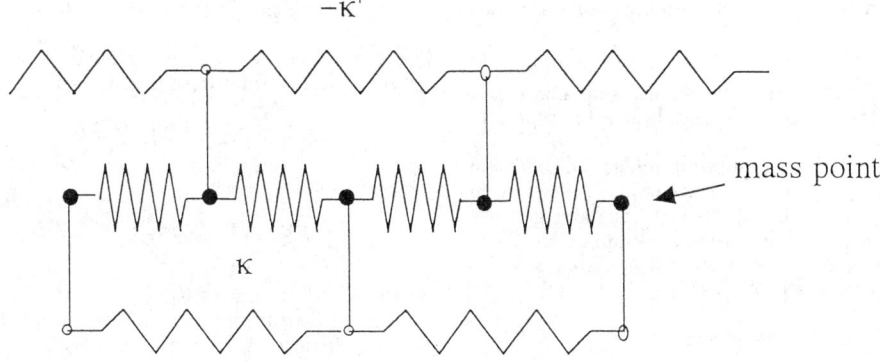

Figure 2. Nearest and next-nearest neighour interactions.

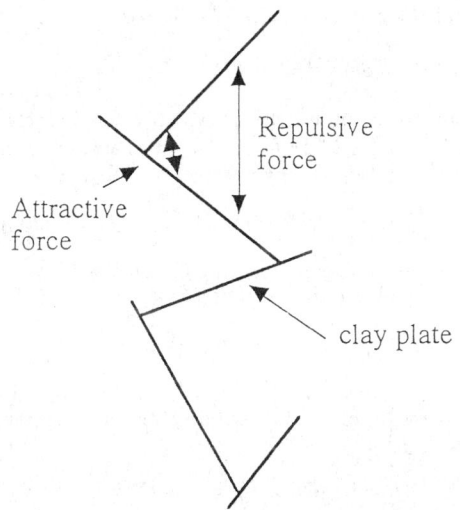

Figure 3. Card house structure of clay.

From the above derivation, we get the gradient dependent model as:

$$\sigma = a\varepsilon - b\varepsilon_{xx} - c\varepsilon^2 \qquad (18)$$

where a, b and c are positive material constants.

Let's discuss about the two-neighbor interactions by the card-house structure of clay (Push[12], Mitchell [13], Shirozu[14]). In the card-house structures, the attractive force is dominant around the points through electrical double layer shown in Figure 3. On the other hand, between the two points far from the point, the repulsive force is significant. This a typical example for the two neighbor interactions introduced in Eqs.(7) and (8). We call this model the second gradient model in the following sections. In the second model, the gradient dependent term works as a stabilizer because of the positive sign of its term and the last term of Eq.(18) plays a softening role. As has been seen in this section, it is evident that the introduction of the gradient term into the constitutive model has a physical nature of the interactions between particles.

3 LOCALIZED SOLUTIONS OF DEFORMATIONS IN GRADIENT DEPENDENT ELASTIC MEDIA

Coleman and Hodgdon[8] and Zbib and Aifantis[15] obtained the analytical solution of shearing motions for gradient dependent material model. If the unloading part is neglected, their model is similar to the second model mentioned in Section 2. Their model is the model for the plastic behavior. However, the models derived in the present study are elastic one because of the loss of loading-unloading conditions. In reality, the considering the unloading region leads to the complicated formulation of the problem. In the present study, we only treat the elastic model for simplicity. The second model considered in the present study is effective for the prediction of the equilibrium solution of viscoplastic model for clay proposed by Oka et al.[9] The overstress type viscoplastic model by Oka et al. can be expressed in one-dimensional form as

$$\dot{\varepsilon}^{vp} = \mu[\sigma - (a\varepsilon^{vp} - b\varepsilon_{xx}^{vp} - 3b\varepsilon^{vp2})] . \qquad (19)$$

Since at equilibrium state, $\dot{\varepsilon}^{vp} = 0$, the solution at the equilibrium state can be obtained by solving the second model (Eq.(18)).

The different method from Coleman and Hodgdon's method, and Zbib and Aifantis' one is used for finding the analytical solutions. The results for the second model is equivalent to the Coleman and Hodgdon's result[8] and the Zbib and Aifantis's one[15].

In this section, we focus on a special class of the

second model in one dimensional form expressed by

$$\sigma = a\varepsilon - b\varepsilon_{xx} - 3b\varepsilon^2 \qquad (20)$$

where a, b are material constants and subscript x denotes the partial differentiation with respect to x.

Now we derive the solution under the condition where stress $,\sigma = c_0$, is constant. This condition corresponds to the one where the homogeneous material is loaded in tension or compression. Since c_0 is constant, the equilibrium equation is automatically satisfied. We set

$$\varepsilon' = \varepsilon - e_0 \qquad (21)$$

in which e_0 is a constant to be determined by the values a, b and c. Upon substitution of ε' into Eq.(20),

$$A\varepsilon' - b(\varepsilon'_{xx} + 3\varepsilon'^2) = 0 \qquad (22)$$

where

$$c_0 = ae_0 - 3be_0^2 \qquad (23)$$
$$A = a - 6be_0 \qquad (24)$$

As for the nonlinear theory on the partial differential equations, Hirota ([16],[17],[18]) developed a powerful method to get the soliton solution in the non-linear wave propagation theory. In the present study, Hirota's derivative is used to solve the governing equations made of the gradient dependent constitutive model.

By introducing the logarithmic transformation and Hirota's differential operator, Eq.(22) can be rewritten as

$$(AD_x^2 - bD_x^4)(f \cdot f) = 0 \qquad (25)$$

where the logarithmic transformation,

$$\varepsilon = 2\frac{\partial}{\partial x^2}\log_e f \qquad (26)$$

is used. In Eq.(25), Hirota's operator is defined by

$$D_x^n(f \cdot g) = (\frac{\partial}{\partial x} - \frac{\partial}{\partial x'})^n f(x)g(x')|_{x=x'} \qquad (27)$$

in which f and g are functions of x.

The feature of Hirota's operator is shown in Appendix. Next, in order to solve Eq.(25), we formally expand the function f as

$$f = 1 + \epsilon f^{(1)} + \epsilon^2 f^{(2)} + \epsilon^3 f^{(3)} + ... \qquad (28)$$

where ϵ is an arbitrary small constant. By substituting Eq.(28) into Eq.(25), we obtain

$$(AD_x^2 - bD_x^4)(1 + \epsilon f^{(1)} + \epsilon^2 f^{(2)} + \epsilon^3 f^{(3)} + ...)\cdot$$
$$(1 + \epsilon f^{(1)} + \epsilon^2 f^{(2)} + ...) = 0 . \qquad (29)$$

From the 0th order terms of ϵ, we get

$$o(1); \quad (AD_x^2 - bD_x^4)1 = 0 \qquad (30)$$

It is easily seen that Eq.(30) is always satisfied. From the 1st order term of ϵ, we get

$$2(A\partial_x\partial_x - b\partial_x^4)f^{(1)} = 0 \qquad (31)$$

The solution of $f^{(1)}$ is

$$f^{(1)} = e^{\eta_1}, \quad \eta_1 = k_1 x + \eta_0 \qquad (32)$$

where k_1 and η_0 are constants.

Upon substitution of Eq.(32) into Eq.(31), the following relation is obtained.

$$k_1 = \pm\sqrt{\frac{A}{b}} \qquad (33)$$

From the 2nd order term of ϵ, we get

$$2(A\partial_x\partial_x - b\partial_x^4)f^{(2)} = -(AD_x^2 - bD_x^4)(f^{(1)} \cdot f^{(1)}) = 0 \qquad (34)$$

From the definition of Hirota's operator, the relation $-(AD_x^2 - bD_x^4)(f^{(1)} \cdot f^{(1)}) = 0$ is always satisfied. From Eq.(31), we get

$$f^{(2)} = e^{\eta_1} \quad or \quad 0 \qquad (35)$$

Using the similar manner, we obtain the relations for the higher order term of ε,

$$f^{(n)} = e^{\eta_1} \quad or \quad 0, \quad (n \geq 2) \qquad (36)$$

From the above derivations, f is obtained by taking the value of ϵ as 1.0 because of the arbitrariness of ϵ.

$$f = 1 + Ce^{\eta_1}, \quad \eta_1 = k_1 x, e^{\eta_0} = C \qquad (37)$$

Substituting Eq.(37) into Eq.(26), ε is obtained as

$$\varepsilon = \frac{2Ck_1^2 e^{k_1 X}}{(1 + Ce^{k_1 X})^2} + e_0 = \frac{2Ck_1^2}{C^2 e^{k_1 x} + 2C + e^{-k_1 x}} + e_0$$

$$= \frac{1}{2}k_1^2 e^{\eta_0} \operatorname{sech}^2[\frac{1}{2}k_1 x + \eta_0] + e_0 \qquad (38)$$

Let's discuss the behavior of the solution expressed by Eq.(38). At $x = 0$,

$$\varepsilon = \frac{2Ck_1^2}{(1+C)^2} + e_0 \qquad (39)$$

On the other hand, at infinity, the strain ε is constant,

$$\varepsilon \to e_0 \quad as \quad x \to \pm\infty$$

Figure 4(a) shows a numerical example of the solution by Eq.(38). Comparing two figures Figure 4(a) and Figure 4(b), the width of localized

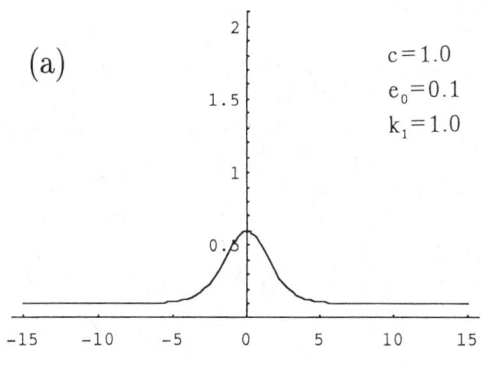

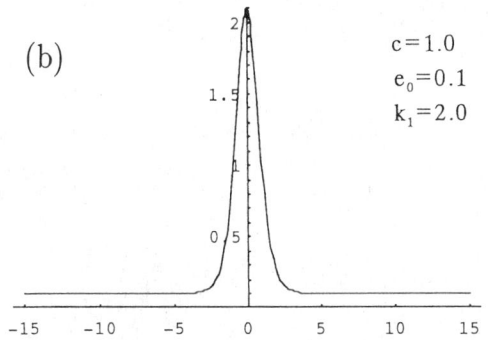

Figure 4. Numerical examples of the solution by Eq.(38).

zone depends on the value of k_1. With increase of k_1, the width of localized zone decreases. This indicates that the parameter k_1 is a measure of the width of shear bands, and the width is determined by the material constants contained in the model. This type of behavior can be called "localized or soliton" solution.

From Eq.(33), k_1 is rewritten by

$$k_1 = \sqrt{\frac{\kappa^2 - \kappa'^2 m^2}{Bh^2} - 6e_0} \quad (40)$$

From Eq.(40), it is evident that the width of localized zone decreases with the decrease of h, the distance of particles. Indirectly, this may show that the localized zone depends of the particle size of the materials. In addition, from Eq.(36), it is evident that the gradient model with only second order gradient term has only one localization solution. It goes without saying that homogeneous solutions satisfy the field relations with gradient dependent model.

4 ACKNOWLEDGEMENT

We wish to express sincere thank to Japanese Society of Promotion of Science for the support of Dr. Mühlhaus's stay in Gifu University in 1993 as a visiting Professor through the Invitation Fellowship Program.

REFERENCES

[1] Aifantis,E.C. (1984), On the microstructural origin of certain inelastic models, ASME, J. Engr. Materials and Tech., 106, 326-330.

[2] Aifantis,E.C. (1987), The physics of plastic deformation, International Journal of Plasticity, 3, 211- 247.

[3] Vardoulakis,I. and Aifantis,E.C. (1991), A gradient flow theory of plasticity for granular materials, Acta Mechanica 87, 197-217.

[4] Mühlhaus,H.-B. and Aifantis,E.C. (1991), The influence of microstructure-induced gradients on the localization of deformation in viscoplastic materials, Acta Mechanica (in Press).

[5] Zbib,H.M. and Aifantis,E.C. (1989), A gradient dependent flow theory of plasticity: Application to metal and soil instability, Appl. Mech. Rev. 42, 295-304.

[6] Thomas,T.Y. (1961), Plastic flow and fracture of solids, Academic Press.

[7] Hill,R. (1962), Acceleration waves in solids, J. Mech. Phy. Solids, 10, 1-16.

[8] Coleman,B.D. and Hodgdon,M.L. (1985), On shear bands in ductile materials, Arch. Rat. Mech. Anal., 90, 219-249.

[9-a)] Oka,F., Yashima,A., Adachi,T. and Aifantis,E.C. (1991), A gradient dependent viscoplastic model for clay and its implication to FEM consolidation analysis, Proc. of 3rd International Conference on Constitutive Laws for Engineering Materials - Theory and Application, ed. by C.S. Desai et al., ASME Press, 313-316.

[9-b)] Oka,F., Yashima,A., Adachi,T. and Aifantis,E.C. (1992), Instability of gradient dependent viscoplastic model for clay saturated with water and FEM analysis, Appl. Mech. Rev., ASME, 45, 3(2), 103-111.

[10] Fermi,E. (1965), Collected Papers of Enrico Fermi, Vol.II, Chicago Univ. Press, 978.

[11] Kunin,I.A. (1982), Elastic Media with Microstructure I, One-Dimensional Models, Solid-State Sciences 26, Springer-Verlag, Berlin-New York.

[12] Push,R. (1970), Microstructural changes in soft quick clay at failure, Canadian Geotechnical Journal, 7(1), 1-7.

[13] Mitchell,J.K. (1976), Fundamentals of Soil Behavior, John Wiley & Sons.

[14] Shirozu,H. (1988), Introduction to Clay Mineralogy, Asakura Pub. Co. (in Japanese).

[15] Zbib,H.M. and Aifantis,E.C. (1988), On the localization and Postlocalization Behavior of Plastic deformation. II. On the Evolution and Thickness of Shear Bands, Res. Mechanica, 23, 279-292.

[16] Hirota,R. (1971), Exact solution of the Korteweg-de Vries equation for multiple collisions of solitons, Phys. Rev. lett., 27, 1192-1194.

[17] Hirota,R. (1976), Direct methods of finding exact solutions of nonlinear evolution equations, in Bäcklund Transformations, R.M. Miura, ed., Lecture Notes in Mathematics 515, Springer-Verlag, New-York.

[18] Ablowitz,M.J. and Segur,H. (1985), Solitons and The Inverse Scattering Transform, SIAM.

Appendix

The feature of the Hirota's operator is as follows: For example, in the case of $n = 2$,

$$D_x^2(f \cdot g) = (\frac{\partial}{\partial x} - \frac{\partial}{\partial x'})^2 f(x)g(x') |_{x=x'}$$
$$= \frac{\partial^2 f}{\partial x^2}g - 2\frac{\partial f}{\partial x}\frac{\partial g}{\partial x} + f\frac{\partial^2 g}{\partial x^2}.$$

6 Numerical modelisation
  Modélisation numérique

Numerical modelling for the behaviour of an elastic medium in the presence of a discontinuity for geotechnical applications

Modélisation numérique du comportement d'un milieu élastique en présence de discontinuités, pour application dans le cadre de la géotechnique

E. Sakellariadi & G. Scarpelli
University of Ancona, Italy

ABSTRACT: The behaviour of stiff overconsolidated clays is often determined not so much by the constitutive properties of the material itself, but rather by the presence of discontinuities. Furthermore the porous nature of the soil can influence the way it interacts with existing discontinuities. In this paper a model adequate for such materials is presented, in which a rigorous mathematical analysis based on the theory of dislocations is adopted for analysing the behaviour of a discontinuity and its interaction with the continuous medium, while the soil mass itself is modelled using straightforward finite element techniques and a linear elastic constitutive law. The two formulations are coupled by means of the principle of superposition. An incremental step-by-step procedure is introduced, through which some aspects of poro-elastic behaviour can be taken into account. Results are obtained in terms of soil displacement field, on the basis of which it seems possible to sustain that a simple linear elastic constitutive law may be adequate for modelling the behaviour of stiff clay traversed by discontinuities, provided that these are correctly taken into account. Moreover the algorithm here proposed is advantageous from a numerical-computational point of view. Two applications of the model are presented, namely a direct shear box modelisation and a well-known real excavation problem.

Dans bien des cas, le comportement des argiles raides surconsolidées n'est pas tant déterminé par les propriétés rhéologiques du matériau lui-même, que par par la présence de discontinuités. De plus, la façon dont le matériau interagit avec les discontinuités existantes peut être influencée par sa nature poreuse. On présente dans cet article un modèle adéquat pour de tels matériaux, dans lequel on adopte une analyse mathématique rigoureuse basée sur la théorie des dislocations pour analyser le comportement d'une discontinuité et son interaction avec le milieu continu; la masse du sol elle-même est modélisée en utilisant la méthode des éléments finis classique, et une loi élastique linéaire. Les deux formulations sont couplées par le principe de superposition. Une procédure incrémentale pas à pas est introduite, qui permet de prendre en compte quelques aspects de comportement poro-élastique. Les résultats sont obtenus en terme de champs de déplacements, sur la base desquels il paraît possible de considérer qu'une loi de comportement linéaire élastique simple peut convenir pour modéliser le comportement d'une argile raide traversée par des discontinuités, à condition qu'on tienne compte correctement de ces discontinuités. L'algorithme proposé ici comporte en outre des avantages d'un point de vue numérique. On présente deux applications du modèle, à savoir une modélisation d'une boîte de cisaillement direct et un problème d'excavation.

1 INTRODUCTION

The behaviour of stiff overconsolidated clays is often determined not so much by the constitutive properties of the material itself, but rather by the presence of discontinuities. In fact, it is well known from experimental evidence and everyday practice that in such materials the existence of planes of discontinuity, either entirely contained in the medium or intersecting the boundary, constitutes a constraint for displacements. Once these planes have formed the behaviour of the entire structure is determined by the relative displacements that can develop along these surfaces and depends on the way the discontinuities themselves behave, for example whether they propagate or not; the remaining part of the soil appears hardly to be deforming at all. Strain localisation, developing for some reason along certain planes in the medium, can be thought of as such a discontinuity.

Existing literature and engineering design methods adopted in these cases come under two separate groups; these are on the one hand the methods based on limit equilibrium techniques, and on the other the recently developed sophisticated models based on finite element discretizations. In the first case, the existence of failure surfaces or slip bands is postulated, together with the meeting of some strength criterion along these surfaces, while the rest of the soil is assumed to be rigid or elastic. Classical soil mechanics is greatly based on such approaches. In the second case, the evolution of computer-based numerical analyses has rendered possible new techniques based on discretization processes with the use of very sophisticated constitutive laws. While maintaining a continuum approach for a solid containing discontinuities, these methods attempt to correctly model its behaviour through the use of particular constitutive or numerical algorithms that are capable of reproducing the effects of strain localisation.

The approach presented in this work is intermediate to those mentioned above. The discontinuity surfaces are directly modelled as such, by means of a surface integral formulation in which the theory of dislocations is used in order to represent the discontinuity as a mathematically equivalent distribution of edge dislocations. The resulting equations are solved numerically by an algorithm which provides the exact analytical solution for certain collocation points. The soil mass is modelled by means of the finite element method, using a simple linear elastic constitutive law; this discretized model is numerically coupled to the surface integral model for the discontinuity, thus obtaining a global solution in which the behaviour of the soil mass and that of the discontinuity are analysed in separate ways but in such a fashion that the one takes the other correctly into account.

The algorithm has been implemented into a computer code which is in part derived from the existing code MULTIFRAC, developed at the M.I.T. in relation to hydraulic fracturing of inhomogeneous rock layers. In its current form the computer program can tackle the stress and deformation analysis of an elastic medium containing one or more discontinuity surfaces. The medium can be either an infinite or a finite one, and it can be loaded by any kind of surface or volume loads, including self-weight. The discontinuity surfaces can be modelled either as opening fractures or as shear bands, and both an internal (hydraulic) pressure and a relative shear strength across the surface can be taken into account; the transmission of compressive normal stresses across the two surfaces is also correctly modelled. The discontinuities can be arbitrarily shaped and can be either fully contained in the medium or intersect its boundary. The porous nature of the medium can be taken into account by means of appropriate fundamental solutions used in the discontinuity modelisation; in this case an incremental step-by-step procedure is performed.

Fracture Mechanics concepts are contemplated, in that stress intensity factors are correctly computed, and their usefulness in comparing different configurations has been checked, although currently no calculations are made regarding the propagation of discontinuity surfaces. In fact, the object of the applications of the algorithm presented in this paper is to correctly compute the displacements of a continuum containing discontinuities, the aim being to investigate into the possibility of assuming very simple constitutive laws for a deforming medium on condition that the discontinuities are correctly taken into account. An affirmative answer would imply that sophisticated constitutive laws may not be indispensable when modelling overconsolidated clays traversed by fissures, shear bands, or other forms of discontinuities, provided that an adequate model can be used for the latter.

Moreover the algorithm here presented, apart from giving accurate results regarding displacement calculations, is also advantageous from a numerical-computational point of view. In fact, though the discontinuity may propagate or change its relative configuration, the finite element mesh remains the same and therefore computation time is reduced to a minimum.

The general methodology of the model is illustrated in the next two sections; in section 2 an outline of the analytical formulation adopted for a discontinuity is given and a possible algorithm for its solution is suggested. Both elastic and poro-elastic media are contemplated. Section 3 illustrates how this formulation is coupled with the finite element method. Two examples regarding respectively a direct shear box test and an excavation problem are illustrated in section 4. Finally in section 5 a few concluding remarks and considerations regarding future intentions are made. It is to be kept in mind throughout that the main aim of this work is to formulate a model which can be effectively used for the analysis of actual engineering problems.

2 ANALYTICAL FORMULATION; NUMERICAL SCHEME FOR DISCONTINUITIES

2.1 *Analytical formulation*

A displacement discontinuity surface can be modelled as a continuous distribution of edge dislocations (Bilby and Eshelby 1968; Rice 1968). On these grounds a singular integral equation can be formulated, the solution of which provides a function linked to the dislocation density, from which the relative displacements along the discontinuity surface can be computed by integration. If the appropriate influence functions describing stress fields and displacements due to unit edge dislocations are known, it is possible to determine the way in which a discontinuity surface modifies the response of a continuum to applied external loads.

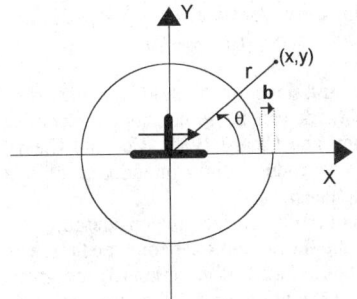

Figure 1: Edge dislocation in the X direction, at the origin of coordinates.

An edge dislocation singularity (see Figure 1) is quantified by a vector **b** called the Burgers vector. In modelling a discontinuity surface as a continuous distribution of edge dislocation singularities, the relative displacement of a point on the discontinuity is described by a vector **b**, or by its components b_β; for plane analyses, as performed in this work, β takes on values 1 and 2 representing X and Y directions respectively.

A continuum containing M discontinuities is considered, as in Figure 2. For a point **x'** on one of these surfaces, the strength of dislocation over an infinitesimal segment $d\Gamma_m$ is given by:

$$db_\beta = \delta_\beta(\mathbf{x'}) \, d\Gamma_m \qquad (1)$$

where $\delta_\beta(\mathbf{x'})$ represents the β-th component of dislocation density at point **x'**.

Assuming that fundamental solutions for the elementary problem of an edge dislocation placed in an infinite medium are known, it is possible to obtain the stress state generated at any point **x** of the medium by the presence of the system of M discontinuities by integrating over these surfaces, as follows:

$$\sigma_\alpha(\mathbf{x}) = \sum_{m=1}^{M} \int_{\Gamma_m} \Psi_\sigma^\delta(\mathbf{x},\mathbf{x'}) \, \delta_\beta(\mathbf{x'}) \, d\Gamma_m \qquad (2)$$

where:
Ψ : influence functions, given by the fundamental solutions for the problem of an edge dislocation at a point in an elastic or in a poro-elastic medium;
$\delta_\beta(\mathbf{x'})$: β-th component of dislocation density at point **x'**. Note that for **x'** very near the tip of the discontinuity $\delta(\mathbf{x'})$ tends to infinity;
$\sigma_\alpha(\mathbf{x})$: α-th component of stress at point **x**;
Γ_m : surface of m-th discontinuity;
x : point of the medium (internal, or on the boundary);
x' : point on a discontinuity surface.

For the sake of simplicity, the assumption M=1 can be made. Equation (2) becomes:

$$\sigma_\alpha(\mathbf{x}) = \int_\Gamma \Psi_\sigma^\delta(\mathbf{x},\mathbf{x'}) \, \delta_\beta(\mathbf{x'}) \, d\Gamma \qquad (3)$$

Using equation (3) it is therefore possible to calculate the stress state caused by a discontinuity surface at any point of the medium, providing the functions Ψ and $\delta(\mathbf{x'})$ are known. Although this is true for the influence functions Ψ, the dislocation density function $\delta(\mathbf{x'})$ is an unknown. To calculate this function it is necessary to write equation (3) with reference to points **x** where the stresses are known or can be imposed a priori; in this way equation (3) becomes an integral equation with $\delta(\mathbf{x'})$ as the unknown function. Such points are all points lying on the discontinuity surface: in fact, at these points normal stress components are zero if tensile (corresponding to a mode I or opening mode of loading) while are left unaffected by the discontinuity if compressive; shear stress components are more complex to express, but still they can be determined by means of a constitutive law linked to the intrinsic behaviour of the discontinuity. Such a constitutive law may for example be the one proposed by Palmer and Rice (1973) in which the shear components depend on the relative shear displacement of points on opposite sides of the discontinuity surface. For simplicity in this work the shear components of stress on the discontinuity surface are assigned a constant value, corresponding to the material's residual shear strength. A zero value implies a freely slipping discontinuity, corresponding to mode II (sliding mode of loading). If a more complex constitutive behaviour was taken into account, and in particular one in which the shear stress values depend on the solution (i.e. the relative displacements), a suitable iterative procedure to solve the non-linear problem would be

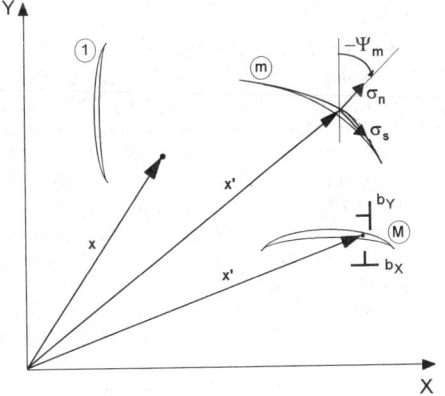

Figure 2: Plane system of M discontinuities.

required. Moreover it is to be noted that, even in the Palmer and Rice model for τ-δ behaviour, shear stresses are constant and equal to a residual shear strength for the larger part of the shear band, while they depend on relative displacement only at points lying on a small zone (called the end zone) near the tip of the band.

The equation for determining the dislocation density function $\delta(x')$, obtained by writing equation (3) for a certain number of points lying on the discontinuity surface, is a singular integral equation due to a singularity of the type $(x-x')^{-1}$ presented by the Ψ influence functions and of the type $1/(r^{1/2})$ presented near the tip (point where $r = 0$) by the unknown dislocation density function. Section 2.3 deals with the numerical solution of this equation.

It is interesting to note that the singular equation which was here obtained through a physical consideration of the problem and a surface integral approach, is formally similar to the equation obtained as the result of the mathematical formulation of general boundary value problems using the boundary element method in its indirect approach. In fact it is known that, while in the direct approach of the boundary element method the unknown functions correspond to the actual physical variables of the problem, in the indirect approach these unknown functions are represented by some fictitious source densities, while the values of the physical variables are consequently obtained by integration (Aliabadi and Rooke 1991, Brebbia and Dominguez 1989). More specifically, the indirect approach uses the Somigliana identity and, by defining an integral equation for the reference state and one for the complementary domain, arrives at a formulation of the problem by means of the singular integral equation:

$$t_1^i = \int_\Sigma p_{lk}^* \mu_k \, d\Sigma \qquad (4)$$

in which:
t_1^i: traction components on the boundary;
p_{lk}^*: fundamental solution for the point force problem;
$\mu_k \, d\Sigma$: fictitious source densities which can be interpreted as point dislocations.

The above equation is valid for the solution of general boundary value problems for a continuous medium and is not associated specifically to discontinuity problems. In fact the boundary element method cannot be directly applied for the formulation of fracture problems, in that a singularity arises making the problem ill-posed, as first demonstrated by Cruse (1972).

2.2 Influence functions; formulation of the problem for elastic and poro-elastic media

The analytical expressions for the influence functions Ψ used in this work can be found in the literature (Greene, Annigeri and Cleary 1983; Rice and Cleary 1976); they are reproduced in Appendix C of this work for completeness.

When the discontinuity surface is considered to be interacting with an elastic homogeneous medium, the influence functions $\Psi(x,x')$ of equation (3) are given by the fundamental solution expressing stress at point x of an elastic medium caused by a unit edge dislocation at point x'.

When the discontinuity surface is considered to be interacting with an elastic porous medium, such as described by Biot's theory (Biot 1940), the influence functions $\Psi(x,x',t)$ are given by the fundamental solution expressing the generalised stress field at point x of a poro-elastic medium at time t, caused by a unit singularity that appeared at point x' at time $t = 0$ and was kept constant for all time $t > 0$. In this case by singularity either an edge dislocation $b(x')$ or a point fluid source $q(x')$ can be meant (see Figure 3). In this way both the geometrical configuration of the discontinuity and the pore fluid flow across it are modelled. Denoting by $\delta(x')$ the density of dislocations and by $\theta(x')$ the density of fluid sources, and considering a generalised stress field which has as its components shear and normal stresses σ_α and pore water pressure p, the following system of integral equations is formulated in place of equation (3):

$$\begin{cases} \sigma_\alpha(x,t) = \int_\Gamma \left[\Psi_\sigma^\delta(x,x',t) \, \delta_\beta(x') + \Psi_\sigma^\theta(x,x',t) \, \theta(x') \right] d\Gamma \\ p(x,t) = \int_\Gamma \left[\Psi_p^\delta(x,x',t) \, \delta_\beta(x') + \Psi_p^\theta(x,x',t) \, \theta(x') \right] d\Gamma \end{cases} \qquad (5)$$

In this case the influence functions have different orders of singularity, and precisely Ψ^δ are singular as $(x-x')^{-1}$, while Ψ^θ only as $\ln(x-x')$. Numerical solution of this system of equations is more complex. In fact, while there are examples of solutions for coupled singular integral equations all with the $(x-x')^{-1}$ singularity kernels, and a number of solutions can be found in the literature on the numerical solution of integral equations with a $\ln(x-x')$ kernel, there appears to be little practical experience with coupled equations of the kind of equation (5) (Cleary 1978). Furthermore, the numerical scheme adopted in this work is based on the possibility of exploiting the known $r^{-1/2}$ singularity of the density δ near the discontinuity tip, a technique which proved to be extremely convenient regarding both the computation

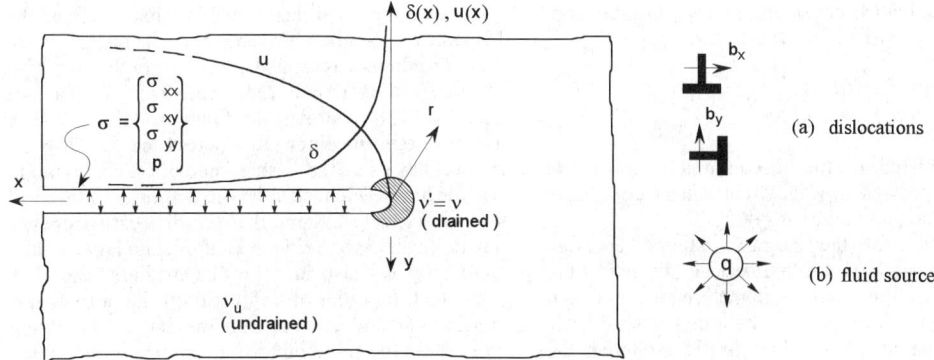

Figure 3: Discontinuity in a poro-elastic medium (adapted from Cleary 1978).

of the stress intensity factors and the accurate solution of the equation (see section 2.3). In the case of θ(x') however there is no well-known singularity as r→0; it is not therefore possible to use the same numerical scheme for equation (5).

Consequently, and in accordance with the mathematical analyses proposed by Cleary (1978), in the present work we concentrate on a class of problems where the fluid source density has a lesser degree of importance than dislocation density. The reduced version of equation (5) then takes the form:

$$\sigma_\alpha(\mathbf{x},t) = \int_\Gamma \Psi_\sigma^\delta(\mathbf{x},\mathbf{x}',t) \delta_\beta(\mathbf{x}',t) \, d\mathbf{x}' \qquad (6)$$

similar to equation (3) but with an additional time dependence. Formulating the problem by means of equation (6) amounts to the assumption that pore pressure or rate of fluid flow across the discontinuity is to be accepted as whatever equation (5b) implies when θ(x') = 0 (Cleary 1978). The physics of the discontinuity in this case is such that there are no fluid sources along its sides and therefore any fluid exchange occurring between the discontinuity and the continuum is neglected.

Assuming σ(x) constant in time for points on the discontinuity surface, as would be for all cases of imposed tractions as discussed in section 2.1, an incremental step-by-step procedure is formulated for the numerical solution of equation (6). The only assumption made is that the values of δ evolve smoothly in time i.e. as a continuous function. This should be so, considering the physical meaning of the evolution of δ: assuming that the response of the medium is not instantaneous implies that, as the effects of the discontinuity are being felt at the points of the medium, the displacement configuration (hence the values of δ) change to accommodate imposed stresses.

At the first step, relative to time $t = t_0$ when the discontinuity (or discontinuity segment) appears, the following integral equation must be solved:

$$\sigma_\alpha(\mathbf{x},t_0) = \int_\Gamma \Psi_\sigma^\delta(\mathbf{x},\mathbf{x}',t_0-t_0) \cdot \delta_\beta(\mathbf{x}',t_0) d\Gamma \qquad (7)$$

obtaining values of $\delta_\beta(\mathbf{x}',t_0)$. From these the relative displacements along the discontinuity surface and stress intensity factors can be calculated, corresponding to time $t = t_0$; these should be equal to the values that would result from the solution of integral equation (3) when the elastic properties of the soil are appropriate to describe its behaviour in undrained conditions.

At the next steps, relative to times $t = t_N$ (N=1,2,....), the integral equation becomes:

$$\sigma_\alpha(\mathbf{x},t_N) = \int_\Gamma \Psi_\sigma^\delta(\mathbf{x},\mathbf{x}',t_N-t_0) \cdot \delta_\beta(\mathbf{x}',t_0) d\Gamma +$$
$$+ \sum_{i=1}^{N} \int_\Gamma \Psi_\sigma^\delta(\mathbf{x},\mathbf{x}',t_N-t_i) \cdot \Delta^{t_i}\delta_\beta(\mathbf{x}') d\Gamma$$
$$(N=1,2,...) \qquad (8)$$

where $\Delta^{t_i}\delta_\beta(\mathbf{x}') = \delta_\beta(\mathbf{x}',t_i) - \delta_\beta(\mathbf{x}',t_{i-1})$.

Values for $\Delta^{t_N}\delta_\beta(\mathbf{x}')$ are obtained at each time step t_N, having calculated $\Delta^{t_i}\delta_\beta(\mathbf{x}')$ for $i = 1,2,...,N-1$ at the previous steps. Consequently, at the generic time $t = t_N$ the values of the dislocation density function can be calculated from:

$$\delta_\beta(\mathbf{x}',t_N) = \delta_\beta(\mathbf{x}',t_0) + \sum_{i=1}^{N} \Delta^{t_i}\delta_\beta(\mathbf{x}'). \qquad (9)$$

Relative displacements and stress intensity factors for $t = t_N$ can be therefore computed. After a certain number of time steps (i.e. for t→∞) these values should tend to those resulting from the solution of integral equation (3) considering elastic properties of the soil as appropriate for drained conditions.

2.3 *Numerical implementation*

Variables related to the discontinuity surface are hereafter expressed in a local curvilinear coordinate system s, such that at the tip $s = \pm 1$.

The solution of the singular integral equation requires some kind of numerical discretization. Following a convenient scheme (Narendran and Cleary 1984), where the known order of singularity of the dislocation density function is exploited, the unknown function $f(x')$ is defined by means of:

$$\delta_\beta(s') = \frac{f_\beta(s')}{\sqrt{1-s'^2}} \qquad (10)$$

In this way the unknown function is no longer singular at the discontinuity tips. The integral equation (equation (3) or (6)) written for a point lying on the discontinuity surface transforms into:

$$\sigma_\alpha(s) = \int_{-1}^{1} \Psi_\sigma^\delta(s,s') \frac{f_\beta(s')\,ds'}{\sqrt{1-s'^2}} \qquad (11)$$

Keeping in mind that the influence functions Ψ present a singularity of the order $1/(s-s')$, this equation can be expressed as an equation of the type:

$$\sigma_\alpha(s) = \int_{-1}^{1} \frac{f_\beta(s')\,ds'}{\sqrt{1-s'^2} \cdot (s-s')} \qquad (12)$$

Like any other integral equation, also this equation could be solved numerically using collocation techniques and suitable interpolation functions. The accuracy of the solution obtained would then depend on the choice of those functions and of collocation points. However there exists in literature (Erdogan and Gupta 1972) a particular method for the solution of singular integral equations specifically of the type of equation (12). Proceeding by this method, which is still based on collocation techniques, it can be proven that the particular choice of collocation points and interpolating functions is such that the solution obtained at those points corresponds to the exact analytical solution of the problem. This method, based on certain properties of the Chebyshev orthogonal polynomials, is consistently adopted in this work in tackling the surface integrals related to formulations of discontinuity problems; its details are illustrated in Appendix B. Therefore, if equation (10) is assumed as accurate, the solution obtained for the dislocation density function is mathematically exact, i.e. there is no error linked to the adopted discretization technique.

It should be noted that, while in what concerns the fundamental solution for an edge dislocation in an elastic medium it is immediate to verify that the type of integral equation that emerges is that of equation (12), regarding the fundamental solution in a poro-elastic medium (Rice and Cleary 1976) the applicability of the same numerical integration scheme has been checked by the authors. To that end it is sufficient to assume that the dislocation density function will have the same kind of singularity at discontinuity tips also in a porous medium, and then prove that the order of singularity of the poro-elastic influence functions is the same as that of the elastic ones; these too therefore can be approximated taking a series of Chebyshev polynomials as interpolation functions (see Appendix B).

Using the Chebyshev scheme the singular integral equation (11) is therefore transformed into the linear algebraic system:

$$\mathbf{C}f = \mathbf{T} \qquad (13)$$

where:
C : matrix of coefficients derived through the application of the Chebyshev scheme method;
T : known tractions on the discontinuity surface;
f : values of the unknown function at N points on the discontinuity surface.

No boundary conditions have yet been mentioned regarding matrix equation (13). Moreover, boundary conditions become necessary by the fact that the Chebyshev scheme here adopted is in such a way formulated that it supplies one equation less than the number of unknowns (see Appendix B). Narendran and Cleary (1984) provide a review of boundary conditions that should be used; these are the "closure" and the "matching" conditions, regarding respectively cracks entirely contained in a medium and the intersection of two crack segments.

In the present work, two boundary conditions have been used. The first, regarding discontinuities that are entirely contained in the medium (i.e. do not intersect the finite boundary), is the "closure condition" of Narendran and Cleary (1984); this condition, expressed by means of:

$$\int_{-1}^{1} \delta_\beta(s')\,ds' = 0 \qquad (\beta=1,2) \qquad (14)$$

which transforms into:

$$\sum_{k=1}^{N} f_\beta(s'_k) = 0 \qquad (\beta=1,2) \qquad (15)$$

assumes that there should be no net "entrapped" dislocation and as a consequence expresses a mass conservation principle. The second, adopted when a discontinuity surface intersects the boundary of the medium, in accordance with the theory of dislocations (Bilby and Eshelby 1968), is expressed by means of:

$$f(0) = 0 \qquad (16)$$

and imposes that the dislocation density should be zero at the point where the discontinuity intersects the boundary.

Once the values of f are calculated at the collocation points, the dislocation densities are computed at these points via equation (10), and hence, by integration, the relative displacements of the points on the discontinuity surface. The values obtained for the dislocation density function can also be used in equation (3) (or (6)) in order to obtain stresses at any point of the medium due to the presence of the discontinuity.

2.4 Stress intensity factors

Stress intensity factors can be easily calculated, simply by multiplying the value assumed by the unknown function f at the tip of the discontinuity by a constant which is a function of material properties and of the length of the discontinuity. In fact, while it is well known that the stress intensity factors give a measure of the stress field singularity, which is of the $r^{-1/2}$ type, it is also clear from equation (10) that the f function measures the strength of the dislocation density singularity at the tip of the discontinuity, singularity which is of the same type as that of the stress field. In other words (see Narendran and Cleary 1984), expressing the stress intensity factors in terms of normal and shear stresses as follows:

$$K_\alpha(\mathbf{x}_m) = \lim_{(x \to x_m)} \left[\sqrt{2\pi |\mathbf{x} - \mathbf{x}_m|} \; \sigma_\alpha(\mathbf{x}) \right] \qquad (17)$$
$$(\alpha = 1, 2)$$

where $\mathbf{x}_m$: tip of m-th discontinuity,

these are related to the values of the unknown functions at the tip by:

$$\begin{Bmatrix} K_I(\mathbf{x}_m) \\ K_{II}(\mathbf{x}_m) \end{Bmatrix} \cong C \cdot \begin{bmatrix} -\cos\psi_m & -\sin\psi_m \\ \sin\psi_m & -\cos\psi_m \end{bmatrix} \begin{Bmatrix} f_1(s_m) \\ f_2(s_m) \end{Bmatrix} \qquad (18)$$

in which C contains material constants and the length of the discontinuity. Rotation of reference systems by the angle ψ_m is necessary because while stress intensity factors K_I and K_{II} refer to opening and sliding directions in the discontinuity local reference system, f_1 and f_2 are computed in the global X and Y directions (see Figure 2). Stress intensity factors are therefore computed by means of equation (18). The value of f at the collocation point nearest to the tip is used. This is in fact very accurate, even when working with a very small number of nodes, because the choice of collocation points as the zeroes of Chebyshev polynomials is such that an accumulation of these points near the tip results.

It should be noted that in the case of a poro-elastic medium the constant C is computed assuming effective stress values for the material constants E and ν; this implies that the infinitesimal zone immediately ahead of the tip is assumed to be always in drained conditions (Figure 3). Such an assumption, originally formulated by Cleary (1978), is necessary in order to calculate a value for the stress intensity factor, seems reasonable and the results obtained have not brought up any evidence against it.

Benchmark problems dealing with opening (mode I) fractures have been solved (Annigeri 1984) using the above formulation; the stress intensity factors obtained show excellent accordance with theoretical results, when such are available.

It should be noted however, that although in classical fracture mechanics and in mode I loading, on which all the benchmark cases are based, the stress intensity factors depend only on the geometry of the configuration and on external loading, in the type of problems that are representative of discontinuity surfaces in clays and that are investigated in this work there are more parameters to be taken into consideration. Namely, when shear strength across the discontinuity surface is accounted for and compressive stresses can be transmitted, the value of the stress intensity factors will change. In particular, K_{II} should decrease, and in the limit would tend to zero; this would correspond to the fact that stresses across the discontinuity surface, acting as internal constraints, have impeded all relative displacements.

Problems of this kind have been tackled (Aliabadi and Rooke 1991) using contact problem techniques; it should be borne in mind however that, in contact problems, the fracture mechanics basics lack physical sense. The algorithm has proven capable of handling this kind of problem correctly. This is illustrated in section 4 with reference to an example of a direct shear box modelisation.

In conclusion, the scheme illustrated is capable of correctly modelling a medium containing one or more discontinuities; these can be of any shape and arbitrarily collocated. Their mutual influence is also correctly accounted for. Coupling of this model with a finite element scheme, as illustrated in the next section, also permits to correctly model the mutual influence between the discontinuities and the external boundary surfaces of the medium.

3 FORMULATION OF COMPLETE ALGORITHM

In the previous section an algorithm capable of modelling a discontinuity surface in an infinite medium was discussed. In order to deal with real problems this algorithm, based on a surface integral method, is coupled with a finite element scheme. This is appropriate, in that both methods deal with the computation of the displacements of certain points in the continuum, when the applied loads or stresses are known. It should be kept in mind however that there is a difference in the degree of approximation of the two procedures: in particular, in the surface integral formulation there is no discretization of the medium into elements, and the exact fundamental solution is used. This fundamental solution is derived on the assumption of elastic behaviour for the medium. Finite elements on the other hand rely on discretization of the continuum and the results obtained can depend on the mesh and type of element adopted.

There are various examples in the literature of coupling techniques between the finite element and the boundary element methods, the latter being, as has been shown, formally linked to the surface integral formulation used here. These techniques rely on the definition of an interface between boundary integral and finite element domains. However, numerical inaccuracies can often arise when the necessary continuity conditions for stresses and compatibility for displacements across this interface are imposed. The algorithm developed in this paper eliminates this difficulty by using an alternative technique for the coupling of the two formulations, based on the principle of superposition, and thus avoiding the stumbling block of further requirements on stresses and strains. Applicability of the principle of superposition is justified within a linear elastic context, which in turn is a basic assumption of the formulation.

Figure 4 illustrates how the principle of superposition is applied in order to couple the surface integral to the finite element formulation. The actual problem of a body containing discontinuity surfaces is resolved into two sub-problems which are: 1) the same medium containing no discontinuities and 2) the discontinuities alone in an infinite medium. Application of the principle of superposition, as originally discussed by Annigeri (1984), is illustrated in Appendix A. The final resulting coupled matrix equation is the following:

$$\begin{bmatrix} K & G \\ S & C \end{bmatrix} * \begin{Bmatrix} U \\ f \end{Bmatrix} = \begin{Bmatrix} F \\ T \end{Bmatrix} \quad (19)$$

where:
K : stiffness matrix relative to the finite element discretization;
C : coefficient matrix resulting from the numerical discretization of the surface integral formulation using the Chebyshev scheme, as illustrated in the previous section and in Appendix B;
G : coupling matrix obtained through a similar numerical discretization of equation (3) (which in this case is not a singular integral equation);
S : coupling coefficient matrix, based on the finite element constitutive law;
U : displacements of the finite element nodes;
f : values at the collocation points of the unknown functions related to the discontinuity, defined by means of equation (10);
F : consistent nodal forces relative to the finite element discretization;
T : constant stresses, known or imposed, acting on the discontinuity surface. In the present formulation these do not depend on the solution, i.e. on the relative displacements of the discontinuity surface.

In the case when the interaction between the medium and the discontinuity is modelled as purely elastic (instant response), it is sufficient to solve the matrix equation (19) once. The solution, in terms of displacements of the finite element nodes and dislocation densities on the discontinuity surface, is directly obtained. Numerical integration of the latter gives the relative displacements on the discontinuity surface. The coefficient matrix of equation (19) is not symmetric, and a general diagonalisation technique appropriate for non-symmetric matrices is adopted.

Alternatively, in the case when the interaction between the medium and the discontinuity is

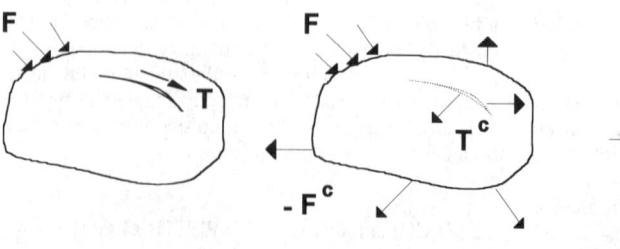
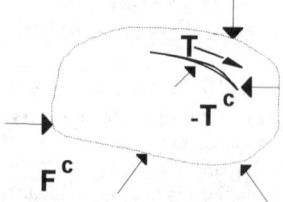

Actual problem ≡ Finite body without discontinuity + Discontinuity in infinite medium

Figure 4: Application of the principle of superposition.

modelled as poro-elastic (time dependant response), a step-by-step procedure based on the same assumptions as those discussed in detail in section 2.2 is adopted. In matrix form, the equation to be solved at each time step t_N is the following:

$$\begin{bmatrix} K & G_N \\ S & C_N \end{bmatrix} * \begin{Bmatrix} \Delta^{t_N} U \\ \Delta^{t_N} f \end{Bmatrix} = \begin{Bmatrix} F \\ T \end{Bmatrix}_{t_N} - \begin{bmatrix} K & G_0 \\ S & C_0 \end{bmatrix} * \begin{Bmatrix} U \\ f \end{Bmatrix}_{t_0} -$$

$$- \sum_{i=1}^{N-1} \begin{bmatrix} K & G_i \\ S & C_i \end{bmatrix} * \begin{Bmatrix} \Delta^{t_i} U \\ \Delta^{t_i} f \end{Bmatrix} \quad (20)$$

where G_X, C_X: coefficient matrices based on influence functions of the type $\Psi(x,x',t_N-t_X)$.

Displacements of the finite element nodes and relative displacements on the discontinuity can be obtained at each time t_N by summing the differences these quantities have undergone between time t_{N-1} and time t_N to the values calculated in relation to time t_{N-1}. The solution for $t = t_N$ is obtained through:

$$\begin{cases} \delta_\beta(x',t_N) = \delta_\beta(x',t_0) + \sum_{i=1}^{N} \Delta^{t_i}\delta_\beta(x') \\ U(t_N) = U(t_0) + \sum_{i=1}^{N} \Delta^{t_i} U \end{cases} \quad (21)$$

Relative displacements and stress intensity factors for $t = t_N$ can be therefore computed. For $N = 0$ ($t = t_0$) the solution corresponding to undrained conditions is obtained, while for $N \to \infty$ the displacements tend towards constant values, corresponding to drained conditions. The coefficient matrix at each time step is the same, while the right hand side vector has to be adjusted each time.

As already mentioned, it should be emphasised that this is an extremely simplified analysis in what regards the poro-elastic nature of the medium, in that this aspect is taken into account only regarding the time dependent effects of the discontinuity on the global stress-strain state, while the analysis of the medium itself disregards all diffusion processes. However, being the objective that of correctly taking into account the effect of the presence of discontinuities on the global stress-strain state, and in particular the role played in this effect by the pore pressure and local diffusion processes, and having obtained plausible results throughout, the analysis is considered by the authors as a successful first step towards a correct numerical modelisation of the problem.

In the applications illustrated in the next section this behaviour is shown by means of plots of the evolution in time of the stress intensity factor and of the relative displacement at the discontinuity origin, for a discontinuity surface in an infinite medium and for one originating at the lateral surface of a direct shear box.

4 APPLICATIONS

Regarding numerical implementation, the code MULTIFRAC originally developed at the M.I.T. for analysing hydraulic fracturing of deep rock layers was used. The code has been modified by the authors, in order to handle correctly the analysis of discontinuity surfaces in soil masses. Principal modifications regard: tractions on the discontinuity surface i.e. the possibility to transmit shear and compressive stresses; interaction of the discontinuity surface with the finite element boundary limiting surface via the closure condition mentioned in section 2.3; poro-elasticity, having included the influence functions of a poro-elastic medium and developed the incremental step-by-step procedure. Calculation of stress intensity factors was also modified.

MULTIFRAC also treats propagation problems linked to internal pressure and pumped fluid flow in a fracture; these parts of the code have not been used in the present work.

Two applications are illustrated, dealing with an excavation problem and with a direct shear box modelisation.

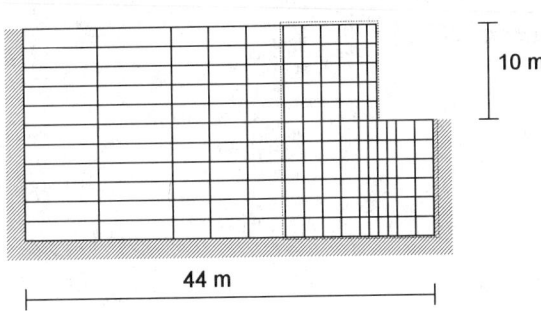

Figure 5a: Finite element mesh for excavation problem.

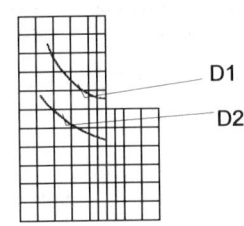

Figure 5b: Position of discontinuities.

4.1 Excavation

The basic geometry for this example was taken from an excavation actually performed in Chicago and which has been monitored and analysed by various authors (Finno, Atmatzidis and Perkins 1989; Finno and Nerby 1989). The choice of this particular example was based on the fact that available measurements indicate the presence and location of zones of shear localisation, which can be modelled as discontinuity surfaces. This example was analysed using only the elastic scheme.

The finite element mesh adopted is shown in Figure 5a. The excavation is 10 metres deep and the material constants are the following: $\gamma = 20$ kN/m³, $E = 2 \cdot 10^4$ kPa, $\nu = 0.2$. As shown in Figure 5b the discontinuity surfaces, denoted by D1 and D2, are such that one intersects the boundary and one is contained in the soil mass; their lengths are 9.40 and 9.10 metres respectively. Both D1 and D2 are modelled as freely slipping ($\tau_{res}=0$) or interacting by residual shear strength of 30 or 60 kPa. Analyses were performed for the following configurations: 1) no discontinuity in soil mass (nofrac); 2) D1 discontinuity, with residual strength $\tau_{res} = 0$ (d1-t0), $\tau_{res} = 30$kPa (d1-t3) and $\tau_{res} = 60$ kPa (d1-t6); 3) D2 discontinuity, with residual strength $\tau_{res} = 0$ (d2-t0) and $\tau_{res} = 30$kPa (d2-t3). Figures 6 to 9 show the results obtained, in terms of plots of ground surface settlements and horizontal displacements of the vertical excavation surface; Figures 6 and 7 refer to the presence of discontinuity D1, Figures 8 and 9 to that of discontinuity D2. In all plots the heavy line represents displacements when no discontinuity is present, calculated by means of linear elastic finite elements.

These results should be regarded as more qualitative than quantitative in nature, even though they do quite closely estimate the measured displacements; in fact, quantitative comparison with values measured at the Chicago site is not appropriate, in that in the present application several elements of the actual configuration have been oversimplified. The significant feature is that the proposed model does reproduce some aspects of behaviour that even quite sophisticated finite element analyses have not managed to compute correctly. In particular it can be seen in Figures 6 and 8 that when the discontinuity surface is freely slipping ($\tau_{res}=0$) localisation effects are more evident. These were correctly computed as a concentration of ground surface settlement at a small zone near the excavation, while no settlement is computed just a few metres further. A localisation effect is also clearly seen as a step in the settlement plot (Figure 8) when discontinuity D2 is considered, which is entirely contained in the soil mass and therefore no visible discontinuity in displacement is seen on the excavation vertical face.

On the other hand, concerning these horizontal displacements, the model correctly predicts a discontinuity in the case of the D1 surface, which intersects the boundary (Figure 7). In this particular diagram a validation for the coupling of the finite element and surface integral models can be seen: in fact a continuous and smooth displacement diagram is obtained by plotting the relative displacement computed at the origin of the discontinuity (points indicated by the small arrows) and the displacements computed at the finite element nodes.

When there is a residual strength of 30 kPa on the discontinuity surface, all displacement diagrams present the same general pattern as when freely

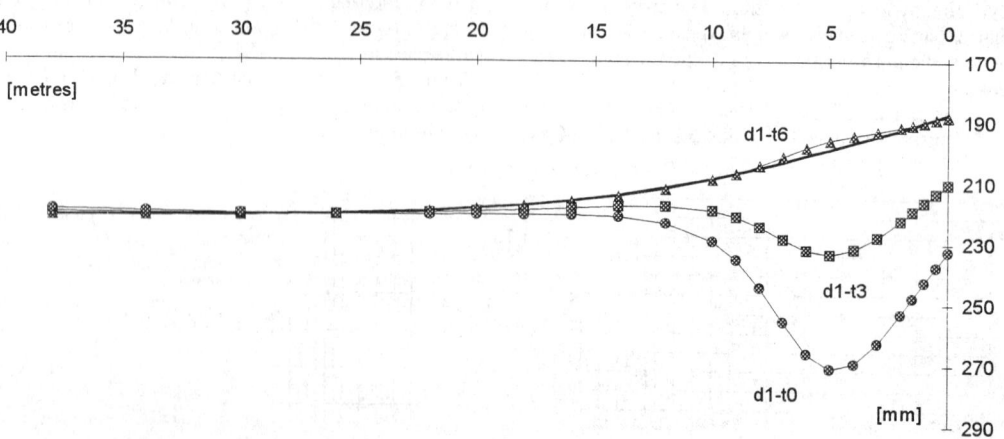

Figure 6: Soil mass containing D1 discontinuity: ground surface settlements.

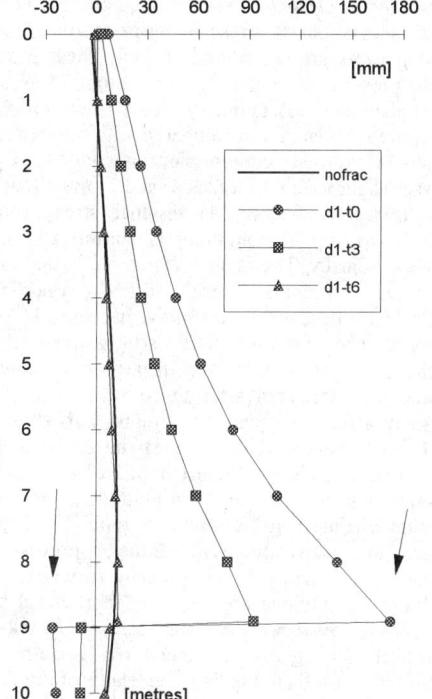

Figure 7: Soil mass containing D1 discontinuity: horizontal displacements of vertical excavation face.

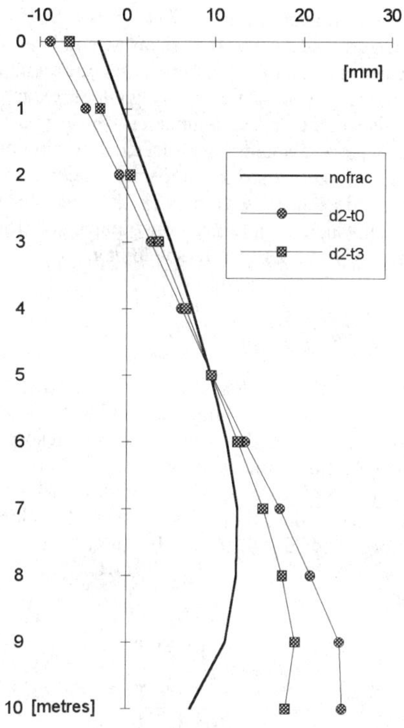

Figure 9: Soil mass containing D2 discontinuity: horizontal displacements of vertical excavation face.

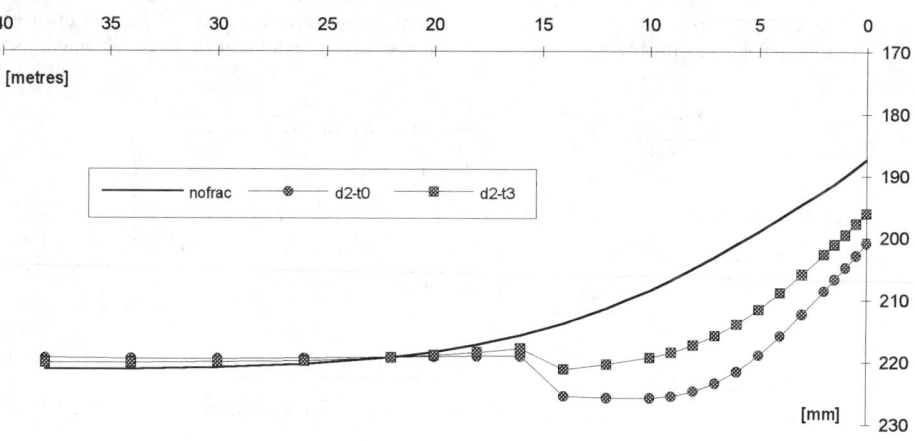

Figure 8: Soil mass containing D2 discontinuity: ground surface settlements.

slipping, but smaller values. This shows that the model correctly interprets this shear strength as an internal constraint on deformation. When the value of the shear strength is sufficiently high (see example d1-t6) displacement patterns similar to those relative to the absence of discontinuity surfaces are obtained and near to zero stress intensity factors are calculated; this correctly translates the fact that a high enough value of internal constraint can "stop" the relative displacement from developing.

4.2 Direct shear box

The mesh used for this example is shown in Figure 10; a 1.50 cm long discontinuity which intersects one side of the shear box was assumed. This configuration was analysed using both the elastic and the poro-elastic schemes. Material constants are the following: $E = 12240$ kPa, $v = 0.02$, $v_u = 0.5$, $B = 0.98$, $c = 10^{-7}$ m^2/s.

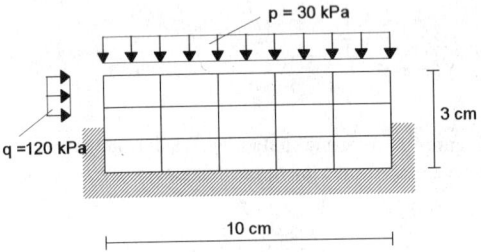

Figure 10: Finite element mesh for direct shear box problem.

Analyses by the elastic scheme aimed at clarifying the correct calculation of stress intensity factors. The following cases are considered: 1) only shear force applied, with no discontinuity considered; 2) shear force applied, with discontinuity present: values of K_I and K_{II} are calculated and compared with theoretical results; 3) normal compression is added: K_{II} decreases slightly. In both cases 2) and 3) the discontinuity is freely slipping; 4) residual strength of 20 kPa across the discontinuity is considered: K_{II} decreases radically. This is as it should be, in that the discontinuity becomes less critical regarding propagation. These results are shown in Figure 11.

Analyses using the poro-elastic scheme are useful for checking the step-by-step procedure. Results obtained are compared with those relative to the analysis of a freely slipping discontinuity surface in a semi-infinite medium. The K_{II} stress intensity factor and the relative sliding at the origin of the discontinuity obtained in this last case are plotted against the logarithm of time t in Figure 12; a good "S" shape results, which correctly recalls diffusion processes, and the (known) values corresponding to undrained and drained conditions are correctly reproduced by the analysis. However these results can only be taken qualitatively as regards the time scale, because in dealing with an infinite medium the length of the discontinuity becomes a parameter of no significance. Figure 13 on the other hand shows the evolution of the same two parameters for the discontinuity in a shear box (Figure 10); in this case the values can be retained also quantitatively indicative. Here the "S" shape of the semilogarithm plots is not so apparent for the relative displacement at the origin, and not at all for the stress intensity factor; furthermore, in neither case are the final "drained" values equal to those which would have been obtained using the

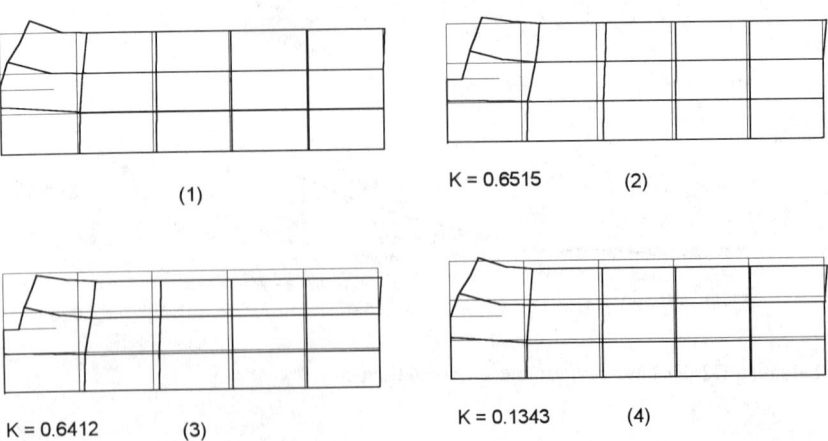

Figure 11: Various analyses of direct shear box.

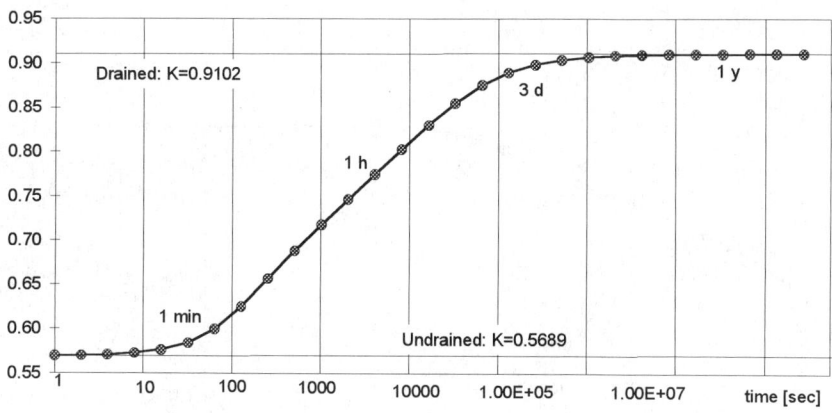
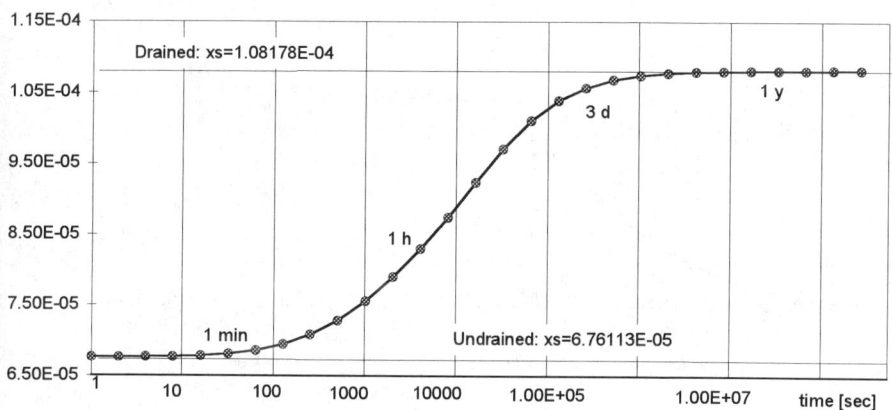

Figure 12: Evolution in time of K_{II} and x_S, for a discontinuity in an infinite medium.

elastic model and effective stress parameters for the material. This could be a result of the interaction of the discontinuity surface with the finite boundary of the medium, but can also (or partly) be due to the fact that the material's response to loading, analysed by means of simple linear elastic finite elements, is considered as instantaneous and no diffusion processes in the material are coupled with the assumed time-dependent poro-elastic response of the same material to the presence of the discontinuity.

5 CONCLUSIONS

The strong point that emerges from the application of the proposed model is the possibility to arrive at a correct prediction of displacements both in a local and in a global sense, even if a rudimental linear elastic constitutive law is used for the material, provided that an adequate modelling of discontinuities is available and correctly introduced.

It is however pointed out that, although the formulation of the algorithm renders it extremely flexible and suitable to be used in various contexts, it is currently being applied only for the deformation analysis of a continuum containing discontinuity surfaces, in which the interaction of these surfaces with each other and with the boundary is correctly taken into account. No analysis regarding the propagation or the initiation of discontinuities is attempted; to that end a greater insight into the material's intrinsic behaviour is required. Likewise, though the porous nature of the medium can be taken into account in what regards the effects of the discontinuity surface on the stress-strain state, no provision has yet been made to incorporate a complete analysis of the diffusion process in the entire medium; this is due to numerical difficulties regarding the solution of certain singular integral equations connected with the model, and remains to be further investigated.

The authors believe that a suitable propagation

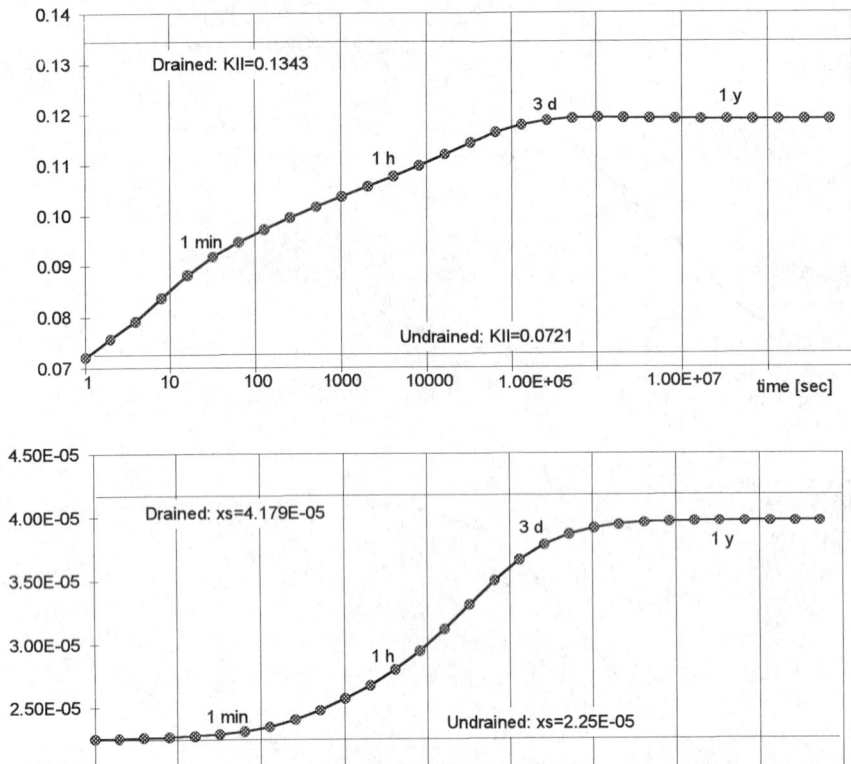

Figure 13: Evolution in time of K_{II} and x_s, for a discontinuity in a direct shear box.

criterion, even if based on fracture mechanics concepts, should in some way take correctly into account pore pressure effects. This is why much attention has been paid to local diffusion processes, their correct modelisation and exact analytical solution of the resulting integral equations, at the cost of disregarding global poro-elastic behaviour.

A drawback of the method, as of all formulations based on fracture mechanics, is the necessity of the presence of some kind of discontinuity to start with, even as an initial notch; stability analysis would seem necessary in order to model initiation of shear bands and overcome this limitation. However, it is also worth mentioning that in most geotechnical applications it is the nature of the structure itself that suggests the most probable location for shear band initiation.

REFERENCES

Abramowitz, M. and Stegun, I.A. (Eds.) 1964. *Handbook of Mathematical Functions*, National Bureau of Standards, Dover, New York, Appl. Math. Series, 55.

Aliabadi, M.H. and Rooke, D.P. 1991. *Numerical Fracture Mechanics*, Computational Mechanics publications - Kluwer Academic Publishers.

Annigeri, B.S. 1984. 'Surface integral finite element hybrid method for localized problems in continuum mechanics', Sc. D. Thesis, M.I.T..

Bilby, B.A. and Eshelby, J.D. 1968. 'Dislocations and the theory of fracture', in *Fracture: an advanced treatise*, ed. Liebowitz, H. Vol. 1, pp. 99-182, Academic Press.

Biot, M.A. 1940. 'General Theory of Three-Dimensional Consolidation' *Journal of Applied Physics*, Vol. 12, pp. 155-164.

Brebbia, C.A. and Dominguez, J. 1989. *Boundary Elements - An Introductory Course*, Computational Mechanics Publications, McGraw-Hill Book Company.

Broek, D. 1986. *Elementary Engineering Fracture Mechanics* 4th edition, Martinus Nijhoff Publishers.

Broek, D. 1989. *The Practical Use of Fracture Mechanics*, Kluwer Academic Publishers.

Cleary, M.P. 1978. 'Moving singularities in elasto-diffusive solids with applications to fracture propagation' *Int. J. Solids and Structures*, Vol. 14, pp. 81-97.

Cruse, T.A. 1972. 'Numerical evaluation of elastic stress intensity factors by the boundary-integral equation method', in *The surface crack: Physical problems and computational solutions*, ed. Swedlow, J.L., pp. 153-170, ASME, New York.

Erdogan, F. and Gupta, G.D. 1972. 'On the numerical integration of singular integral equations' *Quart. Appl. Math.*, pp. 525-534.

Finno, R.J., Atmatzidis, D.K. and Perkins, S.B. 1989. 'Observed performance of a deep excavation in clay' *Journal of Geotechnical Engineering*, Vol. 115, No. 8, pp. 1045-1064.

Finno, R.J., Harahap, I.S. and Sabatini, P.J. 1991. 'Analysis of braced excavations with coupled finite element formulations' *Computers and Geotechnics*, pp. 91-114.

Finno, R.J. and Nerby, S.M. 1989. 'Saturated clay response during braced cut construction' *Journal of Geotechnical Engineering*, Vol. 115, No. 8, pp. 1065-1084.

Greene, T.H., Annigeri B.S. and Cleary, M.P. 1983. 'Computation of stress and displacement using the surface integral method' in *Reports of Research in Mechanics and Materials*, Department of Mechanical Engineering, M.I.T., REL-83-1.

Narendran, V.M. and Cleary, M.P. 1984. 'Elastostatic interaction of multiple arbitrarily shaped cracks in plane inhomogeneous regions' *Engineering Fracture Mechanics*, Vol. 19, No. 3, pp. 481-506.

Palmer, A.C., and Rice, J.R. 1973. 'The growth of slip surfaces in the progressive failure of overconsolidated clay' *Proc. Roy. Soc. London*, **A.** Vol. 332, pp. 527-548.

Rice, J.R. 1968. 'Mathematical analysis in the mechanics of fracture', in *Fracture: an advanced treatise*, ed. Liebowitz, H. Vol. 2, pp. 191-311, Academic Press.

Rice, J.R. and Cleary, M.P. 1976. 'Some basic stress diffusion solutions for fluid-saturated elastic porous media with compressible constituents' *Reviews of Geophysics and Space Physics*, Vol. 14, No. 2, pp. 227-241.

Scarpelli, G. and Iannozzi, F. 1988. 'Il metodo delle discontinuità per lo studio della rottura progressiva nei terreni coesivi' (in Italian) pp. 177-196, in the *Proceedings of conference of the national co-ordination group for geotechnical engineering studies*, Monselice, Italy.

APPENDICES

Appendix A *Coupling the finite element method to the surface integral formulation by means of the principle of superposition.*

As already mentioned in the text, application of the principle of superposition is carried out by dividing the original problem into two sub-systems, as illustrated in Figure 4. These two sub-problems are modelled using finite element and surface integral methods respectively. The principle of superposition implies that the stress and strain fields of the two sub-systems, summed, will result in those of the original problem, providing that the sum of the boundary conditions of the two sub-systems gives the boundary conditions of the original problem. To achieve this last requirement a load correction $\mathbf{F}^C$ in the finite element formulation and a traction correction $\mathbf{T}^C$ on the discontinuity surfaces are required. These are respectively computed as:

1) tractions on the finite body's boundary due to the relative displacements of the discontinuity surface:

$$\mathbf{F}^C = \mathbf{G}^* \cdot f \tag{A1}$$

where:

f : vector of unknowns in sub-problem 2);

$\mathbf{G}^*$: coefficient matrix derived through numerical discretization of appropriate surface integral equations; its elements represent loads at the boundary finite element nodes due to unit dislocation density amplitudes at the collocation points along the discontinuity surface.

2) tractions on the discontinuity surface as a result of displacements at the surrounding finite element nodes:

$$\mathbf{T}^C = \mathbf{S} \cdot \mathbf{U}^{FE} \tag{A2}$$

where:

$\mathbf{U}^{FE}$: vector of unknown finite element nodal displacements;

$\mathbf{S}$: coefficient matrix, its elements representing stresses at the collocation points along the discontinuity surface due to unit displacements of the surrounding finite element nodes.

Consequently, the formulation of the two individual problems is as follows:

For sub-problem 1):

$$\mathbf{K} \cdot \mathbf{U}^{FE} = \mathbf{F} - \mathbf{F}^C \quad (A3)$$

where:
K: finite element stiffness matrix (symmetric, banded);
$\mathbf{U}^{FE}$: vector of finite element nodal displacements;
F: external nodal load vector;
$\mathbf{F}^C$: correction load vector, given by equation (A1).

For sub-problem 2):

$$\mathbf{C}^* \cdot f = \mathbf{T} - \mathbf{T}^C \quad (A4)$$

where:
C: surface integral coefficient matrix, derived through the Chebyshev numerical discretization scheme (non symmetric, full);
f: vector of unknown functions at the discontinuity surface nodal points;
T: vector of known traction conditions on discontinuity surface;
$\mathbf{T}^C$: correction traction vector, given by equation (A2).

Combination of equations (A1) to (A4) gives a complete system of coupled equations, which can be assembled to obtain the following partitioned matrix expression:

$$\begin{bmatrix} \mathbf{K} & \mathbf{G}^* \\ \mathbf{S} & \mathbf{C}^* \end{bmatrix} * \begin{Bmatrix} \mathbf{U}^{FE} \\ f \end{Bmatrix} = \begin{Bmatrix} \mathbf{F} \\ \mathbf{T} \end{Bmatrix} \quad (A5)$$

This equation needs to be reformulated in terms of total nodal displacements **U**, given by:

$$\mathbf{U} = \mathbf{U}^{FE} + \mathbf{U}^{SI} \quad (A6)$$

where:
$\mathbf{U}^{FE}$: displacement field at finite element nodes resulting from finite element discretization of domain without discontinuity surface (first sub-problem);
$\mathbf{U}^{SI}$: displacement field at finite element nodes resulting from the surface integral discretization of the infinite domain containing discontinuities (second sub-problem).

The $\mathbf{U}^{SI}$ vector can be calculated at the finite element nodes by means of:

$$\mathbf{U}^{SI} = \mathbf{L} \cdot f \quad (A7)$$

where:
L: displacements of the finite element nodes due to unit dislocation densities at the collocation points along the discontinuity surfaces. Elements of the **L** matrix are computed using the displacement influence function for an edge dislocation and integrating along the discontinuity surface.

Substituting (A6) into (A5) we obtain:

$$\begin{cases} \mathbf{K} \cdot (\mathbf{U} - \mathbf{U}^{SI}) + \mathbf{G}^* \cdot f = \mathbf{F} \\ \mathbf{S} \cdot (\mathbf{U} - \mathbf{U}^{SI}) + \mathbf{C}^* \cdot f = \mathbf{T} \end{cases} \quad (A8)$$

and hence, using equation (A7):

$$\begin{cases} \mathbf{K} \cdot \mathbf{U} + (\mathbf{G}^* - \mathbf{KL}) \cdot f = \mathbf{F} \\ \mathbf{S} \cdot \mathbf{U} + (\mathbf{C}^* - \mathbf{SL}) \cdot f = \mathbf{T} \end{cases} \quad (A9)$$

Finally, defining:

$$\begin{cases} \mathbf{G} = \mathbf{G}^* - \mathbf{KL} \\ \mathbf{C} = \mathbf{C}^* - \mathbf{SL} \end{cases} \quad (A10)$$

the following matrix equation is obtained:

$$\begin{bmatrix} \mathbf{K} & \mathbf{G} \\ \mathbf{S} & \mathbf{C} \end{bmatrix} * \begin{Bmatrix} \mathbf{U} \\ f \end{Bmatrix} = \begin{Bmatrix} \mathbf{F} \\ \mathbf{T} \end{Bmatrix} \quad (A11)$$

This is the same as equation (19) and is the governing matrix equation for the problem of a finite medium containing discontinuities. Arbitrary force and displacement boundary conditions can be imposed. Details on obtaining the various submatrices and on applicability of the procedure and convergence criteria can be found in the original references.

Appendix B *The Chebyshev numerical integration scheme*

An integral equation of the kind in equation (3) (or (6)) can be tackled by means of interpolation functions and suitable collocation techniques. Namely, the unknown function $\delta(\mathbf{x}')$ would be written in the form:

$$\delta(\mathbf{x}') = \sum_{t=1}^{N} \delta_t \cdot d_t(\mathbf{x}') \quad (B1)$$

where $d_t(\mathbf{x}')$ are suitable interpolation functions.

The stress $\sigma(\mathbf{x})$ would then be imposed at a discrete number of points along the discontinuity surface and, denoting by σ_k ($k = 1,...,N$) the values at these points, the original integral equation would transform into a linear matrix equation:

$$\sigma_k(\mathbf{x}_k) = \sum_{t=1}^{N} W_t \cdot d_t(\mathbf{x}_t) + \Re \quad (B2)$$

where:
W_t : generic weight function;
$\mathbf{x}_k$: points on the discontinuity surface at which tractions are imposed;
$\mathbf{x}_t$: collocation points on the discontinuity surface, at which the value of the unknown function is computed;
$\Re$: error between the solution thus obtained and the exact analytical one; $\Re$ should become smaller when N increases.

As demonstrated by Erdogan and Gupta (1972), the particular interpolation functions and collocation points adopted when using the Chebyshev scheme result in obtaining the exact analytical solution. In particular, as in equation (10), the dislocation density is expressed as:

$$\delta(s) = \frac{f(s)}{\sqrt{1-s^2}} \quad (B3)$$

and equation (3) is written as a singular integral equation of the type:

$$\sigma(s) = \int_{-1}^{1} \frac{f(s')\,ds'}{\sqrt{1-s'^2} \cdot (s-s')} \quad (B4)$$

(This, as already mentioned, exploits the $r^{-1/2}$ singularity exhibited by the dislocation density function $\delta(r)$ near the discontinuity tip, and is useful in determining the strength of that singularity, directly linked to the stress intensity factor).

Interpolating the unknown function by means of the Chebyshev polynomials of the first kind T_j:

$$f(s') = \sum_{j=1}^{\infty} B_j T_j(s') \cong \sum_{j=1}^{P} B_j T_j(s') \quad (B5)$$

where:
T_j : Chebyshev polynomials of the first kind;
B_j : weight functions, which actually do not come into the procedure at all, as will be seen,

equation (B4) becomes:

$$\sigma(s) = \int_{-1}^{1} \frac{\sum_{j=1}^{P} B_j T_j(s')}{\sqrt{1-s'^2}\,(s-s')}\,ds' = \sum_{j=1}^{P} \left[B_j \int_{-1}^{1} \frac{T_j(s')}{\sqrt{1-s'^2}\,(s-s')}\,ds' \right] \quad (B6)$$

At this point the formula:

$$\int_{-1}^{1} \frac{T_n(y)\,dy}{(x-y)\sqrt{1-y^2}} = \pi \cdot U_{n-1}(x) \quad (B7)$$

where:
T_n: Chebyshev polynomial of first kind of order n;
U_{n-1}: Chebyshev polynomial of the second kind of order n-1;

can be used (cfr. Abramowitz and Stegun 1964). Equation (B6) becomes:

$$\sigma(s) = \sum_{j=1}^{P} \left[B_j \cdot \pi \cdot U_{j-1}(s) \right] = \pi \cdot \sum_{j=1}^{P} \left[B_j \, U_{j-1}(s) \right] \cdot \quad (B8)$$

In the literature (Erdogan and Gupta 1972) the following is demonstrated:

$$\sum_{k=1}^{N} \frac{T_j(t_k)}{N(t_k - x_r)} = U_{j-1}(x_r) \quad 0 < j < N \quad (B9)$$

where t_k are the zero points of T_N, given by:

$$t_k = \cos \frac{\pi \cdot (2k-1)}{2N} \quad (B10)$$

and x_r are the zero points of U_{N-1}, given by:

$$x_r = \cos \frac{\pi \cdot r}{N} \quad (B11)$$

Substituting these in (B8) we obtain:

$$\sigma(s_r) = \pi \cdot \sum_{j=1}^{P} \left[B_j \sum_{k=1}^{N} \frac{T_j(s_k)}{N(s_k - s_r)} \right] = \sum_{k=1}^{N} \frac{\pi}{N(s_k - s_r)} \cdot \sum_{j=1}^{P} B_j T_j(s_k) \quad (B12)$$

(N.B. U_{j-1} (and hence σ) now is not written for $s \,\forall$, but for particular points s_r corresponding to the zero points of U_{N-1}.)

which, by virtue of equation (B5), becomes:

$$\sigma(s_r) = \sum_{k=1}^{N} \frac{\pi}{N(s_k - s_r)} \cdot f(s_k) \quad (B13)$$

This last equation, put in matrix form, corresponds to a linear algebraic system in which the matrix of coefficients **C** can be easily calculated by means of:

$$C_{r,k} = \frac{\pi}{N(s_k - s_r)} \quad (B14)$$

where s_k and s_r assume the already mentioned meaning for t_k and x_r, and N is the number of collocation points we choose to assume on the discontinuity surface. The dislocation density values computed for these points will be analytically exact.

The importance of this method is that the exact values for the f function are obtained at certain points (s_k), provided that the tractions on the discontinuity surface are imposed not at arbitrary points, but at those (s_r) corresponding to the zeroes of the Chebyshev polynomial U_{N-1}. Furthermore it is significant to note that the location of these points (both s_k and s_r) in the (-1,1) interval (corresponding to the discontinuity surface, with the tips at -1 and 1) is such that an accumulation of points near the tips is achieved; this is satisfactory in the problems of interest, in that the tip zone is that where the processes influencing global development of configuration are taking place, and in particular it is most advantageous as regards the calculation of the stress intensity factor.

It can also be noted that equation (B13) amounts to an application of the Gauss-Chebyshev integration formula at particular points, such that residual error $R_n = 0$. The procedure here illustrated is equivalent to proof that these particular points are the zeroes of the Chebyshev polynomials U_{N-1}, as illustrated above.

Appendix C *Influence functions for elastic and poro-elastic media; stress fields due to edge dislocation singularities.*

Stress field in an elastic medium due to a unit edge dislocation in the X-direction (Figure 1):

$$\begin{Bmatrix} \sigma_{xx} \\ \sigma_{yy} \\ \sigma_{xy} \end{Bmatrix} = \frac{b_x \overline{E}}{4\pi(1-v^2)} \begin{Bmatrix} -\frac{y(3x^2+y^2)}{(x^2+y^2)^2} \\ \frac{y(x^2-y^2)}{(x^2+y^2)^2} \\ \frac{x(x^2-y^2)}{(x^2+y^2)^2} \end{Bmatrix} \quad (C1)$$

Stress field in an elastic medium due to a unit edge dislocation in the Y-direction (Figure 1):

$$\begin{Bmatrix} \sigma_{xx} \\ \sigma_{yy} \\ \sigma_{xy} \end{Bmatrix} = \frac{b_y \overline{E}}{4\pi(1-v^2)} \begin{Bmatrix} \frac{x(x^2-y^2)}{(x^2+y^2)^2} \\ \frac{x(3y^2+x^2)}{(x^2+y^2)^2} \\ \frac{y(x^2-y^2)}{(x^2+y^2)^2} \end{Bmatrix} \quad (C2)$$

where:

$$\overline{E} = \begin{cases} E : \text{plane stress conditions} \\ \dfrac{E}{4(1-v^2)} : \text{plane strain conditions} \end{cases}$$

E : Young's modulus;
v : Poisson's ratio.

Generalised stress field in a poro-elastic medium due to a unit edge dislocation in the x-direction (Figure 1):

$$\begin{Bmatrix} \sigma_{\theta\theta} \\ \sigma_{r\theta} \\ \sigma_{rr} \\ p \end{Bmatrix} = A\, b_x \begin{Bmatrix} \sin\theta \left\{ 2e^{\frac{-r^2}{4ct}} - \frac{1-v}{v_u-v} - \frac{4ct}{r^2}\left[1 - e^{\frac{-r^2}{4ct}}\right] \right\} \\ \cos\theta \left\{ \frac{1-v}{v_u-v} - \frac{4ct}{r^2}\left[1 - e^{\frac{-r^2}{4ct}}\right] \right\} \\ \sin\theta \left\{ \frac{4ct}{r^2}\left[1 - e^{\frac{-r^2}{4ct}}\right] - \frac{1-v}{v_u-v} \right\} \\ \sin\theta \cdot \eta^{-1}\left[1 - e^{\frac{-r^2}{4ct}}\right] \end{Bmatrix}$$

(C3)

where:

$$A = \frac{G(v_u - v)}{2\pi r(1-v_u)(1-v)}$$

$$\eta = \frac{3(v_u - v)}{[2B(1+v_u)(1-v)]}.$$

In this formulation the material properties are described by means of the five constants G, B, v_u, v and c. These are appropriately defined (Rice and Cleary 1976), and for a saturated medium represent respectively the material's shear modulus, Skempton's constant equal to unity, undrained Poisson's ratio, drained Poisson's ratio and consolidation coefficient or diffusivity.

›
Instability of a viscoplastic model for clay and numerical study of strain localisation

Instabilité d'un modèle élasto-viscoplastique pour l'argile et étude numérique de la localisation de la déformation

Fusao Oka, Atsushi Yashima & Itaru Kohara
Department of Civil Engineering, Gifu University, Japan

Toshihisa Adachi
School of Civil Engineering, Kyoto University, Japan

ABSTRACT: The strain localization phenomenon is currently an attractive subject in geomechanics. It is well-known that shear bands develop in clay specimens during the straining process. Strain localization is closely related to plastic instability. In the present paper, a viscoplastic strain softening model is derived and the instability of the model during the creep process is discussed. It is found that the proposed viscoplastic model is capable of describing plastic instability. Finally, we illustrate the formation of strain localization by a two-dimensional finite element large deformation analysis with the proposed viscoplastic strain softening model. Undrained plane strain compression and extension tests for a normally consolidated clay are simulated by the proposed numerical method, considering a transport of pore water in the material at a quasi-static strain rate.

Le phénomène de Localisation de la déformation est un sujet d'actualité en Géomécanique.
Il est bien connu que des bandes de cisaillement se développent dans les échantillons d'argile au cours de leur déformation. La localisation est liée intimement à l'instabilité plastique. Dans le présent article, on présente une loi de comportement viscoplastique avec adoucissement à la déformation, et étudie son instabilité au cours d'un processus de fluage. On montre que cette loi est en mesure de décrire l'instabilité plastique. Enfin, nous illustrons la formation de la localisation par un calcul aux éléments finis bidimensionnel en grande déformation, mené avec la loi viscoplastique proposée. On modélise ainsi des essais de déformation plane non drainés en compression et en extension, sur une argile normalement consolidée, en prenant en compte l'écoulement du fluide interstitiel dans le matériau en régime quasi-statique.

1 INTRODUCTION

For the last two decades, many constitutive models have been proposed for geomaterials and applied to practical problems. There remain, however, still some outstanding issues to be resolved. One of them is the formation of shear band before failure. The problems of strain localization have been studied in geomechanics as the precursor of the failure phenomenon. The theory was first proposed by Hadamard (1903) and later developed by Thomas (1961), Hill (1962), Mandel (1964) and Rice (1975). For geomaterials, such as sand, Vardoulakis (1979, 1980) treated this problem as a bifurcation. They derived a condition for the initiation of shear bands from the view point of the loss of ellipticity for quasi-static problems and the loss of hyperbolicity for dynamic problems. Two theories are traditionally authoritative in geomechanics to calculate the inclination of shear bands in two-dimensional plane strain problems, namely Mohr-Coulomb theory and the Roscoe theory. Based on the Mandel's work, Vermeer (1982) introduced a compliance approach to derive the angle of shear bands. He succeeded in deriving the inclination of shear bands, which was obtained by Arthur (1977) and Vardoulakis (1980) in the experiments.

The important points involved with the formation of a shear band are predictions of the angle of the shear band, thickness and the post-localization behavior. Both the loss of ellipticity approach and the loss of hyperbolicity approach are very useful in predicting the angle of a shear band. However, these approaches are not appropriate for predicting the thickness of a shear band or simulating the post-localization behavior. Aifantis advocated a new approach for dealing with the localization problem which uses the gradient dependent constitutive model (Aifantis, 1984). Mühlhaus (1986, 1987) took into consideration grain rotations, couple stresses and used the mean grain diameter as a characteristic length based on Cosserat continuum theory. The feature of these methods is the explicit appearance of a characteristic length scale, like the particle size of the soil particle, in the constitutive equation.

As for the problem of post localization, the viscoplastic approach has been used by Loret and Prevost (1990) for dynamic localization problems by introducing the artificial viscocity into rate-independent materials. Generally speaking, the numerical calculation for the post-localization problem can be continued using the elasto-viscoplastic constitutive model, because of the well-posedness of the governing equations. In addition, for dynamic problems, a kind of characteristic length scale is naturally introduced into the analysis as the ratio of wave speed to relaxation time (Needleman, 1988). For static problems, however, defining the characteristic length scale is still an open problem.

In the present paper, we focus on the quasi-static localization problem of water saturated clay. The strain localization problem is treated as an viscoplastic instability of the constitutive model. We first observe that the overstress type constitutive model based on the strain hardening rule cannot describe creep failure phenomena. Then, the elasto-viscoplastic model for clay with strain softening is illustrated. The instability of the viscoplastic model is discussed within the framework of a bifurcation in a wider sense. The critical stress ratio at a minimum viscoplastic strain rate under undrained creep is derived to characterizes the transition between the stable and the unstable regions. Finally, a finite element analysis is carried out for the strain localization and the applicability of the proposed model for simulation to a strain localization problem is shown.

2 FEATURE OF THE OVERSTRESS TYPE VISCOPLASTIC MODEL WITH STRAIN HARDENING

Adachi et al. (1990) made it clear that the overstress viscoplastic model with strain hardening proposed by Adachi and Oka (1982) cannot describe the acceleration creep behavior of clay because of its mathematical structure. Here, we will discuss shortly the structure of our constitutive model in one-dimensional form.

The overstress type viscoplastic constitutive model can be written in one-dimensional form as

$$\dot{\varepsilon}^{vp} = <G(F)> \quad (1)$$

$$F = \sigma - f(\varepsilon^{vp}) \quad (2)$$

where $\dot{()} = \frac{d()}{dt}$, $\dot{\varepsilon}^{vp}$ is the viscoplastic strain rate, σ is the stress and $F = 0$ denotes the static stress-strain relation and $<>$ is the Macauley's bracket.

The function G corresponds to $C\Phi_1$ of Eq.(8) in the followings. The differentiation of the viscoplastic strain rate with respect to time yields the following rate of strain rate as

$$\ddot{\varepsilon}^{vp} = \frac{\partial G}{\partial F}\frac{\partial F}{\partial t}. \quad (3)$$

For the overstress viscoplastic model with strain rate hardening, $\frac{\partial G}{\partial F}$ is positive. If the strain hardening is assumed,

$$\frac{\partial f}{\partial \varepsilon^{vp}} \geq 0. \quad (4)$$

Subsequently, under the creep process where σ is constant,

$$\frac{\partial F}{\partial t} = -\frac{\partial f}{\partial \varepsilon^{vp}}\dot{\varepsilon}^{vp} \leq 0. \quad (5)$$

Eqs.(3) and (5), therefore, produces

$$\ddot{\varepsilon}^{vp} \leq 0. \quad (6)$$

In this case, the viscoplastic strain rate decreases monotonously with time in the creep process. This indicates that the overstress type constitutive model cannot describe the acceleration creep process.

On the other hand, if we assume a strain softening material,

$$\frac{\partial f}{\partial \varepsilon^{vp}} \leq 0. \quad (7)$$

Eventually, it is concluded that the rate of viscoplastic strain rate is always negative in the whole stage of a creep process. This property is inconsistent with the experimental results. The experimental results show the viscoplastic strain rate decreases in the early stage of creep process, then reaches the minimum strain rate, and finally leads to an acceleration creep process. From the above discussions, we conclude that neither viscoplastic model with simple strain softening or the model with strain hardening is a satisfactory model for clay, and non-linear hardening-softening law is necessary for real materials.

3 ELASTO-VISCOPLASTIC SOFTENING CONSTITUTIVE MODEL FOR NORMALLY CONSOLIDATED CLAY

Oka (1981) and Adachi and Oka (1982) developed an elasto-viscoplastic constitutive model for a normally consolidated clay based on the overstress type viscoplastic theory. In this section, Adachi and Oka's model will be generalized by introducing a second material function (Adachi, Oka and Mimura, 1987). The viscoplastic flow rule is given by

$$\dot{\varepsilon}_{ij}^{vp} = C <\Phi_1(F)> \Phi_2(\xi)\frac{\partial f}{\partial \sigma_{ij}} \quad (8)$$

$$F = \frac{f - \kappa_s}{\kappa_s} \quad (9)$$

where $\dot{\varepsilon}_{ij}^{vp}$ is the viscoplastic strain rate, C is a viscoplastic parameter, σ_{ij} is the stress tensor, f is the dynamic yield function, Φ_1 is the material function for the strain rate effect and Φ_2 is the second material function. $F = 0$ denotes the static yield function, κ_s is the hardening parameter.

In the three dimensional case, the function F in Eq.(2) is generalized as Eq.(9). The function F is reformed as a non-dimensional form by introducing the dinominator κ_s in Eq.(9). The second material function is introduced to describe the vanishing of rate dependency at the failure state. Inverting Φ_1, with respect to function F, we obtain

$$f(\sigma_{ij}, \varepsilon_{ij}^{vp}) = $$

$$\kappa_s(\varepsilon_{ij}^{vp})(1 + <\Phi_1^{-1}[\frac{\sqrt{I_2}}{C\Phi_2(\xi)}(\frac{\partial f}{\partial \sigma_{kl}'}\frac{\partial f}{\partial \sigma_{kl}'})^{1/2}]>) \quad (10)$$

in which I_2 is the second invariant of the viscoplastic strain rate tensor expressed by $I_2 = \sqrt{\dot{\varepsilon}_{ij}^{vp}\dot{\varepsilon}_{ij}^{vp}}$ and σ_{kl}' is the Terzaghi's effective stress tensor. Eq.(10) indicates that the yield function depends on the strain rates. On the rate dependency of the constitutive model, Aubry et al. (1985) addressed the fact that the rate dependency vanishes at failure sate. Following this important experimental evidence, a second material function was introduced to describe the vanishing of rate dependency at the failure state. Since the second material function must be infinite at the failure state, the following form for the second material function is adopted:

$$\Phi_2 = 1 + \xi. \quad (11)$$

The internal variable, ξ, expresses the deterioration of the material and satisfies the following evolutional equation:

$$\dot{\xi} = \frac{M^{*2}}{G_2^*(M^* - \eta^*)^2}\dot{\eta}^* \quad (12)$$

where η^* is the stress invariant ratio defined by

$$\eta^* = \frac{\sqrt{2J_2}}{\sigma_m'}. \quad (13)$$

The static yield function is expressed as follows:

$$f = \frac{\sqrt{2J_2}}{M^*\sigma_m'} + ln\frac{\sigma_m'}{\sigma_{me}'} - \kappa_s = 0 \quad (14)$$

where J_2 is the second invariant of the deviatoric part of transformed stress tensor S_{ij}, σ_m' is the mean effective stress, M^{*2} is the value of η^* at the failure state and G_2^* is the material constant. The evolution equation for the hardening parameter is given by

$$\dot{\kappa}_s = \frac{1+e}{\lambda - \kappa}\dot{v}^p. \quad (15)$$

Terzaghi's effective stress concept will be used in the following without concern. Referring the work by Adachi and Oka (1982), material function, $\Phi_1(F)$, is given by

$$\Phi_1(F) = \sigma_m' exp\{m'(\frac{\sqrt{2J_2}}{M^*\sigma_m'} + ln\frac{\sigma_m'}{\sigma_{me}'} - \frac{1+e}{\lambda-\kappa}v^p)\} \quad (16)$$

where m' is a viscoplastic parameter, σ_{me}' is the consolidation stress, λ is the consolidation index, κ is the swelling index, e is the void ratio, σ_{me}' is a initial value of mean effective stress and v^p is the volumetric plastic strain. As for the elastic component of the strain rate tensor, $\dot{\varepsilon}_{ij}^e$, Hooke's isotropic law is used as

$$\dot{\varepsilon}_{ij}^e = \frac{1}{2G}\dot{S}_{ij} + \frac{\kappa}{(1+e)\sigma_m'}\dot{\sigma}_m'\frac{1}{3}\delta_{ij} \quad (17)$$

where G is the elastic shear modulus.

4 INSTABILITY OF THE MODEL

Let's consider the triaxial undrained creep problem. Under the undrained axisymmetric triaxial state ($\sigma_1' > \sigma_2' = \sigma_3'$), the stress states are expressed by q and p' as

$$q = \sigma_1' - \sigma_2', \quad p' = \frac{\sigma_1' + 2\sigma_3'}{3}.$$

Since volumetric strain rate is zero under undrained conditions, the following relation is obtained after the integration of the volumetric viscoplastic strain rate using Eq.(17) and the decomposiotion of total strain rates into the elastic and viscoplastic strain rates.

$$v^p = -\frac{\kappa}{(1+e)}ln\frac{\sigma_m'}{\sigma_{me}'}. \quad (18)$$

From Eq.(8), the axial viscoplastic strain rate under the triaxial stress state is expressed by

$$\dot{\varepsilon}_{11}^{vp} = \sqrt{2/3}\frac{C}{\sigma_m'}\Phi_1(F)\Phi_2(\xi) = \sqrt{2/3}CM\Phi_2(\xi)exp(F) \quad (19)$$

where

$$\Phi_1(F) = \sigma_m'Mexp[m'(\frac{\eta}{M} + ln\frac{\sigma_m'}{\sigma_{me}'} - \frac{(1+e)v^p}{\lambda-\kappa})] \quad (20)$$

$$M = \sqrt{3/2}M^* \quad (21)$$

$$\eta = q/\sigma_m'. \quad (22)$$

By integrating $\dot{\xi}$ under the initial condition that ξ is zero when η is zero, we obtain

$$\Phi_2(\xi) = 1 + \frac{M\eta}{G_2(M-\eta)} \quad (23)$$

where $G_2 = \sqrt{3/2}G_2^*$. The stress-dilatancy equation is obtained from Eqs.(8) and (14) as

$$\frac{dv^p}{d\varepsilon_{11}^p} = M - \frac{q}{\sigma_m'}. \qquad (24)$$

From the creep condition q=constant, and Eq.(18), stress ratio η can be expressed by the viscoplastic volumetric strain under the condition that the viscoplastic volumetric strain v^p is zero at $\eta = \eta_i$ as

$$\eta = \eta_i exp(\frac{(1+e)}{\kappa}v^p). \qquad (25)$$

Eq.(19) can be then rewritten in the following form:

$$\dot{\varepsilon}_{11}^{vp} = \sqrt{2/3}CM \times$$

$$exp[m'(\frac{\eta}{M} + ln\frac{\sigma_m'}{\sigma_{me}'} - \frac{(1+e)v^p}{\lambda-\kappa} + ln(\Phi_2(v^p))/m')]. \qquad (26)$$

From, Eqs.(23) and (25), it is seen that Φ_2 is a positive increasing function with a viscoplastic strain. The third term of the exponential function of Eq.(26) works as a strain hardening term because of its negative sign. On the other hand, the last term in the exponential function in Eq.(26) is a strain softening term due to the positive increasing function of the viscoplastic volumetric strain. This is the reason why we call the model derived here a viscoplastic strain softening model. In other words, the second material function play a role as a strain-softening function.

By differentiating the viscoplastic strain rate $\dot{\varepsilon}_{11}^{vp}$ with respect to time,

$$\ddot{\varepsilon}_{11}^{vp} =$$

$$-b(M-\eta)Z[\dot{\varepsilon}_{11}^{vp}]^2 + \Phi_1(F)\frac{M^2}{G_2(M-\eta)^2}[\frac{\eta(1+e)}{\kappa}]\dot{v}^p \qquad (27)$$

where

$$Z = 1 - \frac{q}{\sigma_m M}(1 - \frac{\kappa}{\lambda}) \qquad (28)$$

Using Eqs.(18) and (24), Eq.(27) can be rewritten as

$$\ddot{\varepsilon}_{11}^{vp} = a(\eta)[\dot{\varepsilon}_{11}^{vp}]^2 \qquad (29)$$

$$a(\eta) =$$

$$[-m'\frac{(\lambda(1+e))}{\kappa(\lambda-\kappa)}(M-\eta)Z + \frac{\eta M^2(1+e)}{\kappa(M\eta + G_2(M-\eta))}]. \qquad (30)$$

In Eq.(39), the sign of the first term in the bracket is negative and increases up to zero at the critical state. On the other hand, the second term is positive monotonically increasing function. Considering that the second material function plays a role as a softening function as mentioned before, the strain-softening enters into $a(\eta)$ through the second term in the bracket of Eq.(30). The stress ratio η_c at the minimum strain rate is given by the equation $a(\eta) = 0$. Since $a(\eta) = 0$ is the cubic equation with respect to η, the solution can be ob-

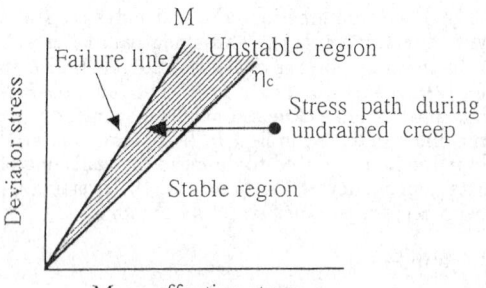

Figure 1. Stress path during undrained creep and η_c at the minimum strain rate.

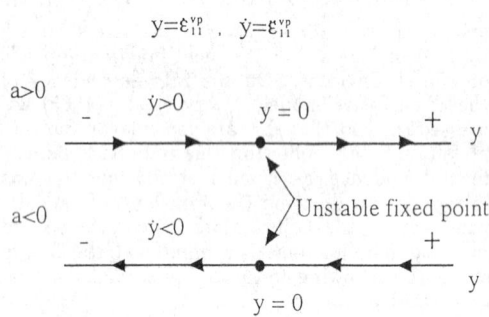

Figure 2. Phase portrait.

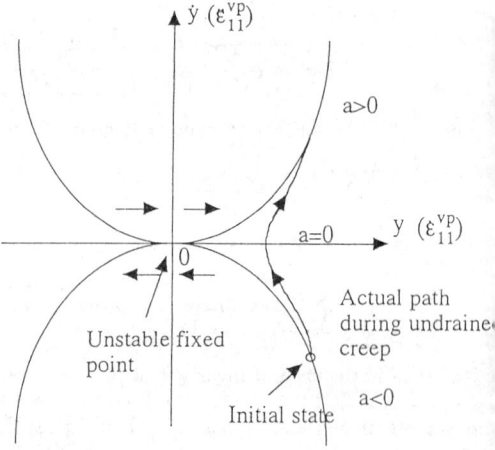

Figure 3. Actual path on the phase portrait.

tained by Cardano's method. The stress ratio η_c at the minimum strain rate is schematically shown in Figure 1. η_c is calculated as $\eta_c = 1.24$ for Osaka alluvial clay in that M (the value of η at the critical state) = $1.29(M^* = 1.05)$.

For simplicity, Eq.(29) can be rewritten as

$$\dot{y} = ay^2 \tag{31}$$

where

$$y = \dot{\varepsilon}_{11}^{vp}. \tag{32}$$

Since the sign of the viscoplastic strain rate is always positive, the positive value of y is only meaningfull.

Firstly, we will examine the mathematical structure of the model expressed in Eq.(31). $y = 0$ is the fixed point in Eq.(31). It is known that the fixed point of this type of differential equation is unstable (Hale and Koçak, 1990). This can be understood from the phase portraits in Figure 2 for different signs of a. The structures of the phase portraits suddenly change at $a = 0$. A system is structurally unstable if the sufficiently nearby vector fields have not always an equivalent phase portrait in the sense of topology. The bifurcation point is defined as the point where the the system is structurally unstable. This means that the point $a = 0$ is a bifurcation point. Figure 3 shows the phase portraits in $\dot{\varepsilon}_{11}^{vp} - \ddot{\varepsilon}_{11}^{vp}$. The initial point has a positive viscoplastic strain rate and a negative rate of viscoplastic strain rate. After a loading, the rate of viscoplastic strain rate increases to the zero value of the rate of viscoplastic strain rate and the viscoplastic strain rate decreases. Finally the rate of viscoplastic strain rate and the viscoplastic strain rate increase up to infinity. a is negative in the stress space where stress ratio $\eta < \eta_c$, while a is positive when $\eta > \eta_c$. By integrating the differential equation under the initial condition that $y = y_0$ when $t = 0$ for a=constant, the solution is obtained as

$$y = \frac{y_0}{(1 - aty_0)}. \tag{33}$$

Eq.(33) shows that y will be unbounded at time $1/ay_0$ if a is positive and the solution of Eq.(31) is unstable. Let's consider two solutions, y_1 and y_2, for two slightly different initial values of y. At $t = t_0$, $y_{02} = y_{01} + \epsilon$, in which y_{01} and y_{02} are positive initial values of y and ϵ is a small positive constant. At time t, the difference between the two solutions can be evaluated at time t by the following equation.

$$|y_1 - y_2| = \frac{\epsilon}{|(1 - aty_{01})(1 - at(y_{01} + \epsilon))|} \tag{34}$$

where y_0 is the positive initial value of y.

If a is negative, the difference becomes zero when $t \to \infty$ and both y_1 and $y_2 \to 0$. This means that the solution is asymptotically stable in the sense of Liapunov. If a is positive, on the other hand, the difference between the two slightly different solutions becomes infinite at the finite time. This indicates that the solution is unstable in regions of the stress space where a is positive and is the reason for the material instability during the acceleration creep process. This case is unstable in the sense of Liapunov.

5 FINITE ELEMENT SOLUTION

In order to numerically study the strain localization of clay, we analyze a rectangular clay specimen under compression test condition at a constant strain rate. The specimen is compressed under the plane strain condition and the boundary of the material is assumed to be impermeable. The transport of the pore water in the specimen, however, is allowed. Horizontal displacements are fixed at the end plates. This restriction on the horizontal displacement is a trigger for the strain localization.

As for the governing equations of water-saturated materials, we adopted Biot's type formulation (Oka et al., 1986). In the finite element analysis, we adopted the updated Lagrangian method for the finite strain. The selection of the finite element type is a very important and difficult task especially in a large deformation analysis. The shear locking of the element can be often eliminated by introducing a reduced integration scheme. It has been, however, demonstrated that the reduced integration leads to a spurious hourglass mode (Irons and Ashmad, 1980), which can easily destroy the real solution. This tendency is more conspicuous in the calculation with reduced integration apllied to four noded and nine noded isoparametric (compatible) elements in plane strain problems than in the calculation with reduced integration applied to eight noded isoparametric (incompatible) element (Bićanić and Hinton, 1979; Kinger and Smith, 1992). In this present study, therefore, an eight-noded quadrilateral element with a reduced Gaussian (2×2) integration is adopted to eliminate the shear locking and to reduce the tendency of appearance of a spurious hourglass mode. On the other hand, the pore water pressures are defined at four corner nodes. The weak form of the continuity equation is integrated with (2×2) full integration. By using this combination of the spatial integration schemes, the effective stresses, the pore water pressures and the strains are calculated at the same integration points in each element. The algorithm of the time integration of the constitutive equation is the type of tangent modulus method by Peierce et

Table 1. Material parameters for Adachi and Oka's model.

Compression index λ	0.355
Swelling index κ	0.0477
Initial mean effect. str. p_0	1.0 (kgf/cm^2)
Initial void ratio e_0	2.0
Viscoplastic parameter m'	12.8
Viscoplastic parameter C	2.85×10^{-8} (1/sec)
Stress ratio at failure M^*	1.45
Poisson's ratio ν	0.33

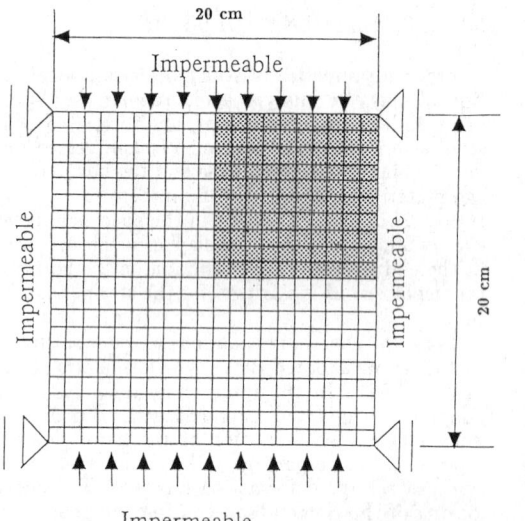

Figure 4. Finite element mesh and boundary conditions.

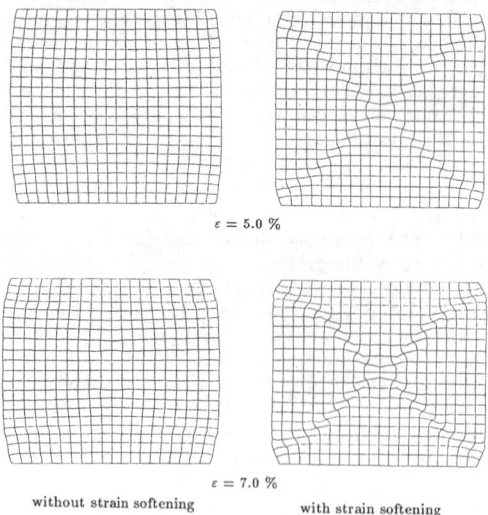

Figure 5. Deformed mesh of specimens with and without strain softening at various stages.

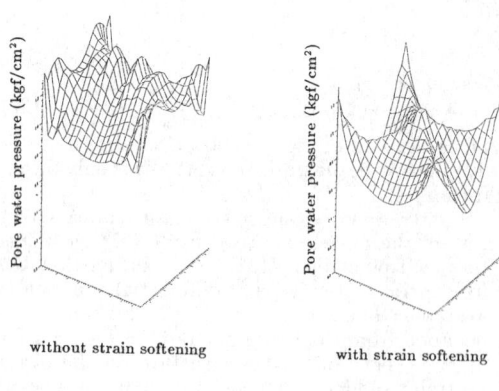

Figure 6. Distributions of the pore water pressure of specimens with and without strain softening at an overall axial compressive strain of 3%.

al.(1984) with θ(interporation parameter) = 0.5. The time increment used in the calculation is 0.6 sec when the strain rate = 1%/min. The material parameters are listed at Table 1.

Effect of strain softening

Firstly, the effect of the strain softening (the vanishing of rate dependency at the failure state) on the strain localization is examined by using material parameters with and without the strain softening term. The finite element mesh used is shown in Figure 4. One quadrant of the whole specimen is employed in the computations due to the symmetry of the loading and boundary conditions. The shaded portion of the finite element mesh corresponds to the analytical area. The algorithm of the time integration of the constitutive equation is the type of tangent modulus method with θ=0.5. The time increment used in the calculation is 0.6 sec when the strain rate = 1%/min. The coefficient of permeability of Eastern Osaka clay was obtained through the consolidation test and $k = 1.16 \times 10^{-8}$cm/sec. For the strain softening term, G_2^*, the value of 0.001 is used in this numerical example.

Figure 5 shows the deformed mesh of specimens with and without strain softening at various stages. In the calculation with the strain softening, the deformations are observed to remain essentially homogeneous up to the maximum load point after which shear bands develop. The deformation mode shifts to one involving bands of highly localized shearing. On the other hand, the numerical calculation without the strain softening finds continued growth of the homogeneous deformation with no tendency exhibited for shear bands to form. The angle of the shear band to the horizontal plane is about 45°.

Figure 6 shows the distributions of the pore water pressure at an overall axial compressive strain of 3%. The positive pore water pressure due to a negative dilatancy develops well along the shear band of the model with the strain softening while the distribution of the pore water pressure is rather uniform in the model without the strain softening.

Figure 7 shows the relationships between the overall deviator stresses and strains with and without strain softening. The overall deviator stress is

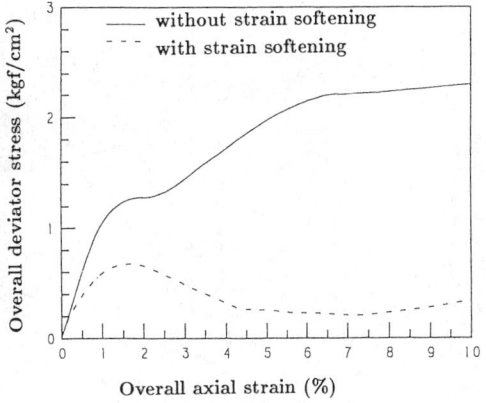

Figure 7. Relationship between the overall deviator stresses and strains of specimens with and without strain softening.

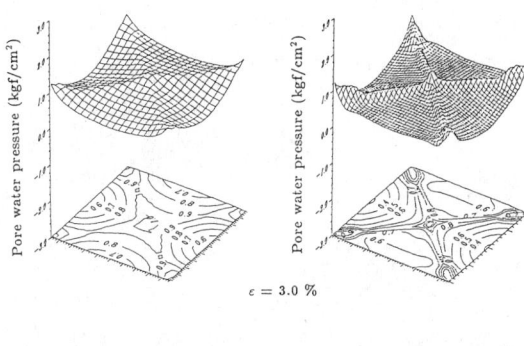

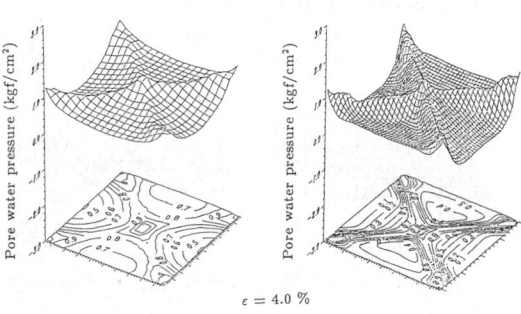

Figure 9. Distributions and contours of the pore water pressure of specimens with a different number of finite elements at various stages.

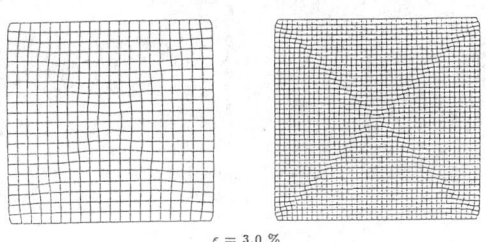

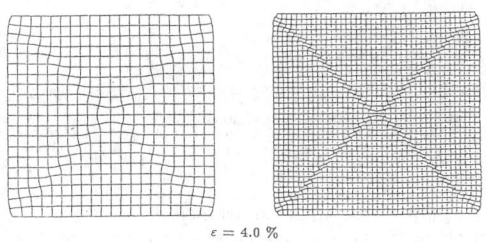

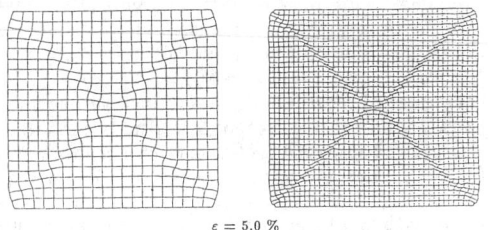

Figure 8. Deformed mesh of specimens with a different number of finite elements at various stages.

calculated by the reaction forces on the end plate. The difference in overall stress behavior exhibited in Figure 7 arises from the difference in deformation modes shown in Figure 5.

Mesh size sensitivity

For dynamic problems with the viscoplastic model, a kind of characteristic length scale is naturally introduced into the analysis as the ratio of wave speed to relaxation time (Needleman, 1988). For static problems discussed in this study, however, defining the characteristic length scale is still an open problem. Therefore, the comprehensive studies on the mesh size sensitivity of the solution have to be carried out. In this study, by the limitation of the computational capacity, only a few different finite element meshes are examined. The discussions based on the calculations with much finer meshes would be a significant further step.

Two different square mesh configurations are prepared to discuss the mesh size sensitivity of the solution. The coarser mesh has 400 elements, while the finer mesh has 1600 elements. One quadrant of the whole specimen is employed in the computations due to the symmetry of the loading and boundary conditions for both cases. Figure 8 shows

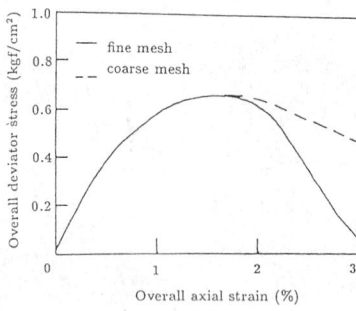

Figure 10. Relationship between the overall deviator stresses and strains of specimens with a different number of finite elements.

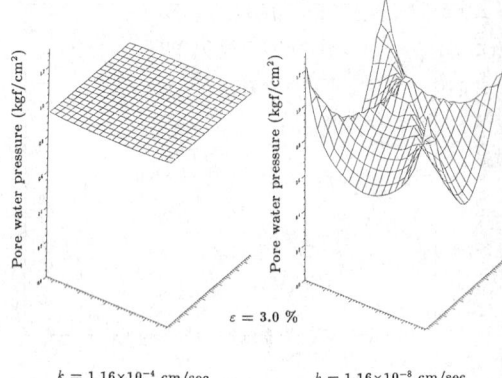

$k = 1.16 \times 10^{-4}$ cm/sec $\qquad k = 1.16 \times 10^{-8}$ cm/sec

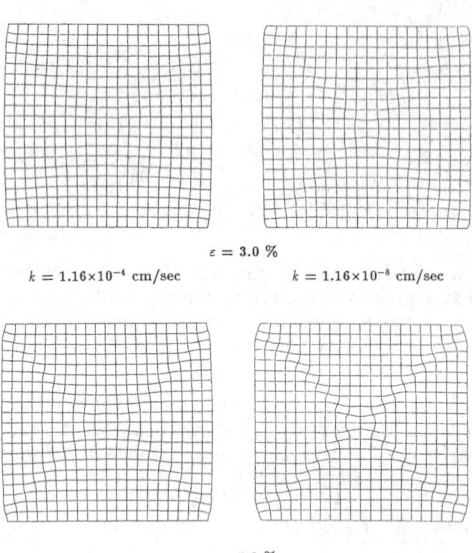

$k = 1.16 \times 10^{-4}$ cm/sec $\qquad \varepsilon = 3.0\% \qquad k = 1.16 \times 10^{-8}$ cm/sec

$\varepsilon = 5.0\%$

Figure 11. Deformed mesh of specimens with a different permeability at various stages.

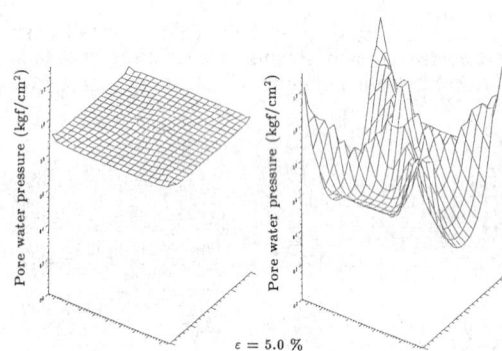

$\varepsilon = 5.0\%$

Figure 12. Distributions of the pore water pressure of specimens with a different permeability at various stages.

in the finer mesh tends to decrease more rapidly than that in the coarse mesh.

Effect of permeability

In the present paper, undrained plane strain compression tests are simulated considering a transport of the pore water in the material. To discuss the effect of permeability, two numerical calculations with a different coefficient of permeability are carried out: k (coefficient of permeability) = 1.16×10^{-4}cm/sec and $k = 1.16 \times 10^{-8}$cm/sec. The coefficient of permeability determined through the consolidation test corresponds to the latter value.

Figure 11 shows the deformed mesh of specimens for both cases at various stages. It is found that the strain localization is moderate for the specimen with a higher permeability and the strain localization is stronger for the specimen with a lower permeability. The same tendencies are observed in the distributions of the pore water pressure shown in Figure 12. The relationships between the overall deviator stresses and strains for two cases are shown in Figure 13. The specimen

the deformed mesh of specimens for both cases at various stages. It is difficult to discuss the difference of the thickness of shear bands in these different mesh configurations. Figure 9 depicts the distributions and contours of the pore water pressure in the specimens. The pore water pressure develops well along shear bands. It is found that the pore water pressure tends to be more localized in the finer mesh system than in the coarser one. For comparison purposes, the relationships between the overall deviator stresses and strains for both cases are shown in Figure 10. Both the overall deviator stresses versus strains coincide up to the peak stress after which the deviator stress

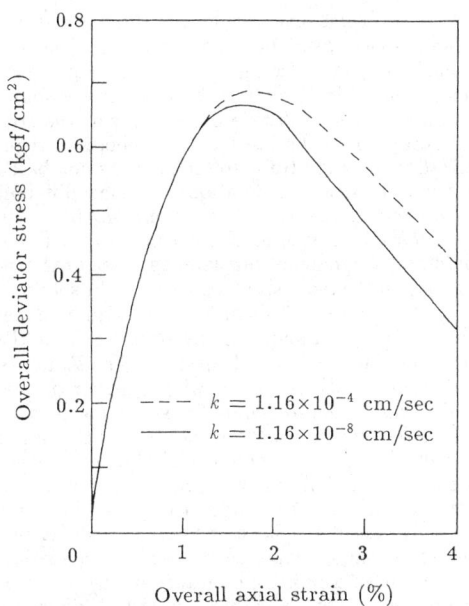

Figure 13. Relationship between the overall deviator stresses and strains of specimens with a different permeability.

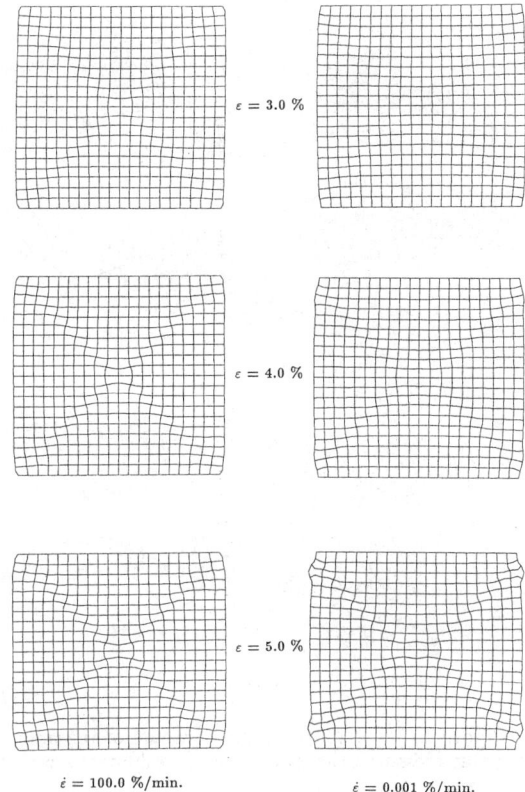

Figure 14. Deformed mesh of specimens under a diferent strain rate condition at various stages.

with a lower permeability shows a softer response than that with a higher permeability.

This tendency obtained here is opposite to the results obtained by Loret and Prévost (1990). They used a Drucker-Prager type of yield function and an associated flow rule in their dynamic analyses of strain localization of fluid-saturated porous media. In their analyses, the positive dilatancy was only introduced due to their constitutive assumption. Rice (1975) also introduced the concept of dilatant hardening for the material with the positive dilatancy. This positive dilatancy retards the generation of shear bands under the undraind condition because the negative pore water pressure generates near the localized area. On the contrary, in this study, the negative dilatancy is only used in the quasi-static analyses of water saturated clay. The positive pore water pressure, therefore, generates in the neighorhood of the localized area. Then, the descent of the effective mean stress accelerates the growth of shear bands. These may be the reasons for the difference between two numerical results on the effect of permeability.

Effect of strain rate

The influence of the strain rate on the behavior of clay can easily be investigated in the calculation. Two numerical calculations with a different strain rate are carried out: $\dot{\varepsilon}$ (overall axial compressive strain rate) = 100%/min and $\dot{\varepsilon}$ = 0.001%/min.

Figure 14 depicts the deformed mesh of specimens for both cases at various stages. It can be seen that the specimen under a higher strain rate condition tends to localize more rapidly than the specimen under a lower strain rate condition. This tendency of the effect of strain rate on the strain localization can be obtained more clearly from the the distributions and contours of the pore water pressure shown in Figure 15. For the case under a higher strain rate condition, the pore water pressure is found to localize significantly along shear bands compared with the case under a lower strain rate condition.

6 CONCLUDING REMARKS

The following conclusions are obtained from the present study.

1. Introducing a second material function into the overstress type viscoplastic flow rule, a viscoplastic softening constitutive model for clay was derived. The second material function controls the vanishing of rate dependency at the critical state.

2. The instability of the proposed model was evaluated in undrained triaxial compression tests.

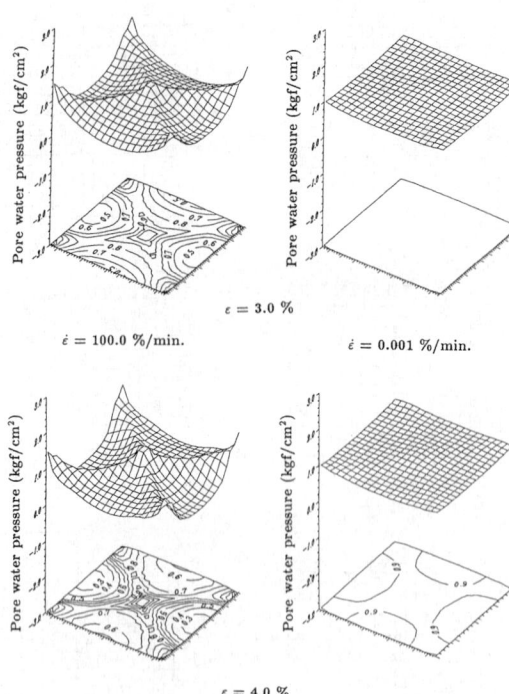

Figure 15. Distributions and contours of the pore water pressure of specimens under a diferent strain rate condition at various stages.

The evolutional equation of viscoplastic strain under undrained creep has critical stress ratio at transition point between the stable and the unstable regions in the sense of Liapunov. In the unstable region, the difference between two initially slightly different solutions becomes very large. At the critical stress ratio, η_c, phase portraits change structurally and the evolutional equation is structurally unstable at this stress ratio. This indicates that the point where the stress ratio is at a critical value of η_c is the bifurcation point. In addition, the fixed point of this evolutional equation is unstable. From the qualitative features mentioned above, it was found that the proposed viscoplastic model has the potential to express the plastic instability of clay.

3. In order to numerically study the strain localization of clay, we analyzed a rectangular clay specimen under compression test condition. The numerical analyses brought out interesting and important features of shear band development from homogeneous deformation states. The pattern of shear band development depended on the incorporation of the strain softening into the constitutive model. For models without strain softening, continued growth of the homogeneous deformation was observed.

4. In problems involving localized shearing there is mesh size dependency of the solution. The viscoplastic regularization has been used to overcome this problem. In this study, however, it was found that the vanishing of rate dependency at the critical state led to the loss of the viscoplastic regularization during strain softening behavior based on the numerical examinations for a few different finite element meshes. The discussions based on the calculations with much finer meshes would be, therefore, a significant further step to lead the conclusions on the mesh size sensitivity of the solution.

5. The course of shear band deformation depended on a transport of the pore water in the specimen and the axial strain rate. The strain localization for the specimen with a higher permeability was moderate, while the strain localization was more significant for the specimen with a lower permeability. This tendency might be partially dependent on the dilatancy properties in the constitutive model. In this study, a negative dilatancy was only considered in the numerical calculations. Therefore, the comparison between the numerical analyses with a positive dilatancy in the constitutive model and those obtained in this study would be useful in this class of problems. It was also found out that the specimen under a higher strain rate condition tends to localize more rapidly than the specimen under a lower strain rate condition.

ACKNOWLEDGEMENTS

The authors wish to acknowledge Mr. K.Ono, a graduate student of Gifu University, for his help with the calculations.

REFERENCES

Adachi,T. and Oka,F. Constitutive equations for normally consolidated clay based on elasto-viscoplasticity, Soils and Foundations, Vol.22, No.4, 57-70 (1982).

Adachi,T., Oka,F. and Mimura,M. An elasto- viscoplastic theory for clay failure, Proc. 8th Asian Regional Conf. on SMFE, Vol.1, 5-8 (1987)

Adachi,T., Oka,F. and Mimura,M. Mathematical structure of an overstress elasto- viscoplastic model for clay, Soils and Foundations, Vol.27, No.3, 31-42 (1987)

Adachi,T., Oka,F. and Mimura,M. Elasto- viscoplastic constitutive equations for clay and its application to consolidation analysis, J. Engineering Materials and Technology, ASME, 112, 202-209 (1990)

Aifantis,E.C. On the microstructural origin of certain inelastic models, J. Mat. Engng. Tech., ASME, Vol.106, 326-330 (1984)

Arthur,J.F.R., Dustan,T., Al-Ani,Q.A.J. and Assadi,A. Plastic deformation and failure in granular media, Géotechnique, Vol.27, 53-74 (1977)

Aubry,D., Kodaissi,E. and Meimon,M. A viscoplastic constitutive equation for clays including damage law, Proc. 5th Int. Conf. on Numerical Meth. in Geomech., Nagoya, edited by Kawamoto and Ichikawa, Vol.1, 421-428 (1985)

Bićanić,N. and Hinton,E. Spurous modes in two-dimensional isoparametric elements, Int. J. Num. Meth. in Engng., Vol.14, 1545-1557 (1979)

Biot,M.A. Mechanics of deformation and accoustic propagation in porous media, Journal of Applied Physics, Vol.33, No.4, 1482-1498 (1962)

Hadamard,J. Lecons sur la propagation des ondes et les équations de l'hydrodynamique, Librairie scientifique, A.Hermann, Paris (1903)

Hale and Koçak. Dynamics and Bifurcation, Springer-Verlag, 308-310 (1990)

Hill,R. Acceleration waves in solids, J. Mech. Phys. Solids, Vol.10, No.1, 1-16 (1962)

Irons,B. and Ashmad,S. Techniques of Finite Elements, Ellis Horwood, Chichester, England (1980)

Kinger,D.J. and Smith,I.M. Eigenvalues of element stiffness matrices. Part I: 2-D plane elements, Engineering Computations, Vol.9, 307-316 (1992)

Loret,B. and Prevost,J.H. Dynamic strain localization in fluid saturated porous media, J. of Engng. Mech., ASCE, Vol.117, No.4, 907-922 (1990)

Mandel,J. Conditions de stabilité et postulat de Drucker, Proc. IUTAM Symp. on Rheology and Soil mechanics, edited by J. Kravtchenko and Siries, P.M., Springer-Verlag, 58-68 (1964)

Mühlhaus,H.B. Shear band analysis in granular materials by Cosserat theory, Ing. Arch., Vol.56, 389-399 (1986)

Mühlhaus,H.B. and Vardoulakis,I. The thickness of shear bands in granular materials, Géotechnique, Vol.37, 271-283 (1987)

Oka,F. Prediction of time dependent behavior of clay, Proc. 10th ICSMFE, Stochholm, Vol.1, 215-218 (1981)

Peirce,D., Shih,C.F. and Needleman,A. A tangent modulus method for rate dependent solids, Computers and Structures, Vol.18, No.5, 875-887 (1984).

Rice,J.R. The localization of plastic deformation, Proc. IUTAM Symp., Theoretical Appl. Mechanics, edited by Koiter,W.T., 207-220 (1975)

Rice,J.R. On the stability of dilatant hardening for saturated rock masses, J. Geophysical Research, Vol.80, No.11 (1975)

Thomas,T.Y. Plastic flow and fracture in solids, Academic Press, New York (1961)

Vardoulakis,I. Bifurcation analysis of the triaxial test on sand samples, Acta Mechanica, Vol.32, 35-54 (1979)

Vardoulakis,I. Shear band inclination and shear modulus of sand in biaxial tests, Int. J. Num. Ana. Meth. in Geomech., Vol.4, 103-119 (1980)

Vermeer,P.A. A simple shear-band analysis using compliance, Proc. IUTAM Conf. on Deformation and Failure of Granular Materials, edited by P.A.Vermeer and H.J.Luger, Balkema, 493-499 (1982)

A strain localisation analysis of frozen sand by elasto-viscoplastic softening model

Modélisation de la localisation de la déformation dans un sable gelé avec un modèle élasto-viscoplastique adoucissant

Toshihisa Adachi
School of Civil Engineering, Kyoto University, Japan

Fusao Oka, Atsushi Yashima & Lim Liong Chu
Department of Civil Engineering, Gifu University, Japan

ABSTRACT: The aim of the present paper is to numerically analyze the behavior of frozen sand by using a viscoplastic constitutive model with strain softening. Two of the authors (Adachi and Oka, 1990) developed this constitutive model introducing the stress history tensor which is a functional of the stress history, with respect to a generalized time measure. It has been shown that Adachi and Oka's model is applicable to the results of triaxial tests on a frozen Toyoura sand at different strain rates. The model is then implemented into a FEM code to numerically simulate the behavior under plane strain conditions. From the numerical results, it is revealed that the formation of shear bands is possible and the characteristics of strain localization, such as shear banding, depend on the strain rates.

Le sujet de cet article est l'analyse numérique du comportement d'un sable gelé avec un modèle viscoplastique adoucissant. Deux des auteurs (Adachi et Oka, 1990) ont développé ce modèle rhéologique qui introduit un tenseur d'histoire de contrainte comme une fonctionnelle de l'histoire de contrainte, par rapport à une mesure de temps généralisé. On montre que le modèle d'Adachi et Oka est applicable aux résultats d'essais triaxiaux sur un sable de Toyoura gelé à différentes vitesses de déformation. Le modèle est intégré dans un code aux éléments finis pour simuler numériquement le comportement en déformation plane. Les résultats numériques montrent que la formation de bandes de cisaillement est possible et que les caractéristiques de la localisation dépendent de la vitesse de déformation.

1 INTRODUCTION

For the last two decades, many constitutive models have been proposed for geomaterials and applied to practical problems. There still remain some outstanding issues which must be resolved. One of them is the formation of shear bands before failure. This problem is closely connected to the strain localization phenomenon and plastic instability. Recently, the importance of simulation of post localization regime has been pointed out. In the present paper, we have numerically analyzed the behavior of frozen sand by using a viscoplastic constitutive model with strain softening. Two of the authors (Adachi and Oka, 1990) developed this constitutive model for frozen sand introducing the stress history tensor that is a functional of stress history, with respect to generalized time measure. They have applied it to the triaxial test results of frozen Toyoura sand at different strain rates.

The model is then implemented into a FEM code to numerically simulate the behavior under plane strain conditions. From the numerical results, it is revealed that the characteristics of the strain localization like shear banding depend on the strain rate.

2 VISCOPLASTIC SOFTENING MODEL FOR FROZEN SAND

Experimental works have revealed that the frozen soil exhibits a rate-sensitive, strain hardening, and at larger strain levels a strain softening behavior (Ladanyi 1981). Many models have been proposed to describe the behavior of frozen soil (e.g., Vialov 1963, Andersland and Al-Nouri 1970, Fish 1980, Assur 1979, and Ting 1983). Most of these studies were conducted under uniaxial loading conditions witout confining presssure. A three dimensional formulation is relevant to problems encountered in geotechnical engineering.

Oka (1985) proposed a new type of viscoplastic model with memory and internal variables. Based on this model, Oka and Adachi (1985) constructed the elasto-plastic constitutive model with strain softening for soft rock introducing a stress history tensor. In this model, a time measure that is similar to the endochornic time by Valanis (1971) is used instead of real time. Adachi, Oka and Poorooshasb (1990) proposed a constitutive model for frozen sand using a generalized measure of time. In this section, the strain softening viscoplastic model for frozen sand is rationally reformulated and then generalized in part. For simplicity, only infinitesimal strain fields are considered in the present paper. A new time measure is introduced given by

$$dz = F(\text{strain rate})dt \quad (1)$$

where dz is an increment of the new time measure, t is the real time and the function F is to be determined experimentally for the particular medium.

The stress history tensor σ_{ij}^* is given as a functional of a reduced stress history with respect to the new time measure, z,

$$\sigma_{ij}^* = \sigma_{ij}^*[\sigma_r^z(z-z')] \quad (2)$$

$$\sigma_r^z = [\sigma_{ij}; 0 < z' \le z] \quad (3)$$

where σ_r^z is the reduced stress history. Here, the stress history tensor is expressed by a single exponential type kernel function as follows:

$$\sigma_{ij}^* = \int_0^z \frac{1}{\tau} exp[(z-z')/\tau](\sigma_{ij}(z') - \sigma_{ij}(0))dz' \quad (4)$$

where τ is a material parameter expressing the retardation of stress with respect to the time measure, $\sigma_{ij}(0)$ is the value of the stress tensor at $z=0$. z is a measure defined by

$$dz = gdt \quad (5)$$

$$g = g(\dot{\varepsilon}_{ij}) . \quad (6)$$

In the present study, only infinitesimal strain fields are dealt with and the total strain rate tensor is decomposed into its elastic and plastic components; i.e.,

$$\dot{\varepsilon}_{ij} = \dot{\varepsilon}_{ij}^e + \dot{\varepsilon}_{ij}^p . \quad (7)$$

The elastic strain rate is given in a linear elastic form given by

$$\dot{\varepsilon}_{ij}^e = \frac{\dot{S}_{ij}}{2G} + \frac{\dot{\sigma}_m}{3K}\delta_{ij} \quad (8)$$

where G is the elastic shear modulus, K is the elastic bulk modulus and σ_m is the mean stress.

It is assumed that the viscoplastic strain rate, $\dot{\varepsilon}_{ij}^{vp}$, is given by the non-associated flow rule

$$d\varepsilon_{ij}^p = H\frac{\partial f_p}{\partial \sigma_{ij}}df_y \quad (9)$$

where f_p is the plastic potential function, f_y is the yield function and H is the loading index. The yield function is assumed to be a function of the stress history tensor and hardening/softening parameter, and is given by

$$f_y = \bar{\eta}^* - \kappa = 0 \quad (10)$$

$$\bar{\eta}^* = (\eta_{ij}^* \eta_{ij}^*)^{1/2} \quad (11)$$

$$\eta_{ij}^* = S_{ij}^*/\sigma_m^* \quad (12)$$

where S_{ij}^* is the deviatoric part of the stress history tensor, σ_m^* is the mean stress history.

Next, a specification of the loading conditions will be given.
Loading:
If $f_y = 0$ and $df_y = (\partial f_y/\partial \sigma_{ij}^*)d\sigma_{ij}^* > 0$, $d\varepsilon_{ij}^p \ne 0$.
Neutral loading:
If $f_y = 0$ and $df_y = 0$, $d\varepsilon_{ij}^p = 0$.
Unloading:
If $f_y = 0$ and $df_y < 0$, $d\varepsilon_{ij}^p = 0$.

It is assumed that the evolutional equation of strain hardening/softening parameter κ is given by the following relation

$$\dot{\kappa} = \frac{G'(M_f^* - \kappa)^2 \dot{\gamma}^p}{M_f^{*2}} \quad (13)$$

in which $\dot{\gamma}^p$ is the second invariant of the deviatoric plastic strain rate tensor, given by the equation

$$\dot{\gamma}^p = (\dot{e}_{ij}^p \dot{e}_{ij}^p)^{1/2} \quad (14)$$

where $\dot{e}_{ij}^p$ is the deviatoric plastic strain rate tensor. Then, κ is given by

$$\kappa = \int_0^t \dot{\kappa} dt . \quad (15)$$

When γ^p is zero at an initial state under proportional loading condition, hyperbolic function is given by integration,

$$\kappa = \frac{M_f^* G' \gamma^p}{M_f^* + G' \gamma^p} \quad (16)$$

in which M_f^* is the value of η^* at the residual strength state and G' is the initial tangent of Eq.(13).

Plastic potential function and overconsolidation boundary surface

The plastic potential function is assumed to be a function of real stress and is given by the following form which is similar to the function for soft rock,

$$f_p = \bar{\eta} + \tilde{M} ln[(\sigma_m + b)/(\sigma_{mb} + b)] = 0 \quad (17)$$

in which b and σ_{mb} are material parameters, and $\bar{\eta}$ is a stress invariant ratio defined by

$$\bar{\eta} = [S_{ij}S_{ij}/(\sigma_m + b)^2]^{1/2} . \quad (18)$$

An overconsolidated boundary surface is also introduced as

$$f_b = \bar{\eta} + \bar{M}_m^* ln[(\sigma_m + b)/(\sigma_{mb} + b)] = 0 \quad (19)$$

where $\bar{M}_m^*$ is the value of stress ratio where the

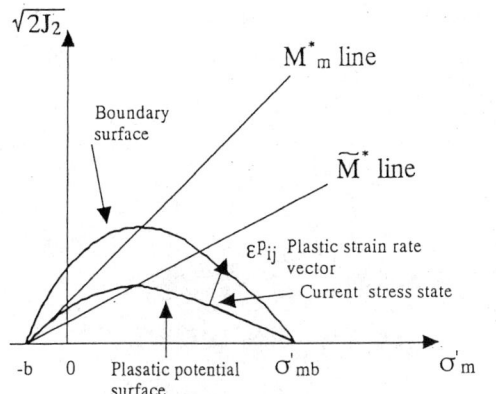

Figure 1. Boundary surface and plastic potential surface.

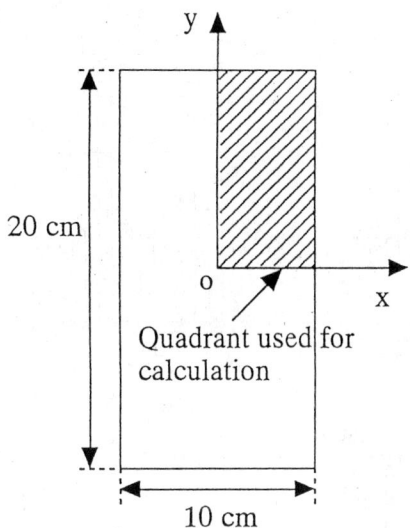

Figure 2. Material configuration.

maximum compression takes place. Inside the boundary surface ($f_b < 0$), $\bar{M}$ is given by

$$\bar{M} = -\bar{\eta}/ln[(\sigma_m + b)/(\sigma_{mb} + b)] . \quad (20)$$

On the other hand, outside the boundary surface ($f_b \geq 0$), $\bar{M}$ is constant

$$\bar{M} = \bar{M}_m^* . \quad (21)$$

The boundary surface and plastic potential function in the case of isotropically consolidated state are shown in Figure 1.

The total strain increment is given by considering the elastic strain increment and Prager's consistency condition as:

$$d\varepsilon_{ij}^p = \Lambda[\frac{\bar{\eta}_{ij}}{\bar{\eta}} + (\bar{M} - \bar{\eta})\frac{\delta_{ij}}{3}][\frac{\eta_{kl}^*}{\eta^*} - \eta^*\frac{\delta_{kl}}{3}]\frac{d\sigma_{kl}^*}{\sigma_m^*} \quad (22)$$

$$\Lambda = \frac{M_f^{*2}}{G'(M_f^* - \eta^*)^2} \quad (23)$$

where

$$\bar{\eta}_{ij} = \frac{S_{ij}}{(\sigma_m + b)} .$$

Table 1. Material parameters

E (Young's modulus)	(MPa)	10,000
K (Bulk modulus)	(MPa)	1,600
b	(MPa)	40
σ_{mb}	(MPa)	100
σ_{m0}	(MPa)	5
$\dot{\varepsilon}_0$	(%/min.)	2.7
τ	(sec.)	130
M_f^*		1.1
M_f^* (imperfection)		1.0
e_0 (initial void ratio)		0.624
a		0.92
G'		200
F_0		1

3 FINITE ELEMENT ANALYSIS BY ELASTO-VISCOPLASTIC SOFTENING MODEL

In the finite element analysis, we used 4-noded linear elements. In order to obtain the stiffness matrix, a square element is divided into four triangular elements, and the stiffness matrix is obtained for the square element by taking the average of the four stiffness matrices of four triangular elements. As for the solution procedure, a first-order self correction method (Stricklin et al. 1973) is used. The calculations are carried out under plane strain conditions with smooth boundaries at a constant average strain rate. The material configuration is shown in Figure 2 and the compression axis is aligned with the y-axis. Material properties, initial and loading conditions are listed in Table 1.

First, the mesh size sensitivity was examined by using different numbers of elements without and with imperfection. The finite element meshes are shown in Figures 3((a),(b),(c)). In Figure 4, average stress-strain relations using three different meshes without imperfection are shown. These results show no effect of the mesh size on the stress-strain relations at different average axial strain rates. The stress-strain relations obtained by the two mesh sizes with an imperfection are shown.

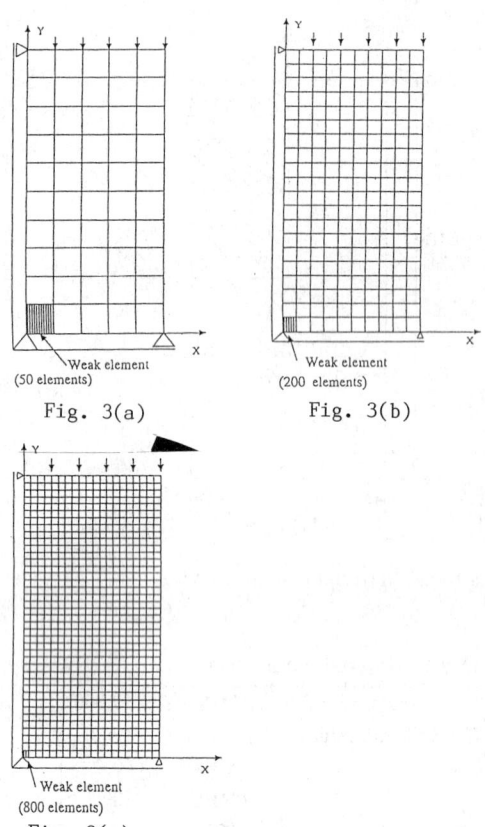

Fig. 3(a) Weak element (50 elements)

Fig. 3(b) Weak element (200 elements)

Fig. 3(c) Weak element (800 elements)

Figure 3. Quarter of finite element mesh.

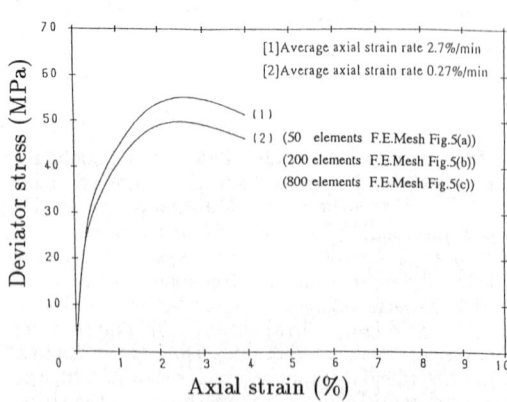

Figure 4. Relations between overall deviator stress and overall axial strain.

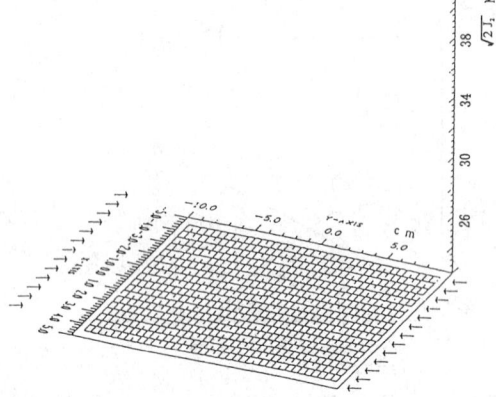

Figure 5(a). Distribution of second invariant of deviatoric stress tensor at axial strain of 0.2 % (axial strain rate = 2.7 %/min.).

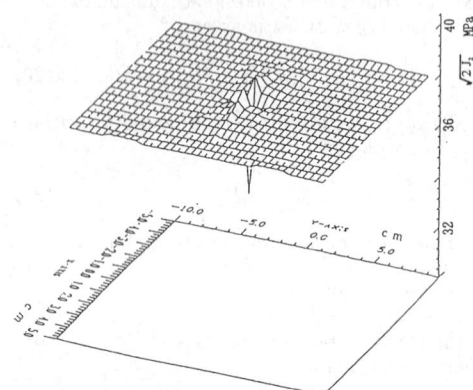

Figure 5(b). Distribution of second invariant of deviatoric stress tensor at axial strain of 2.0 % (axial strain rate = 2.7 %/min.).

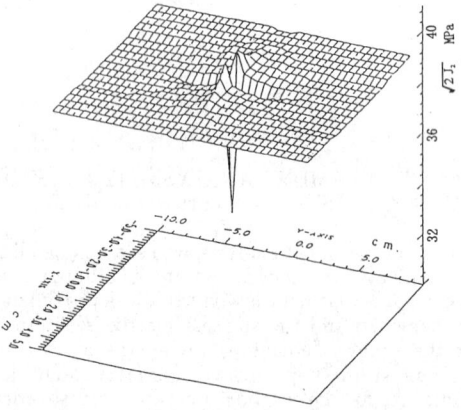

Figure 5(c). Distribution of second invariant of deviatoric stress tensor at axial strain of 2.7 % (axial strain rate = 2.7 %/min.).

An imperfection is placed at the center element of the specimen with a 10 % smaller value of material parameter for failure condition. There is also no significant effect of mesh size on the stress-strain response. This is due to the fact that the proposed model is a rate dependent model and corresponds to the *positive material* defined by Valanis (1984).

The material imperfection is introduced to enhance the growth of inhomogeneity in order to examine the instability of the material. In the defective element is set up at the center of the specimen with 10 % smaller value of material parameter for failure condition.

Figures 5((a), (b), (c)) show the distributions of second invariant of deviatoric stress at different average compressive strains. With increase of average strain, material inhomogeneity grows in a systematic manner. At an average compressive strain of 0.2 %, the distribution of second invariant of the deviatoric stress $\sqrt{2J_2}$ is homogeneous. Then, at the average strain of 2.0 %, the value of $\sqrt{2J_2}$ at the center is low due to the existence of imperfection. Around the center and at the four central parts along the lateral boundaries, it can be seen that the value of $\sqrt{2J_2}$ is relatively high. At the average strain of 3.0 %, the discontinuity line of $\sqrt{2J_2}$ is seen along with the diagonal line from the center, then the last part of this discontinuity line is aligned with y-axis. In addition, at the center part of both lateral surfaces, a slight decrease of $\sqrt{2J_2}$ is observed. From a closer look of Figures 5((a),(b),(c)), we figured out the schematic view of the stress discontinuity line grown up during the compressive deformation. The angle θ and the distance L_c are defined in Figure 6.

Figures 7((a),(b)) show the distributions of $\sqrt{2J_2}$ in the case of the average compressive strain rate of 0.27 %/min. The result in Figures 5 is for the stain rate of 2.7 %/min. Compared with two cases, the value of θ is 45° for Figure 5(c) and 66.8° for Figure 7(b), and L_c is 10mm for Figure 5(c) and 20 mm for Figure 7(b), respectively. This observation indicates that the region is narrower for the case of higher strain rate, where the value of the second invariant of deviatoric stress tensor is relatively low.

Next, we will discuss the distribution of the second invariant of plastic strain increment $\sqrt{\dot{\varepsilon}^p_{ij}\dot{\varepsilon}^p_{ij}}$ in the specimen. Figures 8((a),(b),(c)) show the distributions of the second invariant of plastic strain rate for three different strain rates (0.27 %/min, 2.7 %/min, 5.4 %/min) at the average compressive strain of 2 %. The stress has reached peak value at an overall axial strain of about 2.5 % in Figure 4. From Figure 8(c), it is seen that shear bands have developed before the peak stress where the plastic strain rate is higher than that in other regions. The width of the shear bands is about 10mm. The angle of the bands to the loading axis is about 45°. A comparison of Figures 8((a),(b),(c)) reveals that a more significant local-

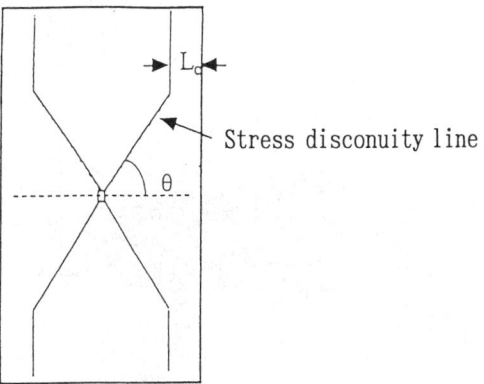

Figure 6. Schematic figure of stress discontinuity line.

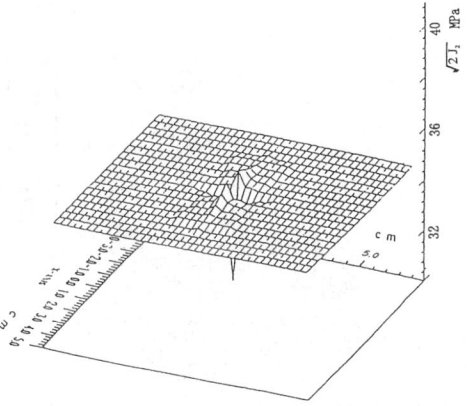

Figure 7(a). Distribution of second invariant of deviatoric stress tensor at axial strain of 2.0 % (axial strain rate = 0.27 %/min.).

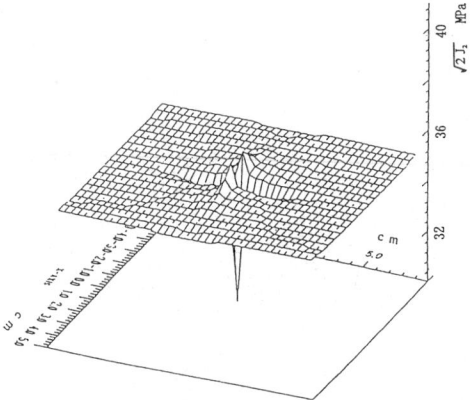

Figure 7(b). Distribution of second invariant of deviatoric stress tensor at axial strain of 3.0 % (axial strain rate = 0.27 %/min.).

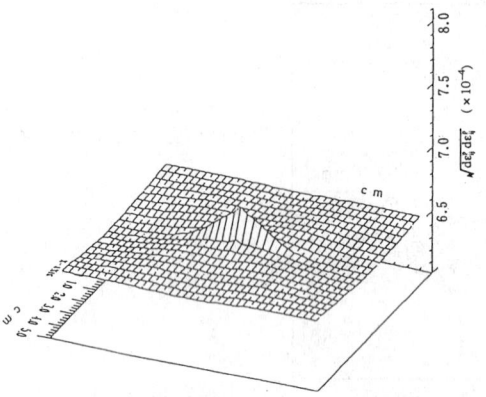

Figure 8(a). Distribution of second invariant of plastic strain increment at axial strain of 2.0 % (axial strain rate = 0.27 %/min.).

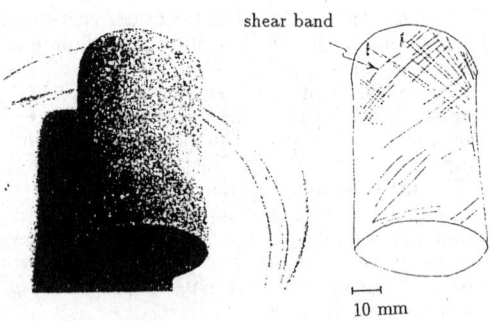

Figure 9. Photograph of specimen of frozen Toyoura sand after triaxial compression test and sketch of shear bands (Shibata et al. 1985)

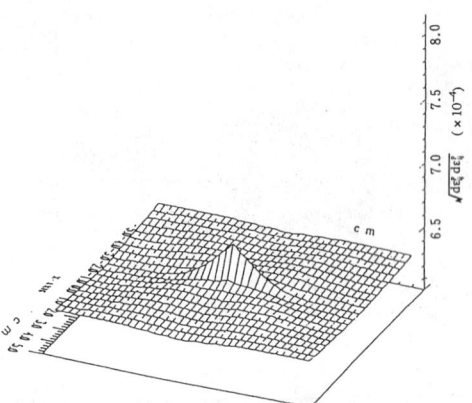

Figure 8(b). Distribution of second invariant of plastic strain increment at axial strain of 2.0 % (axial strain rate = 2.7 %/min.).

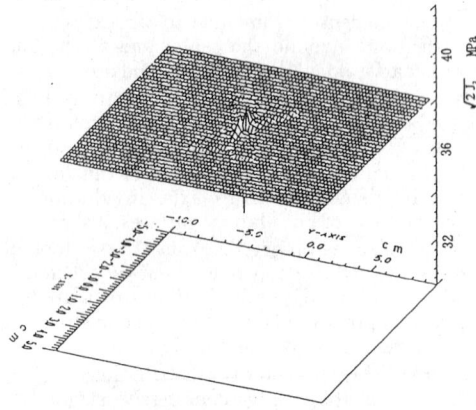

Figure 10(a). Distribution of second invariant of deviatoric stress tensor at axial strain of 2.0 % (axial strain rate = 2.7 %/min.).

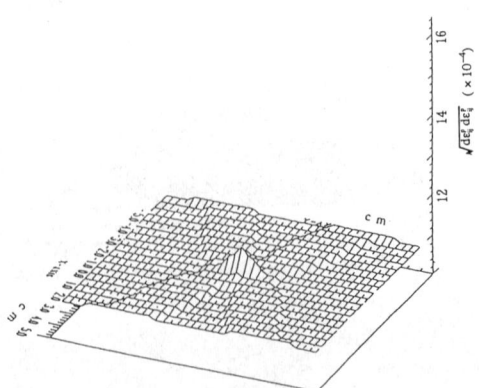

Figure 8(c). Distribution of second invariant of plastic strain increment at axial strain of 2.0 % (axial strain rate = 5.4 %/min.).

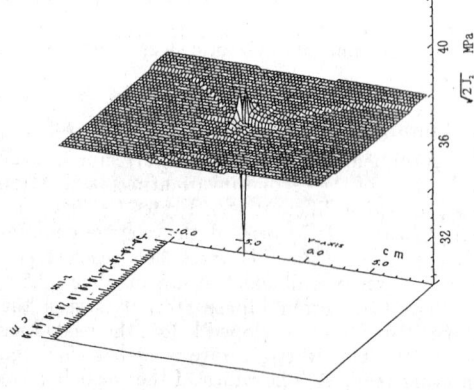

Figure 10(b). Distribution of second invariant of deviatoric stress tensor at axial strain of 3.6 % (axial strain rate = 2.7 %/min.).

ization of plastic strain rate occurs at the higher value of the average compressive strain rate. This is consistent with the fact that strain softening is significant at a higher strain rate (Adachi et al., 1990).

Little research has been done on the formation of shear bands in frozen soil. Thus, the main concerns of the study of mechanical properties of frozen sands only have been the strength and overall stress-strain relations (Sayles, 1988). The development of shear bands in frozen sand was reported by Yashima et al. (1986) and Shibata et al. (1985). They reported that multiple shear bands were observed in the triaxial compression tests on the axisymmetric specimen of frozen Toyoura sand when the stress-strain relation is strain-hardening and strain-softening type. The test specimen, whose material parameters and experimental conditions are similar to those used in the calculation deforms, in a barrel style with multiple shear bands (see Figure 9). Figure 9 shows the photograph of the specimen of frozen Toyoura sand taken after triaxial compression test, in which the temperature was -49°, the confining pressure was 10 MPa and the axial strain rate was 2.7 %/min. The shear bands orientations with respect to the maximum compression direction were about 60°. These trends are consistent with the observations in the calculated results at least on the development of shear bands although the orientation of the shear zone is different. This is partly becuase plane strain conditions were used for the calculation instead of the axisymmetric conditions taken in the experiemnts. It is also worth noting that the shear bands were scarcely observed when the overall stress-strain relation is only strain-hardening type in the experiemnt (Yashima et al.(1986)).

Figures 10((a),(b)) show the results in the case of fine mesh (800 element F.E. Mesh in Figure 3(c)). The tendency is similar to the result in Figures 5((a),(b),(c)). Moreover, a more clear discontinuous line in the distribution of second invariant of deviatoric stress tensor is observed at a strain rate of 3.6 %.

4 CONCLUSIONS

The following conclusions are obtained from the present study.
1. We have applied the viscoplastic softening model for frozen sand to the initial boundary value problem to study the instability (growth of fluctuation) and the strain localization.
2. Through an observation of the distribution of the plastic strain rate, the development of shear bands is plainly seen. The orientation of the shear bands is 45° to the loading axis.
3. The deformation characteristics are less sensitive on the mesh size. From the numerical studies, it is shown that deformation characteristics deeply depend on the strain rate. And from the distribution of the second invariant of the deviatoric stress tensor and the viscoplastic strain rate, it can be said that the higher the strain rate, the clearer the strain localization.

REFERENCES

Adachi,T., Oka,F. and Poorooshasb,H.B. (1990). A constitutive model for frozen sand, J. Energy Resources Technology, ASME, 112, 208-212.

Adachi,T., Oka,F. and Yashima,A. (1991). A finite element analysis of strain localization for soft rock using a constitutive equation with strain softening, Archive of Applied Mechanics, 61, 183-191.

Andersland,O.B. and Al-Nouri,I. (1970). Time dependent strength behavior of frozen soils, Jounal of Soil Mechanics and Foundation Division, ASCE, 96(4), 1249-1265.

Assure,A. (1979). Some promising trends in ice mechanics, Physics and Mechanics of Ice, ed., P.Tyyde, Proc. of the IUTAM Symposium, Springer-Verlag, 1-15.

Fish,A.M. (1980). Kinetic nature of the long-term strength of frozen soils, Proc. 2nd Int. Symp. on Ground Freezing, Tronheim, Norway, 95-108.

Ladanyi,B. (1980). Mechanical behavior of frozen soils, Mechanics of structured media, part B, ed., A.S.P.Selvadurai, 203-245.

May,R.M (1974). Populations with non-overlapping generation stable points, stable cycles, and chaos, Science, 186, 645-647.

Oka,F. (1985). Elasto-viscoplastic constitutive equations with memory and internal variables, Computer and Geotechnics, 1, 1, 59-69.

Oka,F. and Adachi,T. (1985). An elasto-plastic constitutive equations of geologic materials with memory, Proc. of the 5th Int. Conf. on Numerical Methods in Geomechanics, Balkema, 293-300.

Sayles,F.H. (1988). State of the art: Mechanical properties of frozen soil, Proc. 5th Int. Symp. on Ground Freezing, ed. by R.H.Jones and J.T.Holden, 1, 143-165.

Shibata,T., Adachi,T., Yashima,A., Takahashi,T. and Yoshioka,I. (1985). Time dependence and volumetric change characteristic change of frozen sand under triaxial stress condition, Proc. 4th Int. Symp. on Ground Freezing, Sapporo, ed. by S.Kinoshita and M.Fukuda, 173-179.

Stricklin,J.A., Haisler,W.E. and Von Riesemann, W.A. (1973). Evaluation of solution procedure for material and/or geometerically nonlinear structural analysis, AIAA Journal, 11, 3, 292-299.

Ting,J.M. (1983). Tertiary creep model for frozen sands, ASCE Journal of Geotechnical Engineering, 109(7), 932-945.

Valanis,K.C. (1971). A theory of viscoplasticity without a yield surface, Arch. Mech. Stos., 23, 4, 517-533.

Valanis,K.C. (1984). On the uniqueness of the initial value problem in softening materials, J. Applied Mechanics, ASME, 52, 649-653.

Vialov,S.S. (1963). Rheology of frozen soils, Proc. of the 1st Permafrost Conference, Lafayette, Ind., 332-342.

Yashima,A., Shibata,T., Adachi,T., Takahashi,T. and Yoshioka,I. (1986). Mechanical properties of frozen sands under triaxial stress conditions, Proc. Symp. on Freezing of soil, Japanese Society of Soil Mechanics and Foundation Engineering, 43-52. (in Japanese).

Numerical study on localised deformation in a Cosserat continuum
Étude numérique de la déformation localisée dans un milieu continu de Cosserat

J.Tejchman
Institute for Soil Mechanics and Rock Mechanics, Karlsruhe University, Germany

ABSTRACT: Modelling the thickness of shear zones in granular materials by means of many conventional constitutive equations is not possible because they do not include any characteristic length. This shortcoming can be overcome by a polar elastoplastic and a polar hypoplastic model laid down in the frame of a Cosserat continuum. They differ from nonpolar approaches due to the presence of Cosserat rotations and couple stresses using the mean particle diameter as a characteristic length. The formation of shear zones in dense granular body along the wall and inside the material is numerically investigated with a finite element method of a Cosserat type approach. Some different boundary value problems involving the appearance of shear zones are analysed by means of a polar elastoplastic and a polar hypoplastic approach. To simulate the different wall roughness, new boundary conditions are introduced which allow for Cosserat rotations. The numerical calculations show that the Cosserat effect is in shear zones noticeable. The thickness and the kinematics of shear zones are described realistically with a Cosserat approach. The thickness of shear zones is independent of the mesh size. Comparison between the numerical calculations and the experimental results shows acceptable agreement.

Beaucoup des lois de comportement classiques ne permettent pas de modéliser l'épaisseur des bandes de cisaillement, parce qu'elles ne comportent pas de longueur interne. Cette lacune peut être comblée par les modèles polaires élastoplastiques et hypoplastiques développés dans le cadre des milieux de Cosserat. La différence avec l'approche non-polaire réside dans la prise en compte de rotations de Cosserat et de couples de contrainte en utilisant le diamètre moyen des particules comme longueur caractéristique.

On étudie ici la formation des bandes de cisaillement dans un milieu granulaire dense, le long de la frontière et à l'intérieur du matériau, en utilisant une approche de Cosserat dans une méthode aux éléments finis. On considère divers problèmes aux limites avec apparition de bandes de cisaillement, avec des lois de comportement polaires élastoplastiques et hypoplastiques. On peut prendre en compte différentes rugosités du mur en introduisant de nouvelles conditions aux limites qui permettent des rotations de Cosserat. Les résultats montrent que l'effet de Cosserat est notable dans les zones de cisaillement. L'épaisseur et la cinématique de la bande sont décrites de façon réaliste avec une approche de Cosserat; l'épaisseur est indépendante de la taille du maillage. La comparaison de ces résultats avec l'expérience montre un accord acceptable.

1 INTRODUCTION

Modelling the thickness of shear zones in granular materials and other grain size effects by means of many conventional constitutive relations is not possible because they do not include any characteristic length. If the material behaviour involves strain softening, the governing differential equations will change the type, giving rise to ill-posed boundary value problems (Benallal et al. 1991). The numerical results show dependence on the mesh size and the mesh alignment.

The shortcomings of the existing models can be overcome among others by a polar elastoplastic (Mühlhaus 1986, 1990) and a polar hypoplastic model (Tejchman 1994) laid down in the frame of a Cosserat continuum (Schäfer 1962). A Cosserat (polar) continuum differs from a nonpolar continuum owing to the fact that additional rotation, called ω^c, appears in the kinematics (Fig.1). The

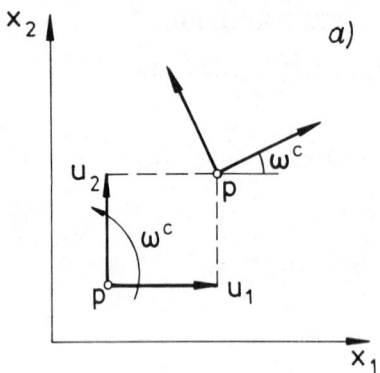

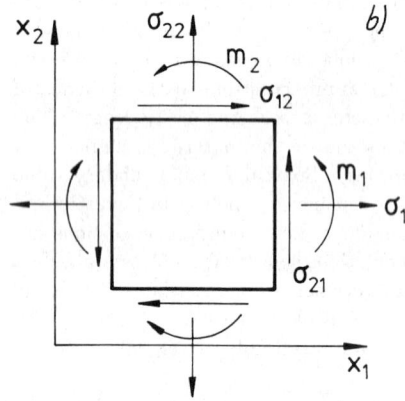

FIGURE 1. Degrees of freedom in a plane Cosserat continuum, u_1, u_2 - horizontal and vertical translation, ω^c - Cosserat rotation. b) Stresses σ_{ij} and couple stresses m_i in a Cosserat element

gradient of rotation $\partial \omega^c / \partial x_i$ corresponds to curvatures κ_i which are associated with couple stresses m_i. As a consequence, the stress tensor σ_{ij} and the deformation tensor ε_{ij} are nonsymmetric, and the constitutive relation is endowed with a characteristic length. From the presence of a characteristic length (mean grain diameter d_{50}), the calculated thickness of shear zones is independent of the spatial discretisation (Tejchman 1989, de Borst 1990). The Cosserat approach admits the localised deformation in the softening region without losing ellipticity of the governing differential equations for static problems (de Borst et al. 1992).

The effectiveness of an elastoplastic Cosserat approach in solving various boundary value problems involving localisation has been demonstrated by Mühlhaus (Mühlhaus 1990), Tejchman et al. (Tejchman 1989, Gudehus and Tejchman 1991, Tejchman and Wu 1993), de Borst et al. (de Borst 1990, de Borst et al. 1992), Vardoulakis et al. (Papanastasiou and Vardoulakis 1992, Vardoulakis and Unterreiner 1993) and Dietsche (Dietsche 1993).

The intention of this paper is to study some different boundary value problems involving localisation by means of a finite element method of a Cosserat type approach using two models: an elastoplastic and a hypoplastic.

2 A POLAR ELASTOPLASTIC APPROACH

A polar elastoplastic model for granular materials with isotropic hardening and softening was laid down by Mühlhaus (Mühlhaus 1986) and was later developed by Mühlhaus and Vardoulakis (Mühlhaus and Vardoulakis 1987) and Mühlhaus (Mühlhaus 1990). It can be summarized as follows:

$$d\varepsilon_{ij} = d\varepsilon^e_{ij} + d\varepsilon^p_{ij}, \qquad d\kappa_i = d\kappa^e_i + d\kappa^p_i, \qquad (1)$$

$$d\varepsilon^e_{ij} = \frac{1}{E}[(1+\nu)d\sigma_{ij} - \nu d\sigma_{kk}], \quad i=j \quad (2)$$

$$d\varepsilon^e_{ij} = \frac{1}{2G}\frac{\partial \tau^2}{\partial \sigma_{ij}}, \quad d\kappa^e_i = \frac{1}{2G}\frac{\partial \tau^2}{\partial m_i}, \quad i \neq j \quad (3)$$

$$d\varepsilon^p_{ij} = \lambda \frac{\partial g}{\partial \sigma_{ij}}, \quad d\kappa^p_i = \lambda \frac{\partial g}{\partial m_i} \qquad (4)$$

$$\tau = (a_1 s_{ij} s_{ij} + a_2 s_{ij} s_{ji} + \frac{a_3}{d_{50}^2} m_i m_i)^{\frac{1}{2}} \qquad (5)$$

$$f = \tau - \mu(e_0, \gamma^p) p \qquad (6)$$

$$g = \tau - \alpha(e_0, \gamma^p) p \qquad (7)$$

wherein τ – the second invariant of the deviatoric stress tensor, s_{ij} – nonsymmetric deviatoric stress tensor ($s_{ij} = \sigma_{ij} - p\delta_{ij}$), p – mean stress, σ_{ij} – stress tensor, m_i – couple stresses, a_1, a_2, a_3 – constants ($a_1 = 3/8$, $a_2 = 1/8$, $a_3 = 1$), d_{50} – mean grain diameter, f, g – yield and potential function, μ, α – mobilized friction and dilatancy factor, e_0 – initial void ratio, γ^p – equivalent plastic shearing, ε_{ij} – strain tensor, κ_i - curvatures (the superposed index e and p designate the elastic and the plastic strain or curvature respectively), λ – proportionality factor, E – elastic modulus. G –

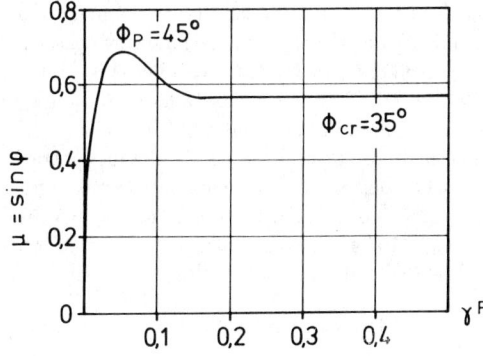
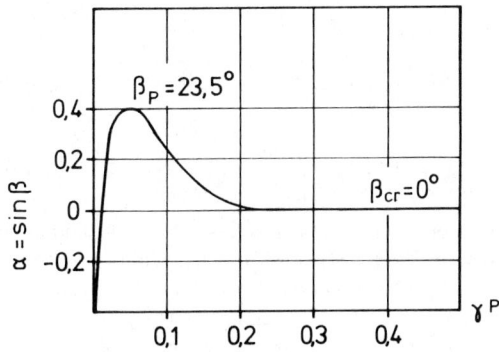

FIGURE 2. Assumed mobilized friction μ and mobilized dilatancy factor α for dense sand (ϕ - angle of internal friction, β - dilatancy angle, γ^p - plastic shearing)

shear modulus, ν – Poisson ratio, δ_{ij} – Kronecker delta.

The meaning of f, g, τ, γ^p, λ is analogous like in a nonpolar plasticity. An expression for τ was derived from consideration of slip and rotation in a random assembly of circular rods with equal diameter representing grains (Mühlhaus and Vardoulakis 1987).

The factors μ and α, associated with the angle of internal friction ϕ and the angle of dilatancy β, were determined on the basis of tests with dry sand in a biaxial apparatus (Vardoulakis and Graf 1985) and in a model silo with parallel walls (Tejchman 1989). The following data for dense sand was used during calculations: $\phi_p = 45°$, $\phi_{cr} = 35°$, $\beta_p = 23.5°$, $\beta_{cr} = 0°$ and $e_o = 0.55$ (subscript $'o'$ denotes the initial value, $'p'$ is the peak, $'cr'$ the residual value), Fig.2. The Poisson's ratio was assumed $\nu = 0.3$ and the elastic modulus E was equal to $5 \div 70$ MPa, in dependence on the stress level in the problem considered.

3 A POLAR HYPOPLASTIC APPROACH

A hypoplastic constitutive relation to describe an inelastic, irreversible deformations in granulates was proposed by Kolymbas (Kolymbas 1977) and was later developed by Kolymbas (Kolymbas 1991), Wu (Wu 1992), Wu and Bauer (Wu and Bauer 1993) and Gudehus (Gudehus 1994). Its main distinguishing feature is that the constitutive equation is of the rate type and is incrementally nonlinear. It consists, namely, of tensorial functions which are linear and nonlinear in strain rate.

A hypoplastic approach differs thus from elastoplastic models that neither yield surface and potential surface nor the elastic range of deformations and switch function are required. Its main advantages are its simplicity and its ability to describe several aspects of the material behaviour covering a wide range of densities, pressures and deformations (Gudehus 1993, 1994). The state is fully described by Cauchy granular stress tensor and void ratio. The constitutive relation has only four material constants which are rather easy to determine (Wu 1992). A hypoplastic approach is, particularly, useful in problems when great fluctuation of void ratio occurs e.g. silo flows and penetration problems (Gudehus 1994).

For numerical calculations, a hypoplastic constitutive relation according to Wu and Bauer (Wu and Bauer 1993), extended by the Cosserat terms ω^c, m_i, κ_i and d_{50}, was used. The extension was performed in a similar way as it was done by Mühlhaus in his polar elastoplastic formulation (Mühlhaus 1990). The constitutive version by Wu and Bauer takes into account the effect of the stress level (so called barotropy) and of the density (so called pyknotropy) on the material behaviour. The polar hypoplastic law has the following form:

$$\overset{\circ}{T} = F(T, D, m/d_{50}, \kappa d_{50}, e) =$$
$$= C_1 \mathrm{tr}(T)D + C_2 \frac{[\mathrm{tr}(TD) + \mathrm{tr}(m\kappa)]T}{\mathrm{tr}T} +$$
$$+ I_e(C_3 \frac{T^2}{\mathrm{tr}T} + C_4 \frac{T^{*2}}{\mathrm{tr}T})\sqrt{\mathrm{tr}D^2 + \mathrm{tr}(\kappa d_{50})^2}, (8)$$

$$\overset{\circ}{m}/d_{50} = F(T, D, m/d_{50}, \kappa d_{50}, e) =$$

$$= C_1 \text{tr}(T)\kappa d_{50} + C_2 \frac{[\text{tr}(TD) + \text{tr}(m\kappa)]m/d_{50}}{\text{tr}T} +$$
$$+ I_e C_m \frac{m}{d_{50}} \sqrt{\text{tr}D^2 + \text{tr}(\kappa d_{50})^2}, \qquad (9)$$

$$I_e = (1-a)\frac{e - e_{min}}{e_{crt} - e_{min}} + a, \qquad (10)$$

wherein $\mathring{T}$ - Jaumann stress rate, $\mathring{m}$ - Jaumann rate of couple stresses, T - nonsymmetric stress tensor, T^* - deviator of the stress tensor T, $\text{tr}T$ - trace of T, D - stretching (velocity strain) tensor, m - couple stresses, κ - rate of curvatures, d_{50} - mean grain size, I_e - density factor, a - pyknotropy paramater, e - void ratio, $e_{min} = 0.53$ - minimal void ratio, e_{crt} - critical void ratio, C_i, C_m - dimensionless constants. The Jaumann stress rate and the Jaumann rate of couple stresses are defined by:

$$\mathring{T} = \dot{T} - WT + TW, \mathring{m} = \dot{m} - Wm + mW. \quad (11)$$

W is spin tensor. A superposed dot stands for time differentiation. D and W are related to the deformation velocity $v = du/dt$:

$$D = (\nabla v + (\nabla v)^T)/2, \qquad (12)$$

$$W = (\nabla v - (\nabla v)^T)/2. \qquad (13)$$

The two first terms with C_1 and C_2 in Eq.8,9 are linear in D and κd_{50}, and the two last terms with C_3 and C_4 in Eq.8,9 are nonlinear in D and κd_{50}. The nonlinear term in the formula for $\mathring{m}/d_{50}$ differs from the nonlinear terms in the formula to calculate $\mathring{T}$ because the behaviour of couple stresses during shearing is quite different than that of stresses according to their skew symmetry and no sign restrictions (Sec.4.1). The void ratio e in Eq.10 is updated during calculations by the formula:

$$\dot{e} = (1+e)\text{tr}D. \qquad (14)$$

The relations between the critical void ratio e_{crt}, the pyknotropy parameter a and the stress level $\text{tr}T$ (in kPa) are approximated by the following exponential functions:

$$e_{crt} = 0.51 + 0.31\exp[-0.00018(\text{tr}T)], \qquad (15)$$

$$a = 1.0 - 0.2\exp[-0.0001(\text{tr}T)]. \qquad (16)$$

The constants C_i can be calibrated with triaxial compression tests (Wu 1992) and the constant C_m can be found with the aid of calculations for shearing of an infinite sand layer between rigid walls (Sec.4.1). The numerical calculations were carried out for dense Karlsruhe sand (within a polar and a nonpolar continuum) with the following constants: $C_1 = -33.5$, $C_2 = 341.4$, $C_3 = -339.7$, $C_4 = 446.5$ (Wu 1992). This set of constants corresponds to an elastic modulus of about 40 MPa, a critical friction angle of 30° and a critical dilatancy angle of 0°. The constant C_m depends on the mean grain diameter and is equal to -145.1 ($d_{50} = 0.5$ mm) and to -167.4 ($d_{50} = 1$ mm).

4 NUMERICAL RESULTS

4.1 Simple shearing of an infinite layer

Simple shearing of an infinite layer of dense sand between rigid walls was studied with a polar elastoplastic and a polar hypoplastic approach. The height of the layer was 2 cm and the length 80 cm. 800 triangular elements were adopted. The horizontal walls were assumed to be very rough so that the boundary condition $w^c = 0$ was prescribed along the boundaries. The deformation was initiated by the horizontal displacement increments prescribed at the nodes along the top. The vertical load $p = 0.1$ MPa and $d_{50} = 1$ mm (polar models) were assumed. The calculations were performed for large deformations and curvatures using an updated Lagrangian-Jaumann stress rate formulation (Bathe 1982). The calculations with an elastoplastic model were carried out with $E = 5$ MPa. Fig.3 presents results with a nonpolar elastic, a polar elastic and a polar elastoplastic approach, and Fig.4-5 show results with a nonpolar and a polar hypoplastic approach along a cross section in the middle of the specimen. The calculations with an elastic approach were carried out without the vertical load p.

Some observations from calculations can be made:

1. the results with a polar elastoplastic and a polar hypoplastic approach are qualitatively the same,

2. the Cosserat effect is in the shear zone significant. The Cosserat rotations and the couple stresses are noticeable in the shear zone. The differences in displacements and in stresses between the results within a polar and a nonpolar are evident,

3. the stress tensor in the frame of a polar continuum is strongly nonsymmetric,

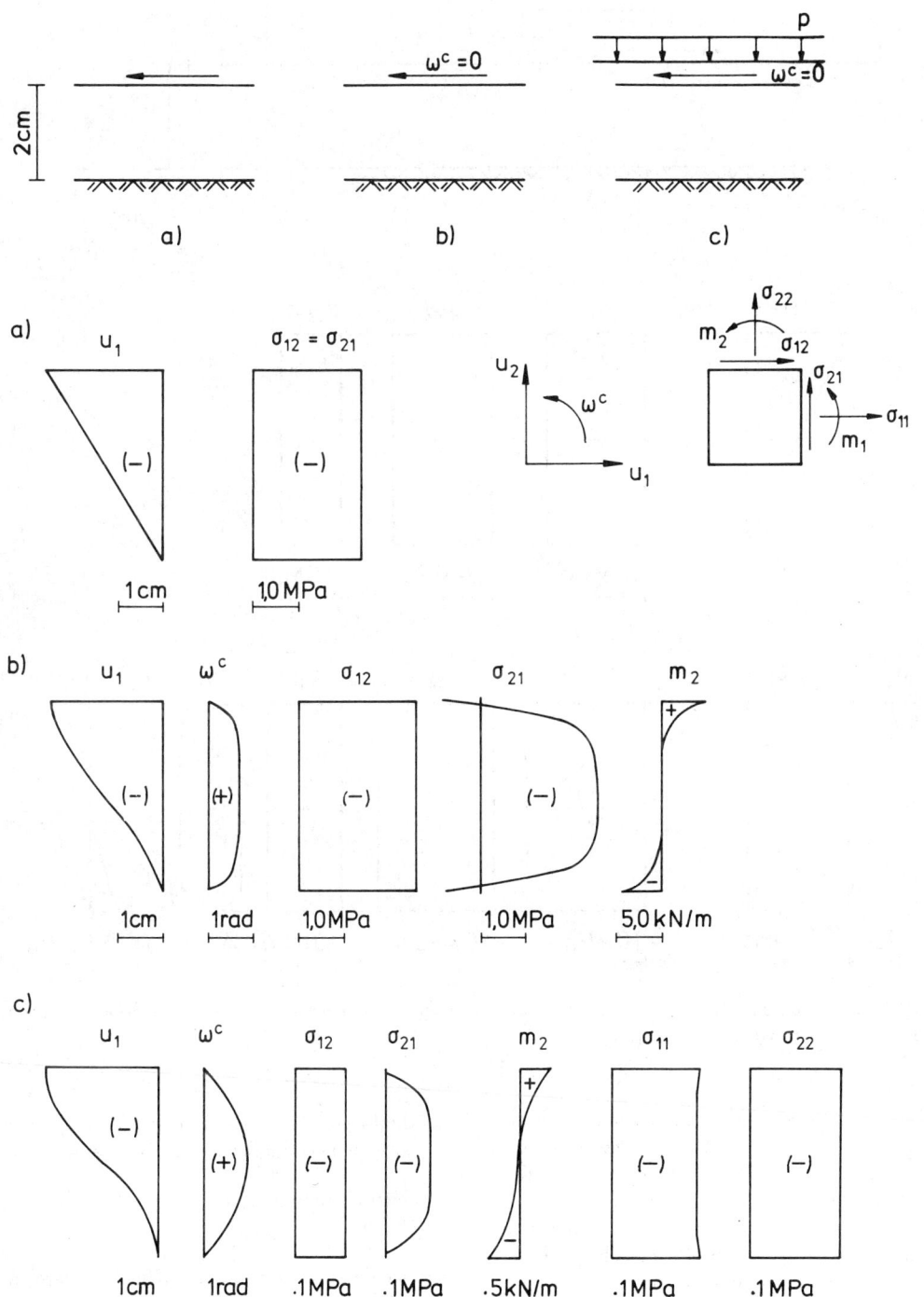

FIGURE 3. Simple shearing of an infinite sand layer: a) nonpolar elastic constitutive approach, b) polar elastic approach ($d_{50} = 1$ mm), c) polar elastoplastic approach ($d_{50} = 1$ mm)

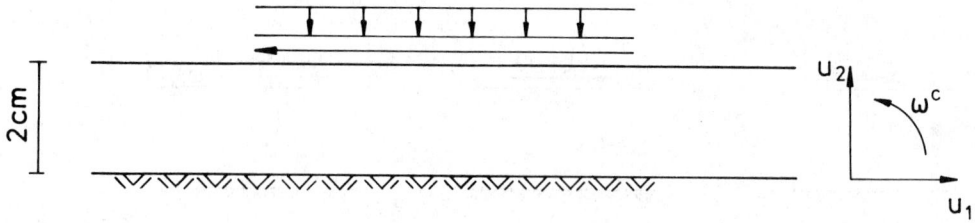

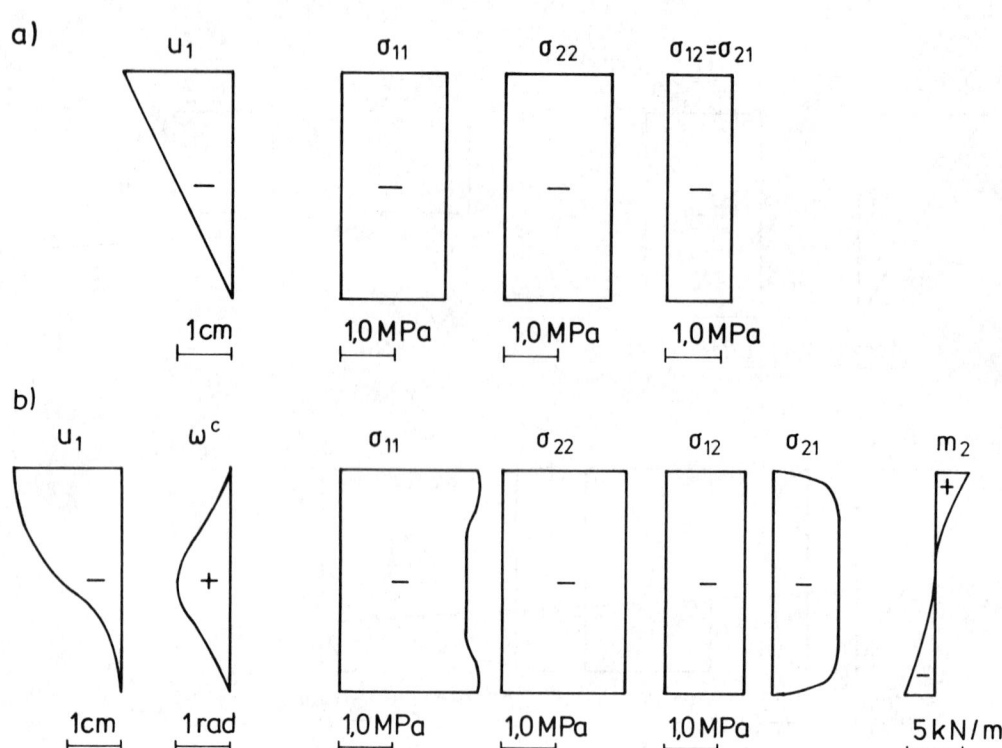

FIGURE 4. Simple shearing of an infinite sand layer: a) nonpolar hypoplastic constitutive approach, b) polar hypoplastic constitutive approach ($d_{50} = 1$ mm)

4. the stresses and the wall friction angle are greater in a polar continuum than in a nonpolar continuum. The maximum wall friction angles are 34.3° (polar hypoplastic approach, $d_{50} = 1$ mm) and 33.3° (nonpolar hypoplastic approach) respectively, Fig.5. The wall friction angles in the residual state are 32.2° (polar hypoplastic approach, $d_{50} = 1$ mm) and 30.9° (nonpolar hypoplastic approach),

5. the bigger the mean grain size is, the bigger are shear stresses and couple stresses,

6. the bigger the mean grain size is, the bigger is the nonsymmetry of the stress tensor,

7. the wall friction angle shows evident softening (Fig.5).

The calculated stresses in the frame of a polar continuum are bigger than in a nonpolar continuum because the virtual work of a Cosserat continuum is greater than this of a nonpolar continuum. It takes in addition into account rotations, curvatures and couple stresses. The Cosserat models are

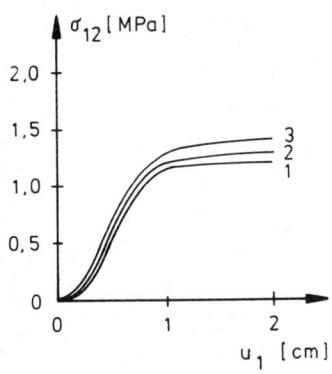 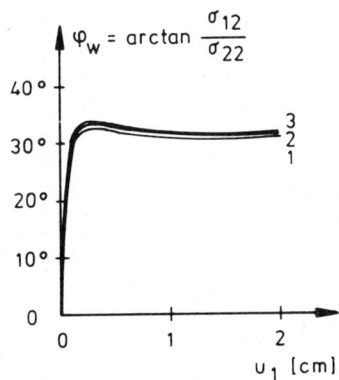

FIGURE 5. Simple shearing of an infinite sand layer (σ_{12} - horizontal shear stress, φ_w - wall friction angle, u_1 - horizontal displacement): 1) nonpolar hypoplastic approach, 2) polar hypoplastic approach ($d_{50} = 0.5$ mm), 3) polar hypoplastic approach ($d_{50} = 1$ mm)

thus much stiffer than nonpolar versions. The calculated Cosserat effect (which causes an increases of stresses and of a wall friction angle) was confirmed in wall friction experiments carried out with a parallelly guided direct shear apparatus (Wernick 1978), in which the thickness of the wall shear zone in dense sand along a very rough wall and the grain diameter of sand were varied (Tejchman 1989, 1993).

4.2 Biaxial compression test

Fig.6 shows the numerical results of a biaxial compression test with an elastoplastic and an hypoplastic constitutive law in the frame of a polar continuum. The deformed meshes and the Cosserat rotations along the height in the middle of a granular specimen are presented. The material parameters (the angle of internal friction ϕ in the case of an elastoplastic approach and the void ratio e in the case of a hypoplastic approach) were distributed stochastically over the sample with a random generator. ϕ and e were increased namely in each element by the value $\phi_1 \cdot r$ and $e_1 \cdot r$ respectively. r is a random number within the range of $(0.01, 0.99)$, $\phi_1 = 2°$ and $e_1 = 0.02$. The specimen was assumed to be 14 cm high and 2 cm wide, and included 1280 triangular elements. The top and the bottom were very smooth. The quasi-static deformations were initiated through the constant vertical displacement increments prescribed to the nodes along the upper edge of the specimen. The mean grain size was assumed 1 mm. The cell pressure was equal to 0.2 MPa and the elasticity modulus using an elastoplastic approach was 70 MPa. The weight of the specimen was taken into account. The calculations were carried out for small deformations and curvatures. The results with two different approaches are qualitatively in good agreement with each other (Fig.6). One shear zone appears in the deformed sample which is marked out by the concentration of displacements, plastic deformations and by the presence of Cosserat rotations. In the remaining region of the specimen, the deformations and rotations are negligible. The rotations are the best criterion to detect the shear zones (Bardet and Proubet 1992, Oda et al. 1982, Uesugi et al. 1988). The thickness of the shear zone estimated on the basis of the Cosserat rotations is about 1.5 cm (i.e. $15 \times d_{50}$) and is bigger than the width of triangular elements. The inclination against the bottom is about 58° which is different from the direction of the mesh alignment - 54.5°. The shear zone comes first into being at the weekest place of the specimen determined by the random distribution of material parameters. The thickness and the inclination of the interior shear zone within a Cosserat continuum depend insignificantly on the mesh if the size of finite elements is not greater than $5 \times d_{50}$ (using triangle finite elements with linear shape functions). The calculations with a homogeneous specimen resulted in no shear zones. The sample had the form of a 'elephant' foot due to the specimen weight growing towards the bottom. The calculated patterning of shear bands in a

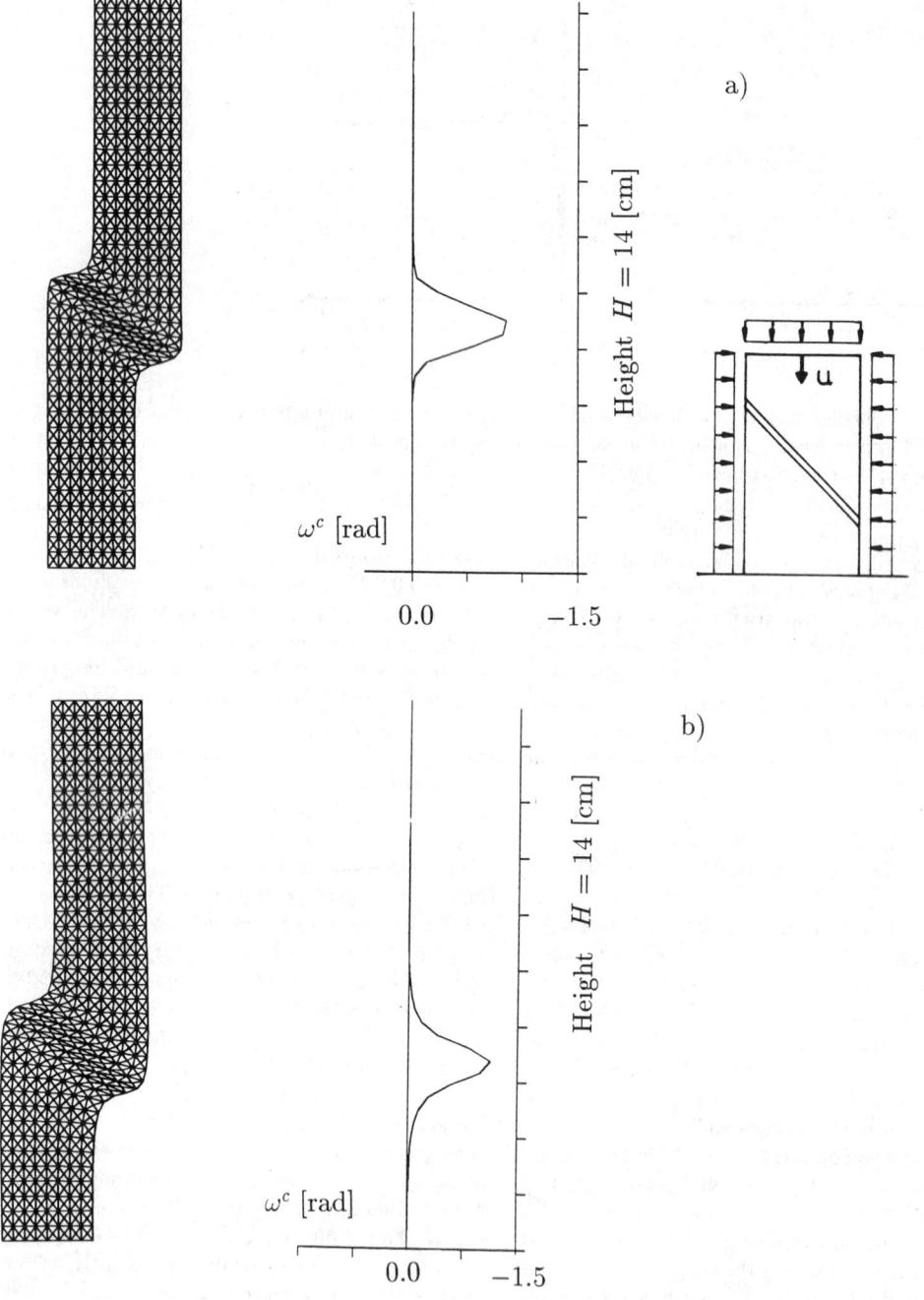

FIGURE 6. Biaxial compression test - deformed mesh and the Cosserat rotation along the height in the middle of a granular specimen ($d_{50} = 1$ mm): a) polar elastoplastic constitutive approach, b) polar hypoplastic constitutive approach

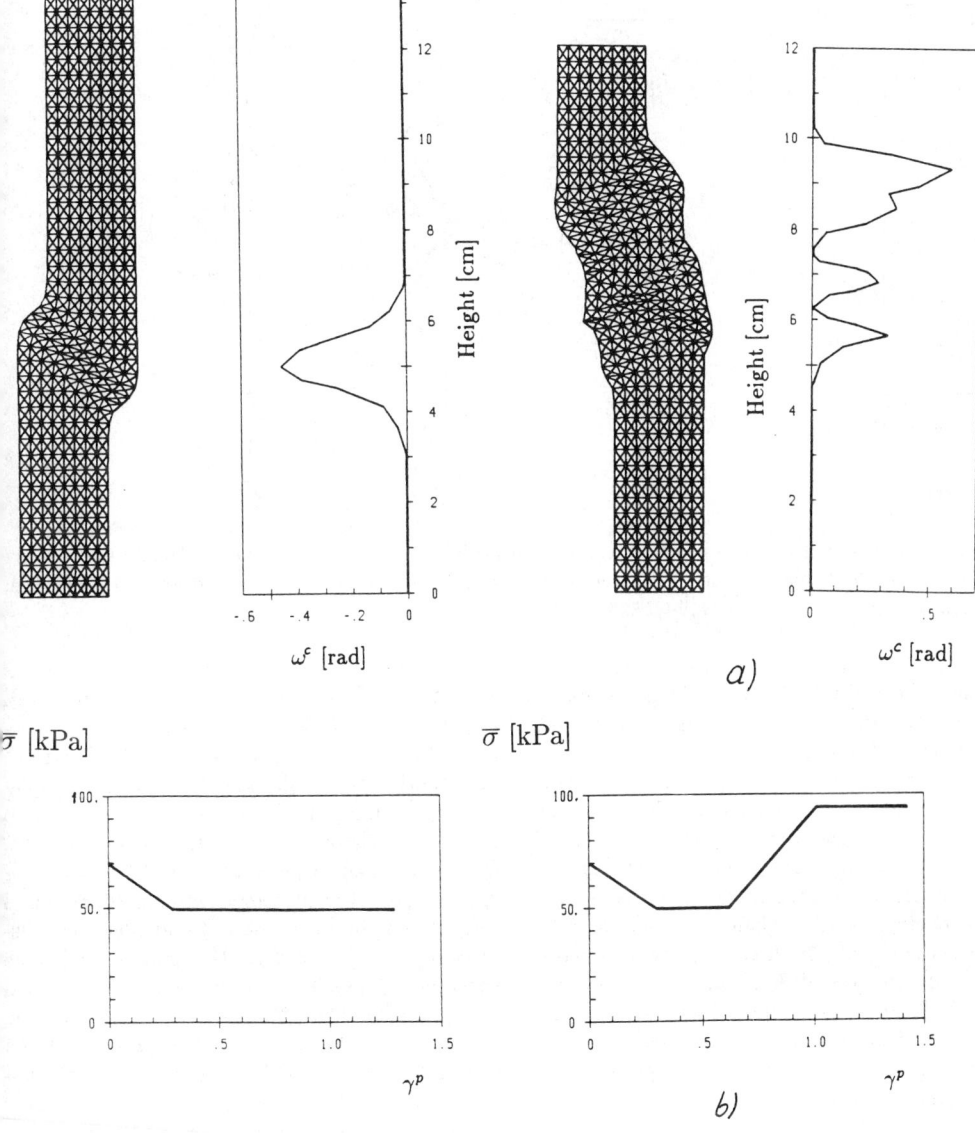

FIGURE 7. a) Biaxial compression test - deformed mesh and the Cosserat rotation along the height in the middle of the von Mises specimen ($L = 1$ mm). b) Yield strength function $\bar{\sigma}$ versus plastic shearing γ^p with softening followed by hardening

von Mises material during a biaxial compression test is presented in Fig.7 (Tejchman and Wu 1993). It was achieved with an elastoplastic approach following Mühlhaus. The deformed mesh, the Cosserat rotations along the height in the middle of the specimen and the assumed yield strength functions versus the plastic shearing are shown. Random distribution of the yield strength acted as a trigger for the localized deformation. The equivalent yield strength function was multiplied in each element with the value r ranging randomly from 0.01 to 0.99. The characteristic length L was assumed to be 1 mm. The calculations with the stress-strain curve including solely softening yielded only

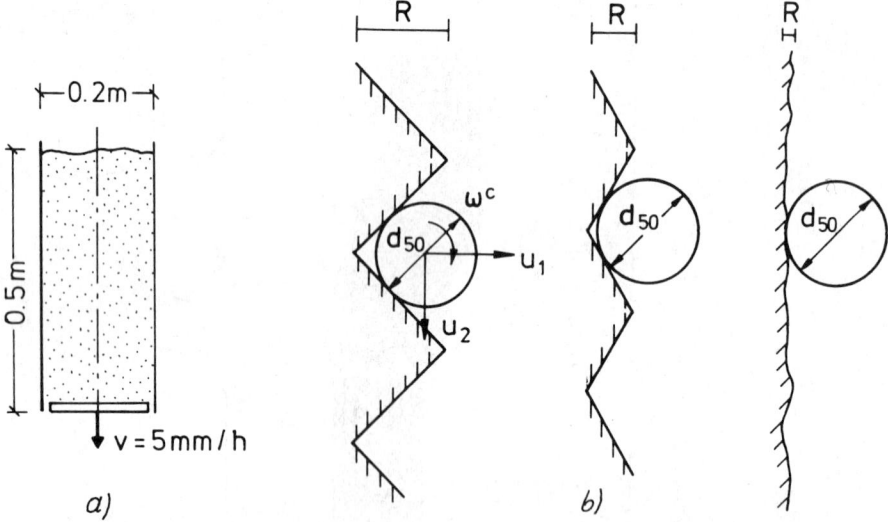

FIGURE 8. a) Controlled confined flow of a granular material in a model silo with parallel walls and a slowly movable bottom. b) Boundary conditions along a silo wall (R - wall roughness): very rough wall ($R \geq d_{50}$), rough wall ($0 < R < d_{50}$), smooth wall ($R \cong 0$)

one single shear band with a thickness of about $15 \times L$. A patterning of shear bands was obtained with the stress-strain curve with softening followed by hardening which is typical for compression tests in metals. The first shear came into being at the weekest place and developed until the plastic shearing reached the hardening region. At this stage, further deformation in the shear zone required an increase of stresses larger than the formation of a further shear zone. The first shear zone became inactive, and the second shear zone began to develop. The whole process repeated successively and the pattern of shear zones could be observed. Rehardening inside shear zones is thus responsible for the pattern formation.

4.3 Controlled confined flow in a silo

Numerical calculations were performed for the onset of a confined controlled mass flow in a plane strain model silo with a slowly movable bottom, Fig.8a (Tejchman 1989). The silo was filled with dry sand. The silo was 50 cm high, 20 cm wide and 60 cm long. 1200 triangular elements were used. The calculations were carried out with very rough walls (roughness R bigger than the mean grain size d_{50}), rough walls (roughness bigger than zero but smaller than the mean grain size) and smooth walls (roughness almost equal to zero), Fig.8b. For modelling very rough walls, full shearing of the material along a rigid wall was assumed (horizontal displacement u_1, vertical displacement u_2 and Cosserat rotation ω^c were equal to zero). In case of rigid smooth and rough walls, the calculations were made with the assumptions: horizontal displacement u_1 equal to zero and the relation between rotations and vertical displacements $\omega^c/u_2 = C_1$. For smooth walls, the constant C_1 should be very small; one assumed $C_1 = (1/d_{50})/10000$. For rough walls one assumed $C_1 = (1/d_{50})/10$. The second boundary condition can be connected to the wall roughness in the following way: $\omega^c/u_2 = R/d_{50}^2$. The wall roughness R is thus $d_{50}/10000$ for smooth surfaces, $d_{50}/10$ for rough surfaces and $R = d_{50}$ for very rough surfaces. The calculated displacements (polar elastoplastic model) in dense sand ($d_{50} = 0.5$ mm) in a silo model after bottom displacement u are shown in Fig.9. During calculations, large deformations and curvatures were taken into account. The wall was very rough, rough and smooth. As the initial stress state, K_o-state was assumed. The elastic modulus was 5 MPa. The calculated thickness of wall shear zone is about $40 \times d_{50}$ for very rough walls $10 \times d_{50}$ for rough walls and $(1-2) \times d_{50}$ for smooth walls. The obtained results are in good agreement with

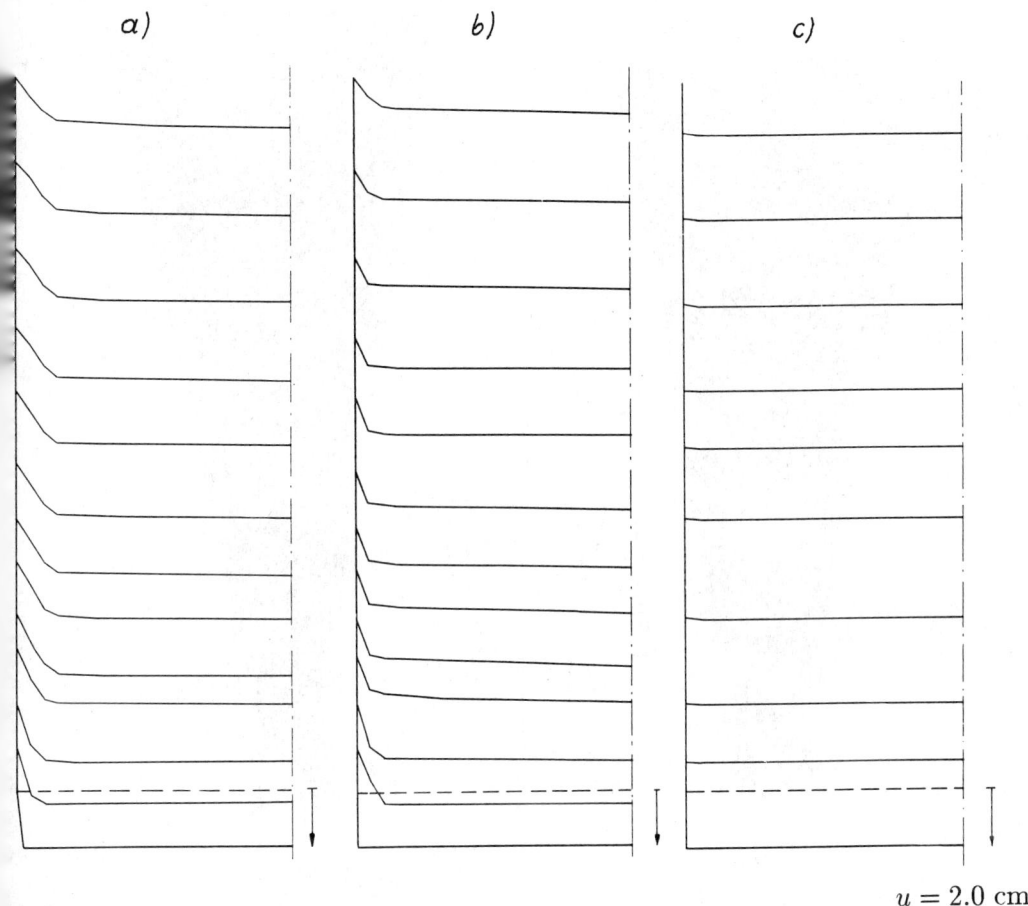

$u = 2.0$ cm

FIGURE 9. Calculated displacements in dense sand ($d_{50} = 0.5$ mm) in a silo model with parallel walls ($H = 0.5$ m, $B = 0.2$ m) after bottom displacement u with a constant velocity: a) very rough walls, b) rough walls, c) smooth walls

experimental results (Tejchman 1989). The calculations with a hypoplastic model resulted in similar results. The calculated thickness of the wall shear zone does not depend on the mesh refinement if the size of finite elements in the wall region is not greater than $8 \times d_{50}$ (very rough walls). With a Cosserat model, the scale effect due to the grain size and to the partial dilatancy constraint in wall shear zones (caused by the stiffness of the neighbouring material) can quantitatively be estimated (Tejchman 1989). The calculations show that the thickness of wall shear zones depends on the wall roughness, the density of the material, the mean particle size, the particle elasticity and the boundary conditions of the whole system (Tejchman 1989, Gudehus and Tejchman 1991).

The calculated displacements in a planar model silo with convergent walls and a slowly movable bottom, filled with a dry sand are presented in Fig.10. The height of the silo was 0.5 m, the wall inclination 79° and the outlet opening 0.10 m. The walls were very rough. A polar elasto-plastic approach was used. The angle of internal friction was stochastically over the whole silo distributed with a random generator. Two different dilatancy functions β were considered. In the first case, the dilatancy function without any volume changes in the region of large shear deformations was assumed. The calculated flow was symmetric. Only two shear zones appeared above the outlet (Fig.10a). In the second case, the dilatancy function with a varying tendency of volume changes

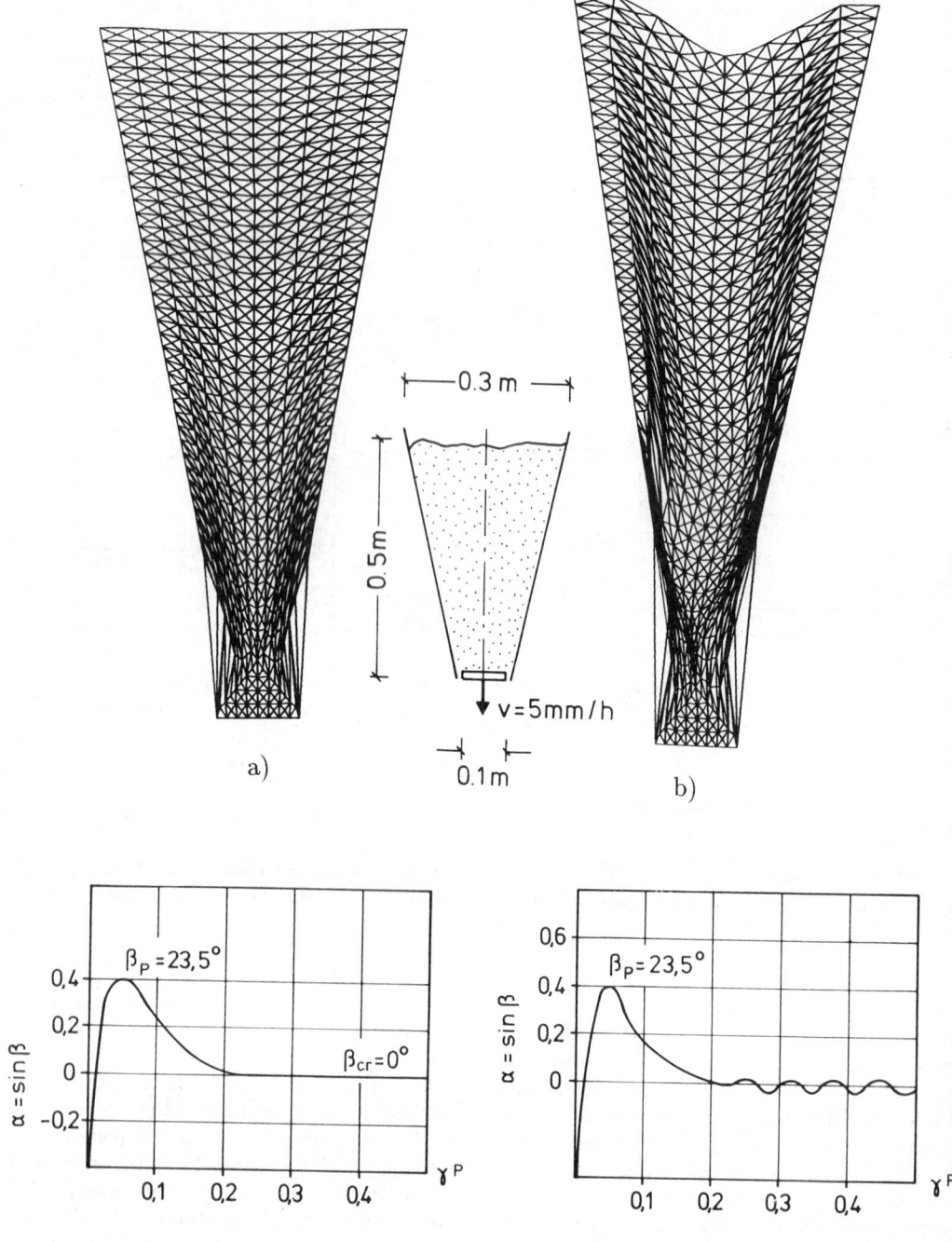

FIGURE 10. Calculated displacements in dense sand ($d_{50} = 0.5$ mm) in a silo model with convergent very rough walls: a) constant dilatancy function β in the residual state, b) fluctuating dilatancy function β in the residual state (γ^p - plastic shearing)

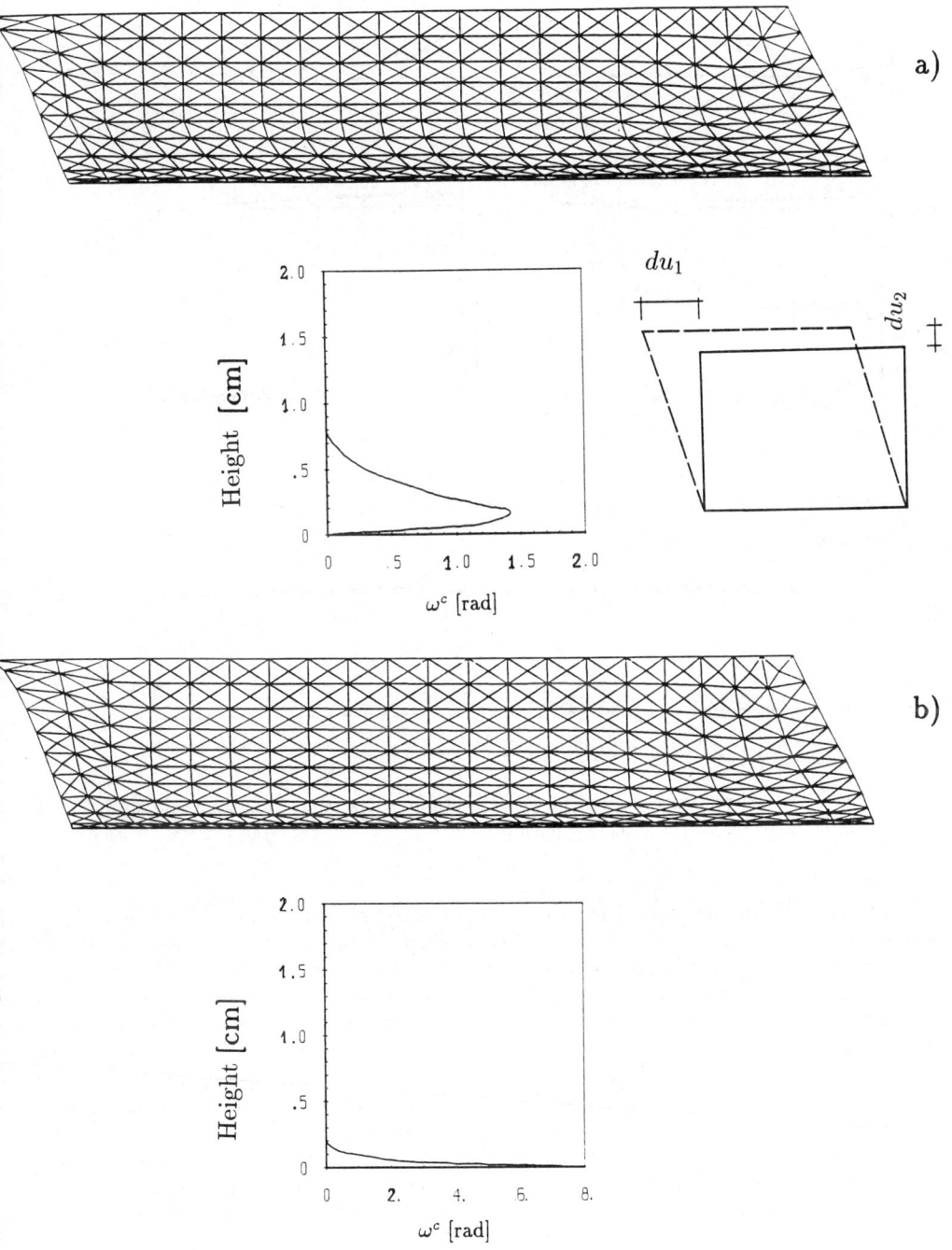

FIGURE 11. Calculated displacements and the Cosserat rotation (along the height in the middle) in a dense sand ($d_{50} = 0.5$ mm) during simple shearing along very rough wall (a) and rough wall (b)

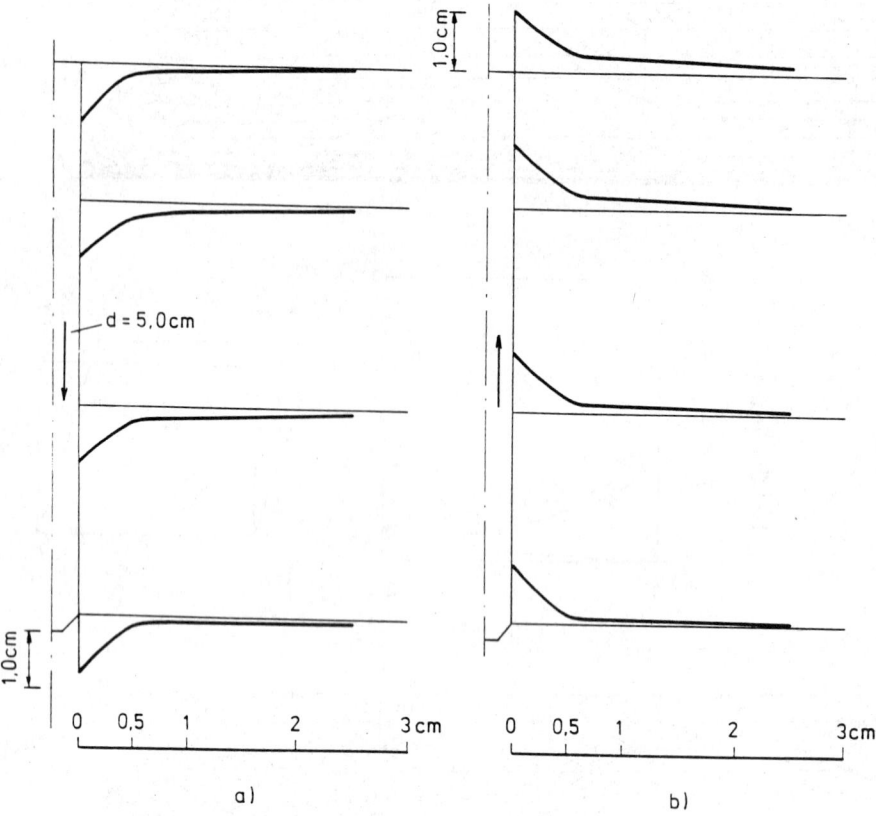

FIGURE 12. Calculated displacements in a dense sand (d_{50} = 0.25 mm) along an axially loaded model pile (L = 0.5 m, D = 0.05 m) during pushing (a) and pulling (b)

in a residual state was assumed. The curve's run was artificially adopted. The calculated displacement field was nonsymmetric and strongly deformed (Fig.10b). It corresponded to experimental observations (Tejchman 1989). The fluctuating behaviour of granular materials is characteristic for many flow and penetration problems. With elastoplastic constitutive relations it seems difficult to describe realistically this property. The hypoplastic approach, on the other hand, creates much more possibilities. The effect of a fluctuating void ratio can come out as a result.

4.4 Simple shear test along a wall with a different roughness

The calculated displacements in dense sand (d_{50} = 0.5 mm) along a very rough and a rough wall during simple shearing and the Cosserat rotations along the height in the middle of the specimen are presented are presented in Fig.11, (Tejchman and Wu 1994). The calculations were performed with a polar elastoplastic model. The specimen was 10 cm long and 2 cm high, and included 800 triangular elements. A quasi-static deformation was initiated through the constant horizontal displacement increments prescribed at the nodes along both sides of the specimen. The horizontal displacements along the upper edge were constrained by the same amount. Large deformations and curvatures were considered. With rough and very rough surfaces, the localized zones with certain thickness can be observed. The localized zone is manifested by the concentration of plastic deformation and by the presence of Cosserat rotations. The calculated thickness of the wall shear zones was found to depend strongly on the wall roughness: 7.5 mm (i.e.

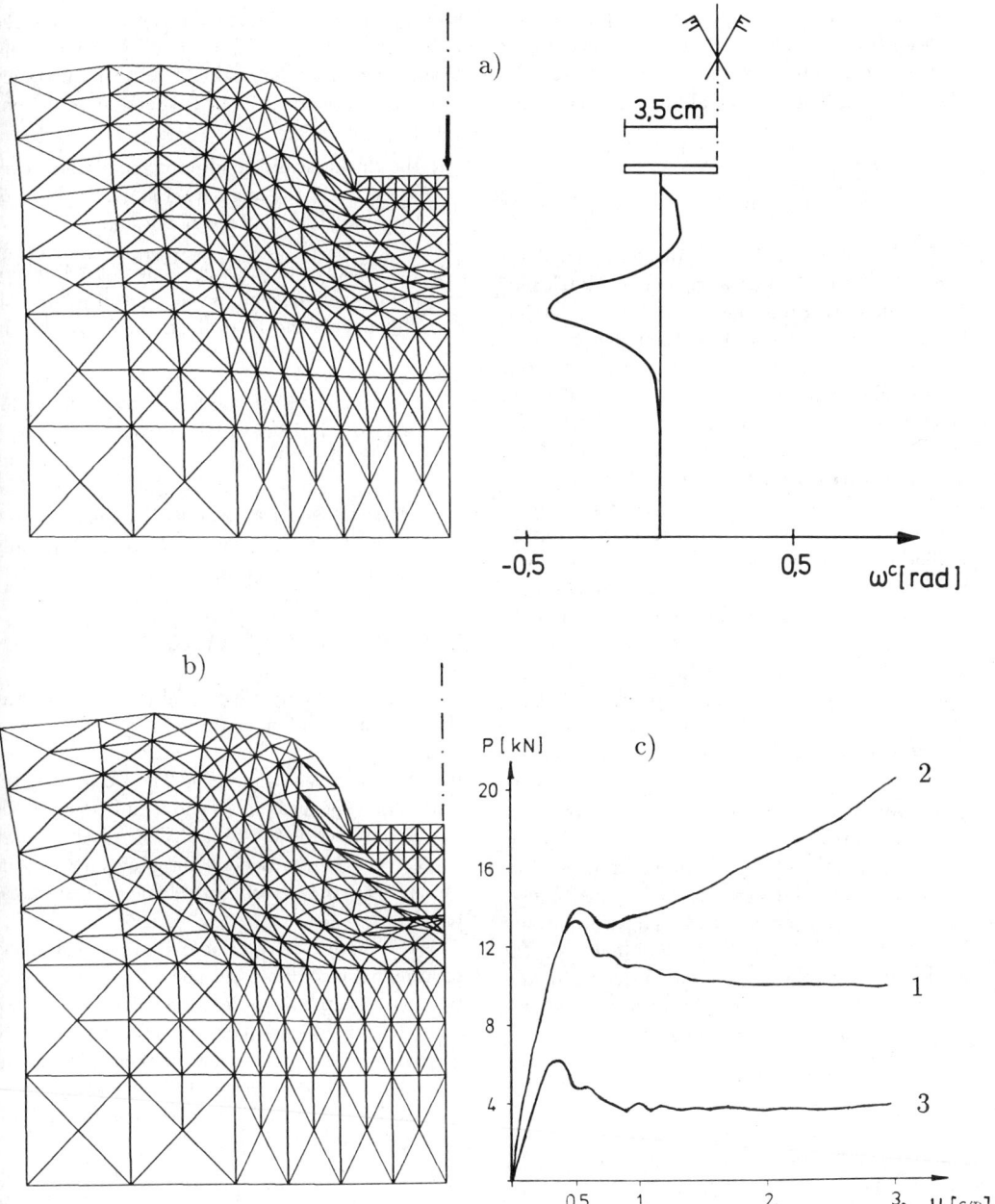

FIGURE 13. a) Calculated displacements in dense sand ($d_{50} = 5$ mm) close to an axially loaded model of strip foundation and the Cosserat rotation under the foundation. b) Calculated displacements under the foundation in the frame of a nonpolar continuum. c) Calculated vertical force P on the foundation versus the vertical displacement u: 1) polar continuum (small deformations), 2) polar continuum (large deformations), 3) nonpolar continuum (small deformations)

$15 \times d_{50}$) for a very rough wall and 1.5 mm (i.e. $3 \times d_{50}$) for a rough wall. The obtained values are realistic as compared to the experimental data (Tejchman 1989, Uesugi et al. 1988, Yoshimi and Kishida 1981).

4.5 Axially loaded piles

Fig.12 shows the calculated displacements in dense sand ($d_{50} = 0.25$ mm) along an axially loaded model pile during pushing and pulling (Tejchman 1989). An elastoplastic polar approach was used for an axial symmetric case with consideration of large deformations and curvatures. The pile (length 0.5 m, diameter 0.05 m) was rigid and very rough. 550 triangular elements were adopted. As the initial stress state, K_o-state was assumed. The elastic modulus was 5 MPa. The deformation was initiated through the constant successive vertical displacement increments prescribed to the nodes along the pile shaft. The calculated thickness of the shear zone along the piles was found to be $20 \times d_{50}$ for $d_{50} = 0.25$ mm (Fig.11) and $7 \times d_{50}$ for $d_{50} = 2.4$ mm. It is in accordance with observed results (Meißner 1983). The unit shaft and the unit base resistance increased both with decreasing pile diameters and increasing grain diameters. This scale effect in model piles is caused mainly by grain rotations and by volume changes in the shear zone which are constrained partially by the surrounding material. The grain rotations depend on the mean grain diameter and the volume changes on the pile diameter (due to radial deformations) and on the mass of the surrounding material. The results from model tests can be transferred to large piles if the ratio between mean grain size and pile diameter is the same in both cases.

4.6 Axially loaded strip foundation

Fig.13 presents the results for an axially loaded model strip foundation on dense sand by means of an elastoplastic approach. Only the deformation field close to the foundation is shown. The calculations were carried out in the frame of a nonpolar and polar continuum taking into account small and large deformations and curvatures. The rigid and very rough model strip foundation was 7 cm wide. Symmetry with respect to the centerline was assumed. 1200 triangular elements were adopted. The earth pressure at rest was assumed as an initial stress state. The elastic modulus was 15 MPa, the mean grain diameter $d_{50} = 5$ mm, the angle of internal friction at the peak 40° and the dilatancy angle at the peak 12° (Fig.2). Some observations can be made:

1. the thickness of the interior shear zone developed under the foundation is, on the basis of Cosserat rotations, about $8 \times d_{50}$,

2. the thickness of the shear zone in the frame of a nonpolar continuum is equal to the breadth of elements and is much smaller than that in a polar continuum,

3. the calculated vertical force acting on the foundation is in a nonpolar continuum much smaller than that in a polar continuum due to formation of a narrower shear zone. The narrower the shear zone is, the smaller is the effect of sand dilatancy constrained partially in this zone on the vertical force.

5 CONCLUDING REMARKS

The presented numerical results show that the thickness and the kinematics of shear zones along wall and inside granular material can be described by a Cosserat type approach. The Cosserat effect is noticeable in shear zones and its influence on the results obtained is substantial. The results within a polar continuum depend insignificantly on the mesh used if the size of finite elements is not greater than $5 \times d_{50}$.

The numerical calculations with the aid of a Cosserat approach will be continued. The main objective is to study and to describe dynamical phenomena (so called silo-quake) appearing during silo emptying. They occur in form of pulsations in granular silo fills and in form of pulsations and shocks in cohesive silo fills. Silo pulsations can realistically be described with an elastoplastic Cosserat model with consideration of inertial forces (Tejchman and Gudehus 1993). Silo shocks however, the phenomena connected to cohesion fluctuation and arching (Tejchman 1989), need to be further investigated.

Acknowledgement — Financial support from the German Research Community (SFB 219 - SILOS) is gratefully acknowledged.

REFERENCES

Benallal, A., Billardon, R. & Geymonat, G. 1991. Localization phenomena at the boundaries and interfaces of solids. *Proc. Int. Conf. Constitutive Laws for Engineering Materials in Tucson*: 387-390.

Bardet, J.P. & Proubet, J. 1992. A numerical investigation of the structure of persistent shear bands. *Géotechnique*. 41: 599-613.

Bathe, K.J. 1982. Finite Element Procedures in Engineering Analysis. Prentice-Hall, Inc., Englewood Cliffs, New Jersey.

de Borst, R. 1990. Simulation of localisation using Cosserat theory. *Proc. 2nd Int. Conf. Computer Aided Analysis and Design of Concrete Structures*: 931-944.

de Borst, R., Mühlhaus, H. B., Pamin, J. & Sluys, L.Y. 1992. Computational modelling of localisation of deformation. *Proc. 3rd Int. Conf. Comp. Plast*. Pineridge Press, Swansea: 483-508.

Dietsche, A. 1993. Lokale Effekte in linear-elastischen und elasto-plastischen Cosserat-Kontinua. Publication Series of the Institute of Mechanics, Karlsruhe University.

Gudehus, G. & Tejchman, J. 1991. Some mechanisms of a granular mass in a silo - model tests a numerical Cosserat approach. *Advances in Continuum Mechanics*, Springer Verlag: 178-193.

Gudehus, G. 1993. Some comments on shear localisation in granular bodies. Int. Workshop on Localisation and Bifurcation Theory for Soils and Rocks in Aussois, France.

Gudehus, G. 1994. A comprehensive equation of state for granular materials. Submitted to *Soils and Foundations*.

Kolymbas, D. 1977. A nonlinear viscoplastic constitutive law for soils. Publication Series of the Institute of Soil Mechanics and Rock Mechanics, Karlsruhe University, No. 77.

Kolymbas, D. 1991. An outline of hypoplasticity. *Archive of Applied Mechanics*, 61: 143-151.

Meißner, H. 1983. Tragverhalten axial oder horizontal belasteter Bohrpfähle in körnigen Böden. Publication Series of the Institute of Soil Mechanics and Rock Mechanics, Karlsruhe University, No. 93.

Mühlhaus, H. B. 1986. Shear band analysis in granular materials by Cosserat theory. *Ing. Arch.*, 56: 389-399.

Mühlhaus, H. B. 1990. Continuum models for layered and blocky rock. *Comprehensive Rock Engineering*, 2.

Mühlhaus, H. B. & Vardoulakis, I. 1987. The thickness of shear bands in granular materials. *Géotechnique*, 37: 271-283.

Oda, M., Konishi, J. & Nemat-Nasser, S. 1982. Experimental micromechanical evaluation of strength of granular materials, effect of particle rolling. *Mech. Mater.*, 1: 269-283.

Papanastasiou, P. & Vardoulakis, I. 1992. Numerical treatment of progressive localization in relation to borehole stability. *Int. J. Num. Anal. Meth. Geomech.*, 16: 389-424.

Schäfer, H. 1962. Versuch einer Elastizitätstheorie des zweidimensionalen ebenen Cosserat-Kontinuums. *Miszellaneen der Angewandten Mechanik*, Berlin, Akademie-Verlag: 277-292

Tejchman, J. 1989. Scherzonenbildung und Verspannungseffekte in Granulaten unter Berücksichtigung von Korndrehungen. Publication Series of the Institute of Soil Mechanics and Rock Mechanics, Karlsruhe University, No. 117.

Tejchman, J. & Wu, W. 1993. Numerical study on shear band patterning in a Cosserat continuum. *Acta Mech.*, 99: 61-74.

Tejchman, J. & Gudehus, G. 1993. Silo-music and silo-quake, experiments and a numerical Cosserat approach. *Powder Technology*, 76: 201-212.

Tejchman, J. 1993. Cosserat effect in the shear zone - experiments on shear resistance between dense sand and very rough walls, Institute for Soil Mechanics, Karlsruhe University.

Tejchman, J. & Wu, W. 1994. Numerical study on sand and steel interfaces. To be published in *Mech. Research Commun.*.

Tejchman, J. 1994. Numerical study on localized deformation in a Cosserat Continuum. *Proc. 8th Inter. Conf. on Comp. Meth. and Advan. in Geomech.* Morgantown, West Virginia.

Uesugi, M., Kishida, H. & Tsubakihara, Y. 1988. Behavior of sand particles in sand-steel friction. *Soils and Foundations*, 28, 1: 107-118.

Wernick, E. 1978. Tragfähigkeit zylindrischer Anker in Sand unter besonderer Berücksichtigung des Dilatanzverhaltens. Publication Series of the Institute of Soil Mechanics and Rock Mechanics, Karlsruhe University, No.75.

Wu, W. & Kolymbas, D. 1991. On some issues in triaxial extension tests. *ASTM. Geotech. Testing Journ.*, 14: 276-287.

Wu, W. 1992. Hypoplastizität als mathematisches Modell zum mechanischen Verhalten granularer Stoffe. Publication Series of the Institute of Soil Mechanics and Rock Mechanics, Karlsruhe University, No.129.

Wu, W. &, Bauer, E. 1993. A hypoplastic models for barotropy and pyknotropy of granular soils. *Proc. Int. Workshop on Modern Approaches to Plasticity, Elsevier*: 365-383.

Vardoulakis, I. & Graf, B. 1985. Calibration of constitutive models for granular materials using data from biaxial experiments. *Géotechnique*, 35: 299-317.

Vardoulakis, I. & Unterreiner, P. 1993. Interfacial localisation in simple shear tests on a granular medium modelled as a Cosserat continuum. *Mechanics of Geomaterials Interfaces.*

Yoshimi, Y. & Kishida, T. 1981. A ring torsion apparatus for evaluating friction between soil and metal surfaces. *Geotech. Test. Journ.*, GTJODJ, 4: 145-152.

Author index
Index des auteurs

Adachi, T. 237, 249

Bigoni, D. 51
Boulon, M. 141
Brinkgreve, R.B.J. 89

Caillerie, D. 35
Chambon, R. 35, 101
Chu, L.L. 249
Cormery, F. 127

Darve, F. 73
Désoyer, T. 127
Desrues, J. 101
di Prisco, C. 59
Dragon, A. 127

Finno, R.J. 189

Gudehus, G. 3

Halm, D. 127

Harris, W.W. 189

Kamegai, Y. 165
Kim, Y.-S. 181
Kohara, I. 237
Kolymbas, D. 13

Maciejewski, J. 19
Mühlhaus, H.-B. 201, 211
Muir Wood, D. 155
Mróz, Z. 19

Niemunis, A. 113
Nova, R. 59

Oka, F. 201, 211, 237, 249

Park, C.-S. 165

Rybicki, S. 41

Sakellariadi, E. 219

Scarpelli, G. 219
Siddiquee, M.S.A. 165
Sikora, Z. 41
Stone, K.J.L. 155
Sulem, J. 141

Tatsuoka, F. 165, 181
Tejchman, J. 257
Tillard, D. 101

Unterreiner, P. 141

Vardoulakis, I. 141
Vermeer, P.A. 89
Viggiani, G. 189

Willis, J.R. 51
Wu, W. 113

Yashima, A. 237, 249
Yoshida, T. 165